Computational Intelligence Paradigms for Optimization Problems Using MATLAB®/SIMULINK®

Computational Intelligence Paradigms for Optimization Problems Using MATLAB®/SIMULINK®

S. Sumathi

L. Ashok Kumar

Surekha P.

CRC Press
Taylor & Francis Group
Boca Raton London New York

CRC Press is an imprint of the
Taylor & Francis Group, an **informa** business

CRC Press
Taylor & Francis Group
6000 Broken Sound Parkway NW, Suite 300
Boca Raton, FL 33487-2742

First issued in paperback 2017

© 2016 by Taylor & Francis Group, LLC
CRC Press is an imprint of Taylor & Francis Group, an Informa business

No claim to original U.S. Government works

ISBN-13: 978-1-4987-4370-9 (hbk)
ISBN-13: 978-1-138-85538-0 (pbk)

Visit the Taylor & Francis Web site at
http://www.taylorandfrancis.com

and the CRC Press Web site at
http://www.crcpress.com

Contents

Preface

Computational intelligence (CI) techniques have attracted several research engineers, decision makers, and practicing researchers in recent years for solving an unlimited number of complex real-world problems, particularly related to the area of optimization. In the indecisive, uncertain, chaotic existence of multiple decision variables, complex constraints, and turmoil environment, classical and traditional approaches are incapable of obtaining complete and satisfactory solutions to optimization problems. CI embraces techniques that use global search optimization, machine learning, approximate reasoning, and connectionist systems. CI techniques involve a combination of learning, adaptation, and evolution used for intelligent and innovative applications. CI is considered as one of the most innovative research directions, since efficient, robust, and easy-to-use solutions to complex real-world problems can be developed.

About This Book

Solutions to constraint-based optimization problems using CI paradigms with MATLAB® m-files and Simulink® models are chosen as the pedestal for this book. This book is written in an attempt to explore the performance of CI in terms of knowledge representation, adaptability, optimality, and processing speed for different real-world problems. This book covers CI paradigms and their role in different engineering applications, such as unit commitment and economic load dispatch, harmonic reduction, load frequency control and automatic voltage regulation, job shop scheduling, multidepot vehicle routing and digital image watermarking. The impact of CI algorithms in the area of power systems, control systems, industrial automation, and image processing is explained with practical implementation using MATLAB/Simulink. Each application is presented with an implementation of CI algorithm, methodology to apply CI algorithms to the problem under consideration, solution approach using MATLAB/Simulink, and an extensive analysis with problem-dependent test data.

A detailed explanation is included for all the topics with the intention that engineers and scientists can benefit from preliminaries on the subject from both application and research points of view. This book is intended for higher-level undergraduate, postgraduate, and research students interested in understanding and implementing CI algorithms for various applications based on MATLAB/Simulink.

Salient Features

The salient features of this book include

- Role of CI paradigms in electrical engineering applications
- Implementation based on MATLAB m-files and Simulink models
- Experimental analysis and results of test systems
- Provides the reader with more practical implementation
- Prepares the reader to be more suitable for an industrial career

Organization of This Book

This book is organized into seven chapters. Chapter 1 describes the basic concepts of CI algorithms and the role of computational intelligence (CI) paradigms in engineering applications. A detailed classification of CI algorithms is presented with pseudocode, application areas, and advantages and disadvantages of each individual algorithm. A survey of application areas is presented, with an introduction to constraint-based problems focused in the later chapters.

In Chapter 2, the mathematical formulation of unit commitment (UC) and economic load dispatch (ELD) problems along with the framework to solve these problems is provided. The implementation of optimization techniques such as genetic algorithms (GA), fuzzy based-radial basis function network (FRBFN), enhanced particle swarm optimization (EPSO), differential evolution with opposition-based learning (DE-OBL), improved DE-OBL (IDE-OBL), artificial bee colony (ABC), and cuckoo search optimization (CSO), in solving UC and economic load dispatch (UC-ELD) problems along with a comparative analysis based on fuel cost, robustness, computational efficiency, and algorithmic efficiency is presented.

Chapter 3 focuses on harmonic effects and methods to eliminate harmonics by applying voltage source inverter (VSI) drives for harmonic measurement in pulp and paper industry. MATLAB/Simulink models are developed for drives applied for harmonic reduction with the application of CI algorithms like GA and bacterial foraging optimization (BFO) for reducing harmonics.

Chapter 4 gives an insight to the importance of load frequency control (LFC) and automatic voltage regulator (AVR) in power generating systems. The LFC and AVR system with PID controller is modeled along with different parameters used for simulation. LFC and AVR are tested with a PID controller tuned by fuzzy logic, GA, ant colony optimization (ACO), and particle swarm optimization (PSO) algorithms. In addition, the fusion of evolutionary algorithms with fuzzy, GA, BF, and the improved versions of PSO algorithm are also applied to test the performance of LFC and AVR in interconnected power systems.

In Chapter 5, the job shop scheduling problem (JSSP) is solved in a fuzzy sense using optimization meta-heuristics such as GA, stochastic PSO (SPSO), ACO, and genetic swarm optimization (GSO). The formulation and mathematical model of JSSP with the implementation of intelligent heuristics are delineated with MATLAB m-file snippets. The algorithms are analyzed by conducting experiments on a standard set of 162 benchmark instances thus evaluating the JSSP performance.

Chapter 6 deals with the basic concepts and mathematical model of the multidepot vehicle routing problem (MDVRP). The problem is implemented with bio-inspired techniques such as GA, modified PSO (MPSO), ABC, GSO, and improved GSO (IGSO). The effectiveness of the proposed techniques is tested and validated on five different Cordeau's benchmark instances. Intelligent heuristics are applied to evaluate the performance of the MDVRP instances.

Chapter 7 enlightens the intelligent digital image watermarking based on GA, PSO, and hybrid PSO (HPSO). The pre-watermarking schemes, such as, image segmentation, feature extraction, orientation assignment, and image normalization, are presented with image processing algorithms. The procedure for watermark embedding and extraction of the preprocessed images in DWT is discussed in detail along with geometric and nongeometric attacks. The step-by-step implementation procedure of the heuristic algorithms along with MATLAB m-file snippets is also elaborated in this chapter.

MATLAB® is a registered trademark of The MathWorks, Inc. For product information, please contact:

The MathWorks, Inc.
3 Apple Hill Drive
Natick, MA 01760-2098 USA
Tel: 508-647-7000
Fax: 508-647-7001
E-mail: info@mathworks.com
Web: www.mathworks.com

Acknowledgments

The authors are always thankful to the Almighty for their perseverance and achievements. The authors are grateful to L. Gopalakrishnan, managing trustee, PSG Institutions, and Dr. R. Rudramoorthy, principal, PSG College of Technology, Coimbatore, India, for their wholehearted cooperation and encouragement in this successful endeavor. The authors extend their heartfelt thanks to Dr. A. Soundarrajan, associate professor, Department of Information Technology, PSG College of Technology, Coimbatore and Dr. Maheswaran, Electrical Department, Tamil Nadu Newsprint and Papers Limited, Karur, for their support in completing this project successfully.

Dr. Sumathi owes much to her daughter, S. Priyanka, who has helped a lot in monopolizing her time on book work and substantially realized the responsibility. She feels happy and proud for the steel frame support rendered by her husband, Sai Vadivel. Dr. Sumathi extends her whole-hearted thanks to her parents, for their help with family commitments and their constant support. She is greatly thankful to her brother M. S. Karthikeyan, who has always been a "stimulator" for her progress. She is pertinent in thanking her parents-in-law for their great moral support.

Dr. L. Ashok Kumar acknowledges those people who helped him in completing this book. This book would not have been possible without the help of his students, department staff, and institute and especially his project staff. He sincerely thanks the management and principal, guru, and mentor Dr. Rudramoorthy for providing a wonderful platform to perform. He is thankful to all his students, who are doing their project and research work with me. But the writing of this book was mainly possible because of the support of his family members, parents, and sisters. Most importantly, he is very grateful to his wife, Y. Uma Maheswari, for her constant support during the writing, without her this work would not have been possible. He expresses his special gratitude to his daughter A. K. Sangamithra. Her smiling face and support have helped immensely in overcoming backlogs and completing this book. He apologizes to her for taking a significant amount of his playtime with her to write this book.

Dr. Surekha P. thanks the Almighty for giving her the courage, energy, and time to complete this book in spite of the busy academic career. The author extends her thanks to her daughter, S. Saisusritha, who has been keeping her active with her cute little smiles and naughtiness. She expresses her sincere gratitude to her husband A. Srinivasan and parents, who shouldered a lot of her responsibilities during the months she spent writing this book. She thanks her brother, Siva Paneerselvam, who has always been helping her during hardships.

The authors thank all their friends and colleagues who have been with them in all their endeavors with their excellent, unforgettable help and assistance in the successful execution of this work.

Authors

Dr. S. Sumathi completed her BE in electronics and communication engineering and ME in applied electronics at the Government College of Technology, Coimbatore, Tamil Nadu. She earned her PhD in the area of data mining and is an assistant professor in the Department of Electrical and Electronics Engineering, PSG College of Technology, Coimbatore. She has 25 years teaching and research experience.

Dr. Sumathi has been a recipient of the prestigious gold medal from the Institution of Engineers Journal Computer Engineering Division—Subject Award for the research paper titled, "Development of New Soft Computing Models for Data Mining" 2002–2003, and also Best Project award for UG technical report titled "Self-Organized Neural Network Schemes: As a Data mining tool," 1999. She received the Dr. R. Sundramoorthy award for Outstanding Academic of PSG College of Technology in the year 2006. She has guided a project that received Best MTech thesis award from the Indian Society for Technical Education, New Delhi.

In appreciation of publishing various technical articles, Dr. Sumathi has received national and international journal publication awards, 2000–2003. She prepared manuals for electronics and instrumentation lab, electrical and electronics lab for the EEE Department, PSG College of Technology, Coimbatore. She organized the second National Conference on Intelligent and Efficient Electrical Systems in 2005 and conducted short-term courses on "Neuro Fuzzy System Principles and Data Mining Applications," November 2001 and 2003. Dr. Sumathi has published about 48 papers in national and international journals/conferences and guided many undergraduate and postgraduate projects. She is guiding 10 PhD candidates, among which 5 have completed their degree successfully. She has also published the following books: *Introduction to Neural Networks with MATLAB* (Tata McGraw Hill, 2006), *Introduction to Fuzzy Systems with MATLAB* (Springer-Verlag, 2007), *Introduction to Data Mining and Its Applications* (Springer-Verlag, 2006), *Computational Intelligence Paradigms: Theory and Applications Using MATLAB* (CRC Press, 2010), *Virtual Instrumentation Systems Using LabVIEW* (ACME Learning, 2011), *Evolutionary Algorithms Using MATLAB* (ACME Learning, 2011), and *Solar PV and Wind Energy Conversion Systems* (Springer-Verlag, 2015). She has reviewed papers in national/international journals and conferences. Her research interests include neural networks, fuzzy systems and genetic algorithms, pattern recognition and classification, data warehousing and data mining, operating systems and parallel computing, etc.

Dr. L. Ashok Kumar completed his graduate program in electrical and electronics engineering. He did his postgraduation with electrical machines as his major. He completed his MBA with a specialization HRD (human resource development) in 2008. He has completed his PhD in wearable electronics. He has been a teaching faculty at PSG College of Technology, India, since 2000. He is working as a professor in the Department of EEE at PSG College of Technology, Coimbatore. After graduation, he joined as project engineer in Serval Paper Boards (P) Ltd., (now renamed ITC PSPD, Kovai), Coimbatore.

He has been actively involved in numerous government-sponsored research projects funded by DRDO, DSIR, CSIR, DST, L-RAMP, Villgro, and AICTE. He has published 82 technical papers in reputed national and international journals and presented 93 research articles in national and international conferences. He visited many countries for presenting his research work. He was selected as a visiting professor at the University of Arkansas, Fayetteville, Arkansas. He was trained in Germany in the installation of grid-connected and stand-alone solar PV systems. He is a registered charted engineer from IE (India). Dr. Kumar is a recipient of many national awards from IE, ISTE, STA, NFED, etc. He is a member of various national and international technical bodies like IE, ISTE, IETE, TSI, BMSI, ISSS, SESI, SSI, CSI, and TAI.

Dr. Kumar has received several awards: the Young Engineer Award, Dr. Triguna Charan Sen Prize from the Institution of Engineers, P. K. Das Memorial Best Faculty Award 2013–EEE, Prof. K. Arumugam National Award from the Indian Society for Technical Education, and the UGC Research Award. He published the book *Solar PV and Wind Energy Conversion Systems* (Springer, 2015). His research areas include wearable electronics, solar PV and wind energy systems, textile control engineering, smart grid, energy conservation and management, and power electronics and drives.

Dr. Surekha P. completed her BE in electrical and electronics engineering in PARK College of Engineering and Technology, Coimbatore, Tamil Nadu, and master's degree in control systems at PSG College of Technology, Coimbatore, Tamil Nadu. She earned her PhD in electrical engineering, Anna University, Chennai, Tamil Nadu. Her research work includes computational intelligence paradigms for engineering applications. She is an associate professor in the Department of Electrical and Electronics Engineering, PES University, Bangalore. She was a rank holder in both BE and ME programs. She has received Alumni Award for best performance in curricular and cocurricular activities during her master's degree program.

She is a member of technical bodies like IET, ISTE, etc. She has been serving as a reviewer for *Journal of Soft Computing*, Elsevier, and for several IEEE-sponsored international conferences. She has published 34 papers in national and international journals and conferences. She has published books such as *Computational Intelligence Paradigms: Theory and Applications Using MATLAB* (CRC Press, 2010), *Virtual Instrumentation Systems Using LabVIEW* (ACME Learning, 2011), *Evolutionary Algorithms Using MATLAB* (ACME Learning, 2011), and *Solar PV and Wind Energy Conversion Systems* (Springer-Verlag, 2015). Her areas of interest include robotics, virtual instrumentation, control systems, smart grid, evolutionary algorithms, and computational intelligence.

1

Introduction

Learning Objectives

On completion of this chapter, the reader will have knowledge on:

- The classification of computational intelligent algorithms.
- Understanding the operators, parameters, applications, advantages and disadvantages of different computational intelligent algorithms.
- A brief introduction to the applications concentrated in this book: "Unit Commitment and Economic Load Dispatch Problem," "Harmonic Reduction," "Load Frequency Control and Automatic Voltage Regulation," "Job Shop Scheduling," "Multidepot Vehicle Routing," and "Digital Image Watermarking."
- A detailed literature survey of the above mentioned applications.

1.1 Computational Intelligence Paradigms

Optimization is an integral part of science and technology. The growing interest in the application of computationally intelligent bio-inspired paradigms to optimization engineering has introduced the potentials of using the state-of-the-art algorithms in several real-time case studies. Computational intelligence (CI) techniques have attracted the attention of several research engineers, decision makers, and practicing researchers in recent years for solving an unlimited number of complex real-world problems particularly related to the research area of optimization. In indecisive, uncertain, or chaotic existence of multiple decision variables, combined with complex constraints, and in presence of a turbulent environment, classical and traditional approaches are incapable of obtaining complete and satisfactory solutions for optimization problems. Therefore, new global optimization methods are required to handle these issues seriously. Such new methods are based on biological behavior, thus providing a generic, flexible, robust, and versatile framework for solving complex global optimization problems.

CI is a successor to artificial intelligence (AI), which relies on heuristic algorithms such as fuzzy systems, neural networks, and evolutionary computation (EC). In addition, CI also embraces techniques that use swarm intelligence, fractals and chaos theory, and artificial immune systems. CI techniques involve a combination of learning, adaptation, and evolution used for intelligent and innovative applications. The major classes of problems in CI are grouped into five categories: control problems, optimization problems, classification problems, regression problems, and nonlinear programming (NP)-complete problems. Optimization problems are undoubtedly the most important class of problem in CI research, since any type of problem can be reframed virtually as an optimization problem. Solving an optimization problem using CI methods requires knowledge about the problem domain and, most importantly, the fitness of a potential solution to the problem. CI is considered as one of the most innovative research directions, since efficient, robust, and easy-to-use solutions to complex real-world problems can be developed on the basis of complementary techniques such as neural networks, fuzzy logic (FL), evolutionary algorithms (EA), and swarm optimization. Thus, solutions to constraint-based optimization problems using CI paradigms are chosen as the pedestal for this research in an attempt to explore the CI characteristics in terms of knowledge representation, adaptability, optimality, and processing speed.

Several real-world applications involve complex optimization processes, which are difficult to solve without advanced computational tools. Because of the increasing challenges of fulfilling optimization goals in current applications, there is a strong drive to advance the development of efficient optimizers. The major challenges of the emerging problems include (1) objective functions that are prohibitively expensive to evaluate, (2) objective functions that are highly multimodal and discontinuous, and (3) dynamic nature of the problems (change with respect to time). Conventional classical optimizers such as linear/nonlinear programming, dynamic programming, and stochastic programming perform poorly and fail to effect any improvement in the face of such challenges. This has motivated researchers to explore the use of CI techniques to augment classical methods.

More specifically, CI techniques are powerful tools offering several potential benefits such as robustness (impose little or no requirements on the objective function), versatility (capable of handling highly nonlinear mappings), hybridization (combination of techniques to improve search), self-adaptation to improve the overall performance, and parallel operation (easier decomposition of complex tasks). However, successful application of CI techniques to real-world problems is not straightforward and requires expert knowledge in obtaining solutions through trial-and-error experiments.

This chapter gives a brief description of the CI paradigms and their detailed classification. Each algorithm in the classification category is presented with the pseudo-code, advantages, disadvantages, and applications. The broad areas engineering to which the CI paradigms can be applied are also listed in later sections of this chapter. An introduction to the applications focused in detail in the following chapters of this book—"Unit Commitment and Economic Load Dispatch," "Harmonic Reduction," "Load Frequency Control and Automatic Voltage Regulation," "Job Shop Scheduling," "Multidepot Vehicle Routing," and "Digital Image Watermarking"—is given in this chapter.

1.2 Classification of Computational Intelligence Algorithms

CI provides a complementary view of the implementation of a combination of methods, such as learning, adaptation, evolution, and optimization, used to create intelligent and innovative applications. CI is closely related to soft computing, a combination of artificial neural networks (ANN), FL and genetic algorithms (GAs), and connectionist systems such

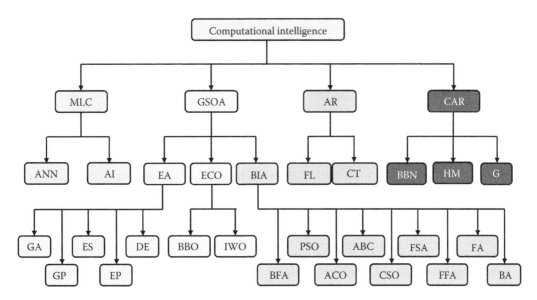

FIGURE 1.1
Computational intelligence paradigms.

as AI and cybernetics. Further, specifically, CI can be thought of as the study of adaptive mechanisms to facilitate intelligent behavior in complex, changing, and uncertain environments. There are also other AI approaches that satisfy both these definitions as well as the requirement of modeling some naturally occurring phenomenon, which do not fall neatly into one of the paradigms mentioned earlier. A more pragmatic approach might be to specify the classes of problems that are of interest without being too concerned about whether or not the solutions to these problems satisfy any constraints implied by a particular definition for CI.

In this chapter, we classify the paradigms of CI into four major groups; global search and optimization algorithms (GSOA) machine learning and connectionist (MLC) systems, approximate reasoning (AR) approaches, and conditioning approximate reasoning (CAR) approaches and connectionist systems, as illustrated in Figure 1.1. The GSOA techniques are classified into groups such as EA, ecology-based algorithms (ECOs), and bio-inspired algorithms (BIAs). Further classifications of these algorithms are presented in the following sections. The widely used ANN and AI belong to the category MLC systems. The major AR approaches using multivalued logic include FL and chaos computing (CC). The CAR approaches include probabilistic models such as hidden Markov models (HMMs), graphical models (GMs), and Bayesian belief networks (BBNs). AI belongs to the category of connectionist systems.

1.2.1 Global Search and Optimization Algorithms

GSOA algorithms belong to the CI group and are able to find the globally optimal solution for a given optimization problem. The major objective of GSOA is to obtain the fastest possible progress toward the best possible solution. Earlier, conventional approaches based on mathematical programming were used to achieve the objective. But the conventional methods search the space in a systematic manner, use models of the objective function, and/or store all the points sampled during the run. The process of maintaining the models and storing sampled points is both time and space consuming. To overcome the difficulties faced by conventional methods, the GSOA algorithms were developed. The behavior of GSOA algorithms is based on biology and nature. Based on their behavior, GSOAs are classified into three major groups: EA, ECO, and BIAs. The following sections give a short description about each of the algorithms in these subgroups.

1.2.1.1 Evolutionary Computation Algorithms

EC algorithms (Fogel et al. 2000) are probabilistic search methods that simulate the process of natural selection based on biological Darwinian evolution. This computing technique is a combination of computer science, natural genetic evolution, and some basic mathematical concepts. These algorithms were widely applied recently in different fields for various reasons. Practical problems that cannot be solved using traditional methods such as gradient methods, hill climbing, clustering, random search, etc., can be solved using EAs. EC involves an iterative search procedure, to arrive at the solution from a population of existing solutions. Upon every iteration, the fit solutions are selected and the poor solutions are neglected. The fit solutions have better characteristics and are selected for future evolutions. The characteristics are estimated in terms of a fitness measure, based on which a recombination is performed. Recombination is a process in which the solutions with high fitness are swapped partially with other solutions in the population to yield new solutions. Sometimes, mutation is also performed to modify the elements of the fit solution. Recombination and mutation are together referred to as the "genetic operators." Over several iterations, the entire population will contain more fit solutions than in the previous generations, and these tend to endorse the survival of the fittest. The basic pseudo-code of an EA is as follows:

```
Initialize population of individuals
While convergence not reached do
      Selection
      Genetic Operators
      Evaluation
Test for convergence
```

The fundamental features of EC, such as intelligence, parallelism, hybridization, computational cost, multiplicity of solutions, statistical property, robustness, and global optimization, have made EC applicable to several problems in all disciplines. As a search algorithm, EC is rising as the fundamental technology in the field of intelligent computation.

1.2.1.1.1 Genetic Algorithms

GAs, formerly known as "genetic reproductive plans," were initially introduced as an optimization tool by John H. Holland in the early 1960s (Goldberg 1989). GA is applied by evaluating the fitness of all of the individuals in the population. Once the fitness function is evaluated, a new population is created by performing operations such as crossover, fitness-proportionate reproduction, and mutation on the individuals. Every time an iteration is performed, the old population is rejected, and the iteration continues using the new population. GAs accentuate the significance of sexual reproduction using crossover as the main operator and mutation as the secondary operator. Sometimes, GAs use probabilistic selection similar to evolutionary programming

[EP; not applicable in evolutionary strategies (ESs)]. The basic operation of a GA is illustrated with the pseudo-code:

```
Initialize population P(0)
Evaluate population P(0)
While fit solution not found do
     For each individual in the population
          t=t+1
          Generate population P(t) from P(t-1)
          Evaluate fitness for each individual in population P(t)
          Selection
          Recombination
          Mutation
          Place new offspring in the population
     End For
End While
```

An initial population is randomly generated consisting of individuals, a set of chromosomes, or strings of characters, which represent all the possible solutions to the given problem. The encoding procedure followed by GAs has been the most significant feature because it uses binary encoding regardless of the type of decision variables of the problem to be solved. Later on, several other types of encoding procedures were also adopted depending on the specific type of optimization problem.

After encoding, the fitness function is applied to each individual in the population in order to determine the quality of the encoded solution. The next step is selection, in which the pair of individuals is chosen to reproduce the next generation offspring. The most common genetic operators, crossover and mutation, are applied on these selected parents to produce a new set of offspring. The new offspring is placed in the population, and the entire process iterates until a fit solution is determined.

1.2.1.1.2 *Genetic Programming*

Genetic programming (GP) is a specific form of GA, which manipulates with variable length chromosomes using modified genetic operators. John Koza introduced GP mainly to solve the problem of encoding in GA. The GP paradigm used a tree-based encoding to represent the individuals in the population (Fogel et al. 2000).

GP was developed to allow automatic programming and program induction. It may be viewed as a specialized form of GA, which manipulates with variable length chromosomes (i.e., with a specialized representation) using modified genetic operators. Unlike GAs, GP does not distinguish between the search space and the representation space. However, it is not difficult to introduce genotype/phenotype mapping for any EA formally (it could be one-to-one mapping in its simplest form).

The GP search space includes not only the problem space but also the space of representation of the problem, that is, GP is able to evolve representations. The theoretical search space of GP is the search space of all possible (recursive) compositions over a primitive set of symbols over which the programs are constructed. The programs are represented either as trees or in a linear form (e.g., machine language instructions). Crossover is considered as a major operator for GP. It interchanges randomly chosen subtrees of the parent trees without disrupting the syntax of the programs. Mutation picks a random subtree and replaces it by a randomly generated one. GP traditionally evolves symbolic expressions in a functional language such as Lisp. However, any useful structure may be utilized today. An evolved program can contain code segments that, when removed from the program, would not alter the result produced by the program (e.g., the instruction a = a + 0), that is, semantically redundant code segments. Such segments are referred to as "introns." The size of the evolved program can also grow uncontrollably until it reaches the maximum tree depth allowed, while the fitness remains unchanged. This effect is known as "bloat." The bloat is a serious problem in GP, since it usually leads to time-consuming fitness evaluation and reduction of the effect of search operators. Once it occurs, the fitness almost always stagnates.

These programs are expressed in GP as parse trees, rather than as lines of code. Thus, for example, the simple program "a + b * c" would be represented as follows:

or, to be precise, as suitable data structures linked together to achieve this effect. Because this is a very simple thing to do in the programming language Lisp, many GP users tend to use Lisp. However, this is simply an implementation detail. There are straightforward methods to implement GP using a non-Lisp programming environment.

The programs in the population are composed of elements from the function set and the terminal set, which are typically fixed sets of symbols selected to be appropriate to the solution of problems in the domain of interest. In GP, the crossover operation is implemented by taking randomly selected subtrees in the individuals (selected according to fitness) and exchanging them. It should be pointed out that GP usually does not use any mutation as a genetic operator.

GP can also employ a number of advanced genetic operators. For instance, automatically defined functions (ADF) allow the definition of subprograms that can be called from the rest of the program. Then the evolution is to find the solution as well as its decomposition into ADFs together. The fitness function is either application-specific for a given environment, or it takes the form of symbolic regression. In all cases, the evolved program must be executed in order to find out what it does. The outputs of the program are usually compared with the desired outputs for given inputs. A terminating mechanism of the fitness evaluation process must be introduced to stop the algorithm. The GP algorithm is presented in the pseudo-code form here:

```
Step 1: Generate an initial population of random compositions of the
functions and terminals of the problem.
Step 2: Perform the following sub steps in an iterative manner until the
termination criteria has been satisfied
Step 2a: Execute each program in the population and assign it a fitness value
Step 2b: Create a new population by applying the following
        Reproduce an existing program by coping it into the new population
        Create two new computer programs from existing programs by
        genetically recombining randomly chosen parts of two existing
        programs using the crossover operation applied at a randomly chosen
        crossover point within each program
Step 3: If the result represents a solution to the problem then stop else
continue Step 2.
```

1.2.1.1.3 *Evolutionary Strategies*

Ingo Rechenberg discovered the ES to solve optimization problems in hydrodynamics that were unable to be solved using traditional optimization techniques. This approach was introduced in 1964 (Fogel et al. 2000). In ES, only one parent was selected from the population, and mutation was performed on that to produce an offspring. The best among the parent and the offspring was selected and the remaining ones were rejected. This was the initial scenario. Over the years, several features were added to improve the performance of the ES to solve optimization problems.

In ES, every individual is composed of a vector of real numbers containing the parametric solution the problem, a vector of standard deviations to control the evolution of the individual, and a vector of angles. The angle vector and the standard deviation vectors are used as the main operators for the mutation operation. From the initial population of parents, offspring are produced according to the crossover and mutation mechanisms. The process of selecting parents for the genetic operation is a deterministic selection process. Based on the fitness function, the solutions are evaluated and the best are iterated for the next generation until a fit solution is obtained. The pseudo-code of ES is shown here:

```
Initialize population P(0)
Evaluate population P(0)
While fit solution not found do
    For each individual in the population
            t=t+1
            Generate population P(t) from P(t-1)
            Evaluate fitness for each individual in population P(t)
            Selection
            Recombination / Crossover
            Mutation
            Substitute (µ-1) worst parents with (µ-1) best offspring
    End For
End While
```

The major advantages of ES are (a) it is less sensitive to initial estimates and (b) it is capable of escaping to the local minima. The only disadvantage is that the ES requires a very long time to obtain a reliable solution.

1.2.1.1.4 *Evolutionary Programming*

The intelligent behavior of organisms proved to be an adaptive behavior in the study of Lawrence J. Fogel during the 1960s. Fogel (Fogel et al. 2000) used finite state machines as predictors to evolve the relation between the parents and the offspring. Since there is no fixed representation for a problem to be solved by EP, it is similar to ES and becoming a challenging task for researchers to distinguish between them.

Here, in EP, a population of individuals is allowed to mutate and produce a collection of offspring. EP uses a stochastic tournament selection method to select the individuals from the population. No crossover operation is performed among the individuals of the population. Since only mutation is performed, each parent will produce an offspring. The participation of an individual as a parent in the mutation process is decided in a probabilistic manner. No encoding is performed in the final stage. The pseudo-code of EP is presented here:

```
Initialize population P(0)
Evaluate population P(0)
While fit solution not found do
     For each individual in the population
          t=t+1
          Generate population P(t) from P(t-1)
          Evaluate fitness for each individual in population P(t)
          Selection
          Mutation
          Place best half of the population and delete the rest
     End For
End While
```

Mutation is the only operator that searches the solution space in an effective manner. It is obvious that mutations performed by EP focus on the probability distribution and try to identify the most suitable individuals for selecting the offspring. EPs are multitalented and very powerful techniques for solving approximation and combinatorial problems. Because of its parallel computational search methodology, researchers can solve problems that are computationally demanding.

1.2.1.1.5 Differential Evolution

Differential Evolution (DE) is a paradigm in the EA group developed by Storn and Price in 1995 (Storn and Price 1997). The behavior of DE is similar to that of GAs due to their searching operation among individuals, thus determining an optimal solution. One of the major differences that makes DE more popular than GA largely arises from the encoding scheme used to represent the permutations as vectors. In addition, the mutation operation in DE is a resultant of arithmetic combinations of individuals. The mutation operator in DE is responsible for both exploration and exploitation. At the beginning of the evolution process, the mutation operator of DE supports exploration, and during the progression of evolution the mutation operator favors exploitation. Therefore, DE repeatedly adapts the mutation increments to the best value based on the iterations of the evolutionary process. Some of the advantages of DE are: the algorithm is easy to implement, requires little parameter tuning, converges faster, and is a reliable, accurate, robust, and fast optimization technique. However, finding the best values for the problem-dependent control parameters used in DE is a time-consuming task, which is a disadvantage. A self-adaptive DE (SDE) algorithm can eliminate the need for manual tuning of the control parameters. The pseudo-code of the DE algorithm is provided here:

```
Initialize population P(0)
Calculate the fitness of the initial population P(0)
While fit solution not found do
     For each parent in the population
          Select three solutions at random
          Create one offspring using the DE mutation operator
```

```
        For each member of the next generation
                If fitness_offspring is better than fitness_parent
                        Replace parent with offspring
                End
            End For
        End For
End While
```

1.2.1.2 Ecology-Based Algorithms

Based on the ecosystems, there are a number of algorithms for designing and solving complex problems. These ecosystems comprises of the biological organisms along with the abiotic environment with which organisms interact, such as air, soil, water, and so on. These species in the ecosystem are sparse and they have different interactions among themselves. ECO algorithms include biogeography-based optimization (BBO) and invasive weed colony optimization (IWO) algorithms. The BBO algorithm includes the study of distribution of species over specified time and space, while IWO is capable of adapting to the environment.

1.2.1.2.1 Biogeography-Based Optimization

BBO was developed Dan Simon in 2008. This is a global search optimization algorithm, which was inspired by mathematical models of biogeography proposed by Robert MacArthur and Edward Wilson. BBO deals with the analysis of distribution of species in environment with respect to time and space. This is also referred to as the immigration and emigration of species between habitats, thus in turn sharing information among candidate solutions. In BBO, every possible solution is an island, and the habitability is characterized by features known as suitability index variables (SIVs). The fitness of each solution is based on the habitability characterization, which in turn is called the habitat suitability index (HSI). Information-sharing takes place from high-HSI solutions to low-HSI solutions during the process of emigration of features to the remaining habitats. During the immigration process, the low-HSI solutions accept new features from high-HSI solutions. During the iterative process, immigration and emigration improve the solutions and thus give a solution to the optimization problem. The value of HSI is considered as the objective function, and the algorithm is intended to determine the solutions that maximize the HSI by the immigrating and emigrating features of the habitats. The two major operators of BBO are migration (including emigration and immigration) and mutation. The habitat similar to population in other algorithms is formed randomly from habitat H, which consists of N (SIVs) integers. Each individual of the population is evaluated for fitness followed by migration and mutation steps to reach global minima. In migration, the information is shared between habitats, which depends on the emigration rates μ and immigration rates λ of each solution. Each solution is modified depending on probability P_{mod}, which is a user-defined parameter. Each individual has its own λ and μ, which are functions of the number of species K in the habitat. Poor solutions accept more useful information from good solutions, which improves the exploitation ability of the algorithm. In BBO, the mutation is used to increase the diversity of the population to get the good solutions. Some of the features of BBO include the following: In BBO, the original population is not discarded after each generation, but is rather modified by migration. Another distinctive feature is that, for each generation, BBO uses the fitness of each solution to determine its immigration and emigration rate.

The pseudo-code of the BBO algorithm is presented here:

```
Step 1: Define the probability of migration and probability of mutation
Step 2: Initialize the habitat
Step 3: Compute the immigration rate and emigration rate of each
candidate in the habitat
Step 4: Based on the immigration rate to choose the island to be modified
Step 5: Apply roulette wheel selection on the emigration rate, and select
an island - the SIV is to be emigrated from the chosen island
Step 6: Randomly select an SIV from the selected island to be emigrated
Step 7: Perform mutation based on the mutation probability of each island
Step 8: Evaluate the fitness of each individual island. If the fitness
criterion is not satisfied go to step 3.
```

1.2.1.2.2 Invasive Weed Optimization

IWO was proposed Mehrabian and Lucas in 2006 (Mehrabian and Lucas 2006). IWO is a numerical stochastic search algorithm inspired by the ecological process of weed colonization and distribution. The algorithm works by adapting itself to the environment and is capable of solving multidimensional linear and nonlinear optimization problems with improved efficiency. The population of weed is increased without any control of external factors in a specified geographical area. Since there is no external control, the population can be large or small. Initially, a certain number of weeds are randomly spread over the entire search range. Invasive weeds cover spaces of opportunity left behind by improper tillage, which is followed by enduring occupation of the field. As the colony becomes dense, there are fewer opportunities of life for the ones with lesser fitness. The behavior of the weed colony is dependent on time. These weeds will eventually grow up and execute the following steps and the algorithm proceeds as follows:

```
Step 1: Initialization—A population of initial solutions is initialized
with random positions in the D dimensional space.
Step 2: Fitness Evaluation—Each member in the population is evaluated for
fitness and ranked based on their fitness.
Step 3: Reproduction—New weeds are produced depending on its own fitness,
colony's lowest and highest fitness. This helps to concentrate on the
highest fitness values in the search domain and hence increases
convergence towards the best value.
Step 4: Spatial Dispersal—All the weeds generated from the reproduction
stage are dispersed over the D dimensional search space by normal
distribution of random numbers with mean equal to zero and varying
variance. The standard deviation (SD), σ, of the random function will
be reduced from a previously defined initial value σ initial, to a
final value, σ final, in every generation, which is given as follows:
```

$$\sigma_{iter} = \frac{(iter_{max} - iter)^n}{(iter_{max})^n} (\sigma_{initial} - \sigma_{final}) + (\sigma_{final})$$

```
Step 5: Selection—The maximum number of plants P_max from the best plants
reproduced is selected. The process repeats until optimal solutions have
reached.
```

1.2.1.3 Bio-Inspired Algorithms

BIAs, also referred to as "nature-inspired heuristic algorithms," are an outcome of the interspecies (between species) or intraspecies interaction (within species) in nature. The nature of these interactions can be cooperative or competitive. Cooperation includes division of labor, and represents the core of sociality. The interaction between the species (interspecies interaction) can be classified into three categories based on the outcome of interaction (positive, negative, neutral) and they are in turn termed as mutualism, parasitism, commensalism, respectively. Some of the examples of cooperative interaction are within species (i.e., homogeneous cooperation, also called social evolution), as in the social foraging behaviors of animal herds, bird flocks, insect groups and bacterial colonies; between species (i.e., heterogeneous cooperation, also called symbiosis), as in the mutualism between human and honey guide.

Nature-inspired heuristic techniques should fulfill several requirements:

- Ability to handle different type of problems
- Ease of use with few control variables
- Good convergence mechanism to the global minimum in consecutive independent trials.

In the previous years, the two main streams of the BIA area were ant colony optimization (ACO) and particle swarm optimization (PSO). In the recent years, new BIAs have appeared, inspired by fish schools as well as different aspects of the behavior of bees, ants, bacteria, fireflies, fruit flies, bats, and cuckoos. Despite swarm inspiration being common to these approaches, they have their own particular way to exploit and explore the search space of the problem. A brief discussion of this category of algorithms such as PSO, artificial bee colony (ABC), fish swarm algorithm (FSA), firefly algorithm (FA), bacterial foraging algorithm (BFA), ACO, cuckoo search optimization (CSO), fruitfly algorithm (FFA), and bat algorithm (BA) is presented in the following section.

1.2.1.3.1 Particle Swarm Optimization

PSO is a population-based stochastic optimization technique developed by Eberhart and Kennedy in 1995, inspired by social behavior of bird flocking (Eberhart and Kennedy 1995). PSO shares many similarities with EC techniques such as GA. The system is initialized with a population of random solutions and searches for optima by updating generations. However, unlike GA, PSO has no evolution operators such as crossover and mutation. In PSO, the potential solutions, called "particles," fly through the problem space by following the current optimum particles. Compared to GA, the advantages of PSO are that PSO is easy to implement and there are few parameters to adjust. PSO has been successfully applied in many areas, such as function optimization, ANN training, fuzzy system control, and areas where GA can be applied.

PSO learns from the scenario and uses it to solve the optimization problems. In PSO, each single solution is a "bird" in the search space. We call it "particle." All of particles have fitness values, which are evaluated by the fitness function to be optimized, and have velocities that direct the flying of the particles. The particles fly through the problem space by following the current optimum particles.

PSO is initialized with a group of random particles (solutions), and then searches for optima by updating generations. In every iteration, each particle is updated by following two "best" values. The first one is the best solution (fitness) it has achieved so far. (The fitness value is also stored.) This value is called pbest. Another "best" value that is tracked by the particle swarm optimizer is the best value obtained so far by any particle in the population. This best value is a global best, and is called gbest. When a particle takes part of the population as its topological neighbors, the best value is a local best, and is called lbest. After finding the two best values, the particle updates its velocity and positions using Equations 1.1 and 1.2.

$$v[] = v[] + c1 * \text{rand}() * (\text{pbest}[] - \text{present}[])$$

$$+ c2 * \text{rand}()(\text{gbest}[] - \text{present}[]) \tag{1.1}$$

$$\text{present}[] = \text{present}[] + v[] \tag{1.2}$$

where v[] is the particle velocity, and present[] is the current particle (solution). pbest[] and gbest[] are defined as before. rand () is a random number in (0,1). c1,c2 are learning factors; usually c1 = c2 = 2.

The pseudo code of the procedure is as follows:

```
For each particle
    Initialize particle
END
While maximum iterations is not attained Do
    For each particle
        Calculate fitness value
        If the fitness value is better than the best fitness value
           (pBest) in history set current value as the new pBest
        End
          Choose the particle with the best fitness value of all the
             particles as the gBest
             For each particle
                   Calculate particle velocity according equation (a)
                   Update particle position according equation (b)
             End
End While
```

Particle velocities on each dimension are clamped to a maximum velocity Vmax. If the sum of accelerations would cause the velocity on that dimension to exceed Vmax, which is a parameter specified by the user, then the velocity on that dimension is limited to Vmax.

1.2.1.3.2 Artificial Bee Colony

Karaboga has described an ABC algorithm based on the foraging behavior of honey bees, for numerical optimization problems (Karaboga and Basturk 2007). In the ABC algorithm, the colony of artificial bees consists of three groups: employed bees, onlookers, and scouts. The first half of the colony consists of the employed artificial bees, and the second half includes the onlookers. For every food source, there is only one employed bee. In other words, the number of employed bees is equal to the number of food sources around the hive. The employed bee whose the food source has been abandoned by the bees becomes a scout. In the ABC algorithm, the position of a food source represents a possible solution to the optimization problem, and the nectar amount of a food source corresponds to the quality (fitness) of the associated solution. The number of the employed bees or the onlooker bees is equal to the number of solutions in the population. The pseudo-code of the ABC algorithm is given here:

```
Initialize the population of solutions xi,j, i = 1 ...SN,j = 1 ...D
Evaluate the population
cycle=1
repeat
Produce new solutions υi,j for the employed bees and evaluate them
Apply selection process based on Deb's method
Calculate the probability values Pi,j for the solutions xi,j
Produce the new solutions υi,j for the onlookers from the solutions xi,j
selected depending on Pi,j and evaluate them
Apply selection process based on Deb's method
Determine the abandoned solution for the scout, if exists, and replace it
with a new randomly produced solution xi,j
Memorize the best solution achieved so far
cycle=cycle+1
until cycle=MCN
```

1.2.1.3.3 *Fish Swarm Algorithm*

The FSA proposed by Li et al. (2002) is a population-based swarm intelligent EC technique that simulates the schooling behavior of fish under water. The fishes in the swarm wish to stay very close to the school in order to protect themselves from predators, and in turn look for food, thus avoiding collisions within the group. Based on the fish behavior, the following operations are identified: random, chasing, swarming, searching, and leaping. Hence, the FSA is a fictitious entity of the behavior of true fish. For solving optimization problems, a "fish" within the school is represented by a point, known as a "candidate solution," and the school is the so-called population that is comprised of a set of points or solutions. The search space is the environment in which the artificial fish moves, thus searching for the optimum solution.

  FSA has a strong ability to achieve global optimization by avoiding local minimum. A fish is represented by its position $X_i = (x_1, x_2, ..., x_k, ..., x_D)$ in the D-dimensional search space. The food search behavior of the fish is represented as the food satisfaction factor FS_i. The relationship between two fish is denoted by their Euclidean distance $d_{ij} = \|X_i - X_j\|$. The searching process is a random search, where the fish searches for food with a tendency toward food concentration. The objective of the behavior is to minimize the FS (food satisfaction) factor. The swarming behavior tries to satisfy the food intake, thus entertaining and attracting new swarm members. A fish located at X_i has neighbors within its visual region. X_c identifies the center position of those neighbors and is used to describe the attributes of the entire neighboring swarm. If the swarm center has greater concentration of food than is available at the fish's current position X_i (i.e., $FS_c < FS_i$), and if the swarm (X_c) is not overly crowded ($n_s/n < \delta$), the fish will move from X_i to next X_{i+1}, toward X_c. When a fish locates food, neighboring individuals follow—this is the significance of the following behavior. In the fish's visual region, some fish will be perceived as finding a greater amount of food than others, and these fish will naturally follow the best one (X_{min}) in order to increase satisfaction ($[FS_{min} < FS_i]$ and less crowding $[n_f/n < \delta]$). Here, n_f represents number of fish within the visual region of X_{min}. The three major parameters involved in FSA are visual distance (visual), maximum step length (step), and a crowd factor. FSA's effectiveness seems primarily to be influenced by the former two (visual and step).

The artificial FSA is shown in the pseudo-code form as follows:

```
Random initialize Fish Swarm.
WHILE (is terminal condition reached)
  FOR (NumFish)
    Measure fitness for Fish.
  DO step Follow
  IF (Follow Fail) THEN
      DO step Swarm
      IF (Swarm Fail) THEN
          DO step Prey
    END
  END
  NumFish+1;
  End FOR
End WHILE
Output optimal solution
```

1.2.1.3.4 Firefly Algorithm

FA, inspired by the flashing behavior of fireflies, was proposed by Yang (Yang 2009). The algorithm consists of a population with agents analogous to fireflies. To solve an optimization problem, the algorithm goes through an iterative procedure with numerous agents communicating with each other. The agents share information with each other using bioluminescent glowing lights. These glowing lights enable them to explore the search space more effectively than in standard distributed random search. Every brighter fire fly attracts its partners, thus exploring the search space effectively and efficiently. Based on the glowing and flashing behavioral characteristics, the FA has the following rules:

- All fireflies are unisex and they will move toward more attractive and brighter ones regardless of their sex.

- The degree of attractiveness of a firefly is proportional to its brightness. Also the brightness may decrease as the distance from the other fire flies increases due to the fact that the air absorbs light. If there is no brighter or more attractive fire fly than a particular one, it will then move randomly.

- The brightness or light intensity of a fire fly is determined by the value of the objective function of a given problem.

The pseudo-code of the FA is shown here:

```
Objective function f(x), x = (x₁,.....xd)ᵀ
Generate Initial population of fireflies xᵢ, for I = 1,2,…,n
Light Intensity Iᵢ at xᵢ is determined by f(xᵢ)
Define light absorption coefficient γ
While (max Generations have not reached)
For i=1:n all n fireflies
    For j=1:n all n fireflies
        If (Iⱼ>Iᵢ)
            Move firefly I towards j in d-dimension
        End If
```

```
        Attractiveness varies with distance r according to exp[-γr]
        Evaluate new solutions and update light intensity
    End For j
End For i
Rank the fireflies and find the current best
End While
Post process results and visualization
```

1.2.1.3.5 *Bacterial Foraging Algorithm*

The BFO algorithm is a new class of biologically inspired stochastic global search technique based on the foraging behavior of the bacterium *Escherichia coli* (Passino 2002). This method is used for locating, handling, and ingesting the food in the form of a group foraging strategy. In the evolutionary process, bacteria search for food such that the energy obtained per unit time is maximized. Communication between every individual bacterium with others is also possible through signals. A bacterium takes foraging decisions after considering two previous factors: the energy consumed per unit time, and the communication in terms of signals. The foraging process involves two major actions: tumbling and swimming. The tumble action modifies the orientation of the bacterium. The swimming action, also referred to as "chemotaxis," is the process in which a bacterium moves by taking small steps while searching for nutrients. This is the key idea of BFO, which mimics the chemotactic movement of virtual bacteria in the problem search space. Chemotaxis is continued until a bacterium goes in the direction of a positive nutrient gradient. After a certain number of complete swims, the best half of the population undergoes reproduction, eliminating the rest of the population. In the elimination–dispersal event, the gradual or unexpected changes in the local environment in which a bacterium population lives may occur because of various reasons such as a significant local rise of temperature, which may kill a group of bacteria that are currently in a region with a high concentration of nutrient gradients. In order to escape local optima, an elimination–dispersion event is carried out where some bacteria are liquidated at random with a very small probability and the new replacements are initialized at random locations of the search space. BFO has already been applied to many real-world problems and proved its effectiveness over variants of GA and PSO. Mathematical modeling, adaptation, and modification of the algorithm might be a major part of the research on BFA in future. The pseudo-code of the BFO algorithm is presented here:

```
Initialize population
Calculate the fitness
While fit solution not found do
    For each chemotactic step for bacterium
        Compute fitness function
        Tumble
        Move in the direction of tumble
        Compute step size
        Swim
    End For
    While all bacterium undergoes chemotaxis do
        Reproduction
    End While
    Elimination-Dispersal
    End For
End While
```

1.2.1.3.6 Ant Colony Optimization

ACO is another popular algorithms that been applied in various fields of optimization after PSO was introduced (Dorigo and Gambardella 1997). It is an optimization algorithm modeled on the actions of an ant colony. ACO is a probabilistic technique useful in problems that deal with finding better paths through graphs based on the behavior of ants seeking a path between their colony and a source of food. Artificial ant (simulation agent) plays important role in locating optimal solutions by moving through a parameter space representing all possible solutions. Natural ants lay down pheromones directing the others to resources while exploring their environment. The simulated "ants" similarly record their positions and the quality of their solutions, so that in later simulation iterations more ants locate better solutions. The ACO algorithm requires the following to define the parameters:

- The problem needs to be represented appropriately, which would allow the ants to incrementally update the solutions through the use of a probabilistic transition rules, based on the amount of pheromone in the trail and other problem-specific knowledge. It is also important to enforce a strategy to construct only valid solutions corresponding to the problem definition.

- A problem-dependent heuristic function η that measures the quality of components that can be added to the current partial solution.

- A rule-set for pheromone updating, which specifies how to modify the pheromone value τ.

- A probabilistic transition rule based on the value of the heuristic function η and the pheromone value τ that is used to iteratively construct a solution.

The pseudo-code of ACO is given here:

```
Create construction graph
Initialize the number of ants, pheromone values
While end criterion false do
    For number of ants
        ant_k is positioned on a starting node;
            For problem_size
                Choose the state according to the probabilistic transition
                rules
            End For
        Update the trail pheromone intensity
        Compare and update the best solution
    End For
End While
```

1.2.1.3.7 Cuckoo Search Optimization

The CSO algorithm draws its inspiration from the breeding process of cuckoos where they lay their eggs in the nest of host birds (Yang and Deb 2009). It is a population-based algorithm in which some female cuckoos imitate the colors and patterns of eggs of a few species which they select. The host birds either throw their eggs away or destroy their nests if they find that the eggs do not belong to them. CS is based on three idealized rules:

1. Each cuckoo lays one egg at a time, and dumps its egg in a randomly chosen nest.
2. The best nests with high-quality of eggs will carry over to the next generation.
3. The number of available host nests is fixed and a host can discover an alien egg with a probability.

The egg in a nest represents a solution, while the cuckoo egg represents a new solution. The main aim is to achieve better solutions by employing a new, better egg, which is from cuckoo to replace not-so-good solutions in the nest.

The pseudo-code of CSO is given here:

```
Begin
define objective function.
generate initial population of host nests.
while (criteria not met)
{
get a cuckoo randomly;
evaluate the fitness of it;
choose a nest from the population randomly;
if(fitness of selected nest is high)
Abandon a fraction of worse nests and build new ones at new locations;
keep the best nests (solutions);
rank the nests and find the current best;
}
post process results and visualization;
End
```

1.2.1.3.8 *Fruit fly Algorithm*

FFA is a novel evolutionary optimization approach that is inspired by the knowledge from the food-finding behavior of fruit flies. As a novel EA, the fruit fly optimization algorithm introduced by Pan (2012) includes two main foraging processes: First, smell the food source by the osphresis organ and fly toward the corresponding location. Second, use sensitive vision to find food and fly toward a better site.

The procedure of the FFA is summarized as follows:

```
Step 1: Randomly initialize the location of the fruit fly swarm.
Step 2: Each individual searches for food in a random direction and
distance around the swarm location using osphresis to generate a new
population.
Step 3: Evaluate all the new individuals.
Step 4: Identify the best fruit fly with the maximum smell concentration
value (i.e. the best objective), and then the fruit fly group flies
towards the best location utilizing vision.
Step 5: End the algorithm if the maximum number of generations is
reached; otherwise, go back to Step 2.
```

1.2.1.3.9 Bat Algorithm

BA was introduced by Yang in 2010 (Yang 2010). This algorithm is based on the echo location behavior of micro-bats. It is based on three important rules.

1. For sensing distance, bat uses its echolocation capacity. It also uses echolocation to differentiate between food and prey and background barriers even in the darkness.
2. Bats fly randomly with some characteristics like a velocity, fixed frequency, and loudness to search for a prey.
3. It also features the variations in the loudness from a large loudness to minimum loudness.

Bats find their prey using varying wavelength and loudness, while their frequency, position, and velocity remain fixed. They can adjust their frequencies according to pulse emitted and pulse rate. The algorithm starts with initialization of the population of bats. Each bat is assigned a starting position, which is an initial solution. The pulse rate and the loudness are defined randomly. Every bat will move from local solutions to global best solutions after each iteration. The values of pulse emission and loudness are updated if a bat finds a better solution after moving. This process is continued till the termination criteria are satisfied. The solution so achieved is the final best solution.

The pseudo code of BA is shown here:

```
define objective function
initialize the population of the bats
define and initialize parameters
while(Termination criterion not met)
{
generate the new solutions randomly
if (Pulse rate (rand) >current)
select a solution among the best solution generate the local solution
around the selected best ones.
end if
generate a new solution by flying randomly
if ( loudness & pulse frequency (rand) < current )
accept the new solutions
increase pulse rate and reduce loudness
end if
rank the bats and find the current best
}
output results and visualization
```

Table 1.1 gives a summary of all the algorithms belonging to the GSOA based on the operators, parameters, applications, advantages, and disadvantages.

1.2.2 Machine Learning and Connectionist Algorithms

Machine learning is a subpart of the AI field concerned with the development of knowledge generalization methods, that is, inductive methods. A typical machine learning method infers hypotheses on a domain from examples of situations. Today, since the development of computers, several algorithms have been developed and successfully used in a range of domains (image analysis, speech recognition, medical analysis, etc.). However, there

TABLE 1.1

Comparison of GSOA Algorithms

GSOA	Operators	Parameters	Applications	Advantages	Disadvantages
GA	Crossover Mutation Selection Inversion	Population size No. of generations Crossover rate Mutation rate	Data mining Web site optimizations Wireless sensor networks Path planning robots Transportation and logistics Scheduling problems Assignment problems Pattern recognition VRP Wireless ad hoc networks Software engineering problems Pollutant emission reduction problem Power system optimization problems Optimal learning path in e-learning Web page classification system Structural optimization Defect identification system Molecular modeling Drug design Personalized e-learning system SAT solvers	GA is useful and efficient when • The search space is large complex or poorly known. • No mathematical analysis is available. • Domain knowledge is scarce to encode to narrow the search space • For complex or loosely defined problems since it works by its own internal rules	GA may have a tendency to converge toward local optima rather than the global optimum of the problem if the fitness function is not defined properly. Operating on dynamic data sets is difficult. For specific optimization problems, and given the same amount of computation time, simpler optimization algorithms may find better solutions than GAs. GAs are not directly suitable for solving constraint optimization problems
GP	Crossover Reproduction Mutation Permutation Editing Encapsulation Decimation	Population size Maximum number of generations Probability of crossover Probability of mutation	Design of image exploring agent Epileptic pattern recognition Automated synthesis of analogue electrical circuits Robotics Data mining Cancer diagnosis Fault classification electrical circuits Remote Sensing Applications	New constraints can be imposed directly into the GP fitness function, GP has short execution times, different classifiers are constructed due to different GP runs, GP based evolved optimal classifiers offer a new form of representation, GP search has a wide search area	Long training times, GP has variable performance from run to run, Difficulty in understanding the evolved solutions

(Continued)

TABLE 1.1 (*Continued*)

Comparison of GSOA Algorithms

GSOA	Operators	Parameters	Applications	Advantages	Disadvantages
ES	Mutation Selection Discrete recombination	Population size Maximum number of generations Probability of crossover Probability of mutation	Parameter estimation Image processing Computer vision system Task scheduling Car automation Structural optimization Gas-turbine fault diagnoses VRPs Clustering	Major advantage is the self-adaptation of strategy parameters, ESs have all properties required for global optimization methods	High computational demand, difficult adjustment of parameters
EP	Mutation Selection	Strategy parameter Correlation coefficient Mutation step size	Training, construction, and optimization of neural networks Optimal routing Drug design Bin packing Automatic control Game theory Pattern clustering and classification	Suitable for complex search spaces, Easy to parallelize, robustness	EP not using crossover also has one major disadvantage—that of speed, mutation is a slow way to search for good solutions
DE	Crossover, mutation, selection	Population size Dimension of problem, scale factor, probability of crossover	Image classification Clustering Digital filter design Optimization of non-linear functions Global optimization Multi-objective optimization.	DE automatically adapts the mutation increments to the best value based on the stage of the evolutionary process. DE is easy to implement, requires little parameter tuning, exhibits fast convergence, generally considered as a reliable, accurate, robust and fast optimization technique	Noise may adversely affect the performance of DE due to its greedy nature Also the user has to find the best values for the problem-dependent control parameters used in DE and this is a time consuming task

(Continued)

TABLE 1.1 (*Continued*)

Comparison of GSOA Algorithms

GSOA	Operators	Parameters	Applications	Advantages	Disadvantages
BBO	Migration (emigration and immigration), mutation	Number of habitats (population size), maximum migration rates, mutation rate	Constrained optimization Sensor selection aircraft engine Health estimation Power system optimization Groundwater detection Satellite image classification Graphical user interface Global numerical optimization State estimation	Highly competitive with other optimization algorithms due to faster convergence	BBO is poor in exploiting the solutions, There is no provision for selecting the best members from each generation, A habitat doesn't consider its resultant fitness while immigrating the features, as a result so many infeasible solutions are generated
IWC	Reproduction, spatial dispersal, selection	Weed population size, modulation index, standard deviations	Tuning of robust controller Antenna configuration optimization Encoding for DNA computing Aperiodic thinned array antennas Multiple task assignment of unmanned aerial vehicles Fractional order proportional integral derivative (PID) Controller Training of Neural Networks Multi-carrier code division multiple access interference suppression	Capable of solving real world practical problems but has not gained much popularity	IWC puts more and more search efforts around the best solution found so far and gradually loses exploration ability, thus resulting in premature convergence
PSO	Initializer, updater and evaluator	Number of particles Dimension of particles, Range of particles Learning factors Inertia weight Maximum number of iterations	Biomedical image registration Iterated Prisoner's Dilemma Classification in multiclass databases Feature selection Web service composition Power System Optimization Edge detection in noisy images Scheduling problems VRPs Prediction of ANN Quality of service in ad hoc multicast Anomaly detection Color image segmentation Sequential ordering problem Machinery fault detection Unit commitment computation Signature verification	PSO algorithm maintains its stochastic behavior capacity while searching for the global optimum value.	Stability problem of PSO restricts the success rate of the algorithm.

(*Continued*)

TABLE 1.1 (*Continued*)

Comparison of GSOA Algorithms

GSOA	Operators	Parameters	Applications	Advantages	Disadvantages
ABC	Reproduction, replacement of bee, selection	Number of food sources which is equal to the number of employed or onlooker bees (SN), the value of limit, the maximum cycle number (MCN)	Scheduling problems Image segmentation Capacitated VRP Assembly line balancing problem Solving reliability redundancy allocation problem Training neural networks Decoder–encoder 3-Bit Parity benchmark problems Pattern classification P-center problem	Has a successful decision mechanism that decides which areas within the search space require to be surveyed for best results. The strategy of the ABC algorithm used to discover new nectar sources within the ABC algorithm and manage the capacity of the nectar sources discovered is also substantially powerful. Simplicity, flexibility and robustness. Use of fewer control parameters compared too many other search techniques. Ease of hybridization with other optimization algorithms Ability to handle the objective cost with stochastic nature. Ease of implementation with basic mathematical and logical operations.	Convergence performance of ABC for local minimum is slow
FSA	Prey Swarm Follow Search	Visual distance, max step length, crowd factor	Function optimization Parameter estimation Combinatorial optimization Least squares support vector machine Geo technical engineering problems	The ability to solve complex nonlinear high dimensional problems. Furthermore, it can achieve faster convergence speed and require few parameters to be adjusted. Whereas the FSA does not possess the crossover and mutation processes used in GA, so it could be performed more easily.	High time complexity Lack of balance between global and local search. Lack of benefiting from the experiences of group members for the next movements.

(*Continued*)

TABLE 1.1 (*Continued*)

Comparison of GSOA Algorithms

GSOA	Operators	Parameters	Applications	Advantages	Disadvantages
FA	Flashing patterns and behavior	Population size, Attractiveness parameter, distance, randomness scaling factor, cooling factor, movement	Multi objective optimization Digital image compression Feature selection Nonlinear problems Multimodal design problems Antenna design optimization Load dispatch problems NP-hard scheduling problems Classifications and clustering Training ANN	Automatic subdivision based on attraction toward brightness and the ability of dealing with multimodality.	Getting trapped into several local optima, performs local search as well and sometimes is unable to completely get rid of them, parameters are fixed and they do not change with the time, FA does not memorize or remember any history of better situation, and they may end up missing their situations.
BFA	Reproduction, chemotaxis Dispersion, elimination	Dimension of the search space, number of bacteria, number of chemotactic steps, number of elimination and dispersal events, number of reproduction steps, probability of elimination and dispersal, location of each bacterium, no: of iterations, step size	Harmonic problem in power systems Optimal power system stabilizers design Tuning the PID controller of an AVR Optimal power flow solution Machine learning Job shop scheduling Transmission loss reduction Constrained economic load dispatch problems Application in the null steering of linear antenna arrays	Excellent global searching.	Sensitive to step size value required by the tumble operator, specific type of constraint-handling mechanism which is difficult to generalize and also the high number of parameter values which must be defined by the user

(*Continued*)

TABLE 1.1 (*Continued*)

Comparison of GSOA Algorithms

GSOA	Operators	Parameters	Applications	Advantages	Disadvantages
ACO	Pheromone Update and measure, trail evaporation	Number of ants, iterations, pheromone evaporation rate, amount of reinforcement	Traveling Salesman Problem Quadratic Assignment problem Job-Shop Scheduling Dynamic problem of data network routing Shortest path problem VRP Graph coloring and set covering Agent-based dynamic scheduling Digital image processing Classification problem in data mining	Inherent parallelism Positive Feedback accounts for rapid discovery of good solutions Efficient for Traveling Salesman Problem and similar problems Can be used in dynamic applications (adapts to changes such as new distances) Distributed computation avoids premature convergence.	Theoretical analysis is difficult. Sequences of random decisions, Probability distribution changes by iteration Research is experimental rather than theoretical Time to convergence uncertain.
CSO	Search mechanism, Random walk using Levy flights	No. of nests Tolerance Discovery rate Search dimension Upper and lower bounds of the search domain Levy exponent Levy step size No. of iterations	Optimizing the parameters of various classifiers including Neural Network, RBF, support vector machine parameters Finding optimizing cluster centers Job scheduling Find optimal path Health sector Wireless sensor network Image processing Unit commitment and economic load dispatch	The generic movement model provided by Levy flights in CSO is an effective approach to solve optimization problems. Due to the random walk in Levy flight mechanism, the algorithm converges much faster, thus reducing the computational time.	The main drawback of this method appears in the number of iterations to find an optimal solution. If the value of a discovery rate is small and the value of Levy step size is large, the performance of the algorithm will be poor and leads to considerable increase in number of iterations. In case of vice-versa, the speed of convergence is high but it may be unable to find the best solution.

(Continued)

TABLE 1.1 (*Continued*)

Comparison of GSOA Algorithms

GSOA	Operators	Parameters	Applications	Advantages	Disadvantages
FFA	Location attractiveness based on smelling, Shooting process, Selection	Smelling radius, Shooting radius, Number of evaluations	Military applications Medicine Management Finance Combined with other data mining techniques	FFA can quickly find the global optimum with the high accuracy without falling into local extremum, thus leading to improved robustness, simple update strategy	Definition of update strategy is a drawback of the FFA
BA	Echolocation behavior of bats, Random walks	Velocity and Position Vectors, Variations of Loudness and Pulse Emission	Engineering design Classifications of gene expression data Training ANN Ergonomic workplace problems Prediction for energy modeling Train standard eLearning datasets	Claimed to provide very quick convergence at a very initial stage by switching from exploration to exploitation, more suitable for classification applications.	When BA switches to exploitation stage too quickly, it may lead to stagnation after some initial stage

is currently no ultimate technique, but many different techniques exist with their own advantages and inconveniences (noise sensitivity, expression power, algorithmic complexity, soundness, completeness, etc). The research in this area is very active, and it includes two major groups: ANN and AI.

1.2.2.1 Artificial Neural Networks

ANNs are machine learning approaches that are modeled after the neural structure of the brain. These biologically inspired methods of computing are relatively insensitive to noise, are algorithmically cheap, and have been proven to produce satisfactory results in several domain areas of engineering. These ANNs try to replicate the basic elements of human brain, which is a complicated, versatile, and powerful organ. The essential processing element of a neural network is a neuron, whose function is to receive inputs, combine them in some way, perform a nonlinear operation, and then output the final result. Based on this concept, the functioning of the artificial neuron is analogous to a neuron in the human brain. Inputs are presented to the network consisting of a set of neurons. Each of these inputs is multiplied by a connection weight. The resulting products are simply summed, fed through a transfer function to generate a result, and hence the output is obtained.

A simple neural network consists of the input, the output, and the hidden layer, as shown in Figure 1.2. The set of neurons that interface to the real world to receive its inputs form the input layer, while neurons that provide the real world with the network's outputs form the output layer. Some of the neurons are hidden in between the input and output layers, which forms the hidden layer. There are networks whose outputs are fed back to the hidden layer for further learning, and there also exists a competition between the nodes or neurons in the output layer. When the architecture of the neural network is framed, then the network is ready for the training or learning process. The initial weights are chosen

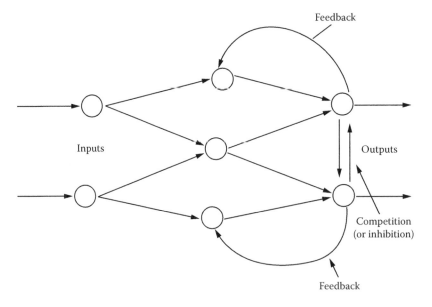

FIGURE 1.2
Simple neural network.

randomly to train a neural network. There are two different categories of training: supervised learning and unsupervised learning. Supervised learning involves a mechanism of providing the network with the desired output either by manually "grading" the network's performance or by providing the desired outputs with the inputs. In unsupervised learning, the network has to make sense of the inputs without outside help. The weights of the neurons are modified during the training process. This does not happen at once but over several iterations, which are determined by the learning rules.

The rate at which ANNs learn using the learning rules depends upon the controlling factor, namely the learning rate. A slower rate takes a long time to perform the learning process. With faster learning rates, however, the network may not be able to make the fine discriminations possible with a system that learns more slowly. Hence a tradeoff is required in deciding the choice of the learning rate. Learning is certainly more complex than the simplifications represented by the learning laws. Some of the learning laws are Hebb's rule, the Hopfield law, the delta rule and the extended Delta rule, the competitive learning rule, the outstar learning rule, the Boltzmann learning law, and the memory-based learning law.-

Several neural network architectures have been developed based on the areas to which they are applied. Most of the applications of ANN fall into four different categories: (i) prediction, (ii) classification, (iii) data association and (iv) data conceptualization. ANNs such as perceptron, back propagation, delta bar delta, extended delta bar delta, directed random search, and higher order neural networks belong to the category of prediction networks. Learning vector quantization, counter-propagation network, and probabilistic neural networks serve as excellent classification algorithms. Data association networks are Hopfield network, Boltzmann machine, Hamming network, and bidirectional associative memory. Adaptive resonance network and self-organizing map belong to the data conceptualization group of networks. A detailed description of these networks including the architecture, training, and testing algorithm can be found in Sumathi and Surekha (2010).

- *Advantages:*
  - Requirement of less formal statistical training
  - Ability to implicitly detect complex nonlinear relationships between dependent and independent variables
  - Ability to detect all possible interactions between predictor variables
  - Availability of multiple training algorithms.
- *Disadvantages:*
  - "Black box" nature of ANN
  - Greater computational burden
  - Proneness to overfitting
  - Empirical nature of model development.
- *Applications:*
  - Image processing
  - Signal Processing
  - Weather prediction and forecast
  - Pattern recognition
  - Classification.

1.2.2.2 Artificial Intelligence

AI comprises the broad area that is focused on the synthesis and analysis of computational agents that behave intelligently (Krishnamoorthy and Rajeev 1996). Any object, biological organism, or a human can be thought of as an agent who can act in the environment in an intelligent manner. An agent can be a person, a dog, a resistor, a thermostat, a bus, a robot, a computer, a city, and so on. An agent is judged by its actions and its response in the form of actions to a particular input. During such actions, it is very important to determine the intelligence involved in the action. Such intelligence of an agent is based on factors such as the following:

- Whether the action carried out by the agent is suitable for satisfying the goals based on the circumstances;
- Whether the agent is adaptable to different environments and different goals;
- Whether the agent is capable of making choices within the computational limitations;
- Whether the agent is capable of learning through experience.

The main objective of an AI algorithm is to construct a machine that demonstrates the intelligence associated with the factors mentioned before. The machine should be able to possess human intelligence, not necessarily the cognitive stages of the human brain. Turing determined whether the machine is capable of possessing intelligence based on a test called the Turing test. In his test, Turing proposed the idea of an "imitation game," in which the communication between the machine and the human would be based on textual messages. According to Turin, if a human cannot question the computer, then it is not fair to call the computer an intelligent machine.

Based on the Turing test, the intelligent behavior of agents is classified as perception involving image recognition and computer vision, reasoning, learning, understanding language involving natural language processing, speech processing, and solving problems. Some of the approaches to AI are as follows:

- **Strong AI**: In the strong AI approach, machines that are capable of really reasoning and solving problems are designed. These kinds of machines should have an intelligence level equal to that of human intelligence. This approach programs machines such that they are capable of maintaining cognitive mental states.
- **Weak AI**: In the process of implementing AI, some machines are unable to realize and solve problems, but they try to act as intelligent agents. In the case of weak AI, suitably programmed machines alone can simulate human cognition.
- **Applied AI**: This approach aims at developing applications for commercial purposes, such as smart systems. This is one of the successful approaches in applications such as smart cards, face recognition, and security systems.
- **Cognitive AI**: This approach develops machines that study the cognitive behavior of the human brain. The theories about machines recognize objects (in case of robotics), or the solution approach to abstract problems belong to this class of AI.

- *Advantages:*
  - No interruption while performing tasks assigned to the AI-based system
  - Hardware advantages such as greater serial speeds and parallel speeds

- Self-improvement advantages like improvement of algorithms, design of new mental modules, and modification of motivational system
- Cooperative advantages such as copyability, perfect cooperation, improved communication, and transfer of skills.
- *Disadvantages:*
  - Limited sensory input
  - Prone to malfunction
  - Lacks the means to store a comparable size of memory
  - Poor speed of knowledge retrieval.
- *Applications:*
  - Biometric recognition (speech, face)
  - Signal processing
  - Game playing
  - Understanding natural language
  - Expert systems
  - Heuristic classification
  - Computer vision.

1.2.3 Approximate Reasoning Approaches

AR is based on the idea of sacrificing soundness or completeness for a significant speedup of reasoning. This is to be done in such a way that the number of introduced mistakes is at least outweighed by the obtained speedup. When pursuing such AR approaches, however, it is important to be critical not only about appropriate application domains but also about the quality of the resulting AR procedures. FL theory and chaos theory are major classes of AR approaches that are becoming very popular.

1.2.3.1 Fuzzy Logic

FL was initiated in 1965 by Lotfi A. Zadeh, Professor of Computer Science, University of California, Berkeley (Ross 2004). Basically, FL is a multivalued logic that allows intermediate values to be defined between conventional evaluations such as true/false, yes/no, high/low, and so on. These intermediate values can be formulated mathematically and processed by computers in order to apply a more human-like way of thinking.

From its inception, FL has been, and to some degree still, an object of skepticism and controversy. In part, skepticism about FL is a reflection of the fact that in English the word "fuzzy" is usually used in a pejorative sense. But, more importantly, for some, FL is hard to accept because by abandoning bivalence it breaks with centuries-old tradition of basing scientific theories on bivalent logic. It may take some time for this to happen, but eventually abandonment of bivalence will be viewed as a logical development in the evolution of science and human thought.

FL forms a bridge between the two areas of qualitative and quantitative modeling. Although the input–output mapping of such a model is integrated into a system as a quantitative map, internally it can be considered as a set of qualitative linguistic rules.

Since the pioneering work of Zadeh in 1965 and Mamdani in 1975, the models formed by FL have been applied to many varied types of information processing including control systems.

Elements of a fuzzy set are taken from a *universe of discourse* or *universe* for short. The universe contains all elements that can come into consideration. The membership function $\mu_A(x)$ describes the membership of the elements x of the base set X in the fuzzy set A, whereby for $\mu_A(x)$ a large class of functions can be taken. Different types of membership functions include the triangular function, the Γ function, the S function, the trapezoidal function, Gaussian function, and the exponential function.

Fuzzy inference is the process of formulating the mapping from a given input to an output using FL. The mapping then provides a basis from which decisions can be made or patterns discerned. The process of fuzzy inference involves all the topics such as fuzzification, defuzzification, implication, and aggregation. Expert control/modeling knowledge, experience, and linking the input variables of fuzzy controllers/models to output variable (or variables) are mostly based on fuzzy rules. There are two basic classes of fuzzy rules: Mamdani fuzzy rules and Takagi–Sugeno fuzzy rules. A fuzzy expert system consists of four components: the fuzzifier, the inference engine, the defuzzifier, and a fuzzy rule base (as shown in Figure 1.3).

In the fuzzy expert system model, a rule base is constructed to control the output variable. The rule base consists of a set of simple IF-THEN rules with a condition and a conclusion (also referred to as the antecedent and consequent). These rules are evaluated based on fuzzy set operations during the inference step. The rules are applied on the fuzzified values obtained from the fuzzification process. The output of the inference step is a set of fuzzy values. These fuzzy values have to be converted to crisp values before giving to the outside world. Hence, this result should be defuzzified to obtain a final crisp output, a process known as defuzzification. Defuzzification is performed according to the membership function of the output variable and there are several methods for defuzzification, including center of gravity method, center of singleton method, maximum methods, and so on.

The FL algorithm can be framed according to the following steps:

1. Define the linguistic variables and terms (initialization).
2. Construct the membership functions (initialization).
3. Construct the rule base (initialization).

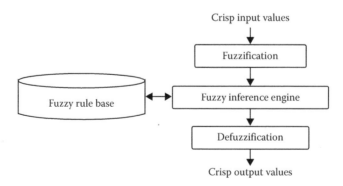

FIGURE 1.3
Fuzzy expert system model.

4. Convert crisp input data to fuzzy values using the membership functions (fuzzification).

5. Evaluate the rules in the rule base (inference).

6. Combine the results of each rule (inference).

7. Convert the output data to non-fuzzy values (defuzzification).

- *Advantages:*
  - Suitable user interface
  - Easy computation and learning
  - Consistency and redundancy of rule base
  - Applicability to solve certain nonlinearities due to universal approximation capabilities
  - Possibility of expansion of rough approximations to fuzzy environment and hence obtaining solutions for real-world problems
  - Applicable in situations where process model of the problem is unavailable.
- *Disadvantages:*
  - No systematic approach to design fuzzy systems: only guidelines are available to choose the fuzzification, inference, and defuzzification methods
  - Suitability of fuzzy logic controls (FLCs) limited to solving trivial problems that do not require high accuracy. Not appropriate for highly complex systems
  - The rule base giving equal importance to all the factors chosen to frame the rules
  - Requires prior knowledge.
- *Applications:*
  - Can be applied to a variety of systems from appliances to domestic devices to really complex problems like waste water management.
  - Several potential areas include
    - Aerospace
    - Automotive
    - Finance
    - Chemical industry
    - Electronics
    - Manufacturing and Industrial automation
    - Image processing, and so on.

1.2.3.2 Chaos Theory

The unpredictable movement of basic components in a system, that is, systems containing simple parts, is called *chaotic behavior* or *chaotic movement*. Such movement is due to the straightforward comparisons, and they can have exceptionally confused arrangements. Though several traditional methods are available, they have failed to analyze the behavior of such chaotic systems. Solutions to the chaotic systems have been possible during recent years through computer-based simulations. Based on an in-depth analysis and resulting perceptions on several simple examples, three properties have risen:

1. The chaotic behavior does not repeat.
2. The movement cannot be predicted since the initial conditions are never known precisely.
3. The return path of the system after the chaotic behavior is complicated.

The properties of chaotic systems are unusual, either taken individually or together; the most efficient way to understand them is by considering particular cases. Examples of chaotic motion of simple systems include body swinging on a pulley, ball bouncing on slopes, and magnetic and driven pendulums.

The term "chaos theory" (CT) appears be to some degree a paradoxical expression. Mathematicians have identified some new and authoritative learning about totally irregular and unfathomable phenomena of chaos systems (Strogatz 2014). However, this is not feasible for all cases. An adequate definition of chaos theory is the qualitative study of unstable, aperiodic behavior in deterministic nonlinear dynamical systems. A dynamical framework may be characterized as a rearranged model for the time-varying behavior of a real framework, and aperiodic behavior is essentially the behavior that happens when no variable depicting the condition of the framework experiences a consistent redundancy of qualities. Aperiodic behavior never repeats, and it keeps on showing the impacts of any little irritation. Consequently, any forecast of a future state in a given framework is aperiodic. Maybe the most identifiable image connected with the butterfly effect is the renowned Lorenz Attractor. Edward Lorenz, an inquisitive meteorologist, was searching for an approach to model the activity of the confused behavior of a chaotic framework.

Hence, he took a few equations from the field of fluid dynamics, simplified them, and got the following three-dimensional system:

$$\frac{dx}{dt} = \text{delta} * (y - x)$$

$$\frac{dy}{dt} = r * x - y - x * z \tag{1.3}$$

$$\frac{dz}{dt} = x * y - b * z$$

Delta represents the "Prandtl number," the ratio of the fluid viscosity of a substance to its thermal conductivity; however, one does not have to know the exact value of this constant; hence, Lorenz simply used 10. The variable r represents the difference in temperature between the top and bottom of the gaseous system. The variable b is the width to height ratio of the box that is being used to hold the gas in the gaseous system. Lorenz used 8/3 for this variable. The resultant x of the equation represents the rate of rotation of the cylinder, y represents the difference in temperature at opposite sides of the cylinder, and z represents the deviation of the system from a linear, vertical graphed line representing temperature. If one were to plot the three differential equations on a three-dimensional plane, using a computer of course, no geometric structure or even complex curve would appear; instead, a weaving object known as the Lorenz Attractor appears. Because the system never exactly repeats itself, the trajectory never intersects itself. Instead, it loops around forever. The attractor will continue weaving back and forth between the two wings, its motion seemingly random, and every action mirroring the chaos that drives the process. Lorenz had obviously made an immense breakthrough in not only chaos theory but life itself. He proved that complex, dynamical systems show order, but they never repeat. Since our world is classified as a dynamical, complex system, our lives, our weather, and our experiences will never repeat; however, they should form patterns.

- *Advantages:*
  - Wide applicability to nonlinear systems
  - Prediction.
- *Disadvantages:*
  - The choice of input parameters is extremely difficult—the approach used to identify and choose the data is highly complex and not an accurate one.
  - Mapping of business environment with chaos theory is complex.
  - It lacks empirical evidence.
- *Applications:*
  - Predicting long-term behavior of biological systems
  - Physical and chemical systems
  - Systems engineering
  - High-dimensional systems

The applications of chaos theory are infinite; seemingly random systems produce patterns of spooky, understandable irregularity. From the Mandelbrot set to turbulence to feedback and strange attractors, chaos appears everywhere. Breakthroughs have been made in the past in the area of chaos theory, and in order to achieve any more colossal accomplishments in the future, they must continue to be made.

1.2.4 Conditioning Approximate Reasoning Approaches

CAR approaches to GM have become an extremely popular tool for modeling uncertainty. These approaches use a standard probability theory to deal with uncertainty. The two most common types of GM are Bayesian networks and Markov models. In real-world applications, there may exist several interdependencies between the states. The independence properties in the distribution can be used to represent such high-dimensional distributions much more compactly. Probabilistic GM provides a general-purpose modeling language for exploiting this type of structure in representation of real-world problems.

1.2.4.1 Hidden Markov Models

HMMs are a formal foundation for making probabilistic models of linear sequence "label-ing" problems. They are commonly applicable to speech signal processing and are gaining much attention in the area of communication systems. Complex models are represented in pictorial form using a conceptual toolkit. The HMM is generally restricted to discrete systems in discrete time domain (Elliot et al. 1995).

To generate a sequence using HMM, consider an example of visiting a state based on the state's emission probability distribution. This example generates two set of information: one set of information is the state path (transition from one state to another), and the other set of information is the *observed sequence*, where residue is emitted from each state in the state path. Here, the state path is referred to as the Markov chain, which is used to decide which state to be visited next. With the available observed sequence, the underlying path is hidden: hence the chain is referred to as HMM.

An HMM is a discrete time process represented as $\{X_k, Y_k\}_k \geq 0$, where $\{X_k\}$ is a Markov chain, which in turn is conditional on $\{X_k\}$. In this case, $\{Y_k\}$ is a sequence of independent random variables (observable sequence) where the conditional distribution of $\{Y_k\}$ depends only on $\{X_k\}$ (hidden sequence). This dependence in HMM can be represented as a graphical model, as shown in Figure 1.4.

An HMM is characterized by the following elements:

1. The number of states in the model, N: There is always a significance to the states of the HMM even though they are hidden. The states are connected in such a manner that any state can be reached from any other state in time t. Let the state at time t be referred to as q_t.

2. The number of distinct observation symbols per state, M: The observation sym-bols correspond to the physical output of the system being modeled. Let the indi-vidual symbols be denoted as $V = \{v_1, v_2, \ldots, v_m\}$.

3. The state transition probability distribution $A = \{a_{ij}\}$, where $a_{ij} = P[q_{t+1} = S_j \mid q_t = S_i]$, $i \geq 1, j \leq N$.

4. The observation symbol probability distribution in state j, $B = \{b_j(k)\}$.

5. The initial state distribution $\pi = \{\pi_i\}$.

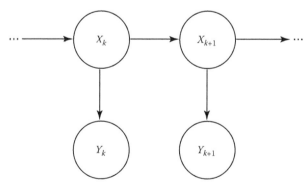

FIGURE 1.4
Dependence structure of HMM.

With appropriate values of N, M, A, B, and π, HMM can be used for generating an observation sequence as follows:

a. Choose an initial state according to state $q_t = S_i$ distribution π.

b. Set $t = 1$.

c. Choose $O_t = v_k$ according to the symbol probability distribution in state S_i.

d. Transit to a new state q_{t+1}.

e. Increment t to $t + 1$.

f. Return to step c if $t < T$; else, terminate the process.

- *Advantages:*
  - HMMs are probabilistic graphic models for which algorithms are known for exact and approximate learning and inference.
  - HMMs are able to represent the variables of a system through probability distributions.
  - HMMs are capable of capturing the dependencies between consecutive measurements.
- *Disadvantages:*
  - HMMs represent the behavior of only discrete systems whereas it fails for continuous-time systems.
  - Because of their Markovian nature, they do not take into account the sequence of states leading into any given state.
  - The time spent in a given state is not captured by the HMM.
- *Applications:*
  - Gene prediction
  - Classification of medical images
  - Decision making in statistics
  - Bioinformatics
  - Facial expression identification from videos
  - Human action recognition from Time Sequential Images

1.2.4.2 Bayesian Belief Networks

A BBN is a graphical representation of a probabilistic dependency model used for understanding and simulating computational models for complex systems in several engineering domain areas (Mittal and Kassim 2007). Though they possess excellent capability for capturing uncertainties, the knowledge required to create and initialize a network is vast. This has prevented the application of BBNs in several applications. BBNs are also represented as graphs with nodes and directional arrows. Nodes in BBN represent variables in the dependency model, while the directional arrows represent the causal relationships between these variables.

The belief in each state of a node is updated whenever the belief in each state of any directly connected node changes. The directed acyclic graph (DAG) $G(U, E)$ is a BBN if and only if it is a minimal I-map of the model M representing the probability distribution $p(U)$, where U is the set of all variables in M. For any ordering X_1, X_2, ..., X_n of the variables in the probability distribution $p(U)$, a DAG in which a minimal set of predecessor variables are designated as parents of each variable X_i, such that

$$p(X_i \mid W); \quad W \in \{X_1, X_2, \dots, X_{i-1}\} \tag{1.4}$$

is a BBN of that probability distribution.

If the distribution is strictly positive, then all of the parent sets are unique for that ordering (Pearl 1988). Although the actual structure of a Bayesian network depends on the particular ordering of the variables used, each such network is an I-map of P, and the resulting DAG represents many of the independencies in M. All orderings that are consistent with the direction of the arrows in a DAG produce the same network topology. For a DAG to be a Bayesian network of M, it is necessary and sufficient that each variable be conditionally independent of all of its non-descendants given its parents, and that no proper subset of its parents also satisfies this condition. Although many different BBNs may be used to represent the same probabilistic model, causation imposes standard orderings on the variables when representing experimental knowledge, producing identical network topologies for the same experimental knowledge.

- *Advantages:*
  - The BBN models are powerful tools for dealing with uncertainties in real-world systems.
  - The graphical representation in BBNs depicts the relationship between variables in a simpler manner.
  - In BBN, networks can be easily redefined—any additional variables can be appended, and mapping can be done without much effort.
  - BBN allows evidences to be added into the network, and the network is updated so that the probabilities can propagate to each node.
  - Qualitative and quantitative analysis is possible in BBN.
  - Researchers can understand the behavior or situation of a system due to the interactive graphical modeling.
- *Disadvantages*:
  - BBN requires considerable knowledge and effort to obtain the parameters for the model and apply them for complex situations.
  - BBN requires a considerable knowledge and effort to construct realistic and consistent graphs, which is quite hard to establish.
  - In situations where domain knowledge is insufficient or inaccurate, the model outcomes are subject to error.
- *Applications:*
  - Medical diagnosis
  - Decision making
  - Social network analysis

1.3 Role of CI Paradigms in Engineering Applications

The significance of creating intelligent artifacts appears in one form or another throughout recorded history. The emergence of digital computers has led to the rise of CI, which gave rise to a range of vital queries like the kind of intelligence that can be processed, the method of storing and accessing the intelligent knowledge in a user-friendly form, and so on. In general, human beings possess two basic extraordinary capabilities. The first capability is the capacity for reasoning, communicating, and making rational decisions in an imprecise, uncertain, vague environment. The second is the capacity to perform a wide variety of physical and mental tasks without any computations. These capacities provide the basis for automation of natural language understanding in different application areas. The application of hybrid CI paradigms in the areas of aerospace, networking, consumer electronics, control systems, power system, robotics, image processing, medicine, and communication systems are discussed in this section.

1.3.1 Aerospace and Electronics

This section presents a complete survey of successful applications of EA and swarm intelligence to various aerospace-related problems. The main intention of this survey is to give the aerospace industry a feel for the power of EA and swarm intelligence and the associated benefits. In the aerospace domain, there is a need for methods to solve complex search and planning problems. These problems often are in a complex domain, which is also often ill-structured. EA and swarm intelligence provide an extensive variety of solutions, allowing us to cope with changing circumstances. Neural networks and AI are capable of providing effective means for resolving several problems encountered in the aerospace industry, especially in nonlinear dynamic control, target feature recognition, classification, and so on. Nowadays, neural networks are a part of hybrid systems, which also utilize FL and expert systems for various aerospace-related applications. FL is a dominant problem-solving methodology, which is similar to human decision making with its ability to work from approximate data and find precise solutions. In aerospace, FL enables very complex real-time problems to be tackled using a simple approach. Bayesian and Markovian inference have been applied in different domains of aerospace engineering, such as identification, classification, searching, maneuvering, tracking, and so on.

1.3.2 Computer Networking

In networking, topology design optimization has been one of the major challenges during recent years. Since there are no exact solving approaches to guarantee an optimal solution, heuristic methods have proven capable of providing good and feasible solutions to applications. The neural network of the brain exhibits the same fundamental structure as that of social or computer networks. Thus neural networks find a wide range of applications in the area of computer networking for switching, routing, network traffic control, and so on. The increasing use of computer networks requires improvements in network technologies and management techniques so that users will receive high-quality service. Implementing AR techniques to networking satisfies such requirements. A major aspect of computer networks that is vital to quality of service is data routing. The most relevant routing information involves various measures that are often obtained in an imprecise or inaccurate manner, thus suggesting that fuzzy reasoning is a natural method to solve computer network problems. Telecommunication traffic control, congestion control, routing, fault localization, and so on are the major applications of communication systems in which probabilistic GM can be applied.

1.3.3 Consumer Electronics

Homes today are filled with a number of high-tech gadgets, from smart phones and PCs to state-of-the-art TV and audio systems, many of them with built-in networking capabilities. A combination of these devices working intelligently forms the building blocks of the smart homes. Though these intelligent systems are wide in their applications, much research-oriented information is not available because of commercial reasons. A wide range of applications covering cameras, stereo systems, gaming systems, smart homes, and so on have been developed over the past few years employing MLC systems. FL occupies a dominant space in the Japanese consumer appliance industry. The nature of CC makes it more effective in the implementation of home automation applications. The applications of probabilistic models in the field of consumer electronics are quite low because of commercial confidentiality reasons.

1.3.4 Control Systems

During the past decade, EAs and SIs have been applied in a wide range of areas, with varying degrees of success, especially in the field of control systems engineering. In the limited space it is not possible to discuss all possible ways in which neural networks have been applied to control system problems. In order to accurately model a plant, the neural network needs to be trained with data, which covers the entire range of possible network inputs. FL and CT are used in system control and in analysis of design, because they shorten the time for engineering development and, sometimes, in the case of highly complex systems, seem to be the only way to solve the problem. Bayesian and Markovian models have been applied in a variety of control system applications such as adaptive control, nonlinear control, dampers, and so on.

1.3.5 Electric Power Systems

Several research papers have been published in the area of power systems based on EA and swarm intelligence. Some of the domains of power systems in which these global and search optimization techniques have been implemented are for solving unit commitment problems, solving economic dispatch (ED) problems, stability analysis, fault analysis, scheduling, and so on. FL and CT have been applied to several power system problems such as stabilization, scheduling, fault diagnosis, load dispatching, and load forecasting. Also, neural networks and probabilistic reasoning approaches are suitable for various power system applications such as load forecasting, scheduling, fault diagnosis, and stabilization of power system. Our focus in this book is on solving unit commitment and economic load dispatch problems, reducing harmonics in power systems, analysis of load frequency control (LFC), and automatic voltage regulator (AVR) using a variety of global search algorithms.

1.3.6 Image Processing

The computational complexity involved in the area of image processing has been overcome with the growth of CI paradigms such as EA and swarm intelligence. During the past 10, years numerous papers in the area of image segmentation, image registration, image regularization, feature extraction, pattern classification, and many more have become available. Neural networks have found a major application in the field of image processing for applications such as classification, segmentation, clustering, pattern recognition, and so on. The computing power required by real-time and embedded applications of image processing such as pattern recognition and shape analysis has improved over the years by the application of neural networks. Image processing in industrial applications deals with subjective concepts like edges, brightness, nonuniformity, and so on. These concepts contain a certain amount of uncertainty, which is solved by FL. Image processing has numerous promising applications using probabilistic GM in domains such as image compression, segmentation, filtering, and so on. This book concentrates on digital image watermarking (DIWM) based on the CI algorithm. This is a promising application in the area of cryptography and authentication.

1.3.7 Medical Imaging

This section provides an overview of EC as applied to problems in medical domains such as diagnosis, prognosis, imaging, signal processing, planning, and scheduling. Neural networks have been applied within the medical domain for clinical diagnosis, image analysis and interpretation, signal analysis and interpretation, and drug development. The current applications of neural networks to medical imaging are reviewed in this section. This trend in medical imaging using artificial neural networks (ANNs) provides an appreciation of the problems associated with implementing neural networks. FL and CT provide a suitable basis for the ability to summarize medical information and extricate from the collections of masses of data related to human diseases that are relevant to performance of the task at hand.

1.3.8 Robotics and Automation

EA and swarm intelligence incorporate principles from biological population genetics to perform search, optimization, and learning. This section discusses issues arising in the application of EA to problems in robotics, such as planning, scheduling, collision avoidance, warehousing, and so on. In this book, we have focused on job shop scheduling problems (JSSPs) and multi-depot vehicle routing problems (MDVRPs). These applications have been gaining more and more attention in industrial automation during recent years. The behavior of biological systems provides both the inspiration and challenge for robotics. The application of ANN to robot control is based on the ability of living organisms to integrate perceptual inputs smoothly with motor responses, even in the presence of novel stimuli and changes in the environment. FL constitutes one of the major trends in the current research on robotics for navigation, path planning, manipulator control, position control, and so on.

1.3.9 Wireless Communication Systems

Several challenges, such as data aggregation and fusion, energy-aware routing, task scheduling, security, optimal deployment, and localization, in the area of wireless communication have been addressed successfully during recent years using the paradigms of CI. CI provides adaptive mechanisms that exhibit intelligent behavior in complex and dynamic environments such as wireless sensor networks. CIs are implemented to wireless networks because of their flexibility, autonomous behavior, robustness against topology changes, communication failures, and scenario changes in the wireless environment. In addition, CI paradigms are more commonly proposed because of their nonconventional approach of solving problems in wireless communication systems.

1.4 Applications of CI Focused in This Book

CI is a relatively young field spanning a range of applications in several areas. The exploration of CI is concerned with subordinate cognitive functions: perceptual experience, object identification, signal analysis, breakthrough of structures in data, simple associations, and control. This section presents an introduction to the applications discussed in this book such as unit commitment and economic load dispatch, harmonic reduction, LFC and automatic voltage regulation, JSS, MDVR, and DIWM.

1.4.1 Unit Commitment and Economic Load Dispatch Problem

Conventional unit commitment and economic load dispatch problems involve the allocation of generated power to different units, with the aim of minimizing the operating cost subject to diverse equality and inequality constraints of the power system. Thus the unit commitment (UC) and economic load dispatch (ELD) problems are considered as large-scale, highly nonlinear, constrained optimization problems. It is therefore of great importance to solve these problems expeditiously and as accurately as possible. Conventional techniques such as gradient-based techniques, Newton methods, linear programming, and quadratic programming offer good results, but when the search space is nonlinear and has discontinuities, these techniques become difficult to solve with a slow convergence ratio and not always seek the global optimal solution. New numerical methods such as lambda iteration and gradient methods are therefore required to cope with these difficulties, especially, those with high-speed search toward the optimal and not being trapped in local minima. In recent years, heuristic optimization algorithms based on the principle of natural evolution and with the ability to solve extremely complex optimization problems have been developed. The growth of such optimization algorithms for solving the UC and ELD problems over the past few years is presented in this section.

Starting from the mid-1980s, efficient power production has been an urgent issue in the electrical power supply industry, with regard to the mode of generation, transmission, and distribution. The uncertainty of demand and outages of generating units led to the concept of dealing with UC problems with imprecise variables. Thus researchers found FL as one of the best alternatives to solve such uncertain problems. Saneifard et al. (1997) demonstrated the FL approach to the UC problem. They found that the need for mathematical models is eliminated and the approach achieved a logical and feasible economical cost of operation of the power system, which is the major objective of UC. Sasaki et al. (1992) explored the possibility of solving the combinatorial optimization problem, in particular to UC, based on the Hopfield neural network. The proposed neural network solved a UC of 30 units over 24 periods, and the obtained results were very encouraging. Walsh and Malley (1997) presented an augmented architecture based on Hopfield network with a new form of interconnection between neurons, giving a more generalized energy function containing both discrete and continuous terms. A comprehensive cost function for the UC problem is developed and mapped to the energy function of the Hopfield network. A new and efficient method based on the radial basis function (RBF) network, which directly gives the optimal value of lambda for a given power demand, is framed for online ED (Aravindhababu and Nayar 2001). Experimental results of two sample systems with 3 and 13 generating units were provided to illustrate the performance of the RBF method.

The growing interest in solving problems based on the principle of evolution and machine learning has paved way to finding best fit solutions for constraint-based optimization problems. The first attempt to apply GA to UC problems has been on both small and large size problems (Sheble et al. 1996). The problem was solved for a 24 hours schedule and a 7 day schedule, and the feasibility of GA was validated over the Lagrangian search technique. Rudolf and Bayrleithner (1999) presented a GA for solving the UC problem of a hydrothermal power system and tested it on a real-scale hydrothermal power system over a period of a day in half-hour time steps for different GA parameters. Swarup and Yamashiro (2002) employed a new strategy for representing chromosomes and encoding the problem search space, which was efficient and could handle large-scale UC. Problem formulation of the UC considered the minimum up- and down-time constraints, start-up cost, and spinning reserve, which was defined as minimization of the total objective function while satisfying the associated constraints. Senjyu et al. (2002) developed a fast solution technique for a large-scale UC problem using GA. The search space was reduced using a unit integration technique with intelligent mutation using a local hill-climbing optimization technique. A hybrid approach to the UC problem based on GA and fuzzy optimization was implemented by Ademovic et al. (2010). In order to obtain a near-optimal solution with less computational time and improve storage requirements, the GA using real-coded chromosomes was employed in place of the more commonly used binary coded scheme. The simulations on a 10-generator system showed satisfactory outcome in terms of total cost compared to the dynamic programming technique.

Huang and Wang (2007) developed a novel technique that combined orthogonal least-squares (OLS) and the PSO algorithms to construct an RBF network for real-time power dispatch. The OLS algorithm was used to determine the number of centers in the hidden layer. With an appropriate network structure, the PSO algorithm was then used to tune the parameters in the network, including the dilation and translation of the RBF centers and the weights between the hidden and output layer. Simulations on an IEEE 30 bus Taipower systems yielded accurate real-time power dispatch solutions. Gaing (2003) presented a PSO approach to solve ED considering generator constraints. Nonlinear characteristics of the generator, such as ramp rate limits, prohibited the operating zone, and nonsmooth cost functions were considered for practical generator operations.

DE is one of the most prominent new-generation EAs, proposed by Storn and Price (1997), to exhibit consistent and reliable performance in nonlinear and multimodal environments, and has proven effective for constrained optimization problems. The application of DE to ELD problems featuring nonsmooth cost functions for discontinuous and nondifferentiable solution spaces was presented by Perez-Guerrero and Cedenio-Maldonado (2005). They concentrated on the applicability of DE to nonsmooth cost ELD problems considering the valve point loading effects, prohibited operating zones, and fuel switching effects. The hybrid differential evolution (HDE) method was presented as a method using parallel processors of the two-member evolution strategy ((1+1)-ES) by Chiou et al. (2007). In order to inspect and accelerate the global search ability of HDE, the concept of the variable scaling factor based on the one-fifth success rule of evolution strategies was embedded in the original HDE. This feature eliminated the need for fixed and random scaling factors in HDE. Wu et al. (2010) presented a multiobjective differential evolution (MODE) algorithm in which the economic emission load dispatch problem was formulated as a nonlinear, constrained, multiobjective problem with competing and noncommensurable objectives of fuel cost, emission, and system loss. In this method, an elitist archive technique was adopted to retain the nondominated solutions obtained during the evolutionary process, and the operators of DE were modified according to the characteristics of multiobjective optimization problems. Moreover, in order to avoid premature convergence, a local random search (LRS) operator was integrated with the proposed method to improve the convergence performance. Yare et al. (2009) developed three heuristic algorithms—the GA, DE, and modified PSO (MPSO)—to solve ED problem for two test systems with 6 and 19 generating units. These heuristic algorithms have been applied in the literature to solve nonconvex ED problems as a replacement for the classical Lagrange-based techniques. Abou El Ela et al. (2010) developed a DE algorithm to solve the emission-constrained economic power dispatch problem. The algorithm attempted to reduce the production of atmospheric emissions such as sulfur oxides and nitrogen oxides caused by the operation of fossil-fueled thermal generation. Such reduction was achieved by including emissions as a constraint in the objective of the overall dispatching problem.

Nayak et al. (2009) presented the application of bio-inspired ABC optimization to the constrained ELD problem. Independent simulations were performed over various systems with different numbers of generating units having constraints such as prohibited operating zones and ramp-rate limits. The basic constraints of the ED problem (such as load demand and spinning reserve capacity) are considered along in addition to practical constraints (such as the ramp-rate limits and the prohibited operating zone).

In Chapter 2, contrary to the literature reports, the UC and ELD problems are solved together. The power is dispatched economically in the ELD problem based on the schedules obtained from the UC problem. The schedules are obtained using GA, and the economic distribution is obtained using fuzzy based-RBF network (FRBFN), enhanced PSO (EPSO), DE with opposition-based learning (DE-OBL), improved DE-OBL (IDE-OBL), ABC optimization, and CSO.

1.4.2 Harmonic Reduction in Power Systems

Technological advancements in the fields of power electronics, automation, and other engineering sectors have fueled industrial growth in countries. Industrial growth and product development gauge the economic development of a nation. Economical variable speed drives (VSDs) occupy an important role in industries. In the pulp and paper industry, 30%–35% of electrical motors are employed for VSD. Variable frequency drives (VFDs) are widely employed in driving three-phase induction and permanent motors because of their high static and dynamic performance. Induction motor drives fed by line-commutated diode and thyristor rectifiers exhibit nonlinear load characteristics and draw nonsinusoidal currents from the supply even when fed from sinusoidal supply voltages. These harmonic currents are injected into the supply systems and disturb the power line, causing power quality problems.

The problems of harmonics in industrial plants are created from a number of sources and produce a variety of undesirable side effects; therefore it is important to understand all solutions that are available. Phase-shifting and detuned filters will be reviewed as concepts for solving certain types of problems related to power quality. Both a theoretical study and a case study were presented by (Zobaa 2004). Asiminoaei et al. (2006)

described the measurements and simulations of harmonics from adjustable-speed drives in a 1.2 MVA heat power station application. The significant amount of line side harmonic currents or voltages existing for different variable-speed DC drives (VSD) was analyzed using a power quality analyzer (PQA). The design of a passive harmonic filter was done for the minimization of these harmonics in MATLAB®/Simulink®. Shirabe et al. (2014) focused on comparing the efficiency of a Si insulated-gate bipolar transistor (IGBT)-based drive with a 6-in-1 GaN module-based drive that was operating at a carrier frequency of 100 kHz with an output sine wave filter. Keypour et al. (2004) proposed active power filters (APFs) that could be employed for harmonic compensation in power systems. Ahmadi et al. (2011) presented a modified four-equation method, which was proposed for selected harmonics elimination for both two-level inverters and multilevel inverters with unbalanced DC sources.

Li et al. (2012) presented a new high-voltage DC (HVDC) system improved by an inductive filtering method, and afterward proposed a unified equivalent circuit model and a corresponding mathematical model for the calculation of the harmonic transfer characteristics of the new HVDC system. Wu et al. (2012) presented a new topology of higher order power filter for grid-tied voltage-source inverters, named the LLCL filter, in which a small inductor is inserted in the branch loop of the capacitor in the traditional LCL filter to compose a series resonant circuit at the switching frequency. Villablanca et al. (2007) described a 12- pulse AC–DC rectifier, which drew sinusoidal current waveforms from the utility. In addition, it could supply either a ripple-free current or voltage to the DC load by using a DC filter.

Different transformer arrangements for 12-pulse-based rectification were studied by Singh et al. (2006a,b). A passive filter is widely used in the field of harmonic suppression and reactive power compensation of power networks because of its low cost. In their publication, Mingwei et al. (2010) proposed a chaotic GA applied to parameter optimization of a passive filter. Verma and Singh (2010) presented a GA-based method to design series-tuned and second-order band-pass passive filters. They demonstrated the ability of the proposed designed passive filters to compensate the current harmonics effectively along with the reduction of the net root-mean squared (RMS) source current. Abraham et al. (2008) presented a simple mathematical analysis of the reproduction step used in the bacterial foraging optimization algorithm (BFOA). Dasgupta et al. (2009) presented a mathematical analysis of the chemotactic step in BFOA from the viewpoint of the classical gradient descent search. They also proposed simple schemes to adapt the chemotactic step size in BFOA with a view to improving its convergence behavior without imposing additional requirements in terms of the number of FEs. Jegathesan and Jerome (2011) presented an efficient and reliable EA-based solution for the selective harmonic elimination (SHE) switching pattern to eliminate the lower order harmonics in a pulse width modulation (PWM) inverter. Sundareswaran et al. (2007) developed the concept of an ant colony systems for a continuous optimization problem of SHE in a PWM inverter.

Salehi et al. (2010) proposed to eliminate low-order harmonics using the optimized harmonic stepped waveform (OHSW) technique. Al-Othman et al. (2013) presented a novel method for output voltage harmonic elimination and voltage control of PWM AC/AC voltage converters using the principle of the hybrid real-coded genetic algorithm-pattern search (RGA-PS) method. RGA is the primary optimizer exploiting its global search capabilities, PS is then employed to fine-tune the best solution provided by RGA in each evolution. In Ray et al. (2009), the harmonic elimination problem in a PWM inverter was treated as an optimization problem, which was solved using the PSO technique. The derived equation for computation of total harmonic distortion (THD) of the output voltage of PWM inverter was used as the objective function in the PSO algorithm. The method was applied to investigate the switching patterns of both unipolar and bipolar cases.

In this application, various harmonics elimination techniques suitable for 2-pulse, 6-pulse, and 12-pulse VFDs were developed and analyzed in terms of power factor improvement and THD reduction, elimination of lower order current harmonics, and minimization of voltage harmonics. These major objectives were achieved through optimization techniques. A GA-based optimization technique that was used to design the passive harmonic filters for single-phase and three-phase diode rectifiers for frontend type VFDs was presented. The method was based a study of the harmonics on pulp and paper industry drives. The success of the method involved accurate representation of the load harmonics. With the harmonics well defined, the harmonic and fundamental frequency drive equivalent circuits were utilized to study the parameters. The GA optimization technique, the filter size, and performance of the filter could be optimized. BFA was also applied for solving the SHE problem. The problem of voltage harmonic elimination together with output voltage regulation was drafted as an optimization task, and the solution was sought through BFA method.

1.4.3 Voltage and Frequency Control in Power Generating Systems

A power generating system has to ensure that adequate power is delivered to the load, both reliably and economically. Any electrical system must be maintained at the desired operating level characterized by nominal frequency and voltage profiles. During transportation, both the active power balance and the reactive power balance must be maintained between generation and utilization of AC power. These two balances correspond to two equilibrium points: frequency and voltage. When either of the balances is broken and reset at a new level, the equilibrium points will float. A good-quality electric power system requires both the frequency and voltage to remain at standard values during its operation. Power systems are subject to constant changes due to loading conditions, disturbances, or structural changes. Controllers are designed to stabilize or enhance the stability of the system under these conditions. The control of frequency is achieved primarily through a speed governor mechanism aided by LFC for precise control. The AVR senses the terminal voltage and adjusts the excitation to maintain a constant terminal voltage. Control strategies such as proportional integral derivative (PID) controllers, decentralized controllers, optimal controllers, and adaptive controllers are adopted for LFC and AVR to regulate the real and reactive power outputs. The daily load cycle changes significantly and hence fixed gain controllers will fail to provide best performance under the wide range of operating conditions. But in general, each controller is designed for a specific situation or scenario and is effective under these particular conditions. Hence, it is desirable to increase the capability of controllers to suit the needs of present-day applications. The main reason to develop better methods to design PID controllers is their significant impact on the performance improvement. The performance indexes adopted for problem formulation are the settling time, overshoot, and oscillations. The primary design goal is to obtain good load disturbance response by optimally selecting the PID controller parameters. Traditionally, the control parameters have been obtained by trial-and-error approach, which consumes a large amount of time in optimizing the choice of gains. To reduce the complexity in tuning PID parameters, EC techniques can be used to solve a wide range of practical problems including optimization and design of PID gains.

The synchronous generator with its controls is one of the most complex devices in a power system. The generation–load imbalance is the main reason behind the instability of voltage and frequency. The dynamic performance of the LFC and AVR control loop decides the quality of a power supply system with respect to frequency and voltage. The fixed-gain conventional controller exhibits poor transient response with increased overshoot, settling time, and oscillations (Mathur and Manjunath 2007). Advances in computer performance have enabled the application of EA to solve difficult, real-world optimization problems, and the solutions have received much attention in control systems problems. The most challenging application of CI is in the area of power system because of the economics and quality of power supply to the consumer. EA, seen as a technique to evolve machine intelligence, is one of the mandatory prerequisites for achieving this goal by means of algorithmic principles that are already working quite successfully in natural evolution. Hence, to improve the characteristics of the frequency and voltage of a power generating system, several intelligent control techniques such as NN, FL, and GA have been employed (Chaturvedic et al. 1999, Zeynelgil et al. 2002).

The first attempt in the area of automatic generation control (AGC) involved the control of the frequency of a power system via the flywheel governor of the synchronous machine (Ibraheem and Kothari 2005). Based on the experiences in implementing AGC schemes, modifications in the design of efficient controllers have been suggested from time to time according to the change in the power system environment (Stankovic et al. 1998). A number of control engineers, for example, Bode, Nyquist, and Black, have established links between the frequency response of a control system and its closed-loop transient performance in the time domain. The investigation carried on conventional approaches resulted in relatively large overshoots and transient frequency deviation. AI techniques such as FL, ANN, and GA have found interesting applications in the LFC and AVR of power generating systems (Ahamed et al. 2002, Mukherjee and Ghosal 2008, Talaq and Al-Basri 1999). El-Hawary (1998) has provided a detailed survey on the applications of FL in power systems. The applications of fuzzy set theory to power systems and the basic procedures for fuzzy set-based methods to solve specific power system problems were explained briefly by Momoh et al. (1995). Masiala et al. (2004) designed an adaptive fuzzy controller for power system load–frequency control to damp frequency oscillations and to track its error to zero at the steady state. The main drawback in the proposed system was that the performance was evaluated only for a small load deviation of 0.01 p.u. Gargoom et al. (2010) proposed voltage and frequency stabilization using FLC for a wind generating system. The simulations results were analyzed for single- and three-phase faults to validate the efficiency of the proposed fuzzy controller. Takashi et al. (1997) proposed an integrated FLC for generator excitation and speed governing control. The simulation results were demonstrated for only small load deviation of 0.5 p.u and the transient oscillations were found to be high. McArdle et al. (2001) proposed an FLC for automatic voltage regulation of a small alternator, and simulation results proved that the proposed controller could outperform the conventional PID control that was employed in the AVR. Nanda and Mangla (2004) proposed a set of fuzzy rules for improving the dynamic performance of an interconnected hydrothermal system.

GA has received great attention in control system in searching for optimal PID controller parameters because of its high potential for global optimization. Oliveira et al. (2002) used a standard GA to make initial estimates for the values of the PID parameters. GA was tested for the PID controller for a nonlinear process and showed robustness and efficiency (Griffin 2003). Rabandi and Nishimori (2001) proposed a new method of time-varying feedback control to improve the stability performance of a complex power control system using GA. Ghosal (2004) developed a new GA-based AGC scheme for evaluating the fitness of GA optimization by selecting a function such as the "figure of merit," which directly depended on transient performance characteristics such as the settling time, undershoot, overshoot, and so on, for a multiarea thermal generating system. Herrero et al. (2002) demonstrated the flexibility of GA in tuning the gains under different operating conditions of the plant. Chile et al. (2008) developed a GA-based optimal PID controller for a turbine speed control system. The simulation results indicated that the proposed controller could reduce the settling time of frequency but transient oscillations and overshoot existed in the system. Milani and Mozafari (2010) proposed an advanced GA-based method to damp the steady-state deviations in frequency with the presence of a step-load disturbance for a fixed-load two-area interconnected power system. Prasanth and Kumar (2008) proposed a new robust load frequency controller for a two-area interconnected power system.

PSO is a CI-based technique that is not largely affected by the size and nonlinearity of the problem, and it can converge to the optimal solution in problems where most analytical methods fail to do so (Valle and Salman 2008). Yoshida and Kenichi (1999) presented a PSO scheme for reactive power and voltage control considering voltage security assessment. Their method was compared with the conventional PID method on practical power system, and showed promising results. Wang et al. (2008b) presented the use of a new PSO-based auto-tuning of a PID controller to improve the performance of the controller. Gozde et al. (2008) selected the optimum parameter value for the integral gain and proportional gain using the PSO algorithm for optimizing the PID values of LFC in a single-area power system. Zareiegovar (2009) proposed a new approach using PSO to tune the parameters of a PID controller for LFC and compared it with a conventional PID controller to validate the effectiveness of the proposed algorithm. Gaing (2004) presented PSO for the optimum design of a PID controller in an AVR system and proved that the proposed method was efficient and robust. Majid Zamani Masoud et al. (2009) proposed a fractional-order PID controller for AVR using PSO and compared the simulation results with those of a conventional controller for performance analysis. Liu et al. (2004) proposed an optimized design based on the PSO algorithm for a PID controller and tested the proposed algorithm by simulation experiments in the common one-order and two-order industrial models.

The interaction of CI techniques and hybridization with other methods such as expert systems and local optimization techniques certainly opens a new direction of research toward hybrid systems that exhibit problem-solving capabilities approaching those of naturally intelligent systems in the future. A novel hierarchical fuzzy genetic information fusion technique was proposed by Buczak and Uhrig (1996), and their algorithm generated satisfactory results with reduced computational complexity. The use of the hybrid algorithm of combining the fuzzy clustering techniques with GA for classification tasks produced better classification results and increased the accuracy of the system in comparison with the conventional GA technique (Gomez-Skarmeta et al. 2001). The EPSO algorithm developed by Fang and Chen (2009) showed that it had stable convergence characteristics and good computational stability for PID control systems. The hybrid approach involving PSO and bacterial foraging (BF) proposed by Biswas et al. (2007) is a new method, and was found to be statistically better since it outperformed in efficiency GA and BF over few numerical benchmarks and in optimizing PID gain parameters. The combined GA and PSO algorithm for PID controller design for an AVR system, proposed by Wong et al. (2009), produced a high-quality solution effectively when applied for real-time control task. Kim et al. (2007) proposed a hybrid GA and BF approach for tuning the gain of a PID controller of an AVR. The performance, which was illustrated for various test functions, clearly showed that the proposed approach was very efficient and could easily be extended for other global optimization problems. To increase the performance characteristics of the controller, Mukherjee and Ghosal (2008), Shayeghi et al. (2008), and Ghosal (2004) developed EA-based algorithms to optimize the PID gains for adaptive control applications. Shi and Eberhart (1999) implemented a fuzzy system to dynamically tune the inertia weight of the PSO algorithm. Their experimental results illustrated that the fuzzy adaptive PSO is a promising optimization method that is especially useful for optimization problems with a dynamic environment. Kim and Cho (2006) proposed a hybrid approach involving GA and BF for tuning the PID controller of an AVR. Simulation results were very encouraging and showed that this novel hybrid model for global optimization could be applied for finding solutions to complex engineering problems.

The major contribution of this application is to use intelligent computing techniques for solving power system optimization problems that were previously addressed by conventional problem-solving methods. Different intelligent computing techniques have been developed for online voltage and frequency stability monitoring. Online stability monitoring is the process of obtaining voltage and frequency stability information for a given operating scenario. The prediction should be fast and accurate so that the control signals can be sent to appropriate locations quickly and effectively. The performance of LFC and AVR has enhanced with the application of FL, GA, PSO, ACO, and hybrid EA optimization techniques.

1.4.4 Job Shop Scheduling Problem

Since the early 1980s, a sequence of new technologies based on AI have been applied to solving JSSPs. These knowledge-based search systems were quite prevalent in the early and mid-1980s because of their four main advantages. First, and perhaps most important, they use both quantitative and qualitative knowledge in the decision-making process. Second, they are capable of generating heuristics that are significantly more complex than the simple dispatching rules used in traditional methods. Third, the selection of the best heuristic can be based on information about the entire job shop including the current jobs, expected new jobs, and the current status of resources, material transporters, inventory, and personnel. Fourth, they capture complex relationships in elegant, new data structures and contain special techniques for powerful manipulation of the information in these data structures.

Fuzzy set theory can be useful in modeling and solving JSSPs with uncertain processing times, constraints, and setup times. These uncertainties can be represented by fuzzy numbers that are described by using the concept of an interval of confidence. Krucky (1994) addressed the problem of minimizing setup times of a medium-to-high product mix production line using FL. The heuristic FL-based algorithm helped in determining how to minimize setup time by clustering assemblies into families of products that shared the same setup by balancing a product's placement time between multiple high-speed placement process steps. Tsujimura et al. (1993) presented a hybrid system that used fuzzy set theory to model the processing times of a flow shop scheduling facility. Triangular fuzzy numbers (TFNs) were used to represent these processing times. Each job was defined by two TFNs, a lower bound and an upper bound, with a branch and bound procedure to minimize the makespan. McCahon and Lee (1990) used triangular and trapezoidal fuzzy numbers to represent imprecise job processing times in job shop production systems. Fuzzy makespan and fuzzy mean flow times were then calculated for greater decision-making information.

A number of approaches have been utilized in the application of GA to JSSPs. Davis (1985) applied GA with blind recombination operators to JSSP. Emphasis was placed on relative ordering schema, absolute ordering schema, cycles, and edges in the offspring, which in turn caused differences in the blind recombination operators. Since Davis' application of GA, numerous implementations have been suggested not only for the job shop problem but also for other variations of the general resource-constrained scheduling problem. In some cases, a representation for one class of problems can be applied to others as well. But in most cases, modification of the constraint definitions requires a different representation. The first attempt in the application of GA to solve JSSP in a real production facility was developed by Starkweather et al. (1992). Their approach was based on dual criteria such as minimization of average inventory in the plant and minimization of the average waiting time for an order to be selected. Husbands (1996) outlined the state of the art in GA for scheduling. He noted the similarity between scheduling and sequence-based problems such as the traveling salesman problem. NP-hard problems such as layout and bin-packing problems, which are similar to the job shop formulation, were also referenced. Cheung and Zhou (2001) developed a hybrid algorithm based on GA and a well-known dispatching rule for sequence-dependent setup time (SDST) job shops where the setup times are separable. The first operations for each of the *m* machines were achieved by GA, while the subsequent operations on each machine were planned by the shortest processing time (SPT) rule. The objective of the approach was to minimize the makespan.

PSO combined with simulated annealing was developed by Wei and Wu (2005) to solve the problem of finding the minimum makespan in the JSS environment. Liao et al. (2007) proposed a type of PSO, based on the discrete PSO, which was effective in minimizing the makespan of a flow shop scheduling problem (FSSP). Lian et al. (2006) applied specific crossover and mutation operators in PSO, which are often used in GA. Liu et al. (2007) proposed an effective PSO-based memetic algorithm (MA) for permutation FSSPs, minimizing the maximum completion time. Ali and Fawaz (2006) combined PSO with taboo search to study a type of fuzzy FSSP with distributed database. Bing and Zhen (2007) were the first to apply fuzzy processing time to FSSPs using PSO. The objective function involved a two-criteria schedule: the fuzzy makespan and robustness of the makespan. The uncertain processing times were represented using TFNs. Niu et al. (2008) addressed the problem of scheduling a set of jobs in a job shop environment with fuzzy processing time combining GA and PSO. The objective was to find a job sequence that minimized the makespan and the uncertainty of the makespan by using an approach for ranking fuzzy numbers.

ACO has been successfully applied to combinatorial optimization problems, such as the traveling salesman problem (Dorigo and Gambardella 1997), for which the method was shown to be very competitive over other meta-heuristics. For the JSSP, Colorni et al. (1994) first applied ACO, but obtained relatively uncompetitive computational results. Zwaan and Marques (1999) developed an ACO for JSSP in which a GA was adopted for fine-tuning the parameters of an ACO approach. Later, Blum and Sampels (2004) developed a state-of-the-art ACO approach to tackle the general shop scheduling problem, including the JSSP and the open shop scheduling problem. Their approach used strong non-delay guidance for constructing schedules with a newly developed pheromone model, where the pheromone values were assigned to pairs of related operations, and applied a steepest descent local search to improve the constructed schedule. Further, Blum (2005) developed a new competitive hybrid algorithm combining ACO with beam search for the open shop scheduling problem. Huang and Liao (2008) presented a hybrid algorithm combining the ACO algorithm with the Tabu search algorithm for the classical JSSP. Instead of using the conventional construction approach to construct feasible schedules, the ACO algorithm employed a novel decomposition method inspired by the shifting bottleneck procedure and a mechanism of occasional reoptimizations of partial schedules. Besides, the Tabu search algorithm was embedded to improve the solution quality in terms of makespan.

This book implements BIA such as GA, stochastic PSO (SPSO), ACO, and genetic swarm optimization (GSO) for solving JSSP with fuzzy processing time in the (λ, 1) interval and ranking based on signed distance.

1.4.5 Multidepot Vehicle Routing Problem

A well-known generalization of the VRP is the MDVRP. In this extended version of VRP, every customer is visited by a vehicle based at one of several depots. In the standard MDVRP, every vehicle route must start and end at the same depot. The possible optimization techniques to solve MDVRPs are heuristics, exact algorithms, and meta-heuristics, among which heuristic/meta-heuristic approaches are more popular. In addition, research activities involving heuristic optimization solutions to MDVRP are comparatively few. Skok et al. (2000) dealt with the MDVRP problem by applying a steady-state GA. Though the procedure adopted was similar to the basic GA, they applied, instead of using a roulette wheel selection, a selection function to select parents for reproduction. Improved results were obtained for cycle crossover and fragment reordering crossover, with a scramble mutation operator. A hybrid GA was later suggested by Jeon et al. (2007), in which they included the following features: (i) production of the initial population by using both a heuristic and a random generation method; (ii) minimization of infeasible solutions instead of eliminating them; (iii) gene exchange process after mutation; (iv) flexible mutation rate; and (v) route exchange process at the end of GA.

To deal with the MDVRP problem more efficiently, two hybrid genetic algorithms (HGAs)—HGA1 and HGA2—were developed by Ho et al. (2008). The major difference between the HGAs was that the initial solutions were generated randomly using HGA1. The Clarke and Wright saving method and the nearest neighbor heuristic were incorporated into HGA2 for the initialization procedure. The results proved that the performance of HGA2 was superior to that of HGA1 in terms of the total delivery time. A GA with indirect encoding and an adaptive inter-depot mutation exchange strategy for the MDVRP with capacity and route-length restrictions was developed by Ombuki-Berman and Hanshar (2009). Efficient routing and scheduling of vehicles was established, proving GAs as competitive algorithms over the Tabu search technique.

Zhang and Lu (2007) offered an algorithm that integrated niche technology with PSO for the vehicle routing and scheduling problem. The inertia weight was decreased linearly to improve the searching ability. Particle position matrix based on goods was established by Wang et al. (2008a) for PSO to solve the VRP model. Vehicle routes to all goods were searched by the PSO to obtain the global solution. A modified PSO algorithm developed by Wenjing and Ye (2010) applied a mutation operator and improved the inertia weight for solving MDVRP.

ABC optimization introduced by Karaboga (2005) has not been applied to solve the MDVRP so far. It has been applied to solve capacitated VRPs, which are also extended versions of the VRPs. Szeto et al. (2011) actually developed a basic and enhanced version of the ABC algorithm for solving the capacitated VRP. The performance of the enhanced heuristic was evaluated on two sets of standard benchmark instances and compared with the original ABC heuristic. The computational results showed that the enhanced heuristic outperformed the original one and could produce good solutions when compared with the existing heuristics. Brajevic (2011) also presented the ABC algorithm for capacitated VRP. Although the global optimality was not guaranteed, the performance of the algorithm was observed to be good and robust. It was also noticed that the algorithm was capable of getting trapped in the local minimum for benchmark instances of small-scale problems.

In this work, the solution to MDVRP is obtained in four stages: grouping, routing, scheduling, and optimization. The performance of the proposed heuristics such as GA, MPSO, ABC optimization, genetic swarm optimization (GSO), and improved GSO (IGSO) on MDVRP benchmark instances was addressed.

1.4.6 Digital Image Watermarking

In response to the increasing demand of distributing multimedia data over the Internet, watermarking technology has received considerable attention in the past few years. Aimed at copyright protection, watermarking is the process of embedding hidden information into a multimedia data. DIWM should be capable of providing qualities such as imperceptibility, robustness, and security of the covered image. A variety of watermarking schemes in the spatial domain, transformation domain, and compression domain have been proposed in the literature to embed a watermark into the host image. Cox and Kilan (1996) pointed that, in order for a watermark to be robust to attack, it must be placed in perceptually significant areas of the image. Kundur and Hatzinakos (1997) embedded the watermark in the wavelet domain, where the strength of watermark was decided by the contrast sensitivity of the original image. Delaigle et al. (1998) generated binary m sequences and then modulated them on a random carrier. A method for casting digital watermarks on images and analyzing its effectiveness was given by Pitas (1996), who also examined the immunity to subsampling.

Lin et al. (2005) proposed a spatial-fragile watermarking scheme in which the original image was divided into several blocks and permuted based on a secret key. The method was capable of restoring the tampered images and was resistant to counterfeiting attacks. Wang and Chen (2009) provided a copyright protection scheme that first extracted the image features from the host image by using the discrete wavelet transform (DWT) and singular value decomposition (SVD). Using the k-means clustering algorithm, the extracted features were clustered, resulting in a master share. The master share and a secret image were used to build an ownership share according to a two-out-of-two visual cryptography (VC) method. Chen and Wang (2009) proposed a fragile watermarking technique based on muzzy C-Means (FCM) to create relationships between image blocks, thus withstanding counterfeiting attacks. A rotation- and scaling-invariant image watermarking scheme in the discrete cosine transform (DCT) domain was presented by Zheng et al. (2009) based on rotation-invariant feature extraction and image normalization. Each homogeneous region was approximated as a generalized Gaussian distribution, with the parameters estimated using expectation maximization.

An innovative watermarking based on GA in the transform domain was developed by Shieh et al. (2004). The genetic-based approach was found to be robust against watermarking attacks. Robustness was achieved because the method used GA to train the frequency set for embedding the watermark. Dongeun et al. (2006) presented a novel watermark extraction algorithm based on DWT and GA. This scheme applied the wavelet domain for watermark insertion and GA for watermark extraction. Wei et al. (2006) also extended image watermarking application based on GA. In order to improve the robustness and imperceptibility of the image, the spread spectrum watermark algorithm, a new approach for optimization in 8 × 8 domain using GA, was presented. Boato et al. (2008)

proposed a new flexible and effective evaluation tool based on GA to test the robustness of DIWM techniques. Given a set of possible attacks, the method finds the best possible unwatermarked image in terms of the weighted peak signal-to-noise ratio (WPSNR). Al-Haj (2007) described an imperceptible and robust watermarking scheme based on combined DWT–DCT DIWM algorithm. Watermarking was done by altering the wavelet coefficients of carefully selected DWT subbands, followed by the application of the DCT transform on the selected subbands.

PSO was also found to be efficient for DIWM optimization. Ziqiang et al. (2007) suggested a method in which the watermark was embedded to the discrete multi-wavelet transform (DMT) coefficients larger than predefined threshold values, and watermark extraction was efficiently performed through the PSO algorithm. Results showed almost invisible difference between the watermarked image and the original image and that it was robust to common image processing operations and JPEG lossy compression. Aslantas et al. (2008) directed a DWT–SVD-based watermarking algorithm that initially decomposed the host image into sub bands; later, the singular values of each subband of the host image were modified by different scaling factors to embed the watermark image. Ishtiaq et al. (2010) investigated a new method for adaptive watermark strength optimization in the DCT domain based on PSO. PSO helped in perceptual shaping of the watermark so that it was less perceptible to the human visual system. A hybrid watermarking scheme based on GA and PSO was presented by Lee et al. (2008). In this hybrid scheme, the parameters of the perceptual lossless ratio (PLR), which were defined by just noticeable difference (JND) for two complementary watermark modulations, were derived.

In the DIWM application discussed in this book, the input image undergoes preprocessing stages such as segmentation, feature extraction, orientation assignment, and image normalization. Intelligent image watermark embedding and extraction is performed in the DWT domain based on GA, PSO, and HPSO, and their performance is evaluated against geometric and nongeometric attacks.

1.5 Summary

This chapter introduced the reader to the concepts of CI paradigms and their classification. The basic concept, working, and pseudo-code of the algorithms were discussed with the advantages and disadvantages of each algorithm, so that an engineer can make a suitable choice of the algorithm. There are numerous applications of CI paradigms in different domains of engineering, for example, aerospace, networking, consumer electronics, control systems, power system, robotics, image processing, medicine, and communication systems. This book provides a set of applications that have been requiring high research attention during the recent years. The applications within the scope of this book include solutions to UC and ELD, harmonic reduction, LFC and automatic voltage regulation, job shop scheduling, multi-depot vehicle routing, and DIWM using CI algorithms. Algorithms such as FL, PSO, ACO, ABC, DE, GA, CSO, and BFA are applied to solve the chosen problems. In addition, these algorithms are either improved to enhance their performance or hybridized for faster convergence. Several combinations of these algorithms are developed for the applications throughout this book using MATLAB/Simulink. The scope of this book is limited to the CI algorithms mentioned here. But owing to the fast-growing and promising research in the area of CI, there are a wide range of problems to which these algorithms can be applied. This book will be paving a path to researchers in the field, thus providing a knowledge base to choose and apply the appropriate CI algorithm to solve constraint-based optimization problems. CI techniques are becoming increasingly common and powerful tools for designing intelligent controllers. The field of CI is rapidly developing owing to its intelligent characteristics and great progress can be expected in the years to come.

References

Abou El Ela, A.A., Abido, M.A., and Spea, S.R., Differential evolution algorithm for emission constrained economic power dispatch problem, *Electric Power Systems Research*, 80(10), 1286–1292, October 2010.

Abraham, A., Biswas, A., Dasgupta, S., and Das, S., Analysis of reproduction operator in bacterial foraging optimization algorithm, in *IEEE Congress on Evolutionary Computation*, Hong Kong, China, June 1–6, 2008, pp. 1476–1483.

Ademovic, A., Bisanovic, S., and Hajro, M., A genetic algorithm solution to the unit commitment problem based on real-coded chromosomes and fuzzy optimization, in *Proceedings of the 15th IEEE Mediterranean Electrotechnical Conference: MELECON 2010*, Valletta, Malta, April 26–28, 2010, pp. 1476–1481.

Ahamed, T.P.I., Rao, P.S.N., and Sastry, P.S., A reinforcement learning approach to automatic generation control, *Electric Power Systems Research*, 63, 9–26, 2002.

Ahmadi, D., Zou, K., Li, C., Huang, Y., and Wang, J., A universal selective harmonic elimination method for high-power inverters, *IEEE Transactions on Power Electronics*, 26(10), 2743–2752, 2011.

Ali, A. and Fawaz, S.A., A PSO and Tabu search heuristics for the assembly scheduling problem of the two-stage distributed database application, *Computers and Operations Research*, 33(4), 1056–1080, 2006.

Al-Haj, A., Combined DWT-DCT Digital watermarking, *Journal of Computer Science*, 3(9), 740–746, 2007.

Al-Othman, A.K., Ahmed, N.A., AlSharidah, M.E., and AlMekhaizim, H.A., A hybrid real coded genetic algorithm—Pattern search approach for selective harmonic elimination of PWM AC/AC voltage controller, *Electrical Power and Energy Systems*, 44, 123–133, 2013.

Aravindhababu, P. and Nayar, K.R., Economic dispatch based on optimal lambda using radial basis function network, *Journal on Electrical Power and Energy Systems*, 24, 551–556, 2001.

Asiminoaei, L., Hansen, S., and Blaabjerg, F., Predicting harmonics by simulations. A case study for high power adjustable speed drive, *Electrical Power Quality and Utilisation Magazine* 2(1), 65–75, 2006.

Aslantas, V., Dogan, A.L., and Ozturk, S., DWT-SVD based image watermarking using particle swarm optimizer, in *Proceedings of the IEEE International Conference on Multimedia and Expo*, Hannover, Germany, June 23–26, 2008, pp. 241–244.

Bing, W. and Zhen, Y., A particle swarm optimization algorithm for robust flow-shop scheduling with fuzzy processing times, in *Proceedings of the IEEE International Conference on Automation and Logistics*, Jinan, China, August 18–21, 2007, pp. 824–828.

Biswas, A., Dasgupta, S., Das, S., and Abraham, A., Synergy of PSO and bacterial foraging optimization—A comparative study on numerical benchmarks, *International Journal on Innovations in Hybrid intelligent Systems*, ASC 44, 255–263, 2007.

Blum, C., Beam ACO - Hybridizing ant colony optimization with beam search: An application to open shop scheduling, *Computers and Operations Research*, 32, 1565–1591, 2005.

Blum, C. and Sampels, M., An ant colony optimization algorithm for shop scheduling problems, *Journal of Mathematical Modeling and Algorithms*, 3, 285–308, 2004.

Boato, G., Conotter, V., and De Natale, F.G.B., GA-based robustness evaluation method for digital image watermarking, in *Digital Watermarking*, Y.Q. Shi, H.-J. Kim, and K. Stefan, Eds. Springer-Verlag, New York, 2008, pp. 294–307.

Brajevic, I., Artificial bee colony algorithm for the capacitated vehicle routing problem, in *Proceedings of the 5th European Conference on European Computing Conference: ECC'11*, Paris, France, April 28–30, 2011, pp. 239–244.

Buczak, A.L. and Uhrig, R.E., Hybrid fuzzy-genetic technique for multisensory fusion, *Information Sciences*, 93(3–4), 265–281, 1996.

Chaturvedi, D.K., Satsangi, P.S., and Kalra, P.K., Load frequency control: A generalized neural network approach, *Journal of Electric Power and Energy Systems*, 21, 405–415, 1999.

Chen, W.C. and Wang, M.S., A fuzzy c-means clustering-based fragile watermarking scheme for image authentication, *Expert Systems with Applications*, 36(2), 1300–1307, 2009.

Cheung, W. and Zhou, H., Using genetic algorithms and heuristics for job shop scheduling with sequence dependant set-up times, *Annals of Operational Research*, 107, 65–81, 2001.

Chile, R.H., Waghmare, L.M., and Lingare, M.J., More efficient genetic algorithm for tuning optimal PID controller, in *Proceedings of Second National Conference on Computing for Nation Development*, New Delhi, India, February 2008. http://www.bvicam.ac.in/news/INDIACom%202008%20Proceedings/pdfs/papers/155.pdf.

Chiou, J.P., Wang, S.K., and Liu, C.W., Non-smooth/non-convex economic dispatch by a novel hybrid differential evolution algorithm, *IET Generation Transmission and Distribution*, 1(5), 793–803, 2007.

Colorni, A., Dorigo, M., Maniezzo, V., and Trubian, M., Ant system for job shop scheduling, *Belgian Journal of Operations Research, Statistics and Computer Science*, 34(1), 39–53, 1994.

Cox, I. and Kilan, J., Secure spread spectrum watermarking for images, audio and video, in *Proceedings of the IEEE International Conference on Image Processing*, Lausanne, Switzerland, September 16–19, 1996, pp. 243–246.

Dasgupta, S., Das, S., Abraham, A., and Biswas, A., Adaptive computational chemotaxis in bacterial foraging optimization an analysis, *IEEE Transactions on Evolutionary Computation*, 13(4), 919–941, 2009.

Davis, L., Job shop scheduling with genetic algorithms, in *Proceedings of the First International Conference on Genetic Algorithms*, Pittsburgh, PA, July 24–26, 1985, pp. 136–140.

Delaigle, J., De Vleeschouwer, C., and Macq, B., Psychovisual approach to digital picture watermarking, *Journal of Electronic Imaging*, 7(3), 628–640, July 1998.

Dongeun, L., Tackyung, K., Seongwon, L., and Joonki, P., Genetic algorithm based watermarking in discrete wavelet transform domain, in *Intelligent Computing*, G. Gerhard, J. Hartmanis, and J.v. Leeuwen, Eds. Springer-Verlag, Berlin, Germany, 2006.

Dorigo, M. and Gambardella, L.M., Ant colony system: A cooperative learning approach to the traveling salesman problem, *IEEE Transactions on Evolutionary Computation*, 1(1), 53–66, 1997.

Eberhart, R. and Kennedy, J., A new optimizer using particles swarm theory, in *Proceedings of the Sixth IEEE International Symposium on Micro Machine and Human Science: MHS '95*, Nagoya, Japan, October 4–6, 1995, pp. 39–43.

El-Hawary, M.E., *Electric Power Applications of Fuzzy Systems*, 2nd edition. New Brunswick, NJ: IEEE Press, 1998.

Elliot, R.J., Aggoun, L., and Moore, J.B., *Hidden Markov Models: Estimation and Control*. Berlin, Germany: Springer, 1995.

Fang, H. and Chen, L., Application of enhanced PSO algorithm to optimal tuning of PID gains, IEEE explore reference 978-1-4244-2723-9/09, pp. 35–39, 2009.

Fogel, D.B., Back, T., and Michalewicz, Z., *Evolutionary Computation 1: Basic Algorithms and Operators*. Bristol, U.K.: Institute of Physics Publishing, 2000.

Gaing, Z.-L., Particle swarm optimization to solving the economic dispatch considering the generator constraints, *IEEE Transactions on Power Systems*, 18(3), 1187–1195, August 2003.

Gaing, Z.-L., A particle swarm optimization approach for optimum design of PID controller in AVR system, *IEEE Transactions on Energy Conversion*, 19(2), 384–391, 2004.

Gargoom, A., Haque, H., and Negnevitsky, M., Voltage and frequency stabilization using PI-like fuzzy controller for the load side converters of the stand alone wind energy systems, in *25th Annual IEEE Applied Power Electronics Conference and Exposition*, Palm Springs, CA, February 21–25, 2010, pp. 2132–2137.

Ghosal, S.P., Application of GA/GA-SA based fuzzy automatic generation control of a multi area thermal generating system, *Electric Power System Research*, 70(2), 115–127, 2004.

Goldberg, D.E., *Genetic Algorithms in Search, Optimization & Machine Learning*. Reading, MA: Addison-Wesley, 1989.

Gomez-Skarmeta, A.F., Valdes, M., Jimenez, F., and Marin-Blazquez, J.G., Approximate fuzzy rules approaches for classification with hybrid-GA techniques, *Information Sciences*, 136(1–4), 193–214, 2001.

Gozde, H., Cengiz, T., Ilhan, K., and Ertugrul, PSO based load frequency control in a single area power system, *Scientific Bulletin*, 2(8), 106–110, 2008.

Griffin, I., On-line PID controller tuning using genetic algorithms, MSc Thesis, School of Electronic Engineering, Dublin City University, Dublin, Ireland, 2003.

Herrero, J.M., Balsco, X., Martinez, J., and Salcedo, V., Optimal PID tuning with genetic algorithms for non-linear process models, in *Proceedings of 15th Triennial World Congress*, Barcelona, Spain, July 21–26, 2002. http://citeseerx.ist.psu.edu/viewdoc/download?doi=10.1.1.4.6291&rep=rep1&type=pdf.

Ho, W., Ho, G.T.S., Ji, P., and Lau, H.C.W., A hybrid genetic algorithm for the multi-depot vehicle routing problem, *Journal of Engineering Applications of Artificial Intelligence*, 21(4), 548–557, 2008.

Huang, C.-M. and Wang, F.-L., An RBF network with OLS and EPSO algorithms for real-time power dispatch, *IEEE Transactions on Power Systems*, 22(1), 96–104, February 2007.

Huang, K.L. and Liao, C.J., Ant colony optimization combined with taboo search for the job shop scheduling problem, *Computers and Operations Research*, 35, 1030–1046, 2008.

Husbands, P., Genetic algorithms for scheduling, *AISB Quarterly*, 89, 38–45, 1996.

Ibraheem, P.K. and Kothari, D.P., Recent philosophies of automatic generation control strategies in power systems, *IEEE Transactions on Power Systems*, 20(1), 346–357, 2005.

Ishtiaq, M., Sikandar, B., Jaffar, M.A., and Khan, A., Adaptive watermark strength selection using particle swarm optimization, *ICIC Express Letters*, 4(5), 1–6, 2010.

Jegathesan, V. and Jerome, J., Elimination of lower order harmonics in voltage source inverter feeding an induction motor drive using evolutionary algorithms, *Expert Systems with Applications*, 38(1), 692–699, 2011.

Jeon, G., Leep, H.R., and Shim, J.Y., A vehicle routing problem solved by using a hybrid genetic algorithm, *Computers and Industrial Engineering*, 53, 680–692, 2007.

Karaboga, D., An idea based on honey bee swarm for numerical optimization, Technical Report TR06, Erciyes University, Kayseri, Turkey, 2005.

Karaboga, D. and Basturk, B., A powerful and efficient algorithm for numerical function optimization: Artificial bee colony (ABC) algorithm, *Journal of Global Optimization*, 39(3), 359–371, 2007.

Keypour, R., Seifi, H., and Yazdian-Varjani, A., Genetic based algorithm for active power filter allocation and sizing, *Electric Power Systems Research*, 71(1), 41–49, 2004.

Kim, D.H. and Cho, J.H., A biologically inspired intelligent PID controller tuning for AVR systems, *International Journal of Control, Automation, and Systems*, 4(5), 624–636, 2006.

Kim, D.H., Abraham, A., and Cho, J.H., A hybrid genetic algorithm and bacterial foraging approach for global optimization, *Elsevier International Journal of Information Sciences*, 177, 3918–3937, 2007.

Krishnamoorthy, C.S. and Rajeev, S., *Artificial Intelligence and Expert Systems for Engineers*. Boca Raton, FL: CRC Press, 1996.

Krucky, J., Fuzzy family setup assignment and machine balancing, *Hewlett-Packard Journal*, June, 51–64, 1994.

Kundur, D. and Hatzinakos, D., A robust digital image watermarking method using wavelet-based fusion, in *Proceedings of the IEEE International Conference on Image Processing*, Santa Barbara, CA, October 26–29, 1997, Vol. 1, pp. 544–547.

Lee, Z.J., Lin, S.W., Su, S.F., and Lin, C.Y., A hybrid watermarking technique applied to digital images, *Applied Soft Computing*, 8(1), 798–808, 2008.

Li, X., Shao, Z., and Qian, J., An optimizing method base on autonomous animates: Fish-swarm algorithm, *Systems Engineering Theory and Practice*, 22, 32–38, 2002.

Li, Y., Luo, L., Rehtanz, C., Yang, D., Rüberg, S., and Liu, F., Harmonic transfer characteristics of a new HVDC system based on an inductive filtering method, *IEEE Transactions on Power Electronics*, 27(5), 2273–2283, 2012.

Lian, Z., Gu, X., and Jiao, B., A similar particle swarm optimization algorithm for job-shop scheduling to minimize makespan, *Applied Mathematics and Computation*, 183, 1008–1017, 2006.

Liao, C.J., Tseng, C.T., and Luarn, P., A discrete version of particle swarm optimization for flow-shop scheduling problems, *Computers and Operations Research*, 34(10), 3099–3111, 2007.

Lin, P.L., Hsieh, C.K., and Huang, P.W., A hierarchical digital watermarking method for image tamper detection and recovery, *Pattern Recognition*, 38(12), 2519–2529, 2005.

Liu, B., Wang, L., and Jin, Y.H., An effective PSO-based memetic algorithm for flow shop scheduling, *IEEE Transactions on Systems, Man and Cybernetics—Part B: Cybernetics*, 37(1), 18–27, 2007.

Liu, Y., Bang, J., and Wang, S., Optimization design based on PSO algorithm for PID controller, in *Proceedings of the 5th World Congress on Intelligent Control and Automation*, Hangzhou, China, June 15–24, 2004, pp. 2419–2422.

Majid Zamani Masoud, K.G., Nazzer, S., and Mostafa, P., Design of a fractional order PID controller for an AVR using particle swarm optimization, *Control Engineering Practice*, 17(12), 1380–1387, 2009.

Masiala, M., Ghnbi, M., and Kaddouri, A., An adaptive fuzzy controller gain scheduling for power system load-frequency control, in *IEEE International Conference on Industrial Technology (ICIT)*, Hammamet, Tunisia, December 8–10, 2004, pp. 1515–1520.

Mathur, H.D. and Manjunath, H.V., Frequency stabilization using fuzzy logic based controller for multi-area power system, *The South Pacific Journal of Natural Science*, 4, 22–30, 2007.

McArdle, M.G., Morrow, D.J., Calvert, P.A.J., and Cadel, O., A hybrid PI and PD type fuzzy logic controller for automatic voltage regulation of the small alternator, in *Proceedings of Power Engineering Society Summer Meeting*, Vancouver, British Columbia, Canada, July 2001, Vol. 3, pp. 1340–1345.

McCahon, C.S. and Lee, E.S., Job sequencing with fuzzy processing times, *Computers and Mathematics with Applications*, 19(7), 31–41, 1990.

Mehrabian, A.R. and Lucas, C., A novel numerical optimization algorithm inspired from weed colonization, *Ecological Informatics* 1, 355–366, 2006.

Milani, A.E. and Mozafari, B., Genetic algorithm based optimal load frequency control in two-area interconnected power systems, *Transaction on Power System Optimization*, 2, 6–10, 2010.

Mingwei, R., Yukun, S., and Xiang, R., A study on optimization of passive filter design, in *Proceedings of the 29th Chinese Control Conference*, Beijing, China, July 29–31, 2010, pp. 4981–4985.

Mittal, A. and Kassim, A., *Bayesian Network Technologies: Applications and Graphical Models*. Hershey, PA: Idea Group Inc, 2007.

Momoh, J.A., Ma, X.W., and Tomsovic, K., Overview and literature survey of fuzzy set theory in power systems, in *Transaction of IEEE PES Winter Power Meeting*, New York, January/February 1995, Paper # 95 WM 208-9 PWRS.

Mukherjee, V. and Ghosal, S.P., Velocity relaxed swarm intelligent tuning of Fuzzy based power system stabilizer, in *IEEE Power Engineering Society Conference*, Pittsburg, PA, July 20–24, 2008, p. 85.

Nanda, J. and Mangla, A., Automatic generation control of an interconnected hydro-thermal system using conventional integral and fuzzy logic controller, in *IEEE International Conference on Electric Utility Deregulation, Restructuring and Power Technologies*, Hong Kong, China, April 5–8, 2004, pp. 372–377.

Nayak, S.K., Krishnanand, K.R., Panigrahi, B.K., and Rout, P.K., Application of artificial bee colony to economic load dispatch problem with ramp rate limits and prohibited operating zones, in *Proceedings of the World Congress on Nature and Biologically Inspired Computing: NaBIC 2009*, Coimbatore, India, December 9–11, 2009, pp. 1237–1242.

Niu, Q., Jiao, B., and Gu, X., Particle swarm optimization combined with genetic operators for job shop scheduling problem with fuzzy processing time, *Applied Mathematics and Computation*, 205(1), 148–158, 2008.

Oliveira, P.M., Cunha, J.B., and Coelho, J.O.P., Design of PID controllers using the particle swarm algorithm, in *Twenty-First IASTED International Conference: Modelling, Identification, and Control (MIC 2002)*, Innsbruck, Austria, February 18–21, 2002. http://www.actapress.com/PaperInfo.aspx?PaperID=26612&reason=500.

Ombuki-Berman, B. and Hanshar, F., Using genetic algorithms for multi-depot vehicle routing, in *Bio-Inspired Algorithms for the Vehicle Routing Problem*, F.B. Pereira and J. Tavares, Eds. Springer Verlag, Berlin, Germany, 2009, pp. 77–99.

Pan, W.-T., A new fruit fly optimization algorithm: Taking the financial distress model as an example, *Knowledge-Based Systems*, 26(2), 69–74, 2012.

Passino, K.M., Biomimicry of bacterial foraging for distributed optimization and control, *IEEE Control Systems Magazine*, 22(3), 52–67, June 2002.

Pearl, J., *Probabilistic Reasoning in Intelligent Systems: Networks of Plausible Inference*. San Francisco, CA: Morgan Kaufmann, 1988.

Perez-Guerrero, R.E. and Cedenio-Maldonado, R.J., Economic power dispatch with non-smooth cost functions using differential evolution, in *Proceedings of the 37th Annual North American Power Symposium*, Ames, IA, October 23–25, 2005, pp. 183–190.

Pitas, I., A method for signature casting on digital images, in *Proceedings of the IEEE International Conference on Image Processing*, Lausanne, Switzerland, September 16–19, 1996, pp. 215–218.

Rabandi, I. and Nishimori, K., Optimal feedback control design using genetic algorithm in multimachine power system, *Electrical Power and Energy Systems*, 23, 263–271, 2001.

Ray, R.N., Chatterjee, D., and Goswami, S.K., An application of PSO technique for harmonic elimination in a PWM Inverters, *Applied Soft Computing*, 9(4), 1315–1320, 2009.

Ross, T., *Fuzzy Logic with Engineering Applications*. Oxford, U.K.: John Wiley and Sons, 2004.

Rudolf, A. and Bayrleithner, R., A genetic algorithm for solving the unit commitment problem of a hydro-thermal power systems, *IEEE Transactions on Power Systems*, 14(4), 1460–1468, 1999.

Salehi, R., Vahidi, B., Farokhnia, N., and Abedi, M., Harmonic elimination and optimization of stepped voltage of multilevel inverter by bacterial foraging algorithm, *Journal of Electrical Engineering & Technology*, 5(4), 545–551, 2010.

Saneifard, S., Prasad, N.R., and Smolleck, H.A., A fuzzy logic approach to unit commitment, *IEEE Transactions on Power Systems*, 12(2), 988–995, 1997.

Sasaki, H., Watanabe, M., and Yokoyama, R., A solution method of unit commitment by artificial neural networks, *IEEE Transactions on Power Systems*, 7(3), 974–981, 1992.

Senjyu, T., Yamashiro, H., Shimabukuro, K., Uezato, K., and Funabashi, T., A fast solution technique for large scale unit commitment problem using genetic algorithm, in *Proceedings of the IEEE/PES Transmission and Distribution Conference and Exhibition*, Yokohama, Japan, October 6–10, 2002, pp. 1611–1616.

Shayegi, H., Shayanfar, H.A., and Malik, O.P., Robust decentralized neural networks based LFC in deregulated power system, *Electric Power Systems Research*, 77, 541–551, 2008.

Sheble, G.B., Maifeld, T.T., Brittig, K., Fahd, G., and Fukurozaki-Coppinger, S., Unit commitment by genetic algorithm with penalty methods and a comparison of Lagrangian search and genetic algorithm economic dispatch example, *International Journal of Electrical Power and Energy Systems*, 18(6), 339–346, 1996.

Shi, Y. and Eberhart, R.C., Empirical study of particle swarm optimization, in *IEEE Congress on Evolutionary Computation*, Piscataway, NJ, July 6–9, 1999, Vol. 3, pp. 1945–1950.

Shieh, C.S., Huang, H.C., Pan, J.S., and Wang, F.H., Genetic watermarking based on transform domain technique, *Pattern Recognition*, 37(3), 555–565, 2004.

Shirabe, K., Swamy, M., Kang, J., Hisatsune, M., Wu, Y., Kebort, D., and Honea, J., Efficiency comparison between Si-IGBT-based drive and GaN-based drive, *IEEE Transactions on Industry Applications*, 50(1), 566–572, 2014.

Simon, D., Biogeography-based optimization, *IEEE Transactions on Evolutionary Computation*, 12(6), 702–713, 2008.

Singh, B., Bhuvaneswari, G., and Garg, V., Harmonic mitigation using 12-pulse AC–DC converter in vector-controlled induction motor drives, *IEEE Transactions on Power Delivery*, 21(3), 1483–1495, 2006a.

Singh, B., Bhuvaneswari, G., Garg, V., and Gairola, S., Pulse multiplication in AC-DC converters for harmonic mitigation in vector-controlled induction motor drives, *IEEE Transactions on Energy Conversion*, 21(2), 342–352, 2006b.

Skok, M., Skrlec, D., and Krajcar, S., The genetic algorithm method for multiple depot capacitated vehicle routing problem solving, in *Proceedings of the 4th International Conference on Knowledge Based Intelligent Engineering Systems and Allied Technologies*, Brighton, U.K., August 30–September 1, 2000, pp. 520–526.

Stankovic, A.M., Tadmor, G., and Sakharuk, T.A., On robust control analysis and design for load frequency regulation, *IEEE Transactions on Power Systems*, 13(2), 449–455, 1998.

Starkweather, T., Whitley, D., and Cookson, B., A genetic algorithm for scheduling with resource consumption, in *Proceedings of the Joint German/US Conference on Operations Research in Production Planning and Control*, Hagen, Germany, June 25–26, 1992, pp. 567–583.

Storn, R. and Price, K.V., Differential evolution a simple and efficient heuristic for global optimization over continuous spaces, *Journal of Global Optimization*, 11(4), 341–359, 1997.

Strogatz, S.H., *Nonlinear Dynamics and Chaos*, 2nd edn. Boulder, CO: Westview Press, 2014.

Sumathi, S. and Surekha, P., *Computational Intelligence Paradigms: Theory & Applications Using MATLAB*. CRC Press, Boca Raton, FL, 2010.

Sundareswaran, K., Krishna, J., and Shanavas, T.N., Inverter harmonic elimination through a colony of continuously exploring ants, *IEEE Transactions on Industrial Electronics*, 54(5), 2558–2565, 2007.

Swarup, K.S. and Yamashiro, S., Unit commitment solution methodology using genetic algorithm, *IEEE Transactions on Power Systems*, 17(1), 87–91, 2002.

Szeto, W.Y., Wu, Y., and Ho, S.C., An artificial bee colony algorithm for the capacitated vehicle routing problem, *European Journal of Operational Research*, 215(1), 126–135, 2011.

Takashi, H., Yoshiteru, U., and Hiroaki, A., Integrated fuzzy logic controller generator for stability enhancement, *IEEE Transactions on Energy Conversion*, 12(4), 1997.

Talaq, J. and Al-Basri, F., Adaptive fuzzy gain scheduling for load frequency control, *IEEE Transactions on Power Systems*, 14(1), 145–150, 1999.

Tsujimura, Y.S., Park, S., Chang, I.S., and Gen, M., An effective method for solving flow shop scheduling problems with fuzzy processing times, *Computers and Industrial Engineering*, 25(1–4), 239–242, 1993.

Valle, Y.D. and Salman, M., Particle swarm optimization: Basic concepts, variants and applications in power systems, *IEEE Transactions on Evolutionary Computation*, 12(2), 171–175, 2008.

Venkata Prasanth, B. and Jayaram Kumar, S.V., Load frequency control for a two area interconnected power system using robust genetic algorithm controller, *Journal of Theoretical and Applied Information Technology*, 1204–1212, 2005–2008.

Verma, V. and Bhim, S., Genetic-algorithm-based design of passive filters for offshore applications, *IEEE Transactions on Industry Applications*, 46(4), 1295–1303, 2010.

Villablanca, M.E., Nadal, J.I., and Bravo, M.A., A 12-pulse AC–DC rectifier with high-quality input/output waveforms, *IEEE Transactions on Power Electronics*, 22(5), 1875–1881, 2007.

Walsh, M.P. and Malley, M.J.O., Augmented hopfield network for unit commitment and economic dispatch, *IEEE Transactions on Power Systems*, 12(4), 1765–1774, 1997.

Wang, M.S. and Chen, W.C., A hybrid DWT-SVD copyright protection scheme based on k-means clustering and visual cryptography, *Computer Standards and Interfaces*, 31(4), 757–762, 2009.

Wang, S., Wang, L., and Wang, D., Particle swarm optimization for multi-depots single vehicle routing problem, in *Proceedings of the Chinese Control and Decision Conference: CCDC 2008*, Yantai, Shandong, July 2–4, 2008a, pp. 4659–4661.

Wang, Y.-B., Peng, X., and Wei, B.-Z., A new particle swarm optimization based auto-tuning of PID controller, in *Proceedings of the Seventh International Conference on Machine Learning and Cybernetics*, Kunming, China, July 12–15, 2008b, pp. 12–15.

Wei, X.J. and Wu, Z.M., An effective hybrid optimization approach for multi-objective flexible job-shop scheduling problems, *Computers and Industrial Engineering*, 48(2), 409–425, 2005.

Wei, Z., Li, H., and Dai, J., Image watermarking based on genetic algorithm, in *Proceedings of the IEEE International Conference on Multimedia and Expo: ICME '06*, Toronto, Ontario, Canada, July 9–12, 2006, pp. 1117–1120.

Wenjing, Z. and Ye, J., An improved particle swarm optimization for the multi-depot vehicle routing problem, in *Proceedings of the International Conference on E-Business and E-Government: ICEE*, Guangzhou, China, May 7–9, 2010, pp. 3188–3192.

Wong, C.-C., Li, S.-A., and Wang, H.-Y., Hybrid evolutionary algorithm for PID controller design of AVR system, *Journal of the Chinese Institute of Engineers*, 32(2), 251–264, 2009.

Wu, L.H., Wang, Y.N., Yuan, X.F., and Zhou, S.W., Environmental/economic power dispatch problem using multi-objective differential evolution algorithm, *Electric Power Systems Research*, 80(9), 1171–1181, 2010.

Wu, W., He, Y., and Blaabjerg, F., An LLCL power filter for single-phase grid-tied inverter, *IEEE Transactions on Power Electronics*, 27(2), 782–789, 2012.

Yang, X.S., Fire fly algorithm for multimodal optimization, in *Proceedings of the Stochastic Algorithms. Foundations and Applications (SAGA 109)*, vol. 5792 of Lecture notes in Computer Sciences Springer, October 2009.

Yang, X.S., A new metaheuristic bat-inspired algorithm, in *Studies in Computational Intelligence*, J. Kacprzyk, Ed. Springer, Berlin, Germany, 2010, pp. 65–74.

Yang, X.-S. and Deb, S., Cuckoo Search via Levy flights, in *Proceedings of the World Congress on Nature and Biologically Inspired Computing: NaBIC 2009*, Coimbatore, India, December 9–11, 2009, pp. 210–214.

Yare, Y., Venayagamoorthy, G.K., and Saber, A.Y., Heuristic algorithms for solving convex and nonconvex economic dispatch, in *Proceedings of the IEEE International Conference on Intelligent System Applications to Power Systems: ISAP'09*, Curitiba, Brazil, November 8–12, 2009, pp. 1–8.

Yoshida, H. and Kenichi, K., A particle swarm optimization for reactive power and voltage control considering voltage stability, in *IEEE International Conference on Intelligent System Applications to Power Systems*, Rio de Janeiro, Brazil, April 4–8, 1999, pp. 1–6.

Zareiegovar, G., A new approach for tuning PID controller parameters of LFC considering system uncertainties, in *Proceedings of International Conference on Environment and Electrical Engineering*, Prague, Czech Republic, March 23–25, 2009, pp. 333–336.

Zeynelgil, H.L., Demiroren, A., and Sengor, N.S., The application of ANN technique for automatic generation control for multi-area power system, *Journal of Electric Power and Energy Systems*, 24, 345–354, 2002.

Zhang, Z. and Lu, C., Application of the improved particle swarm optimizer to vehicle routing and scheduling problems, in *Proceedings of the IEEE International Conference on Grey Systems and Intelligent Services: GSIS 2007*, Nanjing, China, November 18–20, 2007, pp. 1150–1152.

Zheng, D., Wang, S., and Zhao, J., RST invariant image watermarking algorithm with mathematical modeling and analysis of the watermarking processes, *IEEE Transactions on Image Processing*, 18(5), 1055–1068, 2009.

Zheng, D. and Zhao, J., A rotation invariant feature and image normalization based image watermarking algorithm, in *Proceedings of the IEEE International Conference on Multimedia and Expo 2007*, Beijing, China, July 2–5, 2007, pp. 2098–2101.

Ziqiang, W., Xia, S., and Dexian, Z., Novel watermarking scheme based on PSO algorithm, in *Bio-Inspired Computational Intelligence and Applications*, L. Kang, F. Minrui, G.W. Irwin, and S. Ma, Eds. Springer-Verlag, Berlin, Germany, 2007, pp. 307–314.

Zobaa, A.F., Harmonic problems produced from the use of adjustable speed drives in industrial plants: Case study, in *11th International Conference on Harmonics and Quality of Power*, September 12–15, 2004, pp. 6–10.

Zwaan, S. and Marques, C., Ant colony optimization for job shop scheduling, in *Proceedings of the Third Workshop on Genetic Algorithms and Artificial Life: GAAL'99*, 1999.

2

Unit Commitment and Economic Load Dispatch Problem

Learning Objectives

On completion of this chapter, the reader will have knowledge on

- Solving the unit commitment (UC) and economic load dispatch (ELD) problems;
- Application of intelligent algorithms such as genetic algorithm, fuzzy c-means-based radial basis function network, enhanced particle swarm optimization algorithm, improved differential evolution with opposition-based learning, artificial bee colony optimization, and cuckoo search optimization to solve UC and ELD problems;
- Development of MATLAB®/Simulink® m-files for intelligent algorithms;
- Testing the performance of the algorithm on different test cases.

2.1 Introduction

Optimization is an interdisciplinary area providing solutions to nonlinear, stochastic, combinatorial, and multiobjective problems. With the increasing challenges of satisfying optimization goals of current applications, there is a strong drive to improve the development of efficient optimizers. Thus it is important to identify suitable computationally intelligent algorithms for solving the challenges posed by optimization problems. Conventional classical optimizers such as linear/nonlinear programming, dynamic programming, and stochastic programming perform poorly and fail to produce improvement in the face of such challenges. This has motivated researchers to explore the use of computational intelligence (CI) techniques to augment classical methods.

More specifically, CI techniques are powerful tools offering several potential benefits such as robustness (imposing little or no requirements on the objective function), versatility (capable of handling highly nonlinear mappings), hybridization (combination of techniques to improve search), self-adaptation to improve overall performance, and parallel operation (easier decomposition of complex tasks). However, successful application of CI techniques to real-world problems is not straightforward and requires expert knowledge in obtaining solutions through trial-and-error experiments. The growing interest for the application of CI techniques has introduced state-of-the-art methods in solving power system problems. In addition, CI techniques possess an apparent ability to adapt to the nonlinearities and discontinuities in power systems. The motivation of this work is due to the success of CI paradigms to solve unit commitment (UC) and economic load dispatch (UC-ELD) problems with the aim of obtaining optimal schedule for power generating systems. Because of the nonconvex and combinatorial nature of UC and ELD problems, it is difficult to obtain a solution using conventional programming methods such as dynamic programming, mixed integer programming, linear programming and Lagrangian relaxation methods. UC and ELD are nonlinear optimization problems whose solutions can be obtained through CI paradigms since these provide efficient alternatives over analytical methods which suffer from premature convergence, local optimal trapping, and the curse of dimensionality.

The electric power demand is much larger during daytime because of high industrial loads, and during evenings and early mornings because of usage in residential areas. Based on the forecasted power requirements for the successive operating days, the generating units are scheduled on an hourly basis for the next day's dispatch, which in turn is forecasted for a week ahead. The system operators are able to schedule the ON/OFF status and the real power outputs of the generating units to meet the forecasted demand over a time horizon. There may exist large variations in the day-to-day load patterns, so enough power has to be generated to meet the maximum load demand. In addition, it is not economical to run all the units all the time. Hence it is necessary to determine the units of a particular system that are required to operate for given loads. This problem is known as the unit commitment (Rajan 2010) problem.

ELD allocates power to the committed units, thus minimizing the fuel cost. The two major factors to be considered while dispatching power to generating units are the cost of generation and the quantity of power supplied. The relation between the cost of generation and the power levels is approximated by a quadratic polynomial. To determine the economic distribution of load between the various units in a plant, the quadratic polynomial in terms of the power output is treated as an optimization problem with cost minimization as the objective function, and considering equality and inequality constraints.

During the past years, several exact and approximate algorithms have been applied for solving the UC and ELD problems. The exact solutions to these problems can be obtained through numerical calculations, but they cannot be applied for large, practical, real-time systems due to the computational overheads (Vlachogiannis and Lee 2008). Some of these exact methods for solving UC and ELD are the Lambda iteration method (LIM), dynamic programming, mixed integer programming, branch and bound, Newton's method, and Lagrangian relaxation method. The approximate methods include search algorithms such as artificial neural networks (ANNs), genetic algorithms (GAs), tabu search (TS), simulated annealing (SA), evolutionary programming (EP), particle swarm optimization (PSO), ant colony optimization (ACO), artificial bee colony (ABC), differential evolution (DE), bacterial foraging algorithm (BFA), intelligent waterdrop (IWD), and bio-geography-based optimization (BBO) algorithms.

In all the literatures reported, either the UC or the ELD problem is solved individually. Solving UC-ELD problems together using heuristic techniques generates a complete solution for the real-time power system, thereby validating these techniques in terms of optimal solutions, robustness, computational time, and algorithmic efficiency. The purpose of this application is to find out the advantages of the application of the bio-inspired techniques to the UC-ELD problems. An attempt has been made to find out the minimum cost by using intelligent algorithms such as fuzzy-based radial basis function network (FRBFN), enhanced PSO (EPSO), DE with opposition-based learning (DE-OBL), improved DE with OBL (IDE-OBL), ABC optimization, and cuckoo search optimization (CSO). UC and ELD represent a time-decomposed approach to achieve the objective of economic operation and hence they are viewed as two different optimization problems. The UC problem deals with a long time span, typically 24 hours or a week. The ON/OFF timing of the generating units is scheduled to achieve an overall minimum operating cost. ELD is a problem that deals with shorter time span, typically starting from seconds to approximately 20 minutes. It allocates the optimal sharing of generation outputs among synchronized units to meet the forecasted load.

The cost minimization and the rapid response requirement in real-time power systems necessitate this two-step approach. The objective of both the approaches is to minimize the fuel cost with less time of operation, thus meeting the constraints imposed. The units in the system are switched ON/OFF based on an exhaustive search performed by GA. The ON/OFF schedule is then optimized using the heuristics such as FRBFN, EPSO, DE-OBL, IDE-OBL, ABC, and CSO to dispatch power, thus meeting the load demand without violating the power balance and capacity constraints.

The algorithm is evaluated in terms of UC schedules, distribution of load among individual units, total fuel cost, power loss, total power, and computational time. For experimental analysis, four test systems are chosen: the IEEE 30 bus system (6-unit system), 10-unit test system, Indian utility 75-bus system (15-unit system), and the 20-unit test system, including transmission losses, power balance, and generator capacity constraints. The outcome of the experimental results is compared in terms of optimal solution, robustness, computational efficiency, and algorithmic efficiency.

In this chapter, the mathematical formulation of the UC and ELD problems along with the framework to solve these problems is provided. The implementation of the optimization techniques such as GA, FRBFN, EPSO, DE-OBL, IDE-OBL, ABC, and CSO for solving the UC-ELD problem along with a comparative analysis based on fuel cost, robustness, computational efficiency, and algorithmic efficiency is presented.

2.2 Economic Operation of Power Generation

Since an engineer is always concerned with the cost of products and services, the efficient optimum economic operation and planning of electric power generation have always occupied an important position in the electric power industry. With large interconnection of electric networks, the energy crisis in the world, and continuous rise in prices, it is very essential to reduce the running charges of electricity generating units. A saving in the operation of the system even by a small percent represents a significant reduction in the operating cost as well as in the quantities of fuel consumed. The classic problem is the economic load dispatch of the generating systems with minimum operating cost. In addition, there is a need to expand the limited economic optimization problem to incorporate constraints on system operation to ensure the security of the system, thereby preventing the collapse of the system due to unforeseen conditions. However, closely associated with this economic dispatch problem is the problem of the proper commitment of an array of units to serve the expected load demands in an "optimal" manner.

2.3 Mathematical Model of the UC-ELD Problem

To solve problems related to generator scheduling, numerous trials are required to identify all the possible solutions, from which the best solution is chosen. This approach is capable of testing different combinations of units based on the load requirements (Orero and Irving 1995). At the end of the testing process, the combination with the least operating cost is selected as the optimal schedule. While scheduling generator units, the start-up and shut-down time are to be determined along with the output power levels at each unit over a specified time horizon. In turn, the start-up, shut-down, and the running cost are maintained at a minimum. The fuel cost F_i per unit in any given time interval is a function of the generator power output, as given by Equation 2.1:

$$F_T = \sum_{i=1}^{n} F_i(P_i) = \sum_{i=1}^{n} a_i + b_i P_i + c_i P_i^2 \ \ \$/h \tag{2.1}$$

where
    a_i, b_i, c_i represent unit cost coefficients
    P_i denotes the unit power output

The start-up cost (SC) depends upon the down time of the unit, which can vary from a maximum value, when the unit is started from cold state, to a much smaller value if the unit is turned off recently. It can be represented by an exponential cost curve as shown by Equation 2.2:

$$SC_i = \sigma_i + \delta_i * \left\{ 1 - \exp\left(-\frac{T_{off}}{\tau_i} \right) \right\} \tag{2.2}$$

where
    σ_i is the hot start-up cost
    δ_i is the cold start-up cost
    τ_i is the unit cooling time constant
    T_{off} is the time at which the unit has been turned off

The total cost F_T involved during the scheduling process is a sum of the running cost, start-up cost. and shut-down cost given by

$$F_T = \sum_{t=1}^{T} \sum_{i=1}^{N} FC_{i,t} U_{i,t} + SC_{i,t} (1 - U_{i,t-1}) U_{i,t} + SD_{i,t} \tag{2.3}$$

where
    N is the number of generating units
    T is the number of different load demands for which the commitment has to be estimated

The shut-down cost, *SD*, is usually a constant value for each unit, and $U_{i,t}$ is the binary variable that indicates the ON/OFF status of a unit *i* in time *t*. The overall objective is to minimize F_T subject to a number of constraints as follows:

1. System hourly power balance is given by Equation 2.4, where the total power generated must supply the load demand (P_D) and system losses (P_L):

$$\sum_{i=1}^{N} P_{i,t}U_{i,t} = P_D + P_L \tag{2.4}$$

2. Hourly spinning reserve requirements (*R*) must be met. Spinning reserve is the term used to describe the total amount of generation available from all the units synchronized on the system minus the present load plus the losses being incurred. This is mathematically represented as

$$\sum_{i=1}^{N} P_{i,t}^{max}U_{i,t} - (P_D + P_L) = R \tag{2.5}$$

3. Unit rated minimum and maximum capacities must not be violated. The power allocated to each unit should be within their minimum and maximum generating capacity as shown in Equation 2.6:

$$P_{i,t}^{min} \le P_{i,t} \le P_{i,t}^{max} \tag{2.6}$$

4. The initial states of each generating unit at the start of the scheduling period must be taken in to account.

5. Minimum up/down (*MUT/MDT*) time limits of units must not be violated. This is expressed by Equations 2.7 and 2.8, respectively:

$$(T_{t-1,i}^{on} - MUT_i) * (U_{t-1,i} - U_{t,i}) \ge 0 \tag{2.7}$$

$$(T_{t-1,i}^{off} - MDT_i) * (U_{t,i} - U_{t-1,i}) \ge 0 \tag{2.8}$$

where T_{off}/T_{on} is the unit off/on time, while $U_{t,i}$ denotes the unit off/on {0, 1} status.

The principal objective of the ELD problem is to find a set of active power delivered by the committed generators to satisfy the required demand subject to the unit's technical limits at the lowest production cost. The objective of the ELD problem is formulated in terms of the fuel cost, as

$$F_T = \sum_{i=1}^{n} F_i(P_i) = \sum_{i=1}^{n} a_i + b_iP_i + c_iP_i^2 \tag{2.9}$$

The total generated power $\sum_{i=1}^{N} P_i$ should be equal to the sum of total system demand P_D and the transmission loss P_L. This power balance equality constraint is mathematically expressed as

$$\sum_{i=1}^{N} P_i = P_D + P_L \tag{2.10}$$

where P_L is computed using the B coefficients as follows:

$$P_L = \sum_{i=1}^{N} \sum_{j=1}^{N} P_i B_{ij} P_j + \sum_{i=1}^{N} B_{0i} P_i + B_{00} \tag{2.11}$$

where B_{ij}, B_{0i}, and B_{00} are the transmission loss coefficients obtained from the B-coefficient matrix. The generator power P_i should be limited within the range stated by the inequality constraint

$$P_i^{\min} \leq P_i \leq P_i^{\max} \tag{2.12}$$

where $P_i^{\min}$ and $P_i^{\max}$ are the minimum and maximum generator limits corresponding to the ith unit.

2.4 Intelligent Algorithms for Solving UC-ELD

The operation of a modern power system has to incorporate in its mission a strategy that serves to derive the maximum benefits of improved performance and enhanced reliability. The power grid networks have been analyzed using conventional and enumerative techniques for delivering the bulk power, reliably and economically, from power plants to the consumers. The conventional method of solving the generator scheduling problem involves an exhaustive trial of all the possible solutions and then choosing the best among them, which is a complex task. For example, in the UC problem the combination of generating units that produces the least operating cost is taken as the best schedule from several trial runs. In order to alleviate the disadvantages associated with conventional strategies in terms of quality solution and computational time, bio-inspired intelligent techniques are explored in this chapter to solve the UC-ELD problems. The step-by-step procedure of the algorithms applied to optimize the UC and ELD problems using the intelligent heuristics are discussed in detail in this section.

2.4.1 UC Scheduling Using Genetic Algorithm

GAs are adaptive search techniques based on the principles and mechanisms of natural selection and "survival of the fittest" from biological evolution (Goldberg 1989). The algorithm starts with a population of chromosomes, from which a selected group of chromosomes enter the mating pool. Genetic operators are applied to these chromosomes to obtain the best solution based on evaluation of the fitness function. The three prime opera-tors associated with the GA are reproduction, crossover, and mutation. Every generation is made up of a fixed number of solutions that are randomly obtained from the solutions of the previous generation. GA may be phenotypic (operating only on parameters that are placed directly into the fitness function and evaluated) or genotypic (operating on parameters that are used to generate behavior in light of external "environmental" factors) (David Fogel 1995).

In this application, the UC problem is solved using GA, which generates the on/off sta-tus of the generating units. For the UC problem using GA, a chromosome represents the on/off status of each unit for a given load demand. For example, if there are six generating units, the chromosome consists of six genes, and each gene represents the status of one unit. A gene value of 0 represents the "off" status and a gene value of 1 indicates that the unit is on. The step-by-step procedure involved in the implementation of GA for UC prob-lem is explained here:

Step 1: Input data

Specify generator cost coefficients, generation power limits for each unit, and transmission loss coefficients (B-matrix) for the test system. Read hourly load pro-file of the generators for the test system. Initialize the parameters of GA such as number of chromosomes, population size, number of generations, selection type, crossover type, mutation type, crossover probability, and mutation probability, suitably.

Step 2: Initialize GA's population

Initialize population of the GA randomly, where each gene of the chromosomes represents commitment of a dispatchable generating unit. The first step is to encode the commitment space for the UC problem based on the load curve from the load profile. Units with heavy loads are committed (binary 1) and units with lighter loads are de-committed (binary 0). The population consists of a set of UC schedules in the form of a matrix $N \times T$, where N is the number of generators and T is the time horizon.

Step 3: Computation of total cost

The total generation cost for each chromosome is computed as the sum of indi-vidual unit fuel cost.

Step 4: Computation of cost function and fitness function

The augmented cost function for each chromosomes of population is computed using

$$F = \sum_{i=1}^{N} FC_i = a_i * P_i^2 + b_i * P_i + c_i \tag{2.13}$$

where
a_i, b_i, and c_i represents unit cost coefficients
P_i is the unit power output

The fitness function of chromosomes is calculated as the inverse of the augmented cost function.

Step 5: Application of genetic operators

After the computation of the fitness function value for each chromosome of population, crossover and mutation operators are applied to the population and the new generation of chromosomes is generated. A two point crossover technique (Figure 2.1) is applied on two parents to generate two offspring. The offspring are evaluated for fitness, and the best one is retained while the worst is discarded from the population. The mutation operation is performed by selecting a chromosome with specified probability. The chosen chromosome is decoded to its binary equivalent with the unit number, and the time period is selected randomly for the flip bit mutation operation.

Step 6: The algorithm terminates after a specified number of generations have reached. If the termination condition is not satisfied, then it goes to Step 3.

Using the above procedure, the generating units of the test systems are committed/de-committed accordingly and, based on these ON/OFF schedules, the economic dispatch is performed by applying FRBFN, EPSO, DE-OBL, IDE-OBL, ABC, and CSO algorithms.

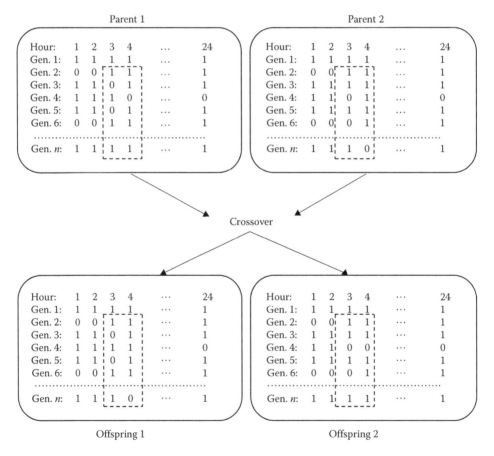

FIGURE 2.1
Crossover operation on UC schedules.

2.4.2 Fuzzy c-Means-Based Radial Basis Function Network for ELD

The methodology of implementing the RBF network to solve the ELD problem is shown in Figure 2.2. The training data based on the selected test systems for different power demands with varying weights are set by the LIM. The values generated should be capable of satisfying all load profiles.

2.4.2.1 Fuzzy c-Means Clustering

Application of clustering methods requires the number of known clusters in advance. There are two options for clustering—validity measures and compatible clustering. The data samples are clustered several times, each time with a different number of clusters $k \in [2,n]$ in validity measures, while in the compatible type of clustering the algorithm starts with a large number of clusters and then proceeds by gradually merging similar clusters to obtain fewer clusters (Meng et al. 2010). In order to validate the nonlinearity of the system, the value of k should be large enough.

The method of selecting the number of hidden units in a neural network is one of the most challenging tasks, requiring more experimentation. In this chapter, an FCM clustering approach is adopted to specify the range of hidden layer neurons in the RBF network. Consider $x_i \in \Re$ is the data patterns in the feature space. Let the initial cluster number be $k = n/2$, and test whether a new center should be added based on the performance of the network. The new cluster center c_{k+1} is added from the remaining samples $[c_1, c_2, \dots, c_k]$. The fuzzy membership matrix is then updated with new centers, and the

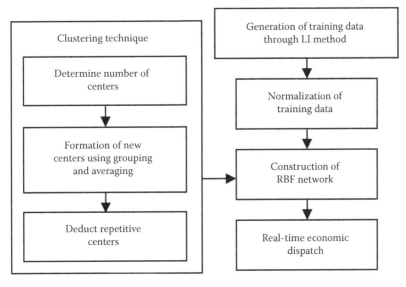

FIGURE 2.2
Schematic of FRBFN methodology.

process is repeated until the condition $k < n$ is satisfied. The clustering algorithm is performed by minimizing the objective function:

$$\min J_m(u,c;x) = \sum_{i=1}^{n} \sum_{j=1}^{k} u_{ji}^m \|x_i - c_j\| \qquad (2.14)$$

where
$\quad m$ is a real number greater than 1
$\quad u_{ji}$ is the degree of membership of x_i in the cluster j
$\quad x_i$ is the ith d-dimensional measured data
$\quad c_j$ is the d-dimensional center of the cluster
$\quad \|*\|$ is the norm expressing the similarity between measured data and the center

The constraints imposed on the degree of membership u_{ji} are given according to Equation 2.15:

$$\begin{cases} u = [u_{ji}], u_{ji} \in [0,1] \\ \sum_{j=1}^{k} u_{ji} = 1, \sum_{j=1}^{k} u_{ji} > 0' \end{cases} \begin{cases} i = 1,2,\ldots,n \\ j = 1,2,\ldots,k \end{cases} \qquad (2.15)$$

The algorithm of the FCM is as follows:

Step 1: For the given dataset, initialize $k \in [n/2, n]$, tolerance $\varepsilon > 0$, initial cluster center c_0, fuzzification constant m, such that $1 < m < \infty$. If $m \to 1$, the membership degrees of the data pattern tend to be either 0 or 1, thus approaching the hard means clustering, and if $m \to \infty$, the membership degrees of the data pattern tend to $1/k$, leading to a high level of fuzziness. Based on experimental analysis conducted by Hathaway and Bezdek (2001), the optimal choice of m is 2.

Step 2: Calculate $u(t) = [u_{ji}(t)]$, where $u_{ji}(t)$ is the membership value of vector x_i to the cluster center c_j; with Euclidean distance $d_{ji} = \|x_i - c_j\|^2$ between x_i and c_j

$$u_{ji}(t) = \frac{1}{\sum_{r=1}^{k} \left\{ \left[d_{ji}(t-1) / d_{ri}(t-1) \right]^{2/(m-1)} \right\}} \qquad (2.16)$$

Step 3: Compute the center $c(t)$, given $c = [c_1, c_2, \ldots, c_k]$ is the array of clusters for $\forall j$:

$$c_j(t) = \frac{\sum_{i=1}^{n} [u_{ji}(t)]^m x_i}{\sum_{i=1}^{n} [u_{ji}(t)]^m} \qquad (2.17)$$

Step 4: Test for stopping condition, else go to step 2. The stopping condition may be the maximum number of iterations or until the condition $\|c(t) - c(t-1) < \varepsilon\|$ is met.

2.4.2.2 Implementation of FCM-Based RBF for the ELD Problem

The major governing parameters for implementing the RBF network are

- Number of centers in the hidden layer
- Position of the RBF centers
- Width of the RBF centers
- Weights applied to the RBF function outputs as they are passed to the summation layer.

The number of hidden neurons, or equivalently radial basis centers, needs to be much larger than the number of clusters in the data. The choice of number of hidden neurons is determined through the FCM algorithm. The output of the hidden neuron is significant only if the Euclidean distance from the cluster center is within a radius of $2\sigma_i$ around the cluster center. The width of the RBF centers is set once the clustering procedure is completed, satisfying the condition that the basis functions should overlap to some extent in order to give a relatively smooth representation of the data. Typically, the width for a given cluster center is set to the average Euclidean distance between the center and the training vectors that belong to that cluster.

The application of RBF network consists of two phases: training and testing. The accuracy of an RBF network model depends on the proper selection of training data. The inputs of the training network are power demand and weights w_1 and w_2, while the outputs constitute the power generated by the generating units. The step-by-step procedure involved in the implementation of ELD using FCM-based RBF network is elaborated here:

Step 1: Divide the dataset into training sets and testing sets to evaluate the network performance.

Step 2: Initialize suitable values for the range of cluster, initial cluster center, tolerance value for FCM, and number of maximum iterations.

Step 3: Compute the membership matrix and update iteratively based on

$$u_{ji}(t) = \frac{1}{\sum_{r=1}^{k}\left\{\left[d_{ji}(t-1)/d_{ri}(t-1)\right]^{2/(m-1)}\right\}} \tag{2.18}$$

Similarly, the clusters center matrix given by Equation 2.17 is computed and updated. If the maximum number of iterations or the specified tolerance level has been reached, then the clustering process stops and algorithm proceeds with Step 6.

Step 4: Compute the cluster radius and weights between the hidden layer and the output layer. The feasible results based on the training and testing data are saved.

Step 5: Based on the current membership matrix, new cluster centers c_{k+1} are determined using the objective function given by Equation 2.19:

$$\min \sum_{1\leq i,j\leq k, i\neq j} (u_{ni} - u_{nj}) \tag{2.19}$$

Go to step 3 to check if the clustering process has completed.

Step 6: The center model that produces minimum error is selected, and the output results are computed based on the testing data.

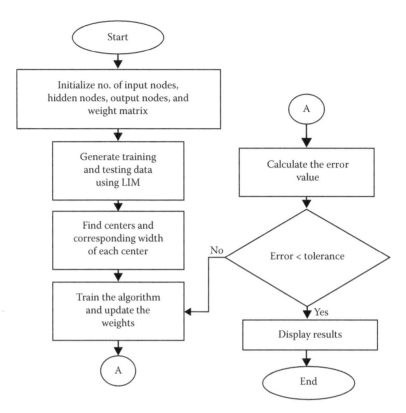

FIGURE 2.3
Flowchart of ELD using RBF network.

Figure 2.3 shows the steps involved in solving the ELD problem using an RBF network. The parameters, such as cost coefficients a_i, b_i, and c_i and minimum and maximum power generated in the ith unit, $P_{Gi\,min}$ and $P_{Gi\,max}$, are given as input to the input nodes. Along with the input parameters, the test data of the inputs are also provided. While propagating along the hidden layers, the weights are updated and the centers are chosen using a random selection method. The network is trained through the algorithm, and the error values are computed. If the difference between the target and trained data is below the tolerance value, the algorithm is stopped and the results are displayed; otherwise the process is repeated until the error converges. The accuracy of the RBF network also depends upon proper selection of the training data. The more uniformly the training data are distributed, the faster the network converges, thus providing the optimal solution.

2.4.3 Solution to ELD Using Enhanced Particle Swarm Optimization Algorithm

PSO is one of the modern heuristic algorithms suitable to solve large-scale nonconvex optimization problems. It is a population-based search algorithm and searches in parallel using a group of particles (Wang et al. 2010). In this chapter, EPSO is applied to solve the ELD problem. The EPSO is an improved version of the conventional PSO, being inspired by the study of birds and fish flocking. In EPSO, a constriction factor is introduced in the velocity update formula to ensure faster convergence. In PSO algorithm, each particle in the swarm represents a solution to the problem and is defined by its position and velocity. Each particle has a position represented by a position-vector x (i is the index of the particle), and a velocity represented by a velocity vector v_i. Each particle remembers its own best position so far in vector $x_i^{\#}$, and its jth dimensional value is $x_{ij}^{\#}$. The best position vector among the swarm so far is then stored in a vector x^*, and its jth dimensional value is x_j^*. During the iteration time t, the update of the velocity from the previous velocity to the new velocity is determined by Equation 2.20. The new position is then determined by the sum of the previous position and the new velocity, given by Equation 2.21.

$$v_{ij}(t+1) = wv_{ij}(t) + c_1 r_1 (x_{ij}^*(t) - x_{ij}(t)) + c_2 r_2 (x_j^*(t) - x_{ij}(t)) \tag{2.20}$$

$$x_{ij}(t+1) = x_{ij}(t) + v_{ij}(t+1) \tag{2.21}$$

where
  w is the inertia weight factor
  r_1 and r_2 are random numbers

that are used to maintain the diversity of the population and are uniformly distributed in the interval [0, 1] for the jth dimension of the ith particle, c_1 is a positive constant called the coefficient of the self-recognition component (cognitive component), and c_2 is a positive constant called the coefficient of the social component. From Equation 2.20, a particle decides where to move next, considering its own experience, which is the memory of its best past position, and the experience of the most successful particle in the swarm.

In EPSO, the acceleration constants c_1, c_2 and the inertia weight w are modified, thus in turn affecting the velocity and position update equations. The constriction factor k is computed using the social and cognitive components according to

$$k = \left| \frac{2}{2 - c - \sqrt{c^2 - 4c}} \right| \tag{2.22}$$

where $c = c_1 + c_2$, such that $c_1 + c_2 \geq 3$.

Similarly, the inertia weight w of the particle is also updated during the iterations in a nonlinear fashion according to Equation 2.23:

$$w = (w_{max} - w_{min}) * \left(\frac{max\_iter - iter}{max\_iter} \right) + w_{min} \tag{2.23}$$

where
  w_{max} is the maximum inertia weight
  w_{min} is the minimum inertia weight
  max_iter is the maximum number of iterations run by the EPSO
  $iter$ is the value of the current iteration

Thus the position and velocity are updated as follows:

$$v_{ij}(t+1) = wkv_{ij}(t) + c_1 r_1 (x_{ij}^*(t) - x_{ij}(t)) + c_2 r_2 (x_j^*(t) - x_{ij}(t)) \tag{2.24}$$

$$x_{ij}(t+1) = x_{ij}(t) + v_{ij}(t+1) \tag{2.25}$$

The algorithm for implementing EPSO to solve the ELD problem is shown here.

Step 1: Initialize the PSO parameters such as population size, maximum inertia weight, minimum inertia weight, initial velocity, initial position, cognitive factor, social factor, error gradient, and maximum number of iterations.

Step 2: To each individual of the population P_g (generator power), employ the B-coefficient loss formula to calculate the transmission loss P_L, according to

$$P_L = \sum_{i=1}^{N}\sum_{j=1}^{N} P_{Gi}B_{ij}P_{Gj} + \sum_{i=1}^{N} B_{0i}P_{Gi} + B_{00} \tag{2.26}$$

where B_{ij}, B_{0i}, and B_{00} are the transmission loss coefficients obtained from the B-coefficient matrix.

Step 3: Calculate the fitness value of each individual in the population using the fitness function given by Equation 2.27:

$$f = \frac{1}{F_{cost} + P_{pbc}} \tag{2.27}$$

where F_{cost} and P_{pbc} are given by

$$F_{cost} = 1 + \text{abs}\frac{\left(\sum_{i=1}^{N} F_i(P_i) - F_{min}\right)}{(F_{max} - F_{min})} \tag{2.28}$$

$$P_{pbc} = 1 + \left(\sum_{i=1}^{N} P_i - P_D - P_L\right)^2 \tag{2.29}$$

where
F_{max} is the maximum generation cost among all individuals in the initial population
F_{min} is the minimum generation cost among all individuals in the initial population
P_i is the power generated by each unit
P_D is the power demand
P_L is the power loss

Step 4: Compare each individual's evaluation value with its *pbest*. The best evaluation value among the *pbests* is denoted as *gbest*.

Step 5: Modify the member velocity of each individual according to Equation 2.24.

Step 6: Modify the member position of each individual according to Equation 2.25.

Step 7: If the evaluation value of each individual is better than the previous *pbest*, the current value is set to be *pbest*. If the best *pbest* is better than *gbest*, the value is set to be *gbest*.

Step 8: If the number of iterations reaches the maximum, then go to Step 9. Otherwise, go to Step 3.

Step 9: The individual that generates the latest *gbest* is the optimal power generation for each unit. Save the computed results.

The aforementioned steps are followed to estimate an optimized solution to the UC-ELD problem for all the test systems considered in this chapter.

2.4.4 Improved Differential Evolution with Opposition-Based Learning for ELD

This section presents DE-OBL algorithm and its improved version IDE-OBL to solve ELD problem with nonsmooth fuel cost curves considering transmission losses, power balance, and capacity constraints. The DE-OBL algorithm varies from the standard differential evolution (SDE) algorithm in terms of three factors: the initial population, mutation, and population size. The IDE-OBL is improved over the DE-OBL by adding an opposition-based generation-jumping step that selects fitter individuals based on opposite points. The concept of OBL and the step-by-step procedure to implement the IDE-OBL for solving ELD are explained later.

2.4.4.1 Concept of OBL

In general, heuristic optimization methods start with a few initial solutions in a population and try to improve them toward optimal solutions during generations. The optimization process terminates when some predefined criteria are satisfied. Without any a priori information about the solutions to the problem under consideration, the optimization starts with a set of random presumptions. The chance of obtaining a fitter solution can be attained through the opposite solution. By monitoring the opposite solution, the presumed fitter solution can be chosen as an initial solution. In fact, according to probability theory, 50% of the time the opposite presumption of the solution is better. Therefore, based on the fitness, two close presumptions have the potential to accelerate convergence. This approach is not only applied to initial solutions but also continuously to each solution in the current population.

2.4.4.1.1 Opposition-Based Population Initialization

Consider a point $P = (x_1, x_2, ..., x_n)$, with D-dimensional space consisting of candidate solutions. Let $f(.)$ be the fitness function used to measure the fitness of the candidate solutions. If $x_i \in [p_i, q_i]$ $\forall i = 1, 2, ..., D$ represents a real number, then the opposite point of x_i (denoted as $\breve{x}_i$) is defined as

$$\breve{x}_i = p_i + q_i - x_i \tag{2.30}$$

Based on Equation 2.30, $\breve{P} = (\breve{x}_1, \breve{x}_2, ..., \breve{x}_n)$ represents the opposite of $P = (x_1, x_2, ..., x_n)$. If $f(\breve{P}) \geq f(P)$, then P can be replaced with $\breve{P}$; otherwise the optimization procedure continues with P. Thus the point and its opposite point are evaluated simultaneously in order to continue the generations with the fitter individuals.

2.4.4.1.2 Opposition-Based Generation Jumping

In IDE-OBL, the evolutionary process can be forced to jump to a new solution candidate that may be fitter than the current one. Based on a jumping rate J_r, after generating a new population by selection, crossover, and mutation, the opposite population is calculated, and the N_p fittest individuals are selected from the union of the current population and the opposite population. Unlike opposition-based initialization, the opposite population for generation jumping is calculated dynamically using the opposite of each variable.

Such dynamic opposition increases the chance to find fitter opposite points, which helps in fine-tuning. By staying within the variables' interval static boundaries, the evolutionary process would jump outside of the shrunken solution space, and the knowledge of current reduced space (converged population) would not be utilized (Rahnamayan et al. 2008). Hence, opposite points are calculated by using the variables' current interval in the population.

2.4.4.2 IDE-OBL for ELD

Though SDE has emerged as one of the most popular technique for solving optimization problem, it has been observed that the convergence rate of SDE does not meet the expectations in case of multiobjective problems. Hence, certain modifications using the concept of opposition-based learning are performed on the SDE. The IDE-OBL varies from the basic SDE in terms of the following factors:

- IDE-OBL uses the concept of opposition-based learning in the initialization phase, while SDE uses uniform random numbers for initialization of population.
- During mutation, DE-OBL chooses the best individual among the three points as the mutant individual, whereas in SDE a random choice is made with equal choice of any of the three being selected. In IDE-OBL, an opposition-based generation phase is added with a jumping factor to determine fitter opposite points.
- SDE uses two sets of population—current population and an advanced population—for the next-generation individuals. Both DE-OBL and IDE-OBL use only one population set throughout the optimization process, which is updated in successive generations with the best individuals found in each generation.

The steps of the algorithm for implementing ELD are explained as follows:

2.4.4.2.1 Initialization

The basic step in IDE-OBL optimization is the creation of an initial population of candidate solutions by assigning random values to each decision parameter of each individual of the population. A population P consisting of N_P individuals is constructed in a random manner such that the values lie within the feasible bounds $X_j^{\min}$ and $X_j^{\max}$ of the decision variable, according to the following rule:

$$X_{i,j}^{(0)} = X_j^{\min} + \text{rand}[0,1] \times \left(X_j^{\max} - X_j^{\min} \right), \quad i = 1,2,\ldots,N_P \quad \text{and} \quad j = 1,2,\ldots,D \quad (2.31)$$

where
  rand [0, 1] represents a uniform random number in the interval [0, 1]
  $X_j^{\min}$ and $X_j^{\max}$ are the lower and upper bounds for the jth component, respectively
  D is the number of decision variables

Each individual member of the population consists of an N-dimensional vector $X_i^{(0)} = \{P_1, P_2, \ldots, P_N\}$, where the ith element of $X_i^{(0)}$ represents the power output of the ith generating unit.

An opposite population P_{add} is constructed using the rule

$$Y_{i,j}^{(0)} = X_j^{\min} + X_j^{\max} - P_{i,j} \quad (2.32)$$

where $P_{i,j}$ denote the points of the population P. The new population P_{new} for the approach is formed by combining the best individuals of both populations P and P_{add} as follows:

$$P_{new} = X_{i,j}^{(0)} \cup Y_{i,j}^{(0)} \tag{2.33}$$

2.4.4.2.2 Mutation

Next-generation offspring are introduced into the population through the mutation process. Mutation is performed by choosing three individuals from the population P_{new} in a random manner. Let X_{ra}, X_{rb}, and X_{rc} represent three random individuals such that $ra \neq rb \neq rc \neq i$, upon which mutation is performed during the Gth generation as

$$V_i^{G+1} = X_{best}^G + F \times \left[X_{rb}^G - X_{rc}^G \right], \quad i = 1, 2, \ldots, N_P \tag{2.34}$$

where

  V_i^{G+1} is the perturbed mutated individual

  X_{best}^G represents the best individual among three random individuals

The difference of the remaining two individuals is scaled by a factor F, which controls the amplification of the difference between two individuals so as to avoid search stagnation and to improve convergence.

2.4.4.2.3 Crossover

New offspring members are reproduced through the crossover operation based on binomial distribution. The members of the current population (target vector) $X_{i,j}^G$ and the members of the mutated individual $V_{i,j}^{G+1}$ are subject to crossover operation, thus producing a trial vector $U_{i,j}^{G+1}$ according to

$$U_{i,j}^{G+1} = \begin{cases} V_{i,j}^{G+1}, & \text{if } rand[0,1] \leq C_r \\ X_{i,j}^G, & \text{otherwise} \end{cases} \tag{2.35}$$

where C_r is the crossover constant that controls the diversity of the population and prevents the algorithm from getting trapped into the local optima. The crossover constant must be in the range of [0 1]. $C_r = 1$ implies the trial vector will be composed entirely of the mutant vector members, and $C_r = 0$ implies that the trial vector individuals are composed of the members of parent vector. Equation 2.35 can also be written as

$$U_{i,j}^{G+1} = X_{i,j}^G \times (1 - C_r) + V_{i,j}^{G+1} \times C_r \tag{2.36}$$

2.4.4.2.4 Selection

Selection is performed with the trial vector and the target vector to choose the best set of individuals for the next generation. In this approach, only one population set is maintained, and hence the best individuals replace the target individuals in the current population. The objective values of the trial vector and the target vector are evaluated and compared. For minimization problems such as ELD, if the trial vector has a better value, the target vector is replaced with the trial vector according to

$$X_i^G = \begin{cases} U_i^{G+1}, & \text{if } f(U_i^{G+1}) \leq f(X_i^G) \\ X_i^G, & \text{otherwise} \end{cases} \quad \text{for } i = 1, 2, \ldots, N_P \tag{2.37}$$

Fitness evaluation: The objective function for the ELD problem based on the fuel cost and power balance constraints is framed as

$$f(x) = \sum_{i=1}^{N} F_i(P_i) + k\left(\sum_{i=1}^{N} P_i - P_D - P_L\right) \tag{2.38}$$

where
    k is the penalty factor associated with the power balance constraint
    $F_i(P_i)$ is the ith generator cost function for output power P_i
    N is the number of generating units
    P_D is the total active power demand
    P_L represents the transmission losses

For ELD problems without transmission losses, setting $k = 0$ is most rational, while for ELD including transmission losses, the value of k is set to 1.

2.4.4.2.5 Generation Jumping

The maximum and minimum values of each variable in current population $[\min_j^p, \max_j^p]$ are used to calculate opposite points instead of using the predefined interval boundaries $[X_j^{\min}, X_j^{\max}]$ of the variables according to Equation 2.39:

$$Y_{i,j}^{(0)} = \min_j^p + \max_j^p - P_{i,j}, \quad i = 1, 2, \ldots, N_p; j = 1, 2, \ldots, D \tag{2.39}$$

The fittest individuals are selected from the new population set $[X_i, Y_{i,j}]$ as the current population.

The pseudocode of the approach is shown here:

```
Generate an initial population P randomly with each individual
representing the power output of the ith generating unit according to
Equation (2.31)
Generate an additional population P_add according to Equation (2.32).
Obtain the new population P_new as per Equation (2.33)
Evaluate fitness for each individual in P_new based on Equation (2.38)
While termination criteria not satisfied
          For i = 1 to N_p
               Mutate random members in P_new to obtain V_i^(G+1)

               Perform crossover on X_i^G and U_i,j^(G+1)

               Evaluate fitness function of X_i^G and U_i^(G+1)

               If f(U_i^(G+1)) ≤ f(X_i^G)
                    Replace existing population with U_i^(G+1)
               End if
          End for
Obtain opposite population for generation jumping (Equation (2.39))
Select the fittest individuals from the set [X_i, Y_i,j] as the current
population
End While
```

2.4.5 ELD Using Artificial Bee Colony Optimization

The ABC optimization algorithm developed by Karaboga and Basturk (2007) is becoming more popular recently because of the foraging behavior of honeybees. ABC is a population-based search technique in which the individuals, known as the food positions, are modified by the artificial bees during the course of time. The objective of the bees in turn is to discover the food sources with high nectar concentration. The colony of artificial bees is grouped into employed bees, onlooker bees, and scout bees. During the initialization phase, the objective of the problem is defined along with the ABC algorithmic control parameters. An employed bee is assigned for every food source available in the problem. In the employed bee phase, the employed bee stays on a food source and provides the neighborhood of the source in its memory. During the onlooker phase, onlooker bees watch the waggle dance of employed bees within the hive to choose a food source. The employed bee whose food source has been abandoned becomes the scout bee. Scout bees search for food sources randomly during the scout phase. Thus the local search is carried out by the employed bees and the onlooker bees, while the global search is performed by the onlooker and the scout bees, thus maintaining a balance between the exploration and exploitation process. The ELD problem is optimized based on the schedules obtained from GA with the application of ABC algorithm, which estimates the power to be shared by each unit that is kept on for the forecasted demand. In this section, the step-by-step procedure to implement ABC technique for ELD is discussed.

Step 1: Initialize ABC's population

Randomly initialize a population of food source positions including the limits of each unit along with the capacity and power balance constraints. Each food source includes the initial schedule of binary bits 0 and 1 obtained from GA, analogous to the chromosomes of the randomly generated population. The population now consists of the employed bees. Initialize all parameters of ABC such as the number of employed bees, number of onlookers, colony size, number of food sources, limit value, and the number of iterations.

2.4.5.1 Employed Bees Phase

Step 2: Evaluation of fitness function

The fitness value of each food source position corresponding to the employed bees in the colony is evaluated using

$$fit(i) = \sum_{i=1}^{N} FC_i + \rho \left(\sum_{i=1}^{N} P_i - P_D - P_L \right) \tag{2.40}$$

where
FC_i represents the fuel cost of the ith generating unit
P_i corresponds to the power of the ith generating unit
P_D denotes the power demand
P_L is the transmission loss
ρ is the penalty factor associated with the power balance constraint

For ELD problems without transmission losses, setting $\rho = 0$ is most rational, while for ELD including transmission losses the value of ρ is set to 1.

The solution feasibility is assessed by comparing the generated power with the load. The generated power should always be greater than the demand of the unit at time j according to

$$\sum_{i=1}^{N} P_{ij} * U_{ij} - P_{Dj} \qquad (2.41)$$

where
  P_{ij} represents the power generated by unit i at time j (24-hour schedule)
  P_{Dj} is the load demand
  U_{ij} represents the on/off status of unit i at time j

Step 3: Choose a food source

The new food source is determined at random by the employed bee by modifying the value of old food source position without changing other parameters, based on Equation 2.42:

$$v_{ij} = x_{ij} + \phi_{ij} * (x_{ij} - x_{kj}) \qquad (2.42)$$

where
  $k \in \{1, 2, ..., n_e\}$
  $j \in \{1, 2, ..., D\}$

Although k is determined randomly, it has to be different from i. $\Phi_{i,j}$ is a random number in the range $\{-1, 1\}$. It controls the production of neighbor food sources around $x_{i,j}$ and represents the comparison of two food positions visually by a bee. In Equation 2.42, as the difference between the parameters $x_{i,j}$ and $x_{k,j}$ decreases, the perturbation on the position $x_{i,j}$ also decreases. Thus, as the search approaches the optimum solution in the search space, the step length is adaptively reduced. This new position is tested for constraints of the ELD problem, and in case of violation, they are set to the extreme limits. The fitness value for the new food position is evaluated using Equation 2.40 and compared with the fitness of the old position. If the fitness of the new food source is better than the old, then the new food source position is retained in the memory. A limit count is also set if the fitness value of the new position is less than the old position. Thus the selection between new and old food positions is based on a greedy selection mechanism.

2.4.5.2 Onlooker Bee Phase

Step 4: Information sharing between employed bee and onlooker bee

Once the searching process is completed by the employed bees, they then share all the food source and position information with the onlooker bees in the dance area. The onlooker bee evaluates the information obtained, and a food source (solution) is chosen randomly based on a probability that is proportional to the quality of the food source according to

$$prob_i = \frac{a * fit(i)}{\max(fit) + b}$$ (2.43)

where
  a and b are arbitrary constants in the range {0, 1}
  $fit(i)$ denotes the fitness of the ith generating unit
  $\max(fit)$ is the maximum fitness value in the population so far

The constants a and b are fixed to 0.9 and 0.1, respectively. The onlookers are now placed into the food source locations based on roulette wheel selection.

Step 5: Modification on the position by onlookers
  Similar to the employed bees, the onlooker bees further produce a modification on the position of the food source in its memory using Equation 2.42. The greedy selection mechanism is repeated to retain the fitter positions in the memory. Again, a limit count is also set if the fitness value of the new position is less than that of the old position.

2.4.5.3 Scout Bee Phase

Step 6: Discover a new food source
  If the solution representing the food source is not improved over a defined number of trial runs (limit > predefined trials), then the food source is abandoned and the scout bee finds a new food source for replacement using

$$P_{ij} = P_{j\min} + \text{rand}[0,1] * (P_{j\max} - P_{j\min})$$ (2.44)

where $P_{j\min}$ and $P_{j\max}$ are the minimum and maximum limits of the parameter to be optimized, that is, the minimum and maximum generation limits of each unit.

Step 7: Memorize best results
  Store the best results obtained so far, and increase the iteration count.

Step 8: Stopping condition
  Increment the timer counter and repeat steps 8–13 for which the 24-hour UC schedules are predetermined through GA. Stop the process if the termination criteria are satisfied; otherwise continue.

2.4.6 ELD Based on Cuckoo Search Optimization

The strength of almost all modern heuristic algorithms is based on biological systems evolved from nature over millions of years (Yang and Deb 2009). These algorithms are governed by two basic principles: search among the current individuals to select the best solutions (intensification or local exploitation), and explore the search space efficiently (diversification or global exploration). In this chapter, a new heuristic technique, the CSO, is used for solving ELD problems. The CSO algorithm is a population-based stochastic algorithm driven by the brood parasitism breeding behavior of certain species of cuckoos. The individuals in this search mechanism are produced through a Levy flight mechanism, which is a special class of random walk with irregular step lengths based on probability distribution. The breeding behavior, Levy flight mechanism, and algorithm for ELD using CSO are discussed in this section.

2.4.6.1 Breeding Behavior

Cuckoos are a tremendously diverse group of birds with regard to breeding systems (Payne et al. 2005). Several species of cuckoos are monogamous, though exceptions exist. The *Anis* and the *Guira* species of cuckoos lay their eggs in communal nests, in the process removing other bird's eggs in the mutual nest. This is a common practice of the cuckoo species in order to increase the probability of hatching their own eggs. Because of the manner of laying eggs in other birds' nests and reproducing offspring, these species are referred to as obligate brood parasites.

The cuckoo species follow three basic types of the brood parasitism: intraspecific, cooperative, and nest takeover. Intraspecific brood parasitism refers to the cuckoos' behavior of laying eggs in another individual's (same species) nest, and further provides no care for the eggs or offspring (Ruxton and Broom 2002). In cooperative breeding, two or more females paired with the same male lay their eggs in the same nest in a cooperative manner and remain mutual throughout the parental care (Gibbons 1986). In nest takeover (Payne et al. 2005), a cuckoo simply occupies another host birds' nest.

During breeding, there is a direct conflict between the host cuckoos and the intruding cuckoos. Once the host bird identifies that the eggs in the nest are alien, it either throw the eggs away or destroys its nest and build a new one elsewhere. Parasitic cuckoos prefer laying their eggs in nests where the host bird has just laid its eggs. Moreover, the eggs laid by these parasitic cuckoos hatch much earlier than the host's eggs. The initial intuition of the cuckoo offspring is to throw out the host eggs, thus increasing its probability of sharing the food provided by the host bird. Ornithology studies have also proved that the cuckoo offspring is also capable of imitating the food call performed by the host offspring to gain more feeding access from the host bird.

2.4.6.2 Levy Flights

Researchers have demonstrated and proved that the behavioral characteristics of different animals and insects are similar to the Levy flight mechanism (Brown et al. 2007, Pavlyukevich 2007, Reynolds and Frye 2007). Studies by Viswanathan et al. (1999) show that several species of birds follow Levy flights during their search for food. The concept of Levy flights was introduced by a French mathematician Paul Levy as a class of random walks with step lengths obtained through probability distribution based on a power-law tail. The distributions that generate such random walks are known as Levy distributions or stable distributions. The Brownian motion in a diffusion process is usually pictured as a sequence of steps or jumps or flights of the walks. The probability of a walk of step size z produces a Gaussian distribution. Levy applied these Brownian motions as a generalized form by considering the distributions for one step, and several steps sharing a similar mathematical form. The Levy distributions decrease as the step size increases according to the power tail law given by

$$P(z) = \frac{1}{|z|^{1+\gamma}}, \quad \text{for } z \to \infty \quad \text{and} \quad \gamma \in [0,2] \text{ is the Levy index} \qquad (2.45)$$

For $\gamma = 2$, Brownian motion can be regarded as the extreme cases of Levy motions, and they do not fall off as rapidly as Gaussian distributions at long walk distances. Levy steps usually do not have a characteristic length scale since the small steps are scattered among the longer steps, leading to the variance of the distribution to diverge.

With $\gamma \leq 0$, the probability distribution given in Equation 2.45 cannot be normalized and hence has no physical meaning. For $0 < \gamma < 1$, the expectation value does not exist (Tran et al. 2004).

Consider a random walk process with step size L; the probability distribution is defined as

$$P(L) = \frac{\gamma}{k(1 + L/k)^{1+\gamma}} \tag{2.46}$$

Equation 2.46 represents a normalized form of the distribution with a Levy scale factor k added to consider the physical dimension of the given problem space.

Levy flights are applied to global optimization problems, in which the behavior of the random walkers is much similar to those in evolutionary algorithms (Balujia and Davies 1998). It is a well-known fact that a good search algorithm should maintain a proper balance between the local exploitation and global exploration. The frequency and lengths of long steps can be tuned by varying the parameters γ and k in the probability distribution Equation 2.46. For optimization applications, Levy flights should be capable of dynamically tuning these two parameters to the best fit landscape. In order to formulate an algorithm with Levy-based steps, Levy flights define a manageable move strategy, with either small steps, long steps, or a combination of both. Later, a single particle (as in greedy, SA, tabu search) or a set of particle(s) (as in GA, evolution strategy, genetic programming, ACO, and scatter search) can be chosen for movement over the search space. The generic movement model provided by Levy flights can also be combined with other known single-solution or population-based meta-heuristics. Such combinations often result in more powerful hybrid algorithms than the original ones.

2.4.6.3 Search Mechanism

The search process in the CSO algorithm is based on three basic principles (Wang et al. 2010):

- Each cuckoo lays one egg at a time and leaves its egg in a randomly chosen host nest.
- The best nest with high quality of eggs will produce offspring carried over to the next generation.
- The number of host nests is predetermined, and the egg laid by the cuckoo is identified by the host bird based on a probability rate (discovery rate) $Dr \in [0, 1]$. In such a situation, upon identification of the cuckoo's egg, the host bird either throws it away, or abandons the nest and builds a new nest.

Each egg in the host nest represents a solution to the optimization problem, while the cuckoo egg represents a new solution. The aim of this search algorithm is to use the new potential cuckoo eggs to replace the less potential eggs in the host's nest. Multiple cuckoo eggs can also be considered in the host's nest, thus leading to optimal solutions at a faster rate. While generating new solutions, a Levy flight is performed according to $x_i^{G+1} = x_i^G + \alpha \otimes Levy(\lambda)$, where $\alpha > 0$ is the step size related to the scales of the problem, $\otimes$ implies entry-wise multiplications, and λ is the expectation of occurrence of an event during a defined interval. The Levy flight provides a random walk with the step length obtained from the probability distribution given in Equation 2.45. The steps form a random walk process following the power law with heavy tail. Levy walk around the best solutions obtained so far picks up new solutions, thus speeding up the local search process.

2.4.6.4 ELD Using CSO

The basic concept of the CSO is constituted by three notions: particle, landscape, and optimizer. The particle is an individual that flies over the landscape, which is defined by all possible solutions to the problem with constraints and objective functions. The movement of the particles in the landscape is controlled by the optimizer. Each particle has its own position and velocity, controlled by a particle manager. The movement of the particles are represented either in the form of real values (continuous) or binary (discrete).

The steps of the CSO algorithm used for searching the optimal solution to the ELD problems are reviewed below:

Step 1: Initialize discovery rate D_r (probability of discovery), number of nests n, search dimension N_d, tolerance, and upper and lower bounds of search dimension to suitable values.

Step 2: Frame the objective function for the economic dispatch problem based on the fuel cost and constraints as

$$fit(x) = \sum_{i=1}^{N} F_i(P_i) + \rho \left(\sum_{i=1}^{N} Pi - P_D - P_L \right) \tag{2.47}$$

where ρ is the penalty factor associated with the power balance constraint. For ELD problems without transmission losses, setting $\rho = 0$ is most rational, while for ELD including transmission losses the value of ρ is set to 1.

Step 3: Each individual of the CS population consists of n host nests $x_i^G = \{P_1, P_2, \ldots, P_N\}$, where N denotes the number of generating units, G denotes the current generation, and the ith element of x represents the power output (P) of the ith generating unit.

Step 4: Obtain a cuckoo solution randomly through Levy flights:
$x_i^{G+1} = x_i^G + \alpha \otimes Levy(\lambda)$, where $\alpha > 0$ is the step size, usually set to 1, $\otimes$ denotes entry-wise multiplications, and $Levy(\lambda) = t^{-\lambda}$, $\lambda \in [1, 3]$.

Step 5: Evaluate the fitness $fit(x_i)$ according to Equation 2.47.

Step 6: Select a nest (j) among the available nests at random, and if ($fit(x_i) > fit(x_j)$), then replace j with the obtained new solution.

Step 7: Based on the discovery rate, the worst nests are replaced with newly built (generated) nests.

Step 8: Retain the best solutions—the nests with high-quality solutions are maintained. The evaluated solutions are ranked in terms of minimum fuel cost, and the current best solution is determined.

Step 9: Test for stopping condition. If the tolerance level has been reached, then stop else continue from Step 4.

2.5 MATLAB® m-File Snippets for UC-ELD Based on CI Paradigms

Experimental analysis is carried out with the goal of verifying or establishing the accuracy of a hypothesis. In this section, the simulation results of the algorithms to optimize the UC and ELD problems are discussed. The main objective of the UC-ELD problem is to obtain the minimum cost solution while satisfying various equality and inequality constraints. The effectiveness of the bio-inspired intelligent algorithms is tested on four test systems: 6-, 10-, 15-, and 20-unit power systems. In all these systems the UC schedules are obtained through GA, and the optimal economic dispatch is performed by FRBFN, EPSO, DE-OBL, IDE-OBL, ABC, and CSO algorithms. A comparative analysis of the these paradigms is performed in order to find the suitable algorithm in terms of fuel cost, standard deviation, computational time, and algorithmic efficiency. The ON/OFF commitment status through GA is implemented in Turbo C ,while the optimal dispatch is executed using MATLAB R2008b on an Intel i3 CPU, 2.53 GHz, 4 GB RAM PC.

2.5.1 Test Systems

Experiment analysis is performed on four test systems: the IEEE 30 bus system (6-unit system), a 10-unit test system, the Indian utility 75-bus system (15-unit system), and a 20-unit test system (referred to as case studies I–IV, respectively), including transmission losses, power balance, and generator capacity constraints. The power demand is varied for 24 hours to determine the schedule in the four test systems. Optimal distribution of load among generating units, fuel cost per hour, power loss, total power, and computational time are computed for each of the test systems using the intelligent algorithms. A brief description of the test systems is given below.

2.5.1.1 Six-Unit Test System

The intelligent algorithms were applied to the IEEE 30 bus system (Labbi and Attous 2010) with six generators located at bus numbers 1, 2, 5, 8, 11, and 13, respectively, and four off-nominal tap ratio transformers in transmission lines 6–9, 6–10, 4–12, and 28–27. All the generating units were valve-point loaded. The load profile of the system over 24 hours was also provided with various demands in the range [117, 435], which is the summation of minimum and maximum power limits. The specifications of the test system data are given in Tables A1–A3 in Appendix A.

2.5.1.2 Ten-Unit Test System

The second case study consisted of a 10-unit test system (Park et al. 2010). The input data included the generator limits, fuel cost coefficients, transmission loss matrix, and load profile for 24 hours. The minimum generating capacity of the system was 690 MW, and the maximum generating capacity was 2358 MW. The load profile and the generator input data and the variation of load for 24 hours per day are given in Tables A4–A6 in Appendix A. The minimum power demand requirement was 1036 MW, and the maximum demand was 2220 MW. The committed schedules of the 10-unit system obtained through GA were further dispatched using the intelligent heuristics.

2.5.1.3 Fifteen-Unit Test System

The Indian utility 75-bus Uttar Pradesh State Electricity Board (UPSEB) system with 15 generating units was chosen as the test system for the analysis of commitment and optimal

TABLE 2.1

GA Parameters for Unit Commitment Problem

S. No.	Parameters	Notations Used	Values
1	No. of chromosomes	n	No. of generators
2	Chromosome size	n_s	24 (hours) × No. of generators
3	No. of generations	N	500
4	Selection method	Sel	Roulette wheel
5	Crossover type	*Cross_type*	Two-point crossover
6	Crossover rate	p_c	0.6
7	Mutation Type	*Mut_type*	Flip bit
8	Mutation rate	p_m	0.001

dispatch (Prabhakar et al. 2009). The load demand, characteristics of generators, and loss coefficients are given in Tables A7–A9 in Appendix A.

2.5.1.4 Twenty-Unit Test System

Case study IV considered in this chapter consisted of 20 generating units. The data were generated based on Abookazemi et al. (2009). The 10-unit system data such as the power limits and cost coefficients were duplicated to obtain the 20-unit system data. The hourly load demand of the 10-unit system was doubled to obtain the hourly profile of the 20-unit system. The load profile, transmission loss coefficients, fuel cost coefficients, and maximum and minimum power are presented in Tables A10–A12 of Appendix A.

2.5.2 GA-Based UC Scheduling

This section discusses the application of GA to solve the UC problem for the test systems with 6, 10, 15, and 20 generating units. The control parameters for GA include the population size, selection type, crossover rate, mutation rate, and total number of generations, as shown in Table 2.1. The population size decides the number of chromosomes in a single generation. A larger population size slows down the GA run, while a smaller value leads to exploration of a small search space. A reasonable range of the population size is in the range {20, 100}, based on the real-valued encoding procedure. In this work, the population size is set to 28. Two-point crossover is used in this work with a crossover probability of 0.6, thus maintaining diversity in the population. The mutation type applied is flip bit, with a mutation rate of 0.001. This value of mutation decreases the diversity of subsequent generations. A flip-bit mutation changes the status of a unit from on to off, or vice versa. The code is developed on Turbo C with the parameters initialized according to Table 2.1. The code snippets for roulette wheel selection is given below:

```
void roulette(int x)// initpop or tempop
{
    sumfitness = 0.0;
    for(i=1;i<=popsize;i++)
    {
        if(x == 1)
        sumfitness = sumfitness + initpop[i].fitness;
        if(x == 2)
        sumfitness = sumfitness + tempop[i].fitness;
    }
}
```

```
for(j=1;j<=popsize;j++)
{
     if(x == 1)
     initpop[j].ratio = sumfitness - initpop[j].fitness;
     if(x == 2)
     tempop[j].ratio = sumfitness - tempop[j].fitness;
}
sumratio1 = 0.0;
sumratio2 = 0.0;
     for(j=1;j<=popsize;j++)
{
     if(x == 1)
     sumratio1 = sumratio1 + initpop[j].ratio;
     if(x == 2)
     sumratio2 = sumratio2 + tempop[j].ratio;
}
int tmp = 0.0;
     for(i=1;i<=popsize;i++)
     {
   if(x == 1)
   {
               tmp = tmp + initpop[i].ratio;
               partition[i] =  tmp;
   }
   if(x == 2)
   {
               tmp = tmp + tempop[i].ratio;
               partition[i] =  tmp;
   }
     }
         int dh = 0;
   int dh2 = 0;
         for(i=1;i<=2;i++)
         {
   if(x == 1)
   {
               sp:
               ranum = random(sumratio1) + 1;
               //cout<<"ranum: "<<ranum;
               if(ranum==0)
                     goto sp;
               //cout<<"\nranum: "<<ranum;
               for(j=1;j<=popsize;j++)
               {
           if(dh == 0 && ranum <= partition[j])
         {
           dh = 1;
         parent[i] = j;
            goto aaa;
         }
                     if(ranum > partition[j] && ranum <= partition[j+1])
                           parent[i] = j + 1;
                           //parent[] simpan index parent yg terpilih
               aaa:
```

```
                          }
            }
            if (x == 2)
            {
                        sp1:
                        ranum = random(sumratio2) + 1;
                        if (ranum==0)
                                goto sp1;
                        for (j=1;j<=popsize;j++)
                        {
                        if (dh2 == 0 && ranum <= partition[j])
                {
                dh2 = 1;
                parent[i] = j;
                goto aab;
                }
                        if (ranum > partition[j] && ranum <= partition[j+1])
                                parent[i] = j + 1;
                aab:
                }
            }
                        }
}
```

2.5.2.1 Six-Unit Test System

The on/off status and the computational time of the six generating units for 24-hour load demand were determined using GA, and are tabulated in Table 2.2. For each hour, the load demand varies and hence the commitment of the units also varies. From Table 2.2, it is clear that the unit P_1 is ON for 24 hours because this unit generates power with minimum fuel cost as the value of coefficient A is minimum for this unit. Units P_5 and P_6 are OFF most of the time because the value of fuel cost coefficient is the maximum for these two units and hence the fuel cost to generate power using these units is more than that of the other units. The UC using GA provides a cost-effective solution by choosing the appropriate units for the forecasted load demand. The computational time required to commit and de-commit the units is recorded, and results show that GA has a much faster convergence rate in solving the UC problem.

2.5.2.2 Ten-Unit Test System

GA is applied to obtain the optimal solution of the UC problem. In each generation, individuals are tested for load demand and unit constraints. These individuals are weighted according to their constraint satisfactions. Worst individuals are weighted with a very high factor so that they will have less chance in competition during the next generation. The process is repeated until a predefined number of generations have been reached. The GA parameters and their settings are shown in Table 2.1. The 10-unit system is solved using GA for obtaining the UC schedules based on the load demand for 24 hours. The forecasted loads, committed schedules, and the computational time are shown in

TABLE 2.2

Commitment of Units Using GA for Six the Unit Test System

Hour	Demand (MW)	Combination of Units						Computational Time (s)
		P_1	P_2	P_3	P_4	P_5	P_6	
1	166	ON	OFF	ON	ON	OFF	ON	1.21
2	196	ON	OFF	ON	ON	ON	ON	1.33
3	229	ON	OFF	ON	ON	ON	ON	1.25
4	267	ON	ON	ON	ON	ON	OFF	1.24
5	283.4	ON	ON	ON	ON	ON	OFF	1.31
6	272	ON	ON	ON	ON	ON	OFF	1.28
7	246	ON	ON	ON	ON	ON	OFF	1.34
8	213	ON	ON	ON	ON	ON	OFF	1.24
9	192	ON	ON	ON	ON	OFF	OFF	1.26
10	161	ON	ON	ON	OFF	OFF	OFF	1.29
11	147	ON	ON	OFF	OFF	OFF	OFF	1.33
12	160	ON	ON	OFF	OFF	OFF	OFF	1.35
13	170	ON	ON	OFF	OFF	OFF	OFF	1.34
14	185	ON	ON	OFF	OFF	OFF	OFF	1.26
15	208	ON	ON	OFF	OFF	OFF	OFF	1.22
16	232	ON	ON	ON	OFF	OFF	OFF	1.27
17	246	ON	ON	ON	OFF	OFF	ON	1.22
18	241	ON	ON	ON	OFF	OFF	ON	1.26
19	236	ON	ON	ON	OFF	OFF	ON	1.37
20	225	ON	ON	ON	OFF	OFF	ON	1.22
21	204	ON	ON	ON	OFF	OFF	ON	1.24
22	182	ON	ON	ON	OFF	OFF	ON	1.29
23	161	ON	ON	ON	OFF	OFF	ON	1.31
24	131	ON	ON	ON	OFF	OFF	OFF	1.26

Table 2.3. In the table, the committed/de-committed status is represented in terms of ON/OFF, and this indicates the number of generating units that are committed during the 24-hour schedule depending upon the demand. It is observed from Table 2.3 that unit P_1 is kept ON throughout the day because this unit generates power with minimum fuel cost as the cost coefficient A is minimum for this unit. Similarly, unit P_8 is the most expensive unit and hence it is kept OFF during most hours of the day. In general, small units can be committed/de-committed at short intervals. The ON times of these units must be minimized since the operational cost of these units is too high. Thus, in UC the order of commitment is determined based on the cost characteristics.

2.5.2.3 Fifteen-Unit Test System

For every hour during the 24-hour schedule, all possible combinations that satisfy the load demand constraints were selected, and these states were allowed to perform the optimal power flow. This procedure was continued for the specified time horizon until all the units in the test system were committed/de-committed. The complete unit commitment

TABLE 2.3

Unit Commitment Schedule for the 10-Unit Test System

Hour	Demand (MW)	Combination of Units										Computational Time (s)
		P_1	P_2	P_3	P_4	P_5	P_6	P_7	P_8	P_9	P_{10}	
1	1036	ON	OFF	OFF	ON	OFF	ON	ON	OFF	OFF	ON	2.08
2	1110	ON	OFF	OFF	ON	ON	ON	OFF	OFF	OFF	OFF	2.15
3	1258	ON	ON	ON	OFF	OFF	OFF	OFF	OFF	OFF	ON	2.31
4	1406	ON	ON	OFF	ON	OFF	ON	OFF	OFF	ON	ON	2.11
5	1480	ON	ON	ON	OFF	ON	OFF	OFF	OFF	ON	OFF	2.35
6	1628	ON	ON	ON	OFF	ON	ON	OFF	OFF	OFF	ON	2.18
7	1702	ON	ON	ON	ON	OFF	ON	OFF	OFF	OFF	ON	2.19
8	1776	ON	ON	ON	ON	ON	OFF	OFF	OFF	OFF	OFF	2.17
9	1924	ON	ON	ON	ON	ON	ON	OFF	OFF	OFF	ON	2.09
10	2072	ON	ON	ON	ON	ON	ON	ON	OFF	OFF	OFF	2.34
11	2146	ON	ON	ON	ON	ON	ON	ON	ON	OFF	OFF	2.26
12	2220	ON	ON	ON	ON	ON	ON	ON	OFF	ON	ON	2.18
13	2072	ON	ON	ON	ON	ON	ON	ON	OFF	OFF	OFF	2.22
14	1924	ON	ON	ON	ON	ON	OFF	OFF	OFF	OFF	ON	2.25
15	1776	ON	ON	ON	ON	ON	OFF	OFF	OFF	OFF	OFF	2.24
16	1554	ON	ON	ON	OFF	ON	OFF	ON	OFF	OFF	OFF	2.31
17	1480	ON	ON	ON	OFF	ON	OFF	OFF	OFF	ON	OFF	2.27
18	1628	ON	ON	ON	OFF	ON	ON	OFF	OFF	OFF	ON	2.29
19	1776	ON	ON	ON	ON	ON	OFF	OFF	OFF	OFF	OFF	2.11
20	2072	ON	ON	ON	ON	ON	ON	ON	OFF	OFF	OFF	2.35
21	1924	ON	ON	ON	ON	ON	ON	OFF	OFF	OFF	ON	2.33
22	1628	ON	ON	ON	OFF	ON	ON	OFF	OFF	OFF	ON	2.24
23	1332	ON	ON	OFF	OFF	ON	OFF	OFF	ON	ON	ON	2.16
24	1184	ON	ON	OFF	OFF	OFF	OFF	OFF	ON	ON	ON	2.06

schedule was obtained, and the power was dispatched in an optimal manner using FRBFN, EPSO, DE-OBL, IDE-OBL, ABC, and CSO algorithms.

The ON/OFF schedule represented in binary form 1/0 for the Indian utility 75-bus 15-unit system is shown in Table 2.4. All units in the system are committed to serve the load demand for the whole day except for units P_7, P_8, and P_{14}. Unit P_6 is not committed at any hour, since operating this unit is too expensive due to the high values of the cost coefficients.

2.5.2.4 Twenty-Unit Test System

The UC solution for the 20-unit system obtained through GA is shown in Table 2.5. It is observed that units P_1 and P_{11} are online throughout the time horizon, and thus they can be referred as base-load units. The cost to generate power by units P_8, P_9, P_{18}, and P_{19} is too high, and hence they are de-committed during most hours of the day. The choice of whether the units should be committed or de-committed is based on the cost characteristics, and hence the units with least cost are committed during 24 hours, while the expensive units are de-committed. Depending upon the power demand prerequisite, the de-committed

TABLE 2.4

Unit Commitment Schedule for the 15-Unit System Using GA

Hour	Demand (MW)	P_1	P_2	P_3	P_4	P_5	P_6	P_7	P_8	P_9	P_{10}	P_{11}	P_{12}	P_{13}	P_{14}	P_{15}	Time (s)
1	3352	1	1	1	1	1	0	1	1	1	1	1	1	1	1	1	3.14
2	3384	1	1	1	1	1	0	0	0	1	1	1	1	1	1	1	3.56
3	3437	1	1	1	1	1	0	1	0	1	1	1	1	1	1	1	3.99
4	3489	1	1	1	1	1	0	1	0	1	1	1	1	1	1	1	3.98
5	3659	1	1	1	1	1	0	1	0	1	1	1	1	1	1	1	3.78
6	3849	1	1	1	1	1	0	1	0	1	1	1	1	1	1	1	3.65
7	3898	1	1	1	1	1	0	1	0	1	1	1	1	1	1	1	3.74
8	3849	1	1	1	1	1	0	1	0	1	1	1	1	1	1	1	3.81
9	3764	1	1	1	1	1	0	1	0	1	1	1	1	1	1	1	3.94
10	3637	1	1	1	1	1	0	1	0	1	1	1	1	1	1	1	3.82
11	3437	1	1	1	1	1	0	1	0	1	1	1	1	1	1	1	3.86
12	3384	1	1	1	1	1	0	0	0	1	1	1	1	1	1	1	3.88
13	3357	1	1	1	1	1	0	0	0	1	1	1	1	1	1	1	3.87
14	3394	1	1	1	1	1	0	0	0	1	1	1	1	1	1	1	3.74
15	3616	1	1	1	1	1	0	0	0	1	1	1	1	1	1	1	3.61
16	3828	1	1	1	1	1	0	0	0	1	1	1	1	1	1	1	3.96
17	3828	1	1	1	1	1	0	0	0	1	1	1	1	1	1	1	3.54
18	3786	1	1	1	1	1	0	0	0	1	1	1	1	1	1	1	3.83
19	3659	1	1	1	1	1	0	0	0	1	1	1	1	1	1	1	3.72
20	3458	1	1	1	1	1	0	0	0	1	1	1	1	1	1	1	3.81
21	3394	1	1	1	1	1	0	0	0	1	1	1	1	1	1	1	3.88
22	3334	1	1	1	1	1	0	0	0	1	1	1	1	1	0	1	3.76
23	3329	1	1	1	1	1	0	0	0	1	1	1	1	1	0	1	3.99
24	3348	1	1	1	1	1	0	0	0	1	1	1	1	1	0	1	3.58

units are committed only when necessary. The computational time for scheduling the generating units for different load demands over 24 hours is also monitored, and the average time is found to be 5.74 s.

2.5.3 ELD Using FRBFN

The accuracy of the RBF network model depends on the proper selection of training data. The inputs of the training network are the power demand and weights w_1 and w_2, while the outputs constitute the power generated by the generating units. Table 2.6 shows the various parameters and their values used in RBFN-based ELD.

The learning rate (α) controls the rate at which the weights are modified due to previous weight updates. It acts as a smoothing parameter that reduces oscillation and helps to attain convergence. This must be a real value between 0.0 and 1.0. In the conducted experiments, the algorithm converged at $\alpha = 0.997$. The step size controls the weights during the training process: the larger the learning rate, the larger the rate of change of weights. Hence to maintain stability in the updation of weights, the value of

TABLE 2.5

Hourly Demand and UC Solution of the 20-Unit System Using GA

Hour	P_D (MW)	Combination of Generating Units																				Time (s)
		1	2	3	4	5	6	7	8	9	10	11	12	13	14	15	16	17	18	19	20	
1	2072	1	0	0	1	0	1	1	0	0	1	1	0	0	1	0	1	1	0	0	1	5.44
2	2220	1	0	0	1	1	1	0	0	0	0	1	0	0	1	1	1	0	0	0	0	5.12
3	2516	1	1	1	0	0	0	0	0	0	1	1	1	1	0	0	0	0	0	0	1	5.78
4	2812	1	1	0	1	0	1	0	0	1	1	1	1	0	1	0	1	0	0	1	1	5.85
5	2960	1	1	1	0	1	0	0	0	1	0	1	1	1	0	1	0	0	0	1	0	5.88
6	3256	1	1	1	0	1	1	0	0	0	1	1	1	1	0	1	1	0	0	0	1	5.64
7	3404	1	1	1	1	0	1	0	0	0	1	1	1	1	1	0	1	0	0	0	1	5.63
8	3552	1	1	1	1	1	0	0	0	0	0	1	1	1	1	1	0	0	0	0	0	5.68
9	3848	1	1	1	1	1	1	0	0	0	1	1	1	1	1	1	1	0	0	0	1	5.77
10	4144	1	1	1	1	1	1	1	0	0	0	1	1	1	1	1	1	1	0	0	0	5.95
11	4292	1	1	1	1	1	1	1	1	0	0	1	1	1	1	1	1	1	1	0	0	5.85
12	4440	1	1	1	1	1	1	1	0	1	1	1	1	1	1	1	1	1	0	1	1	5.99
13	4144	1	1	1	1	1	1	1	0	0	0	1	1	1	1	1	1	1	0	0	0	6.08
14	3848	1	1	1	1	1	1	0	0	0	1	1	1	1	1	1	1	0	0	0	1	5.48
15	3552	1	1	1	1	1	0	0	0	0	0	1	1	1	1	1	0	0	0	0	0	5.77
16	3108	1	1	1	0	1	0	1	0	0	0	1	1	1	0	1	0	1	0	0	0	5.32
17	2960	1	1	1	0	1	0	0	0	1	0	1	1	1	0	1	0	0	0	1	0	5.92
18	3256	1	1	1	0	1	1	0	0	0	1	1	1	1	0	1	1	0	0	0	1	5.99
19	3552	1	1	1	1	1	0	0	0	0	0	1	1	1	1	1	0	0	0	0	0	5.91
20	4144	1	1	1	1	1	1	1	0	0	0	1	1	1	1	1	1	1	0	0	0	5.68
21	3848	1	1	1	1	1	1	0	0	0	1	1	1	1	1	1	1	0	0	0	1	5.64
22	3256	1	1	1	0	1	1	0	0	0	1	1	1	1	0	1	1	0	0	0	1	5.97
23	2664	1	1	0	0	1	0	0	1	1	1	1	1	0	0	1	0	0	1	1	1	5.75
24	2368	1	1	0	0	0	0	0	1	1	1	1	1	0	0	0	0	0	1	1	1	5.64

TABLE 2.6

Parameters of FRBFN

S. No.	Parameters	Notations Used	Values
1	Initial cluster number	k	3
2	Fuzzification constant	m	2
3	Input nodes	Input node	3
4	Output nodes	Output node	No. of generators
5	No. of training patterns	n	456
6	No. of RBF centers	Centers	Problem dependant
7	Momentum factor	m	0.0002
8	Learning rate	α	0.997
9	Step size/tolerance	ε	0.002
10	No. of iterations	$Iter$	500

0.0002 ischosen. For practical implementation, MATLAB m files can be developed similar to the snippet shown below:

```
k=0;
retained=0;
for i=1:centres
    for j=1:npattern
        if newcentres (i, :) ==Input (j, :)
                fprintf(1, '\nCENTRE %d is MATCHED with %d
            %i/p PATTERN\n', i,j);
            retained=retained +1;
            k=0
            break;
        else
            k=1;
            continue;
        end
    end
end
```

The above snippet is applied to check the similarity of newly formed patterns with actual patterns in the clustering process using RBF.

2.5.3.1 Six-Unit Test System

The training data for the FRBFN was generated using LIM. The data was based on the 24-hour ON/OFF status obtained from the UC schedule through GA for various power demands. The generator power limits with transmission losses were also taken into account. A total of 456 training samples were created in this case, and 5.26% of the training data was chosen on a trial-and-error basis as testing data. Figure 2.4 shows the distribution of the initial RBF centers (stars), the selected centers (circles), and the newly formed centers (triangles). The similarity between the newly formed centers and the selected centers was measured, and the repeating centers were deducted in the network. The number of centers became the number of hidden nodes for the FRBFN algorithm. The number of hidden nodes for the six-unit test system based on FCM clustering was found to be 67. The typical relationship between the number of iterations and the error rate for the six-unit generator system is also shown in Figure 2.4. During the training process, the error function is minimized over the given training set by adaptively updating the parameters such as the centers, weights of the centers, and the hidden layer weights.

The optimized results are obtained when the FRBFN converged at the end of 500 iterations. Table 2.7 shows the computational results of the FRBFN for the six-unit generator system for different values of power demand (P_D). The distribution of the load among the six generating units (P_1, P_2, P_3, P_4, P_5, P_6), the optimal fuel cost (*FC*), power loss (P_L), total power (P_T), and the computational time (*CT*) were computed. The final weights w_1 and w_2 were in the range [0.05, 0.95]. From the results, it is observed that the generating unit P_1 is committed to dispatch power during the whole day, while unit P_5 generates power only during certain hours of the day. Maximum power of 283.78 MW is generated during the 5th hour, and minimum power of 131.22 MW is generated during the 24th hour. The total fuel cost consumed by the IEEE 30 bus system during a 24-hour

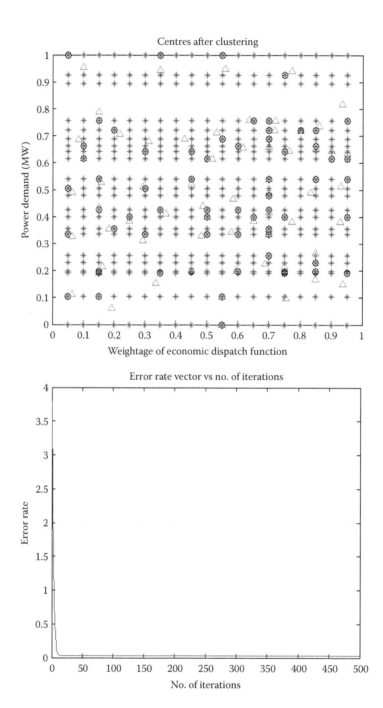

FIGURE 2.4
Clustered centers and error rate for the 6-unit system.

TABLE 2.7

Computational Results of IEEE 30 Bus System Using FRBFN

Hour	P_D (MW)	Distribution of Load among Units (MW)						FC ($/h)	P_L (MW)	P_T (MW)	CT (s)
		P_1	P_2	P_3	P_4	P_5	P_6				
1	166	112.39	0.00	24.37	16.13	0.00	13.41	446.94	0.31	166.31	0.89
2	196	132.23	0.00	19.38	10.02	14.02	20.77	541.97	0.42	196.42	0.78
3	229	173.93	0.00	17.98	10.40	12.05	15.34	649.35	0.69	229.69	0.92
4	267	151.71	80.00	15.26	10.29	10.61	0.00	742.41	0.88	267.88	1.09
5	283.4	200.00	25.96	34.29	12.95	10.58	0.00	799.90	0.38	283.78	0.99
6	272	149.77	58.31	22.78	19.02	22.88	0.00	774.84	0.75	272.75	0.89
7	246	156.98	40.63	18.81	17.43	12.85	0.00	671.28	0.71	246.71	0.99
8	213	144.58	24.62	15.27	13.21	15.87	0.00	561.09	0.55	213.55	1.03
9	192	121.42	40.71	15.45	14.88	0.00	0.00	502.43	0.46	192.46	1.03
10	161	112.55	32.66	16.17	0.00	0.00	0.00	380.84	0.39	161.39	0.85
11	147	117.85	29.55	0.00	0.00	0.00	0.00	354.62	0.39	147.39	0.99
12	160	114.54	45.89	0.00	0.00	0.00	0.00	392.61	0.43	160.43	0.99
13	170	134.46	36.06	0.00	0.00	0.00	0.00	422.57	0.52	170.52	0.98
14	185	155.93	29.71	0.00	0.00	0.00	0.00	468.70	0.64	185.64	1.05
15	208	165.07	43.71	0.00	0.00	0.00	0.00	542.23	0.78	208.78	0.89
16	232	163.15	52.08	17.59	0.00	0.00	0.00	600.30	0.82	232.82	0.99
17	246	175.26	38.96	18.63	0.00	0.00	14.03	647.33	0.88	246.88	0.86
18	241	153.34	54.10	20.03	0.00	0.00	14.31	630.54	0.78	241.78	1.04
19	236	162.80	40.99	18.85	0.00	0.00	14.13	613.78	0.78	236.78	1.11
20	225	155.64	31.74	25.31	0.00	0.00	13.01	578.72	0.70	225.70	0.98
21	204	131.51	37.86	17.09	0.00	0.00	18.07	513.42	0.54	204.54	0.90
22	182	58.96	48.63	40.49	0.00	0.00	34.21	446.92	0.29	182.29	0.96
23	161	109.67	21.00	18.15	0.00	0.00	12.53	387.65	0.34	161.35	0.97
24	131	74.23	37.22	19.77	0.00	0.00	0.00	297.46	0.22	131.22	0.86

schedule is $12,967.90, with a power loss of 13.65 MW. The time taken for the algorithm to compute the results for a single day is 23.02 s.

2.5.3.2 Ten-Unit System

The structural design of the FRBFN is modified for 10-unit test system with 3 input nodes and 10 output nodes. The 10 output nodes correspond to the optimal power generated for each generating unit, and the 3 input nodes represent weights w_1 and w_2 and the power demand. The number of hidden nodes determined using the FCM algorithm was found to be 64 for the 10-unit system. The FRBFN network was trained with 456 patterns generated through the LIM method for 500 iterations with network parameters initialized as shown in Table 2.6. The clustered centers and error rate for the 10-unit test system using FRBFN are shown in Figure 2.5. Among the number of clustered centers, 64 centers were selected (triangles) as hidden units at random with a learning rate of 0.997 and step size of 0.002. The rate of change of error implies that the algorithm converges to a constant error rate of 0.0840 from its initial value of 3.8128.

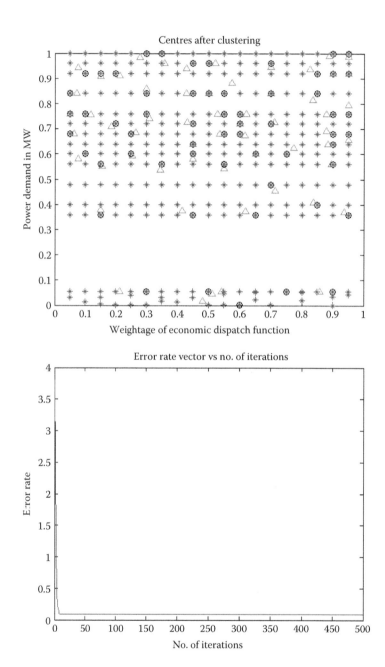

FIGURE 2.5
Clustered centers and error rate for the 10-unit system.

TABLE 2.8

Optimal Dispatch Using FRBFN for the 10-Unit Test System

Hour	P_D (MW)	P_1	P_2	P_3	P_4	P_5	P_6	P_7	P_8	P_9	P_{10}
					Distribution of Load among Units (MW)						
1	1036	432.7	0	0	261.2	0	160	129.6	0	0	55
2	1110	457.9	0	0	252.3	243	160	0	0	0	0
3	1258	406.1	460	340	0	0	0	0	0	0	55
4	1406	428.6	460	0	238.2	0	160	0	0	69.1	55
5	1480	360.7	460	340	0	243	0	0	0	80	0
6	1628	375.9	459.9	340	0	241.5	160	0	0	0	55
7	1702	462.4	460	340	231	0	157.8	0	0	0	55
8	1776	470	425.98	340	300	243	0	0	0	0	0
9	1924	470	361.2	340	300	243	160	0	0	0	55
10	2072	454.3	460	340	298.5	234.8	159.8	130	0	0	0
11	2146	461	460	340	243.7	243	160	130	113.6	0	0
12	2220	470	460	340	300	227.6	160	130	0	80	55
13	2072	449.2	460	340	295.3	243	160	130	0	0	0
14	1924	454.7	436.8	340	267.5	213.4	160	0	0	0	55
15	1776	470	460	340	284.7	226.75	0	0	0	0	0
16	1554	386.8	460	340	0	241.8	0	130	0	0	0
17	1480	470	460	245.4	0	229.3	0	0	0	78.8	0
18	1628	470	452.7	340	0	194.3	119.2	0	0	0	55
19	1776	435.5	460	340	300	243	0	0	0	0	0
20	2072	470	460	340	300	243	134.56	130	0	0	0
21	1924	470	460	340	296.3	222.5	86	0	0	0	55
22	1628	413.6	460	340	0	202	160	0	0	0	55
23	1332	467.6	452.4	0	0	228	0	0	96.8	37	55
24	1184	470	460	0	0	0	0	0	120	80	55

The economic dispatch results using FRBFN for the 10-unit test system, such as the distribution of load among committed units, fuel cost, power generated in each unit, power loss, total power, and computational time, for various values of power demand over the 24-hour schedule are shown in Tables 2.8 and 2.9. From the results, it is seen that unit P_1 generates a maximum power of 10,677 MW, while P_8 generated a minimum power of 330.4 MW. The total fuel cost over the 24-hour time horizon was computed by the FRBFN as \$973,453.4 with a total power loss of 95.29 MW. In general, neural networks take a longer time to train more samples, and in this case study it is observed to be 88.53 s.

2.5.3.3 Fifteen-Unit Test System

The different combinations of the committed/de-committed schedules obtained through GA were dispatched using LIM to determine the fuel cost, individual power generated, power loss, and total power. The training data for the FRBFN was generated based on LIM. A total of 456 training patterns were generated, and among these 48 patterns were

TABLE 2.9

Optimal Dispatch Using FRBFN for the 10-Unit Test System

Hour	P_D (MW)	FC ($/h)	P_L (MW)	P_T (MW)	CT (s)
1	1036	25,241.95	2.5	1,038.5	3.56
2	1110	26,741.91	3.2	1,113.2	3.47
3	1258	30,617.19	3.1	1,261.1	3.44
4	1406	35,175.16	4.9	1,410.9	3.98
5	1480	35,684.98	3.7	1,483.7	3.54
6	1628	39,174.4	4.3	1,632.3	3.58
7	1702	41,318.94	4.2	1,706.2	2.99
8	1776	42,801.05	2.98	1,778.98	4.01
9	1924	46,755.05	5.2	1,929.2	4.08
10	2072	49,141.77	5.4	2,077.4	3.96
11	2146	51,480.3	5.3	2,151.3	3.45
12	2220	54,130.16	2.6	2,222.6	4.05
13	2072	49,130.38	5.5	2,077.5	3.87
14	1924	46,617.37	3.4	1,927.4	3.77
15	1776	42,724.69	5.45	1,781.45	3.21
16	1554	36,669.89	4.6	1,558.6	3.41
17	1480	35,656.16	3.5	1,483.5	3.48
18	1628	39,136.83	3.2	1,631.2	3.89
19	1776	42,768.71	2.5	1,778.5	4.11
20	2072	49,109.59	5.56	2,077.56	3.74
21	1924	46,609.94	5.8	1,929.8	3.56
22	1628	39,136.72	2.6	1,630.6	3.82
23	1332	33,788.15	4.8	1,336.8	3.91
24	1184	33,842.08	1	1,185	3.65

chosen as testing samples. The parameters used for the FRBFN are shown in Table 2.6. The difference between the target and trained input was computed as the error. The algorithm terminated if the error was less than the tolerance value. More accurate results were produced with proper selection of the training data. It was also ensured that the training data were uniformly distributed, so that the network converged faster with optimal and quality solution.

Figure 2.6 shows the chosen centers (triangles) after applying the FCM clustering algorithm and the error rate at which the algorithm converges. The number of centers is found to be 63 after performing the clustering operation, and the error rate at which the clustering converged is observed as 0.0453 (initially it is 2.0715) with respect to the number of iterations.

The experimental results are shown in Tables 2.10 and 2.11. The total power demand requirement for the 15-unit system is 85,470 MW. Unit P_{12} generated a maximum power of 29,227.98 MW and unit P_8 generated a minimum power of 55.7 MW over 24 hours. Unit P_6 is not committed, and hence failed to generate power. The optimal cost for generating the total power including the transmission losses is Rs. 104,601.70.

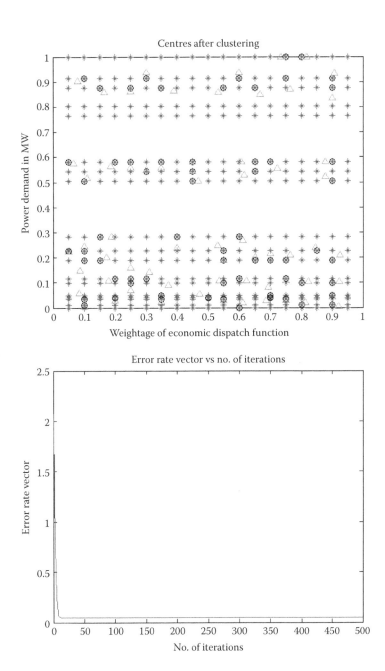

FIGURE 2.6
Clustered centers and error rate for the 15-unit system.

TABLE 2.10

Optimal Dispatch Using FRBFN for the 15-Unit Test System

Hour	P_D (MW)	Distribution of Load among Units (MW)									
		P_1	P_2	P_3	P_4	P_5	P_6	P_7	P_8	P_9	P_{10}
1	3352	429.7	253.3	62.8	107.5	56.5	0	79.9	55.7	201.9	53.2
2	3384	442.2	187.2	145.5	69.1	126.7	0	0	0	152.9	93.6
3	3437	366.8	222.3	163.5	90.3	111.4	0	99.6	0	191	63.6
4	3489	478.8	131.5	49.6	94.8	68.1	0	18.7	0	161.4	95.6
5	3659	100	167	46.8	73.2	34.7	0	100	0	316.3	250
6	3849	618.2	300	44	170	53.2	0	39.8	0	251.5	42.6
7	3898	674.2	222.7	121	127.8	19.3	0	92.3	0	260.4	84.2
8	3849	448.6	191.2	143.3	68.3	205.5	0	58	0	247.2	46
9	3764	636.3	215.9	82.7	55.7	74.9	0	48.2	0	170	244.5
10	3637	215.6	283	200	85.4	191.4	0	41.2	0	278	33.3
11	3437	492.2	285.4	41	87.7	138	0	26.7	0	285.6	69.8
12	3384	218.2	217.8	187.4	43.2	91.4	0	0	0	255.4	123.2
13	3357	204.4	133.1	57.4	94.9	123.5	0	0	0	169.7	133.8
14	3394	548.4	104.5	125.2	53	9.9	0	0	0	107.4	35.5
15	3616	398.4	171.1	166.7	116.2	83.1	0	0	0	154.9	33.9
16	3828	450.5	253.9	172.3	52.6	20.6	0	0	0	507.9	88
17	3828	594.5	115.5	83.8	135.1	182.8	0	0	0	304	41.7
18	3786	100	300	132.9	81.3	144.1	0	0	0	486.9	69.9
19	3659	401.6	162.1	80	167.6	240	0	0	0	228.1	164.1
20	3458	802.16	117.98	765.94	106.87	164.90	0	0	0	183.65	111.28
21	3394	100	300	200	48	156	0	0	0	129	56.1
22	3334	429.9	201	146.3	155.7	83.9	0	0	0	409.5	67.3
23	3329	1177.4	234.7	83.6	64.8	24.5	0	0	0	80.9	57.3
24	3348	100	121.9	115.6	88.4	69.1	0	0	0	405.4	250

2.5.3.4 Twenty-Unit Test System

The GA commitment schedule for the 20-unit system was dispatched economically using the FCM-based RBF network. The training patterns for the network were generated through LIM with a total of 456 samples, and among this 12% of the samples were taken for testing. The number of hidden units was determined based on the centers obtained through FCM clustering algorithm. Figure 2.7 shows the centers obtained after clustering and the rate of change of error attained through the FRBFN. For the 20-unit system, 61 centers were found by FCM. The algorithm was run for 500 iterations. The remaining parameters were set according to Table 2.6. The computed results such as the power generated in each unit, power loss, fuel cost, total power, and computational time for various values of power demand over 24 hours are shown in Tables 2.12–2.14. The results show that a maximum power of 10,324.17 MW was generated by unit P_1, which amounted to 12.87% of the total power generated including a transmission loss of 321.83 MW. The total power loss in the 20-unit system is much higher, which is a major disadvantage of the FRBFN in solving ELD. Though the fuel cost is reasonable, the power loss is relatively high, making the network less suitable for practical, large unit test systems.

TABLE 2.11

Optimal Dispatch Using FRBFN for the 15-Unit Test System

Hour	P_D (MW)	Distribution of Load among Units (MW)					FC (Rs/h)	P_L (MW)	P_T (MW)	CT (s)
		P_{11}	P_{12}	P_{13}	P_{14}	P_{15}				
1	3352	136.8	1300	393.9	22.3	200.3	4056.39	3.95	4092.5	3.35
2	3384	67.4	1300	559	103.7	138.1	4060.09	3.75	4156.4	3.07
3	3437	63.3	1005.7	725.1	137.4	198.6	4241.56	2.41	3110.3	2.83
4	3489	162	1229.3	775.5	37.8	187.9	4160.72	3.83	3719.5	2.91
5	3659	99.7	1292	860.2	91.3	427	4496.63	3.79	3977.2	3.19
6	3849	163.6	1065.1	890.1	39.6	173.5	4779.53	2.28	3407.9	2.96
7	3898	134.7	1265.8	617.4	65.1	215.2	4924.86	4.25	4024.9	2.97
8	3849	153.9	1300	534.3	105.1	350.6	4846.69	3.28	4138.3	3.04
9	3764	98.5	1299.5	790.8	17.3	31.3	4703.51	3.49	4132.8	3.02
10	3637	200	1300	900	150	192.4	4493.39	4.40	3856.6	2.96
11	3437	92.8	1243.7	524.5	75.5	75.4	4220.69	5.58	3917.3	1.37
12	3384	78.2	1300	463.7	65.6	342.5	4018.11	6.58	4199.7	3.13
13	3357	104.3	1300	870.5	90.2	76.8	3966.10	6.16	4083.1	2.99
14	3394	83	1269.5	748.3	103	208.5	3998.45	6.27	3810.6	3.05
15	3616	170.2	1298.5	542.4	126.6	357.1	4453.76	4.39	3870.3	2.85
16	3828	64.3	1275.5	439.6	139.7	366	4759.87	4.56	3883.9	2.71
17	3828	88.8	1292.1	692.5	64	235.4	4756.53	4.34	4047.6	2.77
18	3786	111	1300	900	113.9	203.5	4737.83	4.61	4176.8	2.99
19	3659	54.4	1120.9	698.8	23.6	320.4	4571.74	4.71	3553.5	3.30
20	3458	99.8426	916.7758	690.1196	48.9769	140.1461	4197.87	1.89	2871.8	3.01
21	3394	200	1300	583.8	48	454	4119.15	3.87	4182.4	3.05
22	3334	50.8	1300	900	0	454	3991.05	5.55	4140.4	2.99
23	3329	62.2	653.6	812.9	0	78.5	4018.16	4.66	3814.4	3.04
24	3348	126	1300	564.2	0	265.4	4029.07	4.82	3975.9	2.83

2.5.4 EPSO for Solving ELD

The EPSO parameters to be initialized include particle size, maximum inertia weight, minimum inertia weight, initial velocity, initial position, cognitive factor, social factor, error gradient, and the maximum number of iterations, as shown in Table 2.15. The typical range of population size is [20, 40]. In this case, it was set to a moderate value of 24 to yield better results. The choice of population size was 24 because a smaller population provides a smaller search space, thus resulting in a nonoptimal solution whereas a larger population provides more accurate results but consumes more time.

The maximum inertia weight was 0.9 and the minimum inertia weight was 0.4. This value enables the swarm to fly in a larger area of the search space thus obtaining the best solution. The initial velocity was set to zero and initial position was set at random. Cognitive factor and social factor were set to a constant value of 2, providing equal weight to both social component and cognitive component in order to obtain faster convergence. The constriction factor k for $c_1 = c_2 = 2$ was computed as 0.5. The error gradient value was set to 1×10^{-5} to obtain accurate solutions. The number of iterations for the EPSO was based on the size of the problem. Based on the parameter initialization, the population of particles and their velocities at time zero were initialized. A set

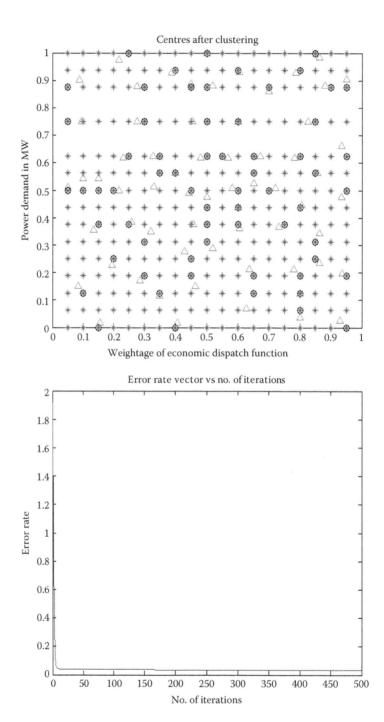

FIGURE 2.7
Clustered centers and error rate for the 20-unit system.

TABLE 2.12

Optimal Dispatch Using FRBFN for the 20-Unit Test System

Hour	P_D (MW)	Distribution of Load among Units (MW)									
		P_1	P_2	P_3	P_4	P_5	P_6	P_7	P_8	P_9	P_{10}
1	2072	470	0	0	231.38	0	131.16	129.92	0	0	55
2	2220	356.39	0	0	300.00	243	160.00	0.00	0	0	0
3	2516	348.85	460	340	0.00	0	0.00	0.00	0	0	55
4	2812	456.58	459.91	0	283.16	0	152.27	0.00	0	33.58	55
5	2960	349.98	460.00	340	0.00	243	0.00	0.00	0	78.37	0
6	3256	294.34	460.00	340.00	0.00	232.33	159.90	0.00	0	0.00	55
7	3404	389.95	455.56	340.00	300.00	0.00	156.63	0.00	0	0.00	55
8	3552	467.95	459.44	338.47	296.79	227.70	0.00	0.00	0	0.00	0
9	3848	442.39	460.00	340.00	300.00	242.62	160.00	0.00	0	0.00	55
10	4144	470.00	460.00	335.33	295.02	242.79	152.29	130.00	0	0.00	0
11	4292	470.00	443.39	340.00	297.56	116.26	160.00	130.00	120.00	0.00	0
12	4440	470.00	460.00	340.00	292.72	243.00	160.00	129.65	0.00	77.64	55
13	4144	430.22	460.00	340.00	300.00	243.00	159.86	130.00	0.00	0.00	0
14	3848	470.00	380.77	340.00	298.45	240.23	160.00	0.00	0.00	0.00	55
15	3552	470.00	460.00	339.64	256.76	242.27	0.00	0.00	0.00	0.00	0
16	3108	470.00	389.89	340.00	0.00	243.00	0.00	130.00	0.00	0.00	0
17	2960	308.04	429.92	340.00	0.00	218.19	0.00	0.00	0.00	79.47	0
18	3256	433.38	460.00	340.00	0.00	206.05	160.00	0.00	0.00	0.00	55
19	3552	406.09	460.00	340.00	300.00	243.00	0.00	0.00	0.00	0.00	0
20	4144	470.00	455.40	340.00	300.00	243.00	160.00	130.00	0.00	0.00	0
21	3848	470.00	460.00	340.00	255.02	243.00	160.00	0.00	0.00	0.00	55
22	3256	470.00	460.00	335.00	0.00	231.38	160.00	0.00	0.00	0.00	55
23	2664	470.00	410.55	0.00	0.00	221.06	0.00	0.00	88.58	46.60	55
24	2368	470.00	460.00	0.00	0.00	0.00	0.00	0.00	120.00	80.00	55

of random population positions bounded by VR were constructed with initial random velocities between *–mv* and *mv*. The initial *pbest* and *gbest* position values were assigned and the objective function was evaluated. A variable was preallocated to keep track of *gbest* for all the iterations to get new velocities and new positions.

```
% get new velocities, positions (this is the heart of the PSO algorithm)
    % each epoch get new set of random numbers
        rannum1 = rand([ps,D]); % for Trelea and Clerc types
        rannum2 = rand([ps,D]);
        if     trelea == 2
        % parameter set 2
          vel = 0.729.*vel...                                % prev vel
                +1.494.*rannum1.*(pbest-pos)...              % independent
                +1.494.*rannum2.*(repmat(gbest,ps,1)-pos);  % social
```

```
elseif trelea == 1
 % parameter set 1
  vel = 0.600.*vel...                                  % prev vel
        +1.700.*rannum1.*(pbest-pos)...                % independent
        +1.700.*rannum2.*(repmat(gbest,ps,1)-pos);     % social
elseif trelea ==3
 % Clerc's Type 1" PSO
  vel = chi*(vel...                                    % prev vel
        +ac1.*rannum1.*(pbest-pos)...                  % independent
        +ac2.*rannum2.*(repmat(gbest,ps,1)-pos));      % social
else
 % common PSO algo with inertia wt
 % get inertia weight, just a linear funct w.r.t. epoch parameter iwe
  if i<=iwe
     iwt(i) = ((iw2-iw1)/(iwe-1))*(i-1)+iw1;
  else
     iwt(i) = iw2;
  end
 % random number including acceleration constants
  ac11 = rannum1.*ac1;    % for common PSO w/inertia
  ac22 = rannum2.*ac2;

  vel = iwt(i).*vel...                                 % prev vel
        +ac11.*(pbest-pos)...                          % independent
        +ac22.*(repmat(gbest,ps,1)-pos);              % social
end

 % limit velocities here using masking
 vel = ( (vel <= velmaskmin).*velmaskmin ) + ( (vel >
 velmaskmin).*vel );
 vel = ( (vel >= velmaskmax).*velmaskmax ) + ( (vel <
 velmaskmax).*vel );
% update new position (PSO algo)
 pos = pos + vel;

% position masking, limits positions to desired search space
% method: 0) no position limiting, 1) saturation at limit,
%         2) wraparound at limit , 3) bounce off limit
 minposmask_throwaway = pos <= posmaskmin;  % these are psXD
 matrices
 minposmask_keep      = pos > posmaskmin;
 maxposmask_throwaway = pos >= posmaskmax;
 maxposmask_keep      = pos < posmaskmax;

 if      posmaskmeth == 1
  % this is the saturation method
  pos = ( minposmask_throwaway.*posmaskmin ) + ( minposmask_
  keep.*pos );
  pos = ( maxposmask_throwaway.*posmaskmax ) + ( maxposmask_
  keep.*pos );
  elseif posmaskmeth == 2
     % this is the wraparound method
```

TABLE 2.13

Optimal Dispatch Using FRBFN for the 20-Unit Test System

Hour	P_D (MW)	Distribution of Load among Units (MW)									
		P_{11}	P_{12}	P_{13}	P_{14}	P_{15}	P_{16}	P_{17}	P_{18}	P_{19}	P_{20}
1	2072	457.63	0	0	299.63	0	118.67	130	0	0	55
2	2220	470.00	0	0	300.00	243	160.00	0	0	0	0
3	2516	470.00	460	339.62	0.00	0	0.00	0	0	0	55
4	2812	469.96	443.28	0.00	205.37	0	159.84	0	0	58.08	55
5	2960	470.00	423.89	340.00	0.00	190.81	0.00	0	0	78.46	0
6	3256	470.00	460.00	340.00	0.00	243.00	154.26	0	0	0.00	55
7	3404	461.28	460.00	328.50	258.77	0.00	159.72	0	0	0.00	55
8	3552	465.26	459.05	311.17	300.00	243.00	0.00	0	0	0.00	0
9	3848	470.00	460.00	340.00	264.01	192.57	85.73	0	0	0.00	55
10	4144	470	460	340	300	242.38	159.93	129.68	0	0.00	0
11	4292	470.00	460.00	340.00	300.00	243.00	160.00	130.00	120.00	0.00	0
12	4440	470.00	460.00	334.16	300.00	243.00	151.10	130.00	0.00	80.00	55
13	4144	470.00	460.00	340.00	300.00	243.00	160.00	130.00	0.00	0.00	0
14	3848	408.9341	460	296.7732	300	241.6739	155.42	0.00	0.00	0.00	55
15	3552	470.00	460.00	324.75	300.00	242.52	0.00	0.00	0.00	0.00	0
16	3108	470.00	460.00	340.00	0.00	243.00	0.00	37.90	0.00	0.00	0
17	2960	470.00	460.00	340.00	0.00	243.00	0.00	0.00	0.00	80.00	0
18	3256	354.06	460.00	340.00	0.00	243.00	160.00	0.00	0.00	0.00	55
19	3552	470.00	460.00	340.00	300.00	243.00	0.00	0.00	0.00	0.00	0
20	4144	470.00	460.00	340.00	300.00	243.00	120.30	130.00	0.00	0.00	0
21	3848	411.33	447.78	340.00	300.00	216.33	109.71	0.00	0.00	0.00	55
22	3256	469.84	459.77	339.37	0.00	237.78	160.00	0.00	0.00	0.00	55
23	2664	470.00	437.98	0.00	0.00	243.00	0.00	0.00	120.00	59.90	55
24	2368	470.00	460.00	0.00	0.00	0.00	0.00	0.00	120	80.00	55

2.5.4.1 Six-Unit Test System

The ELD for the committed schedules of the IEEE 30 bus system was solved using the EPSO algorithm. The input to ELD optimization was the UC ON/OFF schedules obtained through GA, which provides the schedule of each unit along with the generator characteristics. An optimized power generation schedule for each hour was estimated and the process was repeated for different power demands for 24 hours. The results of ELD using EPSO algorithm are shown in Table 2.16. The generating units with a power output of "0" indicate the OFF status of the unit. Since the objective of the UC-ELD problem is to minimize the total fuel cost, the units with smaller cost coefficients are allocated the power first, followed by other online units. In this test system, the most efficient unit is P_1 and hence it operates at its maximum power generating capacity for most of the load requirement. Similarly, the units that are most expensive are allocated minimum power, thus providing improved optimal solutions.

The fuel cost, power loss, total power, and computational time of the ELD using EPSO algorithm were computed. The total fuel cost was computed by summing the fuel cost

TABLE 2.14

Optimal Dispatch Using FRBFN for the 20-Unit Test System

Hour	P_D (MW)	FC ($/h)	P_L (MW)	P_T (MW)	CT (s)
1	2072	50,512.98	6.4	2078.40	3.67
2	2220	53,636.02	12.4	2232.39	3.64
3	2516	61,373.41	12.5	2528.47	3.86
4	2812	70,642.06	20	2832.02	3.67
5	2960	72,099.62	14.5	2974.52	3.84
6	3256	78,534.58	7.83	3263.83	3.67
7	3404	82,791.26	16.4	3420.41	3.64
8	3552	85,724.29	16.8	3568.83	3.84
9	3848	93,468.22	19.3	3867.33	3.89
10	4144	98,496.67	14.6	4187.41	3.92
11	4292	103,361.31	8.21	4300.21	3.75
12	4440	108,545.68	11.3	4451.27	3.69
13	4144	98,530.43	22.1	4166.08	3.66
14	3848	93,585.01	14.2	3862.25	3.72
15	3552	85,580.24	13.9	3565.94	3.75
16	3108	73,550.66	15.8	3123.78	3.84
17	2960	71,728.65	8.6	2968.61	3.72
18	3256	78,392.08	10.5	3266.49	3.73
19	3552	85,644.57	10.1	3562.09	3.67
20	4144	98,440.42	17.7	4161.69	3.69
21	3848	93,355.38	15.2	3863.18	3.87
22	3256	78,393.03	17.7	3433.14	3.62
23	2664	68,335.47	13.7	2677.65	3.67
24	2368	76,343.92	2.09	2370.00	3.70

TABLE 2.15

EPSO Parameters and Settings

S. No.	Parameters	Notations Used	Values
1	Population size	N_s	24
2	Maximum inertia weight	w_{max}	0.9
3	Minimum inertia weight	w_{min}	0.4
4	Initial velocity	$v_{ij}(0)$	0
5	Initial position	$x_{ij}(0)$	Random
6	Cognitive factor	c_1	2
7	Social factor	c_2	2
8	Constriction factor	k	0.5
9	Error gradient	e	1×10^{-25}
10	Maximum number of iterations	max_iter	Problem dependant

TABLE 2.16

ELD Results Using EPSO for the Six-Unit System

Hour	P_D (MW)	Power Generated/Unit (MW)						FC ($/h)	P_L (MW)	P_T (MW)	CT (s)
		P_1	P_2	P_3	P_4	P_5	P_6				
1	166	127.57	0	16.81	10	0	12	440.14	0.36	166.38	2.67
2	196	146.52	0	17.97	10	10	12	536.18	0.47	196.49	2.45
3	229	177.82	0	19.9	10	10	12	647.41	0.69	229.72	2.5
4	267	178.53	49.45	19.96	10	10	0	739.8	0.91	267.94	2.52
5	283.4	190.07	51.95	20.67	10	11.77	0	797.93	1.03	284.46	2.15
6	272	182.05	50.21	20.18	10	10.53	0	757.37	0.95	272.97	2.15
7	246	161.98	45.86	18.94	10	10	0	667.65	0.76	246.78	2.01
8	213	135.99	40.23	17.34	10	10	0	559.69	0.54	213.56	2.12
9	192	127.34	38.35	16.8	10	0	0	492.67	0.47	192.49	2.11
10	161	114.53	31.86	15	0	0	0	380.84	0.39	161.39	2.03
11	147	123.15	24.25	0	0	0	0	355.91	0.4	147.4	2.28
12	160	133.99	26.49	0	0	0	0	393.93	0.48	160.48	2.09
13	170	141.59	28.95	0	0	0	0	423.68	0.54	170.54	2.48
14	185	154.16	31.48	0	0	0	0	469.86	0.64	185.64	2.11
15	208	174.6	34.21	0	0	0	0	543.86	0.81	208.81	2.11
16	232	170.47	43.98	18.4	0	0	0	600.27	0.84	232.85	2.17
17	246	172.05	44.32	18.51	0	0	12	646.54	0.87	246.88	2.12
18	241	168.11	43.47	18.26	0	0	12	630.02	0.84	241.84	2.2
19	236	164.17	42.61	18.02	0	0	12	613.66	0.8	236.8	2.42
20	225	155.5	40.73	17.49	0	0	12	578.19	0.72	225.72	2.14
21	204	138.96	37.15	16.47	0	0	12	512.52	0.58	204.58	2.28
22	182	121.64	33.4	15.41	0	0	12	446.59	0.45	182.45	2.14
23	161	104.62	29.72	15	0	0	12	386.42	0.34	161.34	2.14
24	131	89.75	26.5	15	0	0	0	297.43	0.25	131.25	2.26

of each unit that is in the "ON" state. It was calculated based on the power generated by each unit and its fuel cost coefficients. From Table 2.16 it can be concluded that the total fuel cost incurred per day amounts to $12,918.56 and the average cost per hour is $538.27 for a total power demand of 4953.4 MW. The algorithm produced optimal results at the end of 100 iterations with the total execution time of 55.58 s and mean time per hour of 2.32 s.

2.5.4.2 Ten-Unit Test System

In this case study, 10 units were committed to serve a 24-hour load pattern using GA. The parameters of EPSO were set according to Table 2.15. Based on the GA-committed schedules, the EPSO computes the load distribution among each unit for every hour. The experimental results are shown in Tables 2.17 and 2.18. From the tables, it can be inferred that the units P_1 and P_3 are allocated the maximum load if they are in ON state. This is because the fuel cost coefficiency of these two units is minimum and hence they are highly economical. Unit P_{10} is always allocated 55 MW because its maximum and minimum generator limits are the same. For units P_3, P_6, and P_7, maximum power is allocated for all values of load demand. Analyzing other units such as P_2, P_4, P_5, P_8 and

TABLE 2.17

ELD Results Using EPSO for the 10-Unit System

Hour	Power Generated/Unit (MW)									
	P_1	P_2	P_3	P_4	P_5	P_6	P_7	P_8	P_9	P_{10}
1	470	0	0	223.57	0	160	130	0	0	55
2	470	0	0	240.29	243	160	0	0	0	0
3	406.10	460	340	0	0	0	0	0	0	55
4	470	460	0	245.20	0	160	0	0	20.89	55
5	439.32	460	340	0	224.46	0	0	0	20	0
6	411.56	460	340	0	205.66	160	0	0	0	55
7	470	460	340	221.19	0	160	0	0	0	55
8	470	460	340	267.20	243	0	0	0	0	0
9	470	460	340	201.00	243	160	0	0	0	55
10	470	460	340	274.56	243	160	130	0	0	0
11	469.97	459.98	340	254.67	243	160	130	93.75	0	0
12	470	460	340	300	243	160	130	0	68.38	55
13	469.89	459.82	340	275.06	242.86	159.97	129.95	0	0	0
14	470	460	340	199.38	243	160	0	0	0	55
15	470	460	340	265.62	243	0	0	0	0	0
16	417.15	460	340	0	211.41	0	130	0	0	0
17	438.34	460	340	0	223.84	0	0	0	20	0
18	410.57	460	340	0	205.03	160	0	0	0	55
19	470	460	340	265.62	243	0	0	0	0	0
20	470	460	340	272.98	243	160	130	0	0	0
21	469.98	460	340	199.40	243	160	0	0	0	55
22	410.56	460	340	0	205.03	160	0	0	0	55
23	470	460	0	0	243	0	0	88.50	20.44	55
24	470	460	0	0	0	0	0	120	80	55

P_9 reveals that the load is allocated to these units according to the load demand and the UC schedule. The maximum fuel cost of $53,970.94 is incurred for a load demand of 2,220 MW during the 12th hour and the minimum fuel cost ($25,175.96) is incurred during the 1st hour for a load demand of 1036 MW. The total operating cost for a day amounts to $972,282.3, and the mean cost per hour is $40,511.76. EPSO algorithm takes 59.14 seconds to obtain the ELD results for 24 hours.

2.5.4.3 Fifteen-Unit Test System

The GA committed/de-committed schedules were implemented for economical operation using the EPSO algorithm for the 15-unit test system. The parameters for the EPSO were initialized as shown in Table 2.15. The algorithm terminated at the end of 1500 iterations, and the results of the optimal dispatch were recorded, as shown in Tables 2.19 and 2.20. It is observed that the EPSO is capable of producing better results in terms of optimal fuel cost due to the intrinsic nature of position and velocity updates. Moreover, since the dispatch is based on an hourly basis, the complexity of the EPSO search is reduced to a great extent. From the results, it is observed that a minimum power of 3331.96 MW is

TABLE 2.18

ELD Results Using EPSO for the 10-Unit System

Hour	P_D (MW)	FC ($/h)	P_L (MW)	P_T (MW)	CT (s)
1	1036	25,175.96	2.57	1038.57	2.82
2	1110	26,715.95	3.29	1113.29	2.59
3	1258	30,617.19	3.10	1261.10	2.40
4	1406	35,057.52	5.09	1411.09	2.43
5	1480	35,649.50	3.77	1483.77	2.57
6	1628	39,172.35	4.22	1632.22	2.37
7	1702	41,289.07	4.19	1706.19	2.50
8	1776	42,748.83	4.20	1780.20	2.48
9	1924	46,581.09	5.00	1929.00	2.39
10	2072	49,087.52	5.56	2077.56	2.51
11	2146	51,463.10	5.37	2151.37	2.37
12	2220	53,970.94	6.37	2226.37	2.29
13	2072	49,089.03	5.56	2077.56	2.50
14	1924	46,541.86	3.38	1927.38	2.57
15	1776	42,710.37	2.62	1778.62	2.42
16	1554	36,669.74	4.56	1558.56	2.48
17	1480	35,614.59	2.19	1482.19	2.43
18	1628	39,136.72	2.60	1630.60	2.37
19	1776	42,710.37	2.62	1778.62	2.84
20	2072	49,049.03	3.98	2075.98	2.36
21	1924	46,541.91	3.38	1927.38	2.20
22	1628	39,136.72	2.60	1630.60	2.36
23	1332	33,710.85	4.94	1336.94	2.50
24	1184	33,842.08	4.33	1185.00	2.39

generated at the 23rd hour for a load requirement of 3329 MW. Likewise, the maximum power of 3901.5 MW is generated during the seventh hour for a demand of 3898 MW. Since the transmission losses are considered for this test system, it is observed that the total power loss is 69.86 MW. The fuel cost for the operation of the Indian utility 75-bus for the entire 24 hours is Rs. 101,949.3, and the EPSO algorithm computed these optimal results in a total time of 55.1 s.

2.5.4.4 Twenty-Unit Test System

The most important characteristics of the EPSO algorithm are steady convergence and good computational stability. With the parameter set of the EPSO, the algorithm was run for 1500 iterations. The optimal results were produced at the end of 1200 iterations. The computational results in terms of individual unit power, fuel cost, power loss, total power, and computational time for each value of load demand over the 24-hour time horizon are presented in Tables 2.21–2.23. For the hourly load profile varying in the range [2072, 4440] MW, the total fuel cost was computed by EPSO as $1,957,971.92. The least expensive hour is the first hour with a cost of $50,477.31, and most expensive hour is the 12th hour with cost of $108,546.94. The average computational time required to dispatch power economically is 3.5 s.

TABLE 2.19

Optimal Dispatch Using EPSO for the 15-Unit Test System

Hour	P_D (MW)	Distribution of Load among Units (MW)									
		P_1	P_2	P_3	P_4	P_5	P_6	P_7	P_8	P_9	P_{10}
1	3352	520.53	177.72	51.42	40.02	22.93	0.00	4.39	20.34	202.58	30.08
2	3384	481.37	199.99	40.13	41.51	69.64	0.00	0.00	0.00	224.64	30.03
3	3437	537.82	105.06	40.38	40.00	41.29	0.00	1.14	0.00	199.74	30.01
4	3489	524.79	122.95	75.55	69.71	45.60	0.00	32.99	0.00	224.72	30.07
5	3659	575.71	124.82	61.82	46.16	53.67	0.00	1.16	0.00	232.67	30.00
6	3849	576.15	200.84	53.26	57.24	106.24	0.00	15.28	0.00	232.75	30.38
7	3898	615.11	166.15	42.51	114.41	129.38	0.00	60.95	0.00	236.80	46.81
8	3849	551.32	129.11	118.62	114.20	105.04	0.00	15.22	0.00	205.74	30.00
9	3764	581.25	151.36	67.36	68.82	62.75	0.00	1.00	0.00	235.80	30.01
10	3637	570.56	122.46	98.01	41.70	49.78	0.00	9.30	0.00	213.31	41.60
11	3437	569.43	104.56	54.46	40.37	38.60	0.00	1.68	0.00	217.65	30.00
12	3384	528.03	100.05	40.47	49.21	28.83	0.00	0.00	0.00	198.00	30.51
13	3357	456.28	129.15	40.54	40.65	48.12	0.00	0.00	0.00	228.60	31.50
14	3394	514.08	102.96	40.03	40.00	25.10	0.00	0.00	0.00	174.57	133.82
15	3616	641.96	120.39	60.39	51.32	65.66	0.00	0.00	0.00	250.90	32.87
16	3828	603.42	149.56	75.25	59.37	68.91	0.00	0.00	0.00	218.42	37.14
17	3828	596.65	225.67	63.79	49.43	38.31	0.00	0.00	0.00	246.45	30.01
18	3786	567.32	108.37	94.86	86.82	36.14	0.00	0.00	0.00	329.59	45.27
19	3659	467.61	184.46	107.06	64.72	59.54	0.00	0.00	0.00	250.18	40.09
20	3458	572.33	107.56	51.23	40.09	40.40	0.00	0.00	0.00	203.03	30.32
21	3394	509.97	103.08	48.84	42.35	40.12	0.00	0.00	0.00	198.09	49.48
22	3334	535.68	106.09	40.42	64.88	48.91	0.00	0.00	0.00	191.75	30.06
23	3329	528.56	101.08	40.00	40.00	29.43	0.00	0.00	0.00	201.14	30.00
24	3348	542.02	112.27	50.06	42.13	98.54	0.00	0.00	0.00	202.20	30.00

2.5.5 DE-OBL and IDE-OBL for Solving ELD

The parameters of DE-OBL and IDE-OBL and their settings used for solving the ELD problem are listed in Table 2.24. For optimal parameters, simulations were carried out for 50 trials by varying the basic parameters such as the scale factor (F), crossover rate (C_r), and population size (P). The population size was varied in the range [20,100] according to the test system considered. The parameter F controls the speed and robustness of the search, that is, a lower value of F increases not only increases the convergence rate but also the risk of getting stuck in a local optimum. On the other hand, if $F > 1.0$, then solutions tend to be more time consuming and less reliable. The parameter C_r, which controls the crossover operation, can also be thought of as a mutation rate, that is, the probability that a variable will be inherited from the mutated individual. The role of C_r is to provide a means of exploiting decomposability. In order to select the most suitable {F, C_r} pair, P was fixed, and experimented by varying $F \in [1, 2]$ and $C_r \in [0.1, 1]$ with a step size of 0.2 and 0.1 for F and C_r, respectively. The near-optimum values of F and C_r for most of the case studies were found to be 0.8 for both. To ensure convergence, maximum generations (MAXGEN = 500) was allowed in each experimental run. The parameters and settings

TABLE 2.20

Optimal Dispatch Using EPSO for the 15-Unit Test System

Hour	P_D (MW)	Distribution of Load among Units (MW)					FC (Rs/h)	P_L (MW)	P_T (MW)	CT (s)
		P_{11}	P_{12}	P_{13}	P_{14}	P_{15}				
1	3352	49.47	1299.87	599.71	53.17	282.16	3898.58	2.41	3354.41	2.84
2	3384	45.02	1299.94	608.19	34.55	311.69	3956.39	2.69	3386.69	2.26
3	3437	40.11	1299.99	719.18	64.78	320.39	4018.63	2.89	3439.89	2.29
4	3489	40.01	1300.00	696.63	41.85	286.71	4120.23	2.58	3491.58	2.31
5	3659	41.06	1300.00	753.93	82.41	358.86	4397.81	3.27	3662.27	2.42
6	3849	53.81	1300.00	752.05	97.41	377.06	4742.31	3.46	3852.46	2.31
7	3898	40.95	1300.00	711.81	53.68	382.95	4847.90	3.50	3901.50	2.14
8	3849	127.04	1299.88	777.38	25.35	353.56	4768.43	3.44	3852.44	2.29
9	3764	62.37	1300.00	760.19	84.07	362.35	4583.71	3.34	3767.34	2.34
10	3637	43.04	1300.00	736.70	77.68	335.87	4365.36	3.02	3640.02	2.20
11	3437	40.45	1299.87	703.20	61.79	277.43	4022.77	2.50	3439.50	2.28
12	3384	48.40	1299.99	688.50	58.42	316.41	3931.73	2.82	3386.82	2.23
13	3357	41.86	1299.94	715.27	63.86	263.62	3895.97	2.39	3359.39	2.25
14	3394	41.42	1299.82	660.27	56.81	307.76	3981.33	2.64	3396.64	2.25
15	3616	67.81	1299.79	638.81	87.26	301.43	4342.00	2.60	3618.60	2.18
16	3828	56.50	1300.00	797.10	98.03	367.74	4697.19	3.43	3831.43	2.40
17	3828	103.16	1300.00	774.87	76.51	326.25	4714.57	3.10	3831.10	2.32
18	3786	95.09	1299.99	710.52	62.35	352.85	4645.62	3.18	3789.18	2.14
19	3659	51.76	1300.00	738.89	75.45	322.13	4420.40	2.89	3661.89	2.43
20	3458	41.20	1299.67	688.43	66.87	319.69	4055.53	2.83	3460.83	2.22
21	3394	45.99	1299.93	714.38	79.28	264.88	3955.53	2.39	3396.39	2.23
22	3334	40.03	1299.92	653.29	0.00	325.93	3857.00	2.94	3336.94	2.18
23	3329	40.00	1300.00	701.97	0.00	319.80	3843.44	2.96	3331.96	2.26
24	3348	40.09	1300.00	646.14	0.00	287.12	3886.91	2.59	3350.59	2.32

for DE-OBL and IDE-OBL were the same, except for the jumping factor J_r in IDE-OBL. J_r is an important control parameter in IDE-OBL which, if optimally set, can achieve better results. The experimental analysis reported in Rahnamayan et al. (2008) showed optimal results for $J_r \in [0.3, 0.6]$. Experiments were carried out on the test systems chosen in this work by varying J_r in [0.3, 0.6], and near-optimal solutions were obtained for $J_r = 0.37$. The dimension D varies with respect to the number of generators used in the ELD problem. For 6, 10, 15, and 20 unit test systems, the value of D was set to 5, 9, 14, and 19, respectively.

For implementation using MATLAB, the input variables were checked and initialized along with the population. The population is a matrix of size $NP \times D$. It was initialized with random values between the minimum and maximum values of the parameters. The best member after initialization was evaluated based on the objective function by starting with the first population member. The vectors were selected, and were allowed to enter the new population. This process was performed by checking the cost of the competitor. If competitor was better in the value in the "cost array," then we replaced the old vector with a new one (for new iteration) and saved that value in the "cost array." We updated *bestval* only in case of success to save time. The best member of this iteration was frozen for the coming iteration.

TABLE 2.21

Optimal Dispatch Using EPSO for the 20-Unit Test System

Hour	P_D (MW)	Distribution of Load among Units (MW)									
		P_1	P_2	P_3	P_4	P_5	P_6	P_7	P_8	P_9	P_{10}
1	2072	470.00	0.00	0.00	225.92	0.00	160.00	130.00	0.00	0.00	55.00
2	2220	470.00	0.00	0.00	227.54	243.00	160.00	0.00	0.00	0.00	0.00
3	2516	409.23	460.00	340.00	0.00	0.00	0.00	0.00	0.00	0.00	55.00
4	2812	470.00	460.00	0.00	231.93	0.00	160.00	0.00	0.00	25.47	55.00
5	2960	442.39	460.00	340.00	0.00	225.20	0.00	0.00	0.00	20.00	0.00
6	3256	420.97	451.78	340.00	0.00	215.28	160.00	0.00	0.00	0.00	55.00
7	3404	470.00	460.00	340.00	197.89	0.00	160.00	0.00	0.00	0.00	55.00
8	3552	470.00	460.00	340.00	298.75	243.00	0.00	0.00	0.00	0.00	0.00
9	3848	470.00	460.00	340.00	183.60	242.99	160.00	0.00	0.00	0.00	55.00
10	4144	470.00	460.00	340.00	292.03	243.00	160.00	129.98	0.00	0.00	0.00
11	4292	470.00	459.99	340.00	273.65	242.99	160.00	129.98	78.28	0.00	0.00
12	4440	466.12	460.00	339.93	299.93	242.99	160.00	129.22	0.00	79.94	55.00
13	4144	470.00	460.00	340.00	281.41	243.00	160.00	130.00	0.00	0.00	0.00
14	3848	470.00	460.00	340.00	222.20	243.00	160.00	0.00	0.00	0.00	55.00
15	3552	470.00	460.00	340.00	281.28	243.00	0.00	0.00	0.00	0.00	0.00
16	3108	420.98	460.00	340.00	0.00	212.13	0.00	130.00	0.00	0.00	0.00
17	2960	440.50	460.00	340.00	0.00	223.89	0.00	0.00	0.00	20.00	0.00
18	3256	414.73	460.00	340.00	0.00	203.89	160.00	0.00	0.00	0.00	55.00
19	3552	470.00	460.00	340.00	280.07	243.00	0.00	0.00	0.00	0.00	0.00
20	4144	447.38	459.67	339.96	299.99	242.91	159.66	129.98	0.00	0.00	0.00
21	3848	470.00	460.00	340.00	215.44	243.00	160.00	0.00	0.00	0.00	55.00
22	3256	414.46	460.00	340.00	0.00	203.75	160.00	0.00	0.00	0.00	55.00
23	2664	470.00	460.00	0.00	0.00	243.00	0.00	0.00	94.84	80.00	55.00
24	2368	470.00	460.00	0.00	0.00	0.00	0.00	0.00	120.00	80.00	55.00

```
%-----Select which vectors are allowed to enter the new population-----------
  for i=1:NP
    tempval = feval(fname,ui(i,:),y);   % check cost of competitor
    nfeval  = nfeval + 1;
    if (tempval <= val(i))  % if competitor is better than value in "cost
    array"
        pop(i,:) = ui(i,:);  % replace old vector with new one (for new
        iteration)
        val(i)   = tempval;  % save value in "cost array"
        %----we update bestval only in case of success to save time----------
        if (tempval < bestval)    % if competitor better than the best
        one ever
            bestval = tempval;    % new best value
            bestmem = ui(i,:);    % new best parameter vector ever
        end
    end
  end %---end for imember=1:NP
  bestmemit = bestmem; % freeze the best member of this iteration for the
                    coming
                    % iteration. This is needed for some of the
                    strategies.
```

TABLE 2.22

Optimal Dispatch Using EPSO for the 20-Unit Test System

Hour	P_D (MW)	Distribution of Load among Units (MW)									
		P_{11}	P_{12}	P_{13}	P_{14}	P_{15}	P_{16}	P_{17}	P_{18}	P_{19}	P_{20}
1	2072	470.00	0.00	0.00	226.41	0.00	160.00	130.00	0.00	0.00	55.00
2	2220	470.00	0.00	0.00	259.66	243.00	160.00	0.00	0.00	0.00	0.00
3	2516	409.23	460.00	340.00	0.00	0.00	0.00	0.00	0.00	0.00	55.00
4	2812	470.00	460.00	0.00	299.90	0.00	160.00	0.00	0.00	25.08	55.00
5	2960	442.39	460.00	340.00	0.00	225.20	0.00	0.00	0.00	20.00	0.00
6	3256	420.97	459.98	339.99	0.00	185.77	160.00	0.00	0.00	0.00	55.00
7	3404	470.00	460.00	340.00	252.92	0.00	160.00	0.00	0.00	0.00	55.00
8	3552	470.00	460.00	340.00	244.10	243.00	0.00	0.00	0.00	0.00	0.00
9	3848	470.00	460.00	340.00	228.55	242.95	160.00	0.00	0.00	0.00	55.00
10	4144	469.88	460.00	340.00	268.56	242.87	159.99	130.00	0.00	0.00	0.00
11	4292	469.60	460.00	340.00	248.51	242.92	160.00	129.93	107.71	0.00	0.00
12	4440	466.10	460.00	339.96	299.61	242.67	159.75	129.99	0.00	79.86	55.00
13	4144	470.00	460.00	340.00	278.91	243.00	160.00	130.00	0.00	0.00	0.00
14	3848	470.00	460.00	340.00	183.35	243.00	160.00	0.00	0.00	0.00	55.00
15	3552	470.00	460.00	340.00	255.21	243.00	0.00	0.00	0.00	0.00	0.00
16	3108	421.19	460.00	340.00	0.00	212.02	0.00	130.00	0.00	0.00	0.00
17	2960	440.50	460.00	340.00	0.00	223.89	0.00	0.00	0.00	20.00	0.00
18	3256	414.01	460.00	340.00	0.00	203.81	160.00	0.00	0.00	0.00	55.00
19	3552	470.00	459.99	340.00	256.42	243.00	0.00	0.00	0.00	0.00	0.00
20	4144	447.37	459.91	339.94	299.99	242.96	159.98	129.98	0.00	0.00	0.00
21	3848	470.00	460.00	340.00	190.10	243.00	160.00	0.00	0.00	0.00	55.00
22	3256	414.46	460.00	340.00	0.00	203.75	160.00	0.00	0.00	0.00	55.00
23	2664	470.00	460.00	0.00	0.00	243.00	0.00	0.00	91.92	80.00	55.00
24	2368	470.00	460.00	0.00	0.00	0.00	0.00	0.00	120.00	80.00	55.00

2.5.5.1 Six-Unit Test System

The parameters for DE-OBL and IDE-OBL were set according to the values shown in Table 2.24. The committed schedules obtained through GA were dispatched using DE-OBL and IDE-OBL based on the 24-hour load demand. The heuristic algorithms computed the power to be shared by units P_1 to P_6 for each load demand.

A load demand of "0" indicates that the unit is "OFF." From the results, it is inferred that the operating cost is proportional to the load demand. The computed results of ELD for DE-OBL are shown in Table 2.25. From the results, it is seen that unit P_1 generates a maximum power of 3522.08 MW, while unit P_5 generates a minimum power of 72.73 MW. The total fuel cost over the 24-hour time horizon was computed by DE-OBL as $12,013.69 with a total power loss of 15.03 MW in 13.56 s.

The results for the ELD problem using IDE-OBL algorithm are presented in Table 2.26. The minimum fuel cost is $288.44 for a load demand of 131 MW, and the maximum fuel cost is $774.17 for a load demand of 283.4 MW. Similarly, the total power generated is maximum with 284.48 MW during the 5th hour and minimum with 131.25 MW during the 24th hour. Since the transmission loss is considered in the implementation, the total power loss of the six-unit system over 24 hours was computed as 15.17 MW. The algorithm computed the results for all the 24 hours at the end of 13.2 s.

TABLE 2.23

Optimal Dispatch Using EPSO for the 20-Unit Test System

Hour	P_D (MW)	FC ($/h)	P_L (MW)	P_T (MW)	CT (s)
1	2072	50,477.31	10.32	2082.32	3.79
2	2220	53,592.88	13.20	2233.20	3.15
3	2516	61,371.82	12.47	2528.47	3.51
4	2812	70,366.95	20.43	2872.37	3.24
5	2960	71,466.65	15.18	2975.18	3.17
6	3256	78,532.15	16.96	3264.74	3.65
7	3404	82,783.30	16.82	3420.82	3.14
8	3552	85,703.60	16.84	3568.84	3.20
9	3848	93,406.25	20.05	3868.09	3.57
10	4144	98,447.62	22.31	4166.32	3.10
11	4292	103,192.80	21.57	4313.58	3.35
12	4440	108,546.94	25.65	4466.09	3.24
13	4144	98,446.72	22.31	4166.31	3.46
14	3848	93,248.47	13.55	3861.55	3.56
15	3552	85,548.59	10.49	3562.49	3.42
16	3108	73,541.69	18.33	3126.33	3.20
17	2960	71,326.25	8.79	2968.79	3.20
18	3256	78,388.50	10.44	3266.44	3.49
19	3552	85,548.55	10.49	3562.48	3.56
20	4144	98,394.67	15.69	4159.69	3.45
21	3848	93,248.17	13.55	3861.55	3.37
22	3256	78,388.50	10.44	3266.44	3.32
23	2664	67,659.62	19.75	2802.76	3.29
24	2368	76,343.92	17.32	2370.00	3.14

TABLE 2.24

Parameter Settings of DE-OBL and IDE-OBL

Parameters	Notations Used	Values
No. of members in population	N_P	[20, 100]
Vector of lower bounds for initial population	X_j^{min}	[−2, −2]
Vector of upper bounds for initial population	X_j^{max}	[2, 2]
No. of iterations	Iter	200
Dimension	D	Problem dependant
Crossover rate	C_r	[0, 1]
Step size	F	[1, 2]
Strategy parameter (DE-OBL)	DE/best/2/bin	9
Strategy parameter (IDE-OBL)	DE/rand/1/bin	7
Jumping rate	J_r	0.37
Refresh parameter	R	10
Value to reach	VTR	1.e−6

TABLE 2.25

ELD Results Using DE-OBL for the Six-Unit System

Hour	P_D (MW)	P_1	P_2	P_3	P_4	P_5	P_6	FC ($/h)	P_L (MW)	P_T (MW)	CT (s)
1	166	127.55	0	16.83	10	0	12	440.14	0.36	166.38	0.55
2	196	146.66	0	17.83	10	10	12	536.19	0.47	196.49	0.62
3	229	177.9	0	19	10	10	12.81	647.57	0.7	229.71	0.5
4	267	178.91	48.4	20.27	10	10.35	0	739.83	0.91	267.93	0.56
5	283.4	187.93	52.74	21.6	10	12.18	0	798.02	1.02	284.45	0.56
6	272	181.15	50.75	20.86	10	10.2	0	757.41	0.94	272.96	0.69
7	246	166.3	45.5	15	10	10	0	668.69	0.78	246.8	0.58
8	213	138.63	38.45	16.48	10	10	0	559.82	0.55	213.56	0.53
9	192	127.51	38.3	16.69	10	0	0	492.67	0.48	192.5	0.58
10	161	114.53	31.86	15	0	0	0	380.84	0.39	161.39	0.41
11	147	115.35	32.04	0	0	0	0	354.62	0.38	147.39	0.5
12	160	126.09	34.36	0	0	0	0	392.61	0.46	160.45	0.53
13	170	134.36	36.15	0	0	0	0	422.57	0.52	170.51	0.61
14	185	146.77	38.84	0	0	0	0	468.7	0.61	185.61	0.53
15	208	165.81	42.97	0	0	0	0	542.23	0.78	208.78	0.47
16	232	170.48	43.97	18.4	0	0	0	600.27	0.84	232.85	0.48
17	246	172.3	43.48	19.09	0	0	12	646.58	0.87	246.87	0.55
18	241	167.35	43.71	18.77	0	0	12	630.04	0.83	241.83	0.62
19	236	163.85	43.08	17.86	0	0	12	613.67	0.8	236.79	0.83
20	225	156.83	39.52	17.37	0	0	12	578.23	0.72	225.72	0.61
21	204	139.5	36.93	16.16	0	0	12	512.53	0.58	204.59	0.55
22	182	121.95	33.51	15	0	0	12	446.61	0.45	182.46	0.55
23	161	104.62	29.72	15	0	0	12	386.42	0.34	161.34	0.59
24	131	89.75	26.5	15	0	0	0	297.43	0.25	131.25	0.56

2.5.5.2 Ten Unit Test System

The OBL-based DE and IDE were used to determine the optimal dispatch of the GA-committed schedules for the 10-unit system. The parameters and their values were set according to Table 2.24. The population size for the 10-unit system was set to 40, the crossover probability to 0.6, and step size to 0.8, respectively. Since DE-OBL uses only one set of population during the entire run, the algorithm is capable of attaining a faster convergence rate, thus reducing the computational time. The computed optimal results for DE-OBL and IDE-OBL are shown in Tables 2.27 and 2.28. The total power demand requirement for the 10-unit system is 40,108 MW. Unit P_1 generated a maximum power of 10,574.51 MW and unit P_9 generated a minimum power of 240 MW over 24 hours. The optimal cost for generating the total power including the transmission losses is $973,049.1.

The load distribution among the 10 individual generating units, fuel cost, power loss, total power, and computational time were evaluated through IDE-OBL, and they are shown in Tables 2.29 and 2.30. For an overall power requirement of 40,108 MW, the IDE-OBL algorithm dispatches the power among all the 10 units with a total cost of $972,158.58, including a power loss of 95.3 MW and taking a time of 14.4 s.

TABLE 2.26

ELD Results Using IDE-OBL for the Six-Unit System

Hour	P_D (MW)	P_1	P_2	P_3	P_4	P_5	P_6	FC ($/h)	P_L (MW)	P_T (MW)	CT (s)
1	166	127.57	0	16.81	10	0	12	437.8	0.36	166.38	0.48
2	196	146.52	0	17.97	10	10	12	534.18	0.47	196.49	0.53
3	229	177.82	0	19.90	10	10	12	643.41	0.69	229.72	0.58
4	267	178.52	49.46	19.96	10	10	0	735.8	0.91	267.94	0.58
5	283.4	200	44.64	19.84	10	10	0	774.17	1.08	284.48	0.64
6	272	200	38.05	15	10	10	0	736.77	1.04	273.05	0.59
7	246	162.04	45.83	18.90	10	10	0	657.65	0.76	246.78	0.66
8	213	135.78	40.34	17.43	10	10	0	548.69	0.54	213.56	0.59
9	192	127.41	38.29	16.79	10	0	0	483.01	0.47	192.49	0.5
10	161	114.53	31.86	15	0	0	0	361.73	0.39	161.39	0.51
11	147	115.35	32.04	0	0	0	0	350.96	0.38	147.38	0.55
12	160	126.09	34.36	0	0	0	0	381.92	0.46	160.46	0.47
13	170	134.36	36.15	0	0	0	0	401.42	0.52	170.52	0.53
14	185	146.77	38.84	0	0	0	0	449.7	0.61	185.61	0.55
15	208	165.81	42.97	0	0	0	0	531.98	0.78	208.78	0.51
16	232	170.47	43.98	18.40	0	0	0	591.27	0.84	232.84	0.5
17	246	172.02	44.35	18.50	0	0	12	632.54	0.87	246.87	0.51
18	241	168.26	43.35	18.23	0	0	12	622.02	0.84	241.84	0.5
19	236	164.48	42.48	17.84	0	0	12	604.66	0.80	236.80	0.73
20	225	155.48	40.76	17.48	0	0	12	564.19	0.72	225.72	0.59
21	204	138.88	37.18	16.52	0	0	12	508.52	0.58	204.58	0.48
22	182	121.68	33.35	15.42	0	0	12	438.59	0.45	182.45	0.58
23	161	104.62	29.72	15	0	0	12	381.28	0.34	161.34	0.56
24	131	89.75	26.50	15	0	0	0	288.44	0.25	131.25	0.48

2.5.5.3 Fifteen-Unit Test System

The opposition-based learning concept is introduced into DE and IDE in order to avoid random initial population generation, which is more common in most of the meta-heuristics. Due to this fact, the intricate search process can be avoided, thus leading to a faster convergence rate. The parameters used to obtain the optimal dispatch for the commitment schedule of the Indian utility 75-bus system (15-unit system) are shown in Table 2.24. The economic dispatch results for the 15-unit system using DE-OBL are shown in Tables 2.31 and 2.32. It is noticed from the results that the fuel cost for generating power during the 22nd hour (Rs. 3900.29) is very low compared with that for the remaining hours of the day. The fuel cost is maximum during the seventh hour (Rs. 4873.54), implying that the load requirement is very large during this period. The total power generated by the Indian utility 75-bus system is 85,546.85 MW with a power loss of 76.85 MW, and the algorithm took 16.95 s to produce the optimal results over 24 hours.

The distribution of power among 15 units based on the commitment schedule, fuel cost, power loss, total power and computational time were evaluated using IDE-OBL, and the results are presented in Tables 2.33 and 2.34. It is observed that unit P_{12} generated a maximum power of 31,111.95 MW while unit P_8 generated a minimum power of 27.68 MW.

TABLE 2.27

ELD Results Using DE-OBL for 10 Unit System

Hour	P_1	P_2	P_3	P_4	P_5	P_6	P_7	P_8	P_9	P_{10}
				Power Generated/Unit (MW)						
1	470	0	0	223.57	0	160	130	0	0	55
2	446.87	0	0	263.33	243	160	0	0	0	0
3	406.10	460	340	0	0	0	0	0	0	55
4	436.44	460	0	279.52	0	160	0	0	20	55
5	420.78	460	340	0	243	0	0	0	20	0
6	398.73	460	340	0	218.50	160	0	0	0	55
7	431.71	460	340	259.40	0	160	0	0	0	55
8	455.09	460	340	282.06	243	0	0	0	0	0
9	451.39	460	340	219.56	243	160	0	0	0	55
10	462.22	460	340	300	225.27	160	130	0	0	0
11	458.42	460	340	300	243	160	130	60.07	0	0
12	458.38	460	340	300	243	160	130	0	80	55
13	444.49	460	340	300	243	160	130	0	0	0
14	437.37	460	340	231.93	243	160	0	0	0	55
15	465.94	460	340	269.67	243	0	0	0	0	0
16	385.58	460	340	0	243	0	130	0	0	0
17	419.20	460	340	0	243	0	0	0	20	0
18	408.83	460	340	0	206.77	160	0	0	0	55
19	450.69	460	340	284.86	243	0	0	0	0	0
20	466.27	460	340	276.70	243	160	130	0	0	0
21	455.93	460	340	213.41	243	160	0	0	0	55
22	410.08	460	340	0	205.52	160	0	0	0	55
23	464	460	0	0	243	0	0	94.89	20	55
24	470	460	0	0	0	0	0	120	80	55

The fuel cost computed to generate power for a total load requirement of 85,470 MW is Rs. 102,889.39 with a power loss of 80.38 MW at 16.18 s.

2.5.5.4 Twenty-Unit Test System

The concepts of OBL, DE, and IDE were applied to dispatch power based on the GA's commitment schedule for the 20-unit test system. The various parameters used to implement DE-OBL and IDE-OBL for 20-unit generating system are shown in Table 2.24 except for the dimension D, which was varied based on the size of the problem. Here, $D = 19$ for 20-unit system and the population was usually set based on 10 times the D value. To determine the choice of population size for the 20-unit system, DE-OBL and IDE-OBL algorithms were run with different values for 30 independent trials. The optimal population size was found to be 40, which resulted in minimum mean cost during 28 trials out of 30 trials. For a population size of 40, the crossover probability C_r was increased from 0.1 to 0.9 in steps of 0.1, and the scale factor F was increased from 0 to 1 in steps of 0.2. At the end of 30 trials, the optimal values of C_r and F were found to be 0.6 and 0.8, respectively.

Since the opposition-based generation jumping phase involved in the IDE-OBL, the experimental results were computed with a jumping rate of 0.37. The optimal dispatch

TABLE 2.28

ELD Results Using DE-OBL for the 10-Unit System

Hour	P_D (MW)	FC ($/h)	P_L (MW)	P_T (MW)	CT (s)
1	1036	25,175.97	2.57	1038.57	0.90
2	1110	26,765.92	3.19	1113.19	0.73
3	1258	30,617.19	3.10	1261.10	0.61
4	1406	35,131.00	4.96	1410.96	0.83
5	1480	35,650.05	3.78	1483.78	0.58
6	1628	39,172.61	4.23	1632.23	0.70
7	1702	41,373.02	4.10	1706.10	0.94
8	1776	42,781.70	4.15	1780.15	0.73
9	1924	46,620.78	4.95	1928.95	0.66
10	2072	49,144.65	5.49	2077.49	0.59
11	2146	51,503.09	5.49	2151.49	0.62
12	2220	54,130.94	6.38	2226.38	0.62
13	2072	49,144.65	5.49	2077.49	0.90
14	1924	46,612.04	3.30	1927.30	0.72
15	1776	42,719.26	2.60	1778.60	0.61
16	1554	36,671.32	4.58	1558.58	0.83
17	1480	35,615.17	2.20	1482.20	0.55
18	1628	39,136.72	2.60	1630.60	0.72
19	1776	42,753.02	2.55	1778.55	0.89
20	2072	49,057.26	3.97	2075.97	0.86
21	1924	46,571.75	3.34	1927.34	0.62
22	1628	39,136.72	2.60	1630.60	0.59
23	1332	33,722.23	4.89	1336.89	0.95
24	1184	33,842.08	4.33	1185.00	0.69

results of DE-OBL are shown in Tables 2.35–2.37. For the given load profile of 80,216 MW, the DE-OBL algorithm generated a total power of 80,578.45 MW including the transmission losses with the fuel cost consumption of $1,960,091. The time required for the algorithm to compute results for the 24-hour load profile is 17.58 s.

The experimental results of IDE-OBL for the 20-unit test system are shown in Tables 2.38–2.40. It is seen that a minimum power of 2082.2 MW is generated during 1st hour and a maximum power of 4465.65 MW is generated during 12th hour. The power loss is found to be 380.47 MW, which is due to the large number of generating units. The average time required to dispatch power during each hour is 4.2% of the total computational time.

2.5.6 ABC-Based ELD

The parameters that govern the ABC algorithm are colony size, number of food sources, food source limit, number of employed bees, number of onlooker bees, and maximum number of iterations. The colony size was set to a moderate value of 20, irrespective of the test system. A smaller colony size generates faster solution, but a larger colony size generates more accurate solution but is relatively slow. The number of employed bees and onlooker bees were set to half the value of colony size: in this study it was set to 10. Number of food sources was set to a value of 10, and the food source limit was 100. The number of food sources in the ABC algorithm was equal to number of employed bees. The maximum

TABLE 2.29

Economic Dispatch Using IDE-OBL for the 10-Unit System

Hour	Power Generated/Unit (MW)									
	P_1	P_2	P_3	P_4	P_5	P_6	P_7	P_8	P_9	P_{10}
1	469.99	0	0	223.57	0	160	130	0	0	55
2	470	0	0	240.29	243	160	0	0	0	0
3	406.10	460	340	0.00	0	0	0	0	0	55
4	469.89	460	0	246.20	0	160	0	0	20	55
5	440.19	460	340	0	223.58	0	0	0	20	0
6	411.56	460	340	0	205.66	160	0	0	0	55
7	470.00	460	340	221.19	0	160	0	0	0	55
8	469.96	460	340	267.24	243	0	0	0	0	0
9	470	460	340	201.00	243	160	0	0	0	55
10	469.67	460	340	274.89	243	160	130	0	0	0
11	467.47	460	340	230.78	243	160	130	120	0	0
12	469.86	460	340	300	243	160	130	0	68.51	55
13	455.27	460	340	300	232.22	160	130	0	0	0
14	469.99	460	340	199.39	243	160	0	0	0	55
15	470	460	340	265.62	243	0	0	0	0	0
16	417.15	460	340	0.00	211.41	0	130	0	0	0
17	419.20	460	340	0.00	243	0	0	0	20	0
18	410.56	460	340	0.00	205.04	160	0	0	0	55
19	469.70	460	340	265.92	243	0	0	0	0	0
20	464.54	460	340	278.43	243	160	130	0	0	0
21	470.00	460	340	199.38	243	160	0	0	0	55
22	410.56	460	340	0	205.03	160	0	0	0	55
23	469.84	460	0	0	243	0	0	89.10	20	55
24	470	460	0	0	0	0	0	120	80	55

number of generations was 500, which was chosen based on the convergence of the system. The control parameters of ABC algorithm are given in Table 2.41.

The MATLAB snippets for the phases of the ABC algorithm—employed bee phase, scout bee phase, and onlooker bee phase—are presented below:

```
%%%%%%%%% EMPLOYED BEE PHASE %%%%%%%%%%%%%%%%%%%%%%%%%%%
    for i=1:(FoodNumber)
        %/*The parameter to be changed is determined randomly*/
        Param2Change=fix(rand*D)+1;
        %/*A randomly chosen solution is used in producing a mutant
        solution of the solution i*/
        neighbour=fix(rand*(FoodNumber))+1;
        %/*Randomly selected solution must be different from the solution i*/
            while(neighbour==i)
                neighbour=fix(rand*(FoodNumber))+1;
            end;
        sol=Foods(i,:);
        % /*v_{ij}=x_{ij}+\phi_{ij}*(x_{kj}-x_{ij})  */
        sol(Param2Change)=Foods(i,Param2Change)+(Foods(i,Param2Change)-Foods
        (neighbour,Param2Change))*(rand-0.5)*2;
```

```
    % /*if generated parameter value is out of boundaries, it is
    shifted onto the boundaries*/
     ind=find(sol<lb);
     sol(ind)=lb(ind);
     ind=find(sol>ub);
     sol(ind)=ub(ind);
     %evaluate new solution
     ObjValSol=feval(objfun,sol);
     FitnessSol=calculateFitness(ObjValSol);
    % /*a greedy selection is applied between the current solution i
    and its mutant*/
    if (FitnessSol>Fitness(i)) %/*If the mutant solution is better
    than the current solution i, replace the solution with the mutant
    and reset the trial counter of solution i*/
         Foods(i,:)=sol;
         Fitness(i)=FitnessSol;
         ObjVal(i)=ObjValSol;
         trial(i)=0;
    else
         trial(i)=trial(i)+1; %/*if the solution i can not be
         improved, increase its trial counter*/
    end;
end;

%%%%%%%%%%%%%%%%%%%%%%% ONLOOKER BEE PHASE %%%%%%%%%%%%%%%%%%%%%%%%%%%%%%%
%%%%%%%%%%%%%%%%%%%%%%%

i=1;
t=0;
while(t<FoodNumber)
    if(rand<prob(i))
        t=t+1;
        %/*The parameter to be changed is determined randomly*/
        Param2Change=fix(rand*D)+1;

        %/*A randomly chosen solution is used in producing a mutant
        solution of the solution i*/
        neighbour=fix(rand*(FoodNumber))+1;

        %/*Randomly selected solution must be different from the solution i*/
            while(neighbour==i)
                neighbour=fix(rand*(FoodNumber))+1;
            end;

        sol=Foods(i,:);
        %  /*v_{ij}=x_{ij}+\phi_{ij}*(x_{kj}-x_{ij}) */
        sol(Param2Change)=Foods(i,Param2Change)+(Foods(i,Param2Change)-
        Foods(neighbour,Param2Change))*(rand-0.5)*2;

        %  /*if generated parameter value is out of boundaries, it is
        shifted onto the boundaries*/
         ind=find(sol<lb);
         sol(ind)=lb(ind);
         ind=find(sol>ub);
         sol(ind)=ub(ind);
```

```
        %evaluate new solution
        ObjValSol=feval(objfun,sol);
        FitnessSol=calculateFitness(ObjValSol);

     % /*a greedy selection is applied between the current solution i
     and its mutant*/
     if (FitnessSol>Fitness(i)) %/*If the mutant solution is better
     than the current solution i, replace the solution with the mutant
     and reset the trial counter of solution i*/
         Foods(i,:)=sol;
         Fitness(i)=FitnessSol;
         ObjVal(i)=ObjValSol;
         trial(i)=0;
     else
         trial(i)=trial(i)+1; %/*if the solution i can not be
         improved, increase its trial counter*/
     end;
   end;

   i=i+1;
   if (i==(FoodNumber)+1)
       i=1;
   end;
end;


%/*The best food source is memorized*/
       ind=find(ObjVal==min(ObjVal));
       ind=ind(end);
       if (ObjVal(ind)<GlobalMin)
       GlobalMin=ObjVal(ind);
       GlobalParams=Foods(ind,:);
       end;


%%%%%%%%%%%% SCOUT BEE PHASE %%%%%%%%%%%%%%%%%%%%%%%%%%%%%%%%%%%%%%%%%%%
%%%%%%

%/*determine the food sources whose trial counter exceeds the "limit"
value.
%In Basic ABC, only one scout is allowed to occur in each cycle*/

ind=find(trial==max(trial));
ind=ind(end);
if (trial(ind)>limit)
    Bas(ind)=0;
    sol=(ub-lb).*rand(1,D)+lb;
    ObjValSol=feval(objfun,sol);
    FitnessSol=calculateFitness(ObjValSol);
    Foods(ind,:)=sol;
    Fitness(ind)=FitnessSol;
    ObjVal(ind)=ObjValSol;
end;
```

TABLE 2.30

Economic Dispatch Using IDE-OBL for 10 Unit System

Hour	P_D (MW)	FC ($/h)	P_L (MW)	P_T (MW)	CT (s)
1	1036	25,161.96	2.57	1038.57	0.61
2	1110	26,715.95	3.29	1113.29	0.56
3	1258	30,603.24	3.10	1261.10	0.47
4	1406	35,057.97	5.09	1411.09	0.58
5	1480	35,639.5	3.77	1483.77	0.59
6	1628	39,159.35	4.22	1632.22	0.55
7	1702	41,289.07	4.19	1706.19	0.58
8	1776	42,748.92	4.20	1780.20	0.55
9	1924	46,571.09	5.00	1929.00	0.53
10	2072	49,088.25	5.56	2077.56	0.62
11	2146	51,466.27	5.26	2151.26	0.56
12	2220	53,972.63	6.37	2226.37	0.66
13	2072	49,121.53	5.49	2077.49	0.59
14	1924	46,541.89	3.38	1927.38	0.66
15	1776	42,710.37	2.62	1778.62	0.51
16	1554	36,669.74	4.56	1558.56	0.55
17	1480	35,588.17	2.20	1482.20	0.59
18	1628	39,124.75	2.60	1630.60	0.55
19	1776	42,711.04	2.62	1778.62	0.87
20	2072	49,046.17	3.96	2075.96	0.62
21	1924	46,527.86	3.38	1927.38	0.67
22	1628	39,118.16	2.60	1630.60	0.66
23	1332	33,710.18	4.94	1336.94	0.69
24	1184	33,814.52	4.33	1185.00	0.59

2.5.6.1 Six-Unit Test System

The economic dispatch based on the on/off status of the six generating units for 24-hour load demand was determined, and is shown in Table 2.42. For each hour, load demand varies and hence the commitment of the units also varies. In ELD using ABC algorithm, the load sharing by each unit is uniformly distributed rather than the full load being allocated to a single unit. Thus stress in the generators can be avoided, since none of the units generates to its maximum capacity. Unit P_1 contributes a power of 3516.05 MW, P_2 834.15 MW, P_3 336.13 MW, P_4 90 MW, P_5 72.3 MW, and P_6 120 MW. Thus from the analysis, it is clear that unit P_1 generates the maximum power per day and unit P_5 generates the minimum power. The minimum operating cost is 297.43 $/h for a load demand of 131 MW at the 24th hour. Similarly, the maximum fuel cost (797.9324 $/h) is incurred during the fifth hour for a load demand of 283.4 MW. The total operating cost to generate power from the IEEE 30 bus system per day (24 hours) is $12,912.05. The computation time required to arrive at the solution for 24-hour forecasted load profile is 72.43 s, and the mean time per hour is 3.02 s.

TABLE 2.31

Optimal Dispatch Using DE-OBL for 15 Unit Test System

Hour	P_D (MW)	Distribution of Load among Units (MW)									
		P_1	P_2	P_3	P_4	P_5	P_6	P_7	P_8	P_9	P_{10}
1	3352	408.41	148.83	40	40	193.77	0	6.63	32.19	79.82	88.42
2	3384	592.47	236.24	46.19	40	46.46	0	0	0	68.97	30.90
3	3437	502.10	100	40	170	79.41	0	100	0	115.39	30
4	3489	494.69	100	77.86	170	51.43	0	100	0	172.85	30
5	3659	611.13	105.47	125.38	40	58.33	0	100	0	106.27	30
6	3849	482.47	100	40	101.97	78.16	0	66.01	0	279.27	30
7	3898	529.98	100	80.08	115.34	27.91	0	100	0	247.50	109.30
8	3849	602.67	300	40	40.82	31.27	0	100	0	270.82	41.96
9	3764	466.41	105.78	55.25	170	158.24	0	63.94	0	61.87	69.28
10	3637	472.53	176.42	83.58	50.18	7.77	0	100	0	139.95	81.37
11	3437	446.35	100	129.14	40	27.52	0	6.37	0	60	230.71
12	3384	411.88	100	158.75	40	19.37	0	0	0	158.33	59.80
13	3357	516.11	100	40	95.47	19.58	0	0	0	225.25	79.56
14	3394	593.63	115.07	86.96	170	92.54	0	0	0	189.79	30
15	3616	432.42	152.24	98.95	40	42.64	0	0	0	123.89	38.93
16	3828	431.46	300	40	40.84	64.40	0	0	0	257.72	42.40
17	3828	555.15	100	55.07	170	95.02	0	0	0	194.84	49.43
18	3786	389.96	100	46	73.72	53.26	0	0	0	240.62	65.16
19	3659	397.07	128.52	45.44	44.10	6.84	0	0	0	259.40	83.78
20	3458	362.42	120.36	40	75.18	177.96	0	0	0	70.88	30
21	3394	447.23	100	169.95	64.60	56.23	0	0	0	212.78	41.19
22	3334	387.29	100	51	55.42	26.88	0	0	0	285.23	41.40
23	3329	394.45	100	111.49	89.63	144.35	0	0	0	75.42	31.93
24	3348	529.07	100	138.91	84.73	55.49	0	0	0	307.30	30

2.5.6.2 Ten-Unit Test System

The impact of using ABC for obtaining optimal dispatch of the committed schedules is analyzed in this section. The parameters are set according to Table 2.41. The experimental results for the 10-unit system based on ABC are shown in Tables 2.43 and 2.44.

Unit P_1 is kept ON for the entire day because it has the minimum fuel cost coefficients and hence it also generates the maximum power per day. Unit P_8 is the most expensive unit with a fuel cost coefficient of 0.0048 ($/MW h^2). For units P_2, P_3, P_6, and P_7, the maximum generation limit is allocated for all load demands. For units P_4, P_5, P_8, and P_9, the load sharing is allotted based on the load demand and combination of units in ON state. For each load demand in the 24-hour load profile, the power generated by each unit varies according to their fuel cost function, generating limits, and also the UC schedule. Unit P_1 generates a load of 10,922.16 MW per day, whereas unit P_8 shares a load of only 308.51 MW per day.

It can be concluded that unit P_1 shares the maximum load, which accounts to 27.17% of the total load demand per day, and P_9 shares the minimum power of 0.76% of the total demand.

TABLE 2.32

Optimal Dispatch Using DE-OBL for the 15-Unit Test System

Hour	P_D (MW)	Distribution of Load among Units (MW)					FC (Rs/h)	P_L (MW)	P_T (MW)	CT (s)
		P_{11}	P_{12}	P_{13}	P_{14}	P_{15}				
1	3352	101.53	1300	557.62	119.38	237.41	4010.42	2	3354	0.69
2	3384	40	1142.18	674.25	141.63	327.43	4035.08	2.73	3386.73	0.67
3	3437	52.06	1300	667.70	71.52	210.86	4112.29	2.04	3439.04	0.64
4	3489	40	1300	696.64	10	247.91	4185.53	2.37	3491.37	0.76
5	3659	71.18	1300	687.62	69.94	356.95	4454.05	3.28	3662.28	0.75
6	3849	40	1300	731.30	150	454	4772.53	4.18	3853.18	0.80
7	3898	86.99	1300	695.92	114.54	393.96	4873.54	3.51	3901.51	0.66
8	3849	54.08	1300	556.54	60.97	454	4833.76	4.12	3853.12	0.69
9	3764	40	1300	764.71	150	361.83	4691.42	3.30	3767.30	0.67
10	3637	40	1300	900	10	278.18	4445.41	2.95	3639.95	0.73
11	3437	40	1300	717.71	19.88	322.18	4161.49	2.87	3439.87	0.69
12	3384	56.10	1300	593.30	36.60	454	4003.66	4.14	3388.14	0.75
13	3357	77.51	1212.23	474.58	66.55	454	3986.85	3.84	3360.84	0.78
14	3394	96.47	1300	486.84	30.30	204.21	4055.80	1.81	3395.81	0.76
15	3616	40	1300	900	150	299.87	4383.87	2.94	3618.94	0.72
16	3828	40	1300	900	22.85	392.32	4773.38	3.99	3831.99	0.83
17	3828	47.73	1300	773.92	37.28	454	4738.46	4.44	3832.44	0.67
18	3786	40	1300	900	150	431.45	4680.11	4.16	3790.16	0.69
19	3659	123.89	1300	670.15	150	454	4469.55	4.18	3663.18	0.89
20	3458	40	1300	685.83	105.66	454	4157.26	4.28	3462.28	0.61
21	3394	40	1300	511.55	150	302.86	4027.49	2.39	3396.39	0.69
22	3334	86	1250.68	741.34	0	311.64	3900.29	2.88	3336.88	0.58
23	3329	78.45	1300	708.61	0	297.51	3922.82	2.84	3331.84	0.62
24	3348	40	1300	620.09	0	144.03	3951.91	1.61	3349.61	0.61

The execution time of the algorithm for generating the schedule for 24 hours is 89.93 s, and the average time per hour is 3.75 s.

2.5.6.3 Fifteen-Unit Test System

In this section, the applicability and validity of the ABC algorithm is tested on the 15-unit test system committed through GA. The parameters of ABC were initialized according to Table 2.5. In ABC, the position of a food source represents a possible solution to the optimization problem, and the nectar amount of a food source corresponds to the quality (fitness) of the associated solution.

The simulation results obtained by the ABC algorithm for the economic dispatch at the end of 500 iterations are shown in Tables 2.45 and 2.46. Here, the individual power generated by each unit during each hour, hourly fuel cost, hourly power loss, and computational time are calculated. Unit P_6 has not been committed even for a single hour, and hence no power is generated from this unit. Unit P_7 and P_8 are committed only during certain hours of the day, and thus they generate very low powers of 18.15 and 20 MW, respectively. All the other units are committed throughout the day, and the maximum power of 31,200 MW is generated by unit P_{12}. The ABC algorithm was run for 10 different trials, and the results were consistent during maximum number of trials. This was possible because of the

TABLE 2.33

Economic Dispatch of IDE-OBL for 15 Unit Test System

Hour	P_D (MW)	Distribution of Load among Units (MW)									
		P_1	P_2	P_3	P_4	P_5	P_6	P_7	P_8	P_9	P_{10}
1	3352	333.42	100	83.15	87.33	39.19	0	47.72	27.68	210.03	40.24
2	3384	524.86	162.61	40	62.22	128.50	0	0	0	60	48.27
3	3437	661.68	136.03	41.88	170	44.98	0	23.35	0	147.94	30
4	3489	676.77	100	40	43.21	4.80	0	56.21	0	60	30
5	3659	456.45	115.53	56.92	40	57.80	0	57.80	0	231.06	63.17
6	3849	608.95	100	86.33	40	78.40	0	100	0	167.68	30
7	3898	600.64	100	44.82	73.69	67.89	0	44.36	0	165.86	30
8	3849	467.51	100	40	95.11	52.54	0	53.56	0	283.61	42.59
9	3764	511.30	178.48	40	40	26.86	0	13.70	0	171.01	42.83
10	3637	520.56	129.99	40	141.51	43.77	0	7.07	0	130.95	30
11	3437	506.11	100	40	41.29	2	0	20.17	0	304.29	30
12	3384	347.47	126.02	69.32	40	23.81	0	0	0	152.98	33.44
13	3357	622.06	100	69.97	40	136.97	0	0	0	87.92	30
14	3394	415.26	113.56	55.99	40	24.51	0	0	0	217.28	108.28
15	3616	456.15	100	40	82.08	28.20	0	0	0	226.25	30
16	3828	589.42	100	88.06	70.50	3.62	0	0	0	244.51	87.21
17	3828	591.86	100	40	52.89	138.10	0	0	0	336.28	30
18	3786	657.61	100	40.72	77.55	31.79	0	0	0	249.65	30
19	3659	527.48	100	40	67.53	26.81	0	0	0	254.88	48.67
20	3458	487.55	100	40	40	31.33	0	0	0	60	30
21	3394	359.47	100	40	64.10	9.74	0	0	0	174.42	74.31
22	3334	681.84	109.14	40	40	10.71	0	0	0	236.79	30
23	3329	360.59	188.73	40	79.53	81.05	0	0	0	86.31	55.66
24	3348	689.46	100	40	51.34	39.57	0	0	0	270.55	30

proper tuning of the algorithmic parameters initially. The total fuel cost for the entire day is Rs. 101,642.70 with a power loss of 75.61 MW in 95.68 s.

2.5.6.4 Twenty-Unit Test System

ABC was applied for solving the 20-unit test system in order to determine the optimum solution for each generating unit, thus minimizing the total generation cost. Parameters of the bee's algorithm were heuristically selected, and are given in Table 2.5. Though the algorithm was run for 500 iterations, ABC could effectively find the optimal solutions (at the end of 100 iterations) even before the maximum iterations were reached. The computational results are shown in Tables 2.47–2.49.

Units P_1 and P_2 generate a maximum of 13.56% of the total generated power, while units P_3 and P_4 contribute 12.55%, P_5 and P_6 8.02%, P_7 and P_8 5.47%, P_9 and P_{10} 4.7%, P_{11} and P_{12} 2.97%, P_{13} and P_{14} 1.13%, P_{15} and P_{16} 0.88%, P_{17} contributes 0.36%, P_{18} 0.34%, P_{19} 0.336%, and P_{20} 0.33% of the total power.

The total fuel cost required for generating the power as mentioned above is \$1,958,137.5 with a total power loss of 382.75 MW. ABC generally takes a long computational time, and for this case study it took 100.65 s to dispatch power over 24 hours.

TABLE 2.34

Economic Dispatch of IDE-OBL for the 15-Unit Test System

| Hour | P_D (MW) | Distribution of Load among Units (MW) | | | | | FC (Rs/h) | P_L (MW) | P_T (MW) | CT (s) |
		P_{11}	P_{12}	P_{13}	P_{14}	P_{15}				
1	3352	40	1300	583.37	10	454	3964.69	4.11	3356.11	0.76
2	3384	40	1300	764.90	25.46	229.55	3993.36	2.37	3386.37	0.66
3	3437	42.13	1211.95	647.12	58.90	223.02	4111.61	1.98	3438.98	0.64
4	3489	59.61	1300	658.79	150	312.42	4175.62	2.82	3491.82	0.75
5	3659	40	1300	640.33	150	454	4446.81	4.05	3663.05	0.78
6	3849	52.19	1300	751.62	84.21	454	4768.90	4.39	3853.39	0.55
7	3898	40	1300	900	94.50	440.69	4848.89	4.46	3902.46	0.80
8	3849	40	1300	900	150	327.20	4777.14	3.12	3852.12	0.61
9	3764	40	1300	848.20	150	405.51	4610.10	3.88	3767.88	0.72
10	3637	71.79	1300	802.94	150	271.03	4409.75	2.60	3639.60	0.64
11	3437	103.35	1300	603.57	129.71	258.72	4066.19	2.21	3439.21	0.66
12	3384	105.37	1300	702.02	34.02	454	3994.29	4.44	3388.44	0.61
13	3357	50.74	1300	663.10	24.35	234.15	3939.72	2.25	3359.25	0.61
14	3394	57.02	1300	793.87	49.15	221.29	3997.78	2.22	3396.22	0.58
15	3616	74.49	1300	679.01	150	454	4362.24	4.18	3620.18	0.61
16	3828	74.73	1300	748.37	71.91	454	4720.60	4.32	3832.32	0.62
17	3828	51.82	1300	638.19	144.34	408.04	4735.43	3.52	3831.52	0.78
18	3786	40	1300	658.80	150	454	4650.61	4.12	3790.12	0.59
19	3659	92.72	1300	600.99	150	454	4443.39	4.08	3663.08	0.87
20	3458	40	1300	900	150	281.96	4120.30	2.85	3460.85	0.73
21	3394	40	1300	632.01	150	454	4017.06	4.05	3398.05	0.58
22	3334	69.21	1300	653.06	0	165.10	3891.23	1.84	3335.84	0.59
23	3329	41.18	1300	646.32	0	454	3917.31	4.36	3333.36	0.76
24	3348	40	1300	536.40	0	252.86	3926.38	2.18	3350.18	0.69

2.5.7 CSO-Based ELD

The performance of CSO on ELD is also sensitive to the parameter settings. Compared to the common heuristic algorithms such as GA and EPSO, the number of parameters used in the CSO is small and hence has the potential to solve the ELD at a faster rate. The parameters used in CSO are number of nests or population size (n), tolerance (T), discovery rate (D_r), search dimension (N_d), lower and upper bounds of the search domain (N_{dL} and N_{dU}), the Levy exponent (λ), and the Levy step size (α). The parameters and their settings are shown in Table 2.50.

The two basic parameters that are tuned for optimal solution are n and D_r. The number of nests (n) are varied in [5,30] at intervals of 5, and D_r is set in the range [0,0.5] at intervals of 0.1. For most of the trials executed for the test systems with various n and D_r, much difference was not observed and hence we set n as 15 irrespective of the problem size and D_r as 0.25. The search dimension N_d is problem dependent, and so was set as 6, 10, 15, and 20 based on the number of generating units. The Levy step size was set to 0.01 (usually $L/100$); otherwise, Levy flights may become too aggressive/efficient, which makes new solutions (even) jump outside of the search domain (and thus wasting evaluations). The number of iterations was set to 300 initially, but the optimum results were obtained at the

TABLE 2.35

Optimal Dispatch Using DE-OBL for the 20-Unit Test System

Hour	P_D (MW)	Distribution of Load among Units (MW)									
		P_1	P_2	P_3	P_4	P_5	P_6	P_7	P_8	P_9	P_{10}
1	2072	468.72	0	0	300	0	160	130	0	0	55
2	2220	425.38	0	0	300	243	160	0	0	0	0
3	2516	408.14	460	340	0	0	0	0	0	0	55
4	2812	463.88	337.89	0	300	0	160	0	0	20	55
5	2960	379.21	460	340	0	243	0	0	0	20	0
6	3256	374.59	460	340	0	243	160	0	0	0	55
7	3404	448.10	460	340	300	0	160	0	0	0	55
8	3552	453.04	460	340	259.69	243	0	0	0	0	0
9	3848	352.79	460	340	300	171.49	160	0	0	0	55
10	4144	441.92	460	340	300	243	160	130	0	0	0
11	4292	467.63	460	340	300	243	160	130	120	0	0
12	4440	459.65	460	340	300	243	160	130	0	80	55
13	4144	437.41	460	340	300	235.67	160	130	0	0	0
14	3848	399.26	460	340	283.40	243	160	0	0	0	55
15	3552	406.09	460	340	300	243	0	0	0	0	0
16	3108	371.41	460	340	0	243	0	130	0	0	0
17	2960	312.62	460	340	0	243	0	0	0	80	0
18	3256	355.91	460	340	0	243	160	0	0	0	55
19	3552	462.71	460	340	300	243	0	0	0	0	0
20	4144	425.04	460	340	298.66	243	160	130	0	0	0
21	3848	459.76	460	340	300	243	160	0	0	0	55
22	3256	443.99	460	340	0	213.03	160	0	0	0	55
23	2664	416.81	460	0	0	243	0	0	120	20	55
24	2368	470	460	0	0	0	0	0	120	80	55

end of 30 iterations, proving faster convergence. The MATLAB snippet to develop the CSO algorithm is provided below:

```
while (fmin>Tol),
    % Generate new solutions (but keep the current best)
    new_nest=get_cuckoos(nest,bestnest,Lb,Ub);
    [fnew,best,nest,fitness,P₁,P₁]=
    get_best_nest(nest,new_nest,fitness);
    % Update the counter
     N_iter=N_iter+n;
     % Discovery and randomization
     new_nest=empty_nests(nest,Lb,Ub,pa) ;
     % Evaluate this set of solutions
     [fnew,best,nest,fitness,P₁,P₁]=
     get_best_nest(nest,new_nest,fitness);
     % Update the counter again
      N_iter=N_iter+n;
     % Find the best objective so far
    if fnew<fmin,
        fmin=fnew;
        bestnest=best;
    end
```

```
best_fitness_values(Iter_inc) = fmin;
current_best_solutions(:,Iter_inc) = transpose(bestnest);
Iter_inc = Iter_inc+1;
   if N_iter >= 300  %
   break;
end

end %% End of iterations
```

2.5.7.1 Six-Unit Test System

Using the UC schedule from GA, for each value of power demand, the individual power generated, power loss, total power, fuel cost, and computational time were recorded by applying CSO, as shown in Table 2.51. The algorithm was run for 50 trial runs, among which the best results are presented. For the given load profile of 4,953.4 MW, the CSO algorithm generated a total power of 4,968.02 MW including the transmission losses with the fuel cost consumption of $12,919.66. The entire algorithm for the 24-hour load profile consumed 15.87 s to determine the results.

TABLE 2.36

Optimal Dispatch Using DE-OBL for the 20-Unit Test System

Hour	P_D (MW)	P_{11}	P_{12}	P_{13}	P_{14}	P_{15}	P_{16}	P_{17}	P_{18}	P_{19}	P_{20}
		\multicolumn{10}{c}{Distribution of Load among Units (MW)}
1	2072	470	0	0	153.59	0	160	130	0	0	55
2	2220	470	0	0	231.46	243	160	0	0	0	0
3	2516	410.33	460	340	0	0	0	0	0	0	55
4	2812	470	460	0	300	0	160	0	0	49.58	55
5	2960	470	460	340	0	243	0	0	0	20	0
6	3256	470	460	340	0	155.36	160	0	0	0	55
7	3404	470	460	340	172.60	0	160	0	0	0	55
8	3552	470	460	340	300	243	0	0	0	0	0
9	3848	470	460	340	300	243	160	0	0	0	55
10	4144	470	460	340	288.23	243	160	130	0	0	0
11	4292	470	460	340	169.47	243	160	130	120	0	0
12	4440	470	460	340	300	243	160	130	0	80	55
13	4144	470	460	340	300	243	160	130	0	0	0
14	3848	470	460	340	192.53	243	160	0	0	0	55
15	3552	470	460	340	300	243	0	0	0	0	0
16	3108	470	460	340	0	181.92	0	130	0	0	0
17	2960	470	460	340	0	243	0	0	0	20	0
18	3256	470	460	340	0	167.52	160	0	0	0	55
19	3552	470	460	340	243.73	243	0	0	0	0	0
20	4144	470	460	340	300	243	160	130	0	0	0
21	3848	470	460	340	115.72	243	160	0	0	0	55
22	3256	336.49	460	340	0	243	160	0	0	0	55
23	2664	470	460	0	0	243	0	0	120	20	55
24	2368	470	460	0	0	0	0	0	120	80	55

TABLE 2.37

Optimal Dispatch Using DE-OBL for the 20-Unit Test System

Hour	P_D (MW)	FC ($/h)	P_L (MW)	P_T (MW)	CT (s)
1	2072	50,487.32	10.31	2082.31	0.56
2	2220	53,686.75	12.84	2232.84	0.80
3	2516	61,371.82	12.47	2528.47	0.69
4	2812	70,770.54	19.35	2831.35	0.67
5	2960	71,469.67	15.21	2975.21	0.64
6	3256	78,536.84	16.95	3272.95	0.80
7	3404	82,833.91	16.70	3420.70	0.76
8	3552	85,739.14	16.73	3568.73	0.73
9	3848	93,820.97	19.27	3867.27	0.58
10	4144	98,507.60	22.15	4166.15	0.81
11	4292	103,189.08	21.11	4313.11	0.86
12	4440	108,545.68	25.65	4465.65	0.73
13	4144	98,533.47	22.08	4166.08	0.84
14	3848	93,400.46	13.20	3861.20	0.66
15	3552	85,686.98	10.09	3562.09	0.66
16	3108	73,545.26	18.33	3126.33	0.70
17	2960	71,840.90	8.62	2968.62	0.72
18	3256	78,393.56	10.44	3266.44	0.80
19	3552	85,564.82	10.44	3562.44	1.00
20	4144	98,390.27	15.70	4159.70	0.73
21	3848	93,280.80	13.49	3861.49	0.61
22	3256	78,393.64	10.51	3266.51	0.61
23	2664	67,757.86	18.81	2682.81	0.62
24	2368	76,343.92	17.32	2370.00	1.00

2.5.7.2 Ten-Unit Test System

The parameter setting for the CSO is not a challenging task since the algorithm has a very few of them. The parameters and their settings are chosen according to Table 2.50. For each value of power demand over the 24-hour schedule, the individual power generated, power loss, total power, fuel cost, and computational time (CT) were recorded, as shown in Tables 2.52 and 2.53.

From the results, it is seen that for a minimum load demand of 1,036 MW, the fuel cost is $25,239.66 during the first hour, whereas it is $54,117.43 for the maximum load demand of 2,220 MW during the 12th hour. The total fuel cost to generate a power of 40,108 MW per day is $972,888.5. The computational time of the CSO algorithm for generating the schedule for 24 hours is 19.23 s, and the average time per hour is 0.8 s.

2.5.7.3 Fifteen-Unit Test System

The CSO algorithm based on the brood parasitic behavior of cuckoo birds was tested for optimal dispatch of the committed 15-unit test system. The parameters used by the CSO algorithm are shown in Table 2.50. The major advantage of CSO lies in the small number of parameters during the run of the algorithm. The economic dispatch of the committed schedules was optimized using the CSO algorithm, and the experimental results are shown in Tables 2.54 and 2.55. The CSO algorithm was run for 50 different trials to determine the optimal values of fuel cost, power loss, individual unit power, and computational time.

TABLE 2.38

Economic Dispatch Using IDE-OBL for the 20-Unit Test System

Hour	P_D (MW)	Distribution of Load among Units (MW)									
		P_1	P_2	P_3	P_4	P_5	P_6	P_7	P_8	P_9	P_{10}
1	2072	458.20	0	0	264.93	0	160	130	0	0	55
2	2220	438	0	0	218.94	243	160	0	0	0	0
3	2516	410.41	460	340	0	0	0	0	0	0	55
4	2812	371.74	460	0	300	0	160	0	0	20	55
5	2960	381.37	460	340	0	240.83	0	0	0	20	0
6	3256	427.93	460	340	0	172.43	160	0	0	0	55
7	3404	428.74	460	340	191.87	0	160	0	0	0	55
8	3552	467.50	460	340	245.32	243	0	0	0	0	0
9	3848	435.55	460	340	146.32	243	160	0	0	0	55
10	4144	468.08	460	340	300	243	160	130	0	0	0
11	4292	459.77	460	340	177.30	243	160	130	120	0	0
12	4440	459.65	460	340	300	243	160	130	0	80	55
13	4144	458.06	460	340	272.18	243	160	130	0	0	0
14	3848	469.20	460	340	106.34	243	160	0	0	0	55
15	3552	469.86	460	340	236.63	243	0	0	0	0	0
16	3108	367.91	460	340	0	243	0	130	0	0	0
17	2960	372.82	460	340	0	243	0	0	0	20	0
18	3256	356.26	460	340	0	190.93	160	0	0	0	55
19	3552	460.61	460	340	245.81	243	0	0	0	0	0
20	4144	462.71	460	340	300	243	160	130	0	0	0
21	3848	446.62	460	340	215.21	243	160	0	0	0	55
22	3256	346.69	460	340	0	176.76	160	0	0	0	55
23	2664	440.86	460	0	0	243	0	0	96.37	20	55
24	2368	470	460	0	0	0	0	0	120	80	55

It is seen from the results that the fuel cost for generating power during the 23rd hour (Rs. 3915.73) is very low compared to that for the remaining hours of the day. The fuel cost is maximum during the seventh hour (Rs. 4887.41), implying that the load requirement is very large during this period. The total power generated is 85,533.97 MW with a power loss of 56.98 MW, and the algorithm took 19.69 s to determine the optimal results over the 24 hours.

2.5.7.4 Twenty-Unit Test System

Solving the ELD using CSO is quite simple since the CSO requires very few parameters to be tuned. The parameters and their settings are given in Table 2.50. The search dimension n_d for the CSO depends upon the problem and for case study IV is set to 20. For each power demand, 50 independent trials with 300 iterations per trial have been performed. The results obtained by the CSO at the end of 300 iterations are presented in Tables 2.56 through 2.58.

CSO was able to dispatch power to the 20-unit system with a minimum fuel cost of $50,617.22 for a demand of 2,072 MW and of $108,538.75 for a demand of 4,440 MW. It can be observed from Table 2.58 that the mean fuel cost for the whole day is $81,700.68. In addition, the total power loss is 353.71 MW, and the total cost of dispatching power during the day is $1,960,816.38. The computational time is 17.72 s with the random walk generated by Levy flight mechanism in the CSO algorithm.

TABLE 2.39

Economic Dispatch Using IDE-OBL for the 20-Unit Test System

Hour	P_D (MW)	P_{11}	P_{12}	P_{13}	P_{14}	P_{15}	P_{16}	P_{17}	P_{18}	P_{19}	P_{20}
					Distribution of Load among Units (MW)						
1	2072	470	0	0	199.07	0	160	130	0	0	55
2	2220	470	0	0	300	243	160	0	0	0	0
3	2516	408.06	460	340	0	0	0	0	0	0	55
4	2812	470	460	0	300	0	160	0	0	20	55
5	2960	470	460	340	0	243	0	0	0	20	0
6	3256	399.60	460	340	0	243	160	0	0	0	55
7	3404	470	460	340	300	0	160	0	0	0	55
8	3552	470	460	340	300	243	0	0	0	0	0
9	3848	470	460	340	300	243	160	0	0	0	55
10	4144	470	460	340	262.22	243	160	130	0	0	0
11	4292	470	460	340	300	243	160	130	120	0	0
12	4440	470	460	340	300	243	160	130	0	80	55
13	4144	470	460	340	300	243	160	130	0	0	0
14	3848	470	460	340	300	243	160	0	0	0	55
15	3552	470	460	340	300	243	0	0	0	0	0
16	3108	470	460	340	0	185.42	0	130	0	0	0
17	2960	470	460	340	0	243	0	0	0	20	0
18	3256	446.27	460	340	0	243	160	0	0	0	55
19	3552	470	460	340	300	243	0	0	0	0	0
20	4144	470	460	340	261.20	243	160	130	0	0	0
21	3848	470	460	340	213.59	243	160	0	0	0	55
22	3256	470	460	340	0	243	160	0	0	0	55
23	2664	470	460	0	0	243	0	0	120	20	55
24	2368	470	460	0	0	0	0	0	120	80	55

2.6 Discussion

Based on the experimental results of bio-inspired algorithms for the four test systems, comparative analysis in terms of optimal solution, robustness, computational time, and algorithmic efficiency was performed, as discussed below:

2.6.1 Optimal Fuel Cost and Robustness

For all the test systems considered in this chapter, 20 independent trial runs were performed on each optimization technique. The minimum cost, mean cost, maximum cost, and their standard deviation for specific power demands were evaluated and recorded, as shown in Table 2.59. Since the fuel cost was evaluated based on the combination of UC schedules (GA) and economic dispatch of power, the heuristics are represented as GA-FRBFN, GA-EPSO, GA-DE-OBL, GA-IDE-OBL, GA-ABC, and GA-CSO. In the 15-unit test system, the cost in Indian Rupees is converted to dollars to maintain uniformity in comparison.

Considering the IEEE 30 bus (6-unit) system, the optimal fuel cost obtained by GA-IDE-OBL ($12,660.71) is the least, while the cost computed by GA-FRBFN ($12,967.9) is the highest. The costs computed by GA-EPSO ($12,918.56), GA-DE-OBL ($12,913.69),

TABLE 2.40

Economic Dispatch Using IDE-OBL for the 20-Unit Test System

Hour	P_D (MW)	FC ($/h)	P_L (MW)	P_T (MW)	CT (s)
1	2072	50,502.14	10.20	2082.20	0.70
2	2220	53,660.78	12.94	2232.94	0.51
3	2516	61,359.46	12.47	2528.47	0.62
4	2812	70,575.43	19.74	2831.74	0.53
5	2960	71,431.44	15.20	2975.20	0.69
6	3256	78,433.88	16.96	3272.96	0.51
7	3404	82,801.87	16.61	3420.61	0.62
8	3552	85,708.87	16.82	3568.82	0.62
9	3848	93,485.29	19.87	3867.87	0.62
10	4144	98,451.32	22.30	4166.30	0.61
11	4292	103,176.95	21.07	4313.07	0.61
12	4440	108,526.41	25.65	4465.65	0.61
13	4144	98,472.64	22.24	4166.24	0.53
14	3848	93,262.7	13.54	3861.54	0.56
15	3552	85,550.06	10.49	3562.49	0.59
16	3108	73,523.91	18.33	3126.33	0.61
17	2960	71,329.71	8.82	2968.82	0.58
18	3256	78,375	10.46	3266.46	0.64
19	3552	85,483.18	10.42	3562.42	1.05
20	4144	98,308.42	15.91	4159.91	0.75
21	3848	93,296.18	13.42	3861.42	0.62
22	3256	78,378.57	10.45	3266.45	0.70
23	2664	67,612.67	19.23	2683.23	0.73
24	2368	76,328.92	17.32	2370.00	0.51

TABLE 2.41

ABC Parameters for ELD

S. No.	Parameters	Notations Used	Value
1	Colony size	N_p	20
2	No. of food sources	$N_p/2$	10
3	Food source limit	Limit	100
4	No. of employed bees	N_e	10
5	No. of onlooker bees	N_o	10
6	Maximum no. of iterations	maxCycle	500

GA-ABC ($12,912.05), and GA-CSO ($12,919.66) for the six-unit system are nearly equal. Analysis of the 10-unit system in terms of fuel cost reveals that GA-IDE-OBL produced a minimum cost of $972,158.58, which is less than that of GA-FRBFN ($973,453.4), GA-EPSO ($972,282.3), GA-DE-OBL ($973,049.1), GA-ABC ($972,323.2), and GA-CSO ($972,888.5) algorithms.

In the 15-unit (Indian utility 75-bus) test system, the optimal fuel cost is produced by GA-ABC accounting to $2088.41, but GA-IDE-OBL has a low standard deviation of 0.11827, implying that GA-IDE-OBL is capable of producing optimal fuel cost with stable

TABLE 2.42

Simulation Results of IEEE 30 Bus System Using ABC

Hour	P_D (MW)	Distribution of Load among Units (MW)						FC ($/h)	P_L (MW)	P_T (MW)	CT (s)
		P_1	P_2	P_3	P_4	P_5	P_6				
1	166	127.57	0	16.81	10	0	12	440.14	0.36	166.38	3.04
2	196	146.52	0	17.97	10	10	12	536.18	0.47	196.49	3.01
3	229	177.82	0	19.9	10	10	12	647.41	0.69	229.72	3.03
4	267	178.53	49.45	19.96	10	10	0	739.8	0.91	267.94	3.01
5	283.4	190.07	51.95	20.67	10	11.77	0	797.93	1.03	284.46	3
6	272	182.05	50.21	20.18	10	10.53	0	757.37	0.95	272.97	3.04
7	246	161.98	45.86	18.94	10	10	0	667.65	0.76	246.78	3.03
8	213	135.99	40.23	17.34	10	10	0	559.69	0.54	213.56	3.03
9	192	127.34	38.35	16.8	10	0	0	492.67	0.47	192.49	3.06
10	161	114.53	31.86	15	0	0	0	380.84	0.39	161.39	2.95
11	147	115.35	32.04	0	0	0	0	354.62	0.38	147.39	3.12
12	160	126.09	34.36	0	0	0	0	392.61	0.46	160.45	3.26
13	170	134.36	36.15	0	0	0	0	422.57	0.52	170.51	3.04
14	185	146.77	38.84	0	0	0	0	468.7	0.61	185.61	3.04
15	208	165.81	42.97	0	0	0	0	542.23	0.78	208.78	3.04
16	232	170.47	43.98	18.4	0	0	0	600.27	0.84	232.85	2.79
17	246	172.05	44.32	18.51	0	0	12	646.54	0.87	246.88	3.07
18	241	168.11	43.47	18.26	0	0	12	630.02	0.84	241.84	2.81
19	236	164.17	42.61	18.02	0	0	12	613.66	0.8	236.8	3.35
20	225	155.5	40.73	17.49	0	0	12	578.19	0.72	225.72	3.03
21	204	138.96	37.15	16.47	0	0	12	512.52	0.58	204.58	3.06
22	182	121.64	33.4	15.41	0	0	12	446.59	0.45	182.45	2.85
23	161	104.62	29.72	15	0	0	12	386.42	0.34	161.34	2.92
24	131	89.75	26.5	15	0	0	0	297.43	0.25	131.25	2.85

convergence characteristics. While examining the 20-unit test system, use of GA-EPSO resulted in an optimal fuel cost of $1,957,971.9 for dispatching power over the 24-hour time horizon. But due to the heuristic nature of the GA-EPSO algorithm, unstable variation is observed between the minimum and maximum cost, with a standard deviation of 6.17. Of all the four systems tested, the minimum fuel cost computed by GA-FRBFN is higher than that by GA-EPSO, GA-DE-OBL, GA-IDE-OBL, GA-ABC, and GA-CSO algorithms. This is due to the complexity involved in training and testing of the GA-FRBFN algorithm.

The robustness was evaluated in terms of stability of the bio-inspired algorithms in obtaining the minimum fuel cost. Each algorithm was repeated for 20 trial runs to determine the fuel cost, and the standard deviation (SD) is calculated as shown in Table 2.62. With reference to the six-unit test system, the SD values are 3.191 for GA-FRBFN, 9.953 for GA-EPSO, 0.2247 for GA-DE-OBL, 0.1264 for GA-IDE-OBL, 0.3453 for GA-ABC, and 11.717 for GA-CSO. The SD value of GA-IDE-OBL is less than that for the other algorithms, indicating stability in the optimal solutions. Likewise, of the 10-, 15-, and 20-unit test systems, GA-IDE-OBL resulted in low SD, thus providing stable solutions. Though GA-EPSO, GA-DE-OBL, GA-ABC, and GA-CSO produced minimum fuel cost with marginal differences when compared to GA-IDE-OBL, stability was guaranteed in GA-IDE-OBL due to the smaller values of SD, thus proving its robustness. Based on the observations, it can be

TABLE 2.43

Optimal Dispatch Using ABC for the 10-Unit Test System

Hour	P_D (MW)	Distribution of Load among Units (MW)									
		P_1	P_2	P_3	P_4	P_5	P_6	P_7	P_8	P_9	P_{10}
1	1036	470	0	0	223.57	0	160	130	0	0	55
2	1110	470	0	0	240.29	243	160	0	0	0	0
3	1258	406.10	460	340	0	0	0	0	0	0	55
4	1406	470	460		246.10	0	160	0	0	20	55
5	1480	439.32	460	340	0	224.46	0	0	0	20	0
6	1628	411.56	460	340	0	205.66	160	0	0	0	55
7	1702	470	460	340	221.19	0	160	0	0	0	55
8	1776	470	460	340	267.79	242.41	0	0	0	0	0
9	1924	470	460	340	201.93	242.07	160	0	0	0	55
10	2072	470	460	340	274.56	243.00	160	130	0	0	0
11	2146	469.06	460	340	254.07	238.81	160	130	99.40	0	0
12	2220	469.78	460	340	298.12	243.00	160	130	0	70.48	55
13	2072	470	460	340	274.56	243.00	160	130	0	0	0
14	1924	470	460	340	200.28	242.10	160	0	0	0	55
15	1776	470	460	340	266.21	242.41	0	0	0	0	0
16	1554	417.15	460	340	0	211.41	0	130	0	0	0
17	1480	438.34	460	340	0	223.84	0	0	0	20	0
18	1628	410.56	460	340	0	205.04	160	0	0	0	55
19	1776	470	460	340	266.20	242.42	0	0	0	0	0
20	2072	470	460	340	272.98	243.00	160	130	0	0	0
21	1924	469.91	460	340	200.21	242.26	160	0	0	0	55
22	1628	410.56	460	340	0	205.04	160	0	0	0	55
23	1332	469.83	460	0	0	243.00	0	0	89.11	20	55
24	1184	470	460	0	0	0	0	0	120	80	55

concluded that, among the algorithms tested, GA-IDE-OBL is a suitable one for dispatching power economically to the generating units in the UC-ELD problem.

2.6.2 Computational Efficiency

Computational efficiency was evaluated based on the execution time taken by the algorithms. The sum of the time taken to commit the generating units using GA and the time taken to dispatch the power over the 24-hour schedule using bio-inspired algorithms are shown in Table 2.60. The algorithms are termed GA-FRBFN, GA-EPSO, GA-DE-OBL, GA-IDE-OBL, GA-ABC, and GA-CSO, since the sum of the computational time for UC and ELD problems is considered. In the experiments performed on the test systems, it is found that the time taken by GA-IDE-OBL, GA-DE-OBL, and GA-CSO are lower than with GA-FRBFN, GA-EPSO, and GA-ABC algorithms. For instance, in the six-unit (case study I) test system, the time taken by GA-ABC (72.43 s) to determine the optimal solution is much higher than that of GA-FRBFN (23.02 s), GA-EPSO (55.58 s), GA-DE-OBL (13.56 s), GA-IDE-OBL (13.2 s), and GA-CSO (15.87 s). For the 20-unit system (case study IV), the computational time of GA-ABC is approximately 82% higher than

TABLE 2.44

Optimal Dispatch Using ABC for the 10-Unit Test System

Hour	P_D (MW)	FC ($/h)	P_L (MW)	P_T (MW)	CT (s)
1	1,036	25,175.96	2.57	1,038.57	3.63
2	1,110	26,715.95	3.29	1,113.29	3.67
3	1,258	30,617.19	3.10	1,261.10	3.43
4	1,406	35,057.74	5.09	1,411.09	3.82
5	1,480	35,649.50	3.77	1,483.77	3.76
6	1,628	39,172.35	4.22	1,632.22	3.63
7	1,702	41,289.07	4.19	1,706.19	3.99
8	1,776	42,750.11	4.20	1,780.20	3.39
9	1,924	46,583.04	5.00	1,929.00	3.84
10	2,072	49,087.52	5.56	2,077.56	3.92
11	2,146	51,473.08	5.34	2,151.34	3.92
12	2,220	53,993.48	6.38	2,226.38	3.84
13	2,072	49,087.52	5.56	2,077.56	3.63
14	1,924	46,543.77	3.38	1,927.38	3.99
15	1,776	42,711.67	2.61	1,778.61	3.49
16	1,554	36,669.74	4.56	1,558.56	3.51
17	1,480	35,614.59	2.19	1,482.19	3.92
18	1,628	39,136.72	2.60	1,630.60	3.82
19	1,776	42,711.65	2.61	1,778.61	3.96
20	2,072	49,049.03	3.98	2,075.98	4.07
21	1,924	46,543.62	3.38	1,927.38	3.53
22	1,628	39,136.72	2.60	1,630.60	3.48
23	1,332	33,711.18	4.94	1,336.94	3.98
24	1,184	33,842.08	4.33	1,185.00	3.71

GA-CSO, GA-DE-OBL, and GA-IDE-OBL algorithms. In spite of the capacity to produce optimal solution in terms of fuel cost, GA-ABC takes a longer time to converge, which is a negative aspect of that algorithm. Hence, on the whole, it is observed that GA-IDE-OBL is computationally more efficient than the other heuristics.

2.6.3 Algorithmic Efficiency

Algorithmic efficiency is calculated in terms of the estimated time and calculated time of the algorithm using the BigO notation. The efficiency of GA for solving UC is combined with the efficiency of heuristics applied to solve ELD; hence the algorithms are termed GA-FRBFN, GA-EPSO, GA-DE-OBL, GA-IDE-OBL, GA-ABC, and GA-CSO. The algorithmic efficiency of the intelligent algorithms obtained for the four different case studies is presented in Table 2.61. The variation in algorithm efficiency is due to the increase or decrease in the number of loops in the programming code. For the 15-unit test system (case study III), it is observed that the algorithmic efficiency of GA-FRBFN is 90.35%, GA-EPSO is 89.47%, GA-DE-OBL is 91.18%, GA-IDE-OBL is 90.48%, GA-ABC is 92.19%, and GA-CSO is 97.08%. In all the test cases, significant difference in algorithmic efficiency was observed in GA-ABC and GA-CSO, while not much difference was observed on GA-EPSO, GA-DE-OBL, GA-IDE-OBL, and GA-CSO. On an average, for all the test systems, the algorithmic efficiency is 90.42% for GA-FRBFN, 89.36% for GA-EPSO, 90.78% for GA-DE-OBL, 90.47%

TABLE 2.45

Optimal Dispatch Using ABC for the 15-Unit Test System

Hour	P_D (MW)	Distribution of Load among Units (MW)									
		P_1	P_2	P_3	P_4	P_5	P_6	P_7	P_8	P_9	P_{10}
1	3352	516.96	100.00	40.00	40.00	23.58	0.00	1.00	20.00	192.90	30.00
2	3384	528.45	100.00	40.00	40.00	29.33	0.00	0.00	0.00	200.67	30.00
3	3437	538.59	105.44	40.00	40.00	34.43	0.00	1.00	0.00	207.35	30.00
4	3489	548.65	111.23	40.00	40.00	39.29	0.00	1.00	0.00	214.05	30.00
5	3659	578.38	128.12	54.37	41.53	54.32	0.00	1.00	0.00	233.90	30.00
6	3849	607.08	144.59	68.74	55.88	68.69	0.00	2.76	0.00	253.04	30.00
7	3898	613.91	148.37	72.11	59.25	72.09	0.00	5.65	0.00	257.81	30.53
8	3849	607.10	144.57	68.75	55.88	68.67	0.00	2.74	0.00	253.05	30.00
9	3764	594.50	137.57	62.42	49.58	62.58	0.00	1.00	0.00	244.86	30.00
10	3637	574.76	126.12	52.55	40.00	52.50	0.00	1.00	0.00	231.47	30.00
11	3437	539.90	104.69	40.00	40.00	34.38	0.00	1.00	0.00	208.36	30.00
12	3384	528.48	100.00	40.00	40.00	29.35	0.00	0.00	0.00	200.60	30.00
13	3357	522.08	120.18	40.00	40.00	22.64	0.00	0.00	0.00	193.39	30.00
14	3394	530.47	100.82	40.00	40.00	30.35	0.00	0.00	0.00	201.94	30.00
15	3616	571.21	124.10	50.83	40.00	50.76	0.00	0.00	0.00	229.00	30.00
16	3828	600.72	134.37	91.62	52.51	91.22	0.00	0.00	0.00	246.96	30.00
17	3828	604.35	143.02	67.37	54.51	67.31	0.00	0.00	0.00	251.21	30.00
18	3786	598.06	139.44	64.22	51.37	64.17	0.00	0.00	0.00	247.02	30.00
19	3659	578.53	128.28	54.45	41.60	54.39	0.00	0.00	0.00	233.98	30.00
20	3458	542.82	107.89	40.00	40.00	36.54	0.00	0.00	0.00	210.18	30.00
21	3394	530.52	100.85	40.00	40.00	30.37	0.00	0.00	0.00	201.96	30.00
22	3334	530.15	100.63	40.00	40.00	30.16	0.00	0.00	0.00	201.71	30.00
23	3329	529.07	100.02	40.00	40.00	29.63	0.00	0.00	0.00	200.98	30.00
24	3348	533.16	102.35	40.00	40.00	31.67	0.00	0.00	0.00	203.71	30.00

for GA-IDE-OBL, 91.87% for GA-ABC, and 97.17% for GA-CSO. Since the CSO algorithm has a simple pseudocode with fewer control parameters, it is observed that the efficiency of GA-CSO is the highest among the listed algorithms.

2.7 Advantages of CI Algorithms

The algorithms to solve the UC-ELD test systems with respect to optimal fuel cost, robustness, computational time and algorithmic efficiency are presented in Table 2.62.

For the UC-ELD problem, it can be concluded that, for all the test systems considered, GA-IDE-OBL is computationally more efficient than GA-FRBFN by 78.31%, GA-EPSO by 76.46%, GA-DE-OBL by 10%, GA-ABC by 83.57%, and GA-CSO by 18.73%. The merits and de-merits of the optimization techniques in terms of the computational time are as follows:

- FRBFN: In FRBFN, an FCM clustering approach is adapted to select the range of hidden-layer neurons in the RBF network. Large computational time is required for the learning process, thus making it less suitable for practical implementation of the ELD problem.

TABLE 2.46

Optimal Dispatch Using ABC for the 15-Unit Test System

Hour	P_D (MW)	Distribution of Load among Units (MW)					FC (Rs/h)	P_L (MW)	P_T (MW)	CT (s)
		P_{11}	P_{12}	P_{13}	P_{14}	P_{15}				
1	3352	40.00	1300.00	687.07	50.93	312.32	3879.64	2.77	3354.77	4.20
2	3384	40.00	1300.00	700.16	57.09	321.17	3929.81	2.87	3386.87	4.02
3	3437	40.00	1300.00	711.64	62.45	329.06	4018.32	2.95	3439.95	4.02
4	3489	40.00	1300.00	723.20	67.81	336.80	4105.89	3.03	3492.03	4.32
5	3659	40.00	1300.00	757.00	83.66	360.00	4397.64	3.28	3662.28	3.85
6	3849	50.83	1300.00	789.77	98.94	382.22	4732.64	3.55	3852.55	3.70
7	3898	54.31	1300.00	797.59	102.51	387.50	4820.41	3.62	3901.62	3.98
8	3849	50.84	1300.00	789.70	98.99	382.26	4732.64	3.55	3852.55	3.99
9	3764	44.64	1300.00	775.48	92.22	372.59	4581.70	3.43	3767.43	3.93
10	3637	40.00	1300.00	752.93	81.73	357.18	4359.43	3.25	3640.25	4.82
11	3437	40.00	1300.00	712.22	62.89	326.49	4018.33	2.93	3439.93	3.85
12	3384	40.00	1300.00	700.19	57.06	321.19	3929.81	2.87	3386.87	4.24
13	3357	40.00	1300.00	684.96	51.02	315.53	3886.00	2.82	3359.82	3.87
14	3394	40.00	1300.00	702.43	58.13	322.74	3946.43	2.89	3396.89	3.68
15	3616	40.00	1300.00	748.99	79.91	354.42	4323.03	3.22	3619.22	3.95
16	3828	46.80	1300.00	771.50	92.67	373.03	4697.46	3.42	3831.42	3.93
17	3828	49.47	1300.00	786.66	97.48	380.14	4695.20	3.53	3831.53	4.04
18	3786	46.33	1300.00	779.45	94.14	375.26	4620.59	3.47	3789.47	3.79
19	3659	40.00	1300.00	757.21	83.74	360.11	4397.60	3.29	3662.29	4.02
20	3458	40.00	1300.00	716.57	64.60	332.39	4053.49	2.99	3460.99	3.84
21	3394	40.00	1300.00	702.54	58.14	322.50	3946.43	2.89	3396.89	3.84
22	3334	40.00	1300.00	701.92	0.00	322.41	3851.74	2.99	3336.99	4.02
23	3329	40.00	1300.00	700.72	0.00	321.57	3843.44	2.98	3331.98	4.04
24	3348	40.00	1300.00	705.37	0.00	324.75	3875.05	3.01	3351.01	3.71

- EPSO: In EPSO, a constriction factor is introduced in the velocity update formula to ensure faster convergence. The computational time is high since the fixed values of algorithm parameters cause unnecessary fluctuation of particles, thus prolonging the convergence process.

- DE-OBL and IDE-OBL: In DE-OBL, the concept of OBL is applied to the initialization phase only, while in the IDE-OBL it is applied to both initialization and generation phases. The computational time is comparatively low in IDE-OBL, since opposite presumptions are applied in the initial phase as well as during the algorithmic run.

- ABC: ABC is applied to solve ELD problem since the algorithm uses an effective search process even under complex situations, and the risk of premature convergence is low. The population of solutions increases the computational time due to several iterations involved in the optimization process.

- CSO: The generic movement model provided by Levy flights in CSO is an effective approach to solve optimization problems. Because of the random walk in the Levy flight mechanism, the algorithm converges much faster, thus reducing the computational time.

TABLE 2.47

Optimal Dispatch Using ABC for the 20-Unit Test System

Hour	P_D (MW)	Distribution of Load among Units (MW)									
		P_1	P_2	P_3	P_4	P_5	P_6	P_7	P_8	P_9	P_{10}
1	2072	469.24	0.00	0.00	286.18	0.00	160.00	130.00	0.00	0.00	55.00
2	2220	469.83	0.00	0.00	231.06	243.00	160.00	0.00	0.00	0.00	0.00
3	2516	409.23	460.00	340.00	0.00	0.00	0.00	0.00	0.00	0.00	55.00
4	2812	467.68	460.00	0.00	223.36	0.00	160.00	0.00	0.00	33.00	55.00
5	2960	442.39	460.00	340.00	0.00	225.20	0.00	0.00	0.00	20.00	0.00
6	3256	416.39	460.00	340.00	0.00	205.09	160.00	0.00	0.00	0.00	55.00
7	3404	469.25	460.00	340.00	241.28	0.00	160.00	0.00	0.00	0.00	55.00
8	3552	469.43	460.00	340.00	268.44	241.46	0.00	0.00	0.00	0.00	0.00
9	3848	468.50	460.00	340.00	176.98	240.31	160.00	0.00	0.00	0.00	55.00
10	4144	469.89	460.00	340.00	299.95	243.00	160.00	130.00	0.00	0.00	0.00
11	4292	468.69	460.00	340.00	278.89	234.14	160.00	130.00	92.74	0.00	0.00
12	4440	470.00	460.00	340.00	298.97	243.00	160.00	130.00	0.00	71.68	55.00
13	4144	463.34	460.00	340.00	271.16	241.99	160.00	130.00	0.00	0.00	0.00
14	3848	468.01	460.00	340.00	166.05	237.98	160.00	0.00	0.00	0.00	55.00
15	3552	469.92	460.00	340.00	286.86	241.84	0.00	0.00	0.00	0.00	0.00
16	3108	421.08	460.00	340.00	0.00	212.08	0.00	130.00	0.00	0.00	0.00
17	2960	440.50	460.00	340.00	0.00	223.89	0.00	0.00	0.00	20.00	0.00
18	3256	414.46	460.00	340.00	0.00	203.75	160.00	0.00	0.00	0.00	55.00
19	3552	469.92	460.00	340.00	291.66	241.69	0.00	0.00	0.00	0.00	0.00
20	4144	469.41	460.00	340.00	277.35	243.00	160.00	130.00	0.00	0.00	0.00
21	3848	469.09	460.00	340.00	202.67	240.07	160.00	0.00	0.00	0.00	55.00
22	3256	414.46	460.00	340.00	0.00	203.75	160.00	0.00	0.00	0.00	55.00
23	2664	468.10	458.94	0.00	0.00	243.00	0.00	0.00	51.07	46.65	55.00
24	2368	470.00	460.00	0.00	0.00	0.00	0.00	0.00	120.00	80.00	55.00

TABLE 2.48

Optimal Dispatch Using ABC for the 20-Unit Test System

Hour	P_D (MW)	Distribution of Load among Units (MW)									
		P_{11}	P_{12}	P_{13}	P_{14}	P_{15}	P_{16}	P_{17}	P_{18}	P_{19}	P_{20}
1	2072	469.17	0.00	0.00	167.72	0.00	160.00	130.00	0.00	0.00	55.00
2	2220	469.85	0.00	0.00	256.47	243.00	160.00	0.00	0.00	0.00	0.00
3	2516	409.23	460.00	340.00	0.00	0.00	0.00	0.00	0.00	0.00	55.00
4	2812	467.48	460.00	0.00	270.87	0.00	160.00	0.00	0.00	20.00	55.00
5	2960	442.39	460.00	340.00	0.00	225.20	0.00	0.00	0.00	20.00	0.00
6	3256	416.39	460.00	340.00	0.00	205.09	160.00	0.00	0.00	0.00	55.00
7	3404	469.20	460.00	340.00	211.08	0.00	160.00	0.00	0.00	0.00	55.00
8	3552	469.27	460.00	340.00	278.92	241.29	0.00	0.00	0.00	0.00	0.00
9	3848	467.12	460.00	340.00	245.54	239.55	160.00	0.00	0.00	0.00	55.00
10	4144	469.47	460.00	340.00	261.00	243.00	160.00	130.00	0.00	0.00	0.00
11	4292	464.16	460.00	340.00	257.07	242.65	160.00	130.00	95.13	0.00	0.00
12	4440	469.98	460.00	340.00	300.00	243.00	160.00	130.00	0.00	79.03	55.00
13	4144	469.49	460.00	340.00	298.70	241.57	160.00	130.00	0.00	0.00	0.00
14	3848	465.31	460.00	340.00	257.32	236.77	160.00	0.00	0.00	0.00	55.00
15	3552	469.96	460.00	340.00	252.05	241.84	0.00	0.00	0.00	0.00	0.00
16	3108	421.08	460.00	340.00	0.00	212.08	0.00	130.00	0.00	0.00	0.00
17	2960	440.50	460.00	340.00	0.00	223.90	0.00	0.00	0.00	20.00	0.00
18	3256	414.46	460.00	340.00	0.00	203.75	160.00	0.00	0.00	0.00	55.00
19	3552	469.80	460.00	340.00	247.52	241.88	0.00	0.00	0.00	0.00	0.00
20	4144	469.30	460.00	340.00	277.89	243.00	160.00	130.00	0.00	0.00	0.00
21	3848	467.37	460.00	340.00	212.92	239.37	160.00	0.00	0.00	0.00	55.00
22	3256	414.46	460.00	340.00	0.00	203.75	160.00	0.00	0.00	0.00	55.00
23	2664	468.66	459.99	0.00	0.00	243.00	0.00	0.00	80.40	54.05	55.00
24	2368	470.00	460.00	0.00	0.00	0.00	0.00	0.00	120.00	80.00	55.00

TABLE 2.49

Optimal Dispatch Using ABC for the 20-Unit Test System

Hour	P_D (MW)	FC ($/h)	P_L (MW)	P_T (MW)	CT (s)
1	2072	50,485.36	10.31	2082.31	3.73
2	2220	53,593.41	13.20	2233.20	3.90
3	2516	61,371.82	12.47	2528.47	3.74
4	2812	70,388.68	20.38	2832.38	4.32
5	2960	71,466.65	15.18	2975.18	4.29
6	3256	78,531.71	16.96	3272.96	3.99
7	3404	82,785.78	16.81	3420.81	4.35
8	3552	85,712.18	16.81	3568.81	3.87
9	3848	93,428.51	19.99	3867.99	4.52
10	4144	98,448.61	22.31	4166.31	4.32
11	4292	103,223.29	21.47	4313.47	4.37
12	4440	108,411.39	25.66	465.66	4.57
13	4144	98,467.60	22.25	466.25	4.02
14	3848	93,287.59	13.44	3861.44	4.35
15	3552	85,553.89	10.47	3562.47	3.90
16	3108	73,541.69	18.33	3126.33	3.88
17	2960	71,326.25	8.79	2968.79	3.96
18	3256	78,388.50	10.44	3266.44	4.01
19	3552	85,554.73	10.47	3562.47	4.20
20	4144	98,295.05	15.95	4159.95	4.41
21	3848	93,268.60	13.49	3861.49	4.41
22	3256	78,388.50	10.44	3266.44	4.70
23	2664	67,873.76	19.85	2683.85	4.71
24	2368	76,343.92	17.32	2370.00	4.10

TABLE 2.50

Parameter Settings for Cuckoo Search-Based ELD

S. No.	Parameters	Notations Used	Values
1	No. of nests	n	15
2	Tolerance	T	1.0×10^{-5}
3	Discovery rate	D_r	0.25
4	Search dimension	N_d	Depends on number of generators
5	Upper and lower bounds of the search domain	N_{dL} and N_{dU}	[−1, 1]
6	Levy exponent	λ	[1, 3]
7	Levy step size	α	0.01
8	No. of iterations	N_iter	300

TABLE 2.51

Computational Results of IEEE 30 Bus System Using CSO

Hour	P_D (MW)	Distribution of Load among Units (MW)						FC ($/h)	P_L (MW)	P_T (MW)	CT (s)
		P_1	P_2	P_3	P_4	P_5	P_6				
1	166	123.72	0	15.97	11.48	0	15.19	440.76	0.34	166.36	0.75
2	196	141.91	0	20.21	10.09	10.72	13.54	536.9	0.45	196.47	0.61
3	229	176.01	0	19.27	10.47	10.27	13.69	648.76	0.68	229.71	0.62
4	267	166.89	53	24.5	12.01	11.45	0	740.31	0.84	267.85	0.69
5	283.4	176.52	34.97	37.09	17.59	18.11	0	799.07	0.84	284.28	0.66
6	272	180.85	46.3	21.88	10.12	13.8	0	758.18	0.92	272.95	0.69
7	246	156.8	48.21	18.6	11.98	11.16	0	668.12	0.72	246.75	0.58
8	213	138.78	32.51	18.73	12.13	11.4	0	560.29	0.52	213.55	0.67
9	192	125.26	36.66	19.82	10.73	0	0	493.26	0.46	192.47	0.61
10	161	114.2	31.75	15.44	0	0	0	380.84	0.39	161.39	0.69
11	147	119.95	27.45	0	0	0	0	354.62	0.4	147.4	0.69
12	160	126.03	34.43	0	0	0	0	392.61	0.46	160.46	0.64
13	170	134.29	36.22	0	0	0	0	422.57	0.52	170.51	0.69
14	185	147.23	38.38	0	0	0	0	468.7	0.62	185.61	0.62
15	208	165.81	42.97	0	0	0	0	542.23	0.78	208.78	0.69
16	232	168.35	44.29	20.19	0	0	0	600.27	0.83	232.83	0.81
17	246	161.17	41.2	22.45	0	0	21.97	646.68	0.79	246.79	0.67
18	241	155.99	53.25	19.45	0	0	13.1	630.05	0.79	241.79	0.69
19	236	157.76	48.5	18.41	0	0	12.11	613.8	0.78	236.78	0.64
20	225	149.3	38.32	24.7	0	0	13.35	578.24	0.68	225.67	0.67
21	204	146.06	28.03	17.96	0	0	12.54	512.65	0.6	204.59	0.66
22	182	121.38	33.78	15.24	0	0	12.06	446.64	0.45	182.46	0.61
23	161	102.34	30.75	16.16	0	0	12.09	386.67	0.33	161.34	0.58
24	131	77.24	38.12	15.87	0	0	0	297.44	0.23	131.23	0.64

TABLE 2.52

Optimal Dispatch Using CSO for the 10-Unit Test System

Hour	P_D (MW)	Distribution of Load among Units (MW)									
		P_1	P_2	P_3	P_4	P_5	P_6	P_7	P_8	P_9	P_{10}
1	1036	413.05	0.00	0.00	280.27	0.00	160.00	130.00	0.00	0.00	55.00
2	1110	410.07	0.00	0.00	300.00	243.00	160.00	0.00	0.00	0.00	0.00
3	1258	470.00	460.00	243.71	0.00	0.00	0.00	0.00	0.00	0.00	55.00
4	1406	470.00	147.98	0.00	232.26	0.00	158.93	0.00	0.00	55.07	55.00
5	1480	401.40	460.00	340.00	0.00	243.00	0.00	0.00	0.00	39.35	0.00
6	1628	470.00	460.00	340.00	0.00	89.80	86.29	0.00	0.00	0.00	55.00
7	1702	470.00	460.00	340.00	300.00	0.00	70.34	0.00	0.00	0.00	55.00
8	1776	470.00	460.00	340.00	230.22	243.00	0.00	0.00	0.00	0.00	0.00
9	1924	470.00	268.32	340.00	253.65	243.00	99.29	0.00	0.00	0.00	55.00
10	2072	470.00	328.52	340.00	172.73	188.86	118.19	85.49	0.00	0.00	0.00
11	2146	470.00	460.00	168.47	253.54	134.02	93.45	59.59	71.61	0.00	0.00
12	2220	458.38	460.00	340.00	300.00	243.00	160.00	130.00	0.00	80.00	55.00
13	2072	464.38	460.00	340.00	283.66	240.82	160.00	128.68	0.00	0.00	0.00
14	1924	369.65	460.00	340.00	300.00	243.00	159.57	0.00	0.00	0.00	55.00
15	1776	464.28	460.00	340.00	271.32	243.00	0.00	0.00	0.00	0.00	0.00
16	1554	385.58	460.00	340.00	0.00	243.00	0.00	130.00	0.00	0.00	0.00
17	1480	470.00	275.77	340.00	0.00	155.29	0.00	0.00	0.00	24.60	0.00
18	1628	410.72	460.00	340.00	0.00	204.88	160.00	0.00	0.00	0.00	55.00
19	1776	470.00	322.58	227.10	267.22	171.79	0.00	0.00	0.00	0.00	0.00
20	2072	443.85	460.00	340.00	299.14	242.91	160.00	130.00	0.00	0.00	0.00
21	1924	470.00	460.00	210.28	143.18	224.02	160.00	0.00	0.00	0.00	55.00
22	1628	437.03	460.00	340.00	0.00	178.57	160.00	0.00	0.00	0.00	55.00
23	1332	470.00	421.39	0.00	0.00	218.63	0.00	0.00	112.09	32.63	55.00
24	1184	470.00	460.00	0.00	0.00	0.00	0.00	0.00	120.00	80.00	55.00

TABLE 2.53

Optimal Dispatch Using CSO for the 10-Unit Test System

Hour	P_D (MW)	FC ($/h)	P_L (MW)	P_T (MW)	CT (s)
1	1036	25,239.66	2.32	1,038.32	0.66
2	1110	26,718.75	3.07	1,113.07	0.73
3	1258	30,617.19	3.43	1,228.71	0.67
4	1406	35,081.06	3.15	1,119.24	0.62
5	1480	35,654.55	3.75	1,483.75	0.73
6	1628	39,172.35	3.80	1,501.09	0.69
7	1702	41,294.21	3.97	1,695.34	0.89
8	1776	42,784.72	4.13	1,743.22	0.67
9	1924	46,594.21	3.52	1,729.26	0.69
10	2072	49,087.91	3.63	1,703.79	0.64
11	2146	51,516.20	4.39	1,710.69	0.76
12	2220	54,117.43	6.38	2,226.38	0.89
13	2072	49,105.50	5.53	2,077.53	0.83
14	1924	46,645.15	3.21	1,927.21	0.97
15	1776	42,722.90	2.60	1,778.60	0.90
16	1554	36,669.74	4.58	1,558.58	0.80
17	1480	35,615.36	1.94	1,265.66	0.98
18	1628	39,136.72	2.60	1,630.60	0.90
19	1776	42,713.89	2.22	1,458.70	0.76
20	2072	49,099.79	3.90	2,075.90	0.67
21	1924	46,599.10	3.13	1,722.48	1.03
22	1628	39,136.73	2.60	1,630.60	0.92
23	1332	33,723.31	4.46	1,309.75	0.92
24	1184	33,842.08	4.33	1,185.00	0.90

TABLE 2.54

Optimal Dispatch Using CSO for the 15-Unit Test System

Hour	P_D (MW)	Distribution of Load among Units (MW)									
		P_1	P_2	P_3	P_4	P_5	P_6	P_7	P_8	P_9	P_{10}
1	3352	1005.49	138.22	52.23	56.14	103.55	0.00	37.56	28.62	202.79	62.89
2	3384	661.56	113.51	127.00	49.20	64.77	0.00	0.00	0.00	105.14	36.02
3	3437	446.01	115.46	76.76	75.54	62.52	0.00	9.09	0.00	256.55	33.52
4	3489	478.81	131.52	49.61	94.80	68.06	0.00	18.69	0.00	161.36	95.57
5	3659	421.60	118.39	41.78	60.69	9.11	0.00	100.00	0.00	570.00	196.58
6	3849	618.16	300.00	43.99	170.00	53.19	0.00	39.82	0.00	251.48	42.58
7	3898	473.95	181.31	62.81	80.04	157.12	0.00	11.97	0.00	410.81	134.83
8	3849	534.41	139.17	75.13	82.90	70.47	0.00	44.83	0.00	184.30	203.51
9	3764	611.92	220.11	117.53	47.08	109.66	0.00	77.17	0.00	219.28	141.77
10	3637	506.98	129.98	147.97	41.11	135.50	0.00	10.86	0.00	116.15	80.75
11	3437	100.00	165.52	200.00	124.29	163.63	0.00	43.74	0.00	401.50	69.20
12	3384	936.78	180.38	106.83	40.12	4.43	0.00	0.00	0.00	71.53	101.03
13	3357	204.44	133.10	57.35	94.88	123.54	0.00	0.00	0.00	169.68	133.82
14	3394	548.40	104.46	125.16	53.02	9.87	0.00	0.00	0.00	107.44	35.48
15	3616	392.45	240.54	108.24	115.04	8.13	0.00	0.00	0.00	257.35	69.03
16	3828	1223.47	208.03	83.28	83.63	46.96	0.00	0.00	0.00	332.92	62.74
17	3828	874.38	152.02	99.93	115.65	77.29	0.00	0.00	0.00	90.81	108.04
18	3786	566.20	122.74	99.24	59.16	191.20	0.00	0.00	0.00	165.62	63.96
19	3659	242.99	247.78	200.00	54.26	62.73	0.00	0.00	0.00	155.31	44.90
20	3458	750.14	112.48	99.44	113.18	70.44	0.00	0.00	0.00	234.19	108.12
21	3394	607.82	115.41	60.84	46.24	60.02	0.00	0.00	0.00	332.00	30.57
22	3334	490.70	114.61	165.12	107.93	47.28	0.00	0.00	0.00	139.66	134.01
23	3329	407.87	101.61	67.24	150.32	45.87	0.00	0.00	0.00	196.99	62.34
24	3348	531.75	144.97	97.22	64.38	53.21	0.00	0.00	0.00	211.01	70.54

TABLE 2.55

Optimal Dispatch Using CSO for the 15-Unit Test System

Hour	P_D (MW)	Distribution of Load among Units (MW)					FC (Rs/h)	P_L (MW)	P_T (MW)	CT (s)
		P_{11}	P_{12}	P_{13}	P_{14}	P_{15}				
1	3352	66.55	820.18	454.30	103.45	221.39	3945.94	1.36	3353.36	0.62
2	3384	81.02	1300.00	600.09	55.20	192.35	3983.18	1.87	3385.87	0.87
3	3437	64.47	1300.00	900.00	26.96	71.88	4081.69	1.78	3438.78	0.70
4	3489	162.02	1229.28	775.54	37.84	187.93	4160.72	2.03	3491.03	0.70
5	3659	158.46	1300.00	140.26	91.93	454.00	4494.44	3.80	3662.80	0.72
6	3849	163.60	1065.09	890.12	39.56	173.48	4779.53	2.06	3851.06	2.48
7	3898	139.52	1297.83	667.11	103.82	178.71	4887.41	1.83	3899.83	0.64
8	3849	105.15	1297.93	619.81	141.21	353.16	4775.02	2.97	3851.97	0.64
9	3764	171.12	1300.00	478.62	116.24	155.02	4684.95	1.52	3765.52	0.66
10	3637	56.20	1300.00	749.44	71.65	293.07	4415.34	2.68	3639.68	0.94
11	3437	106.74	1300.00	648.45	31.28	91.09	4149.06	1.45	3445.44	0.81
12	3384	131.10	558.33	765.81	39.47	452.04	3996.89	3.86	3387.86	0.66
13	3357	104.31	1300.00	870.49	90.20	76.83	3966.10	1.64	3358.64	0.66
14	3394	83.01	1269.45	748.25	103.05	208.47	3998.45	2.08	3396.08	0.95
15	3616	77.31	1300.00	705.06	45.18	300.36	4400.67	2.71	3618.71	0.72
16	3828	194.35	1300.00	55.93	121.99	116.04	4729.52	1.35	3829.35	0.69
17	3828	61.57	1300.00	411.70	107.50	432.82	4756.16	3.71	3831.71	0.64
18	3786	63.09	1211.84	861.84	55.01	329.20	4695.02	3.10	3789.10	0.66
19	3659	121.03	1205.10	900.00	95.12	332.95	4456.60	3.19	3662.19	0.67
20	3458	83.09	777.72	828.30	18.33	264.73	4169.39	2.16	3460.16	0.84
21	3394	57.78	1028.38	722.51	21.47	313.52	4018.45	2.55	3396.55	0.75
22	3334	42.97	1262.24	590.65	0.00	240.88	3956.17	2.04	3336.04	1.03
23	3329	149.82	1135.14	707.73	0.00	306.79	3915.73	2.72	3331.72	0.87
24	3348	62.85	1246.67	562.87	0.00	305.07	3925.78	2.53	3350.53	0.78

TABLE 2.56

Optimal Dispatch Using CSO for the 20-Unit Test System

Hour	P_D (MW)	Distribution of Load among Units (MW)									
		P_1	P_2	P_3	P_4	P_5	P_6	P_7	P_8	P_9	P_{10}
1	2072	385.55	0.00	0.00	300.00	0.00	160.00	129.97	0.00	0.00	55.00
2	2220	470.00	0.00	0.00	300.00	233.90	151.94	0.00	0.00	0.00	0.00
3	2516	470.00	460.00	340.00	0.00	0.00	0.00	0.00	0.00	0.00	55.00
4	2812	395.13	457.74	0.00	273.55	0.00	155.44	0.00	0.00	47.78	55.00
5	2960	470.00	197.68	124.97	0.00	96.21	0.00	0.00	0.00	31.84	0.00
6	3256	287.60	460.00	340.00	0.00	243.00	159.51	0.00	0.00	0.00	55.00
7	3404	470.00	228.64	145.88	268.34	0.00	159.97	0.00	0.00	0.00	55.00
8	3552	470.00	427.52	339.44	292.09	243.00	0.00	0.00	0.00	0.00	0.00
9	3848	470.00	459.85	337.78	283.65	124.85	154.02	0.00	0.00	0.00	55.00
10	4144	431.56	460.00	340.00	300.00	243.00	160.00	130.00	0.00	0.00	0.00
11	4292	453.03	459.93	340.00	298.52	128.02	160.00	130.00	120.00	0.00	0.00
12	4440	470.00	460.00	145.48	286.18	106.30	122.48	130.00	0.00	80.00	55.00
13	4144	430.08	460.00	340.00	300.00	243.00	160.00	130.00	0.00	0.00	0.00
14	3848	388.68	460.00	340.00	300.00	235.07	160.00	0.00	0.00	0.00	55.00
15	3552	406.17	460.00	340.00	300.00	242.92	0.00	0.00	0.00	0.00	0.00
16	3108	352.54	460.00	340.00	0.00	200.79	0.00	130.00	0.00	0.00	0.00
17	2960	343.28	460.00	340.00	0.00	193.91	0.00	0.00	0.00	80.00	0.00
18	3256	280.59	460.00	340.00	0.00	243.00	160.00	0.00	0.00	0.00	55.00
19	3552	433.08	460.00	340.00	300.00	243.00	0.00	0.00	0.00	0.00	0.00
20	4144	465.54	460.00	340.00	300.00	243.00	160.00	130.00	0.00	0.00	0.00
21	3848	470.00	445.40	338.59	256.70	175.79	159.97	0.00	0.00	0.00	55.00
22	3256	470.00	459.52	338.00	0.00	243.00	105.28	0.00	0.00	0.00	55.00
23	2664	469.94	460.00	0.00	0.00	120.05	0.00	0.00	79.83	80.00	55.00
24	2368	470.00	460.00	0.00	0.00	0.00	0.00	0.00	120.00	80.00	55.00

TABLE 2.57

Optimal Dispatch Using CSO for the 20-Unit Test System

Hour	P_D (MW)	Distribution of Load among Units (MW)									
		P_{11}	P_{12}	P_{13}	P_{14}	P_{15}	P_{16}	P_{17}	P_{18}	P_{19}	P_{20}
1	2072	444.42	0.00	0.00	271.04	0.00	158.39	122.04	0.00	0.00	55.00
2	2220	457.01	0.00	0.00	274.45	130.07	159.76	0.00	0.00	0.00	0.00
3	2516	385.19	186.85	340.00	0.00	0.00	0.00	0.00	0.00	0.00	55.00
4	2812	451.33	460.00	0.00	300.00	0.00	160.00	0.00	0.00	20.66	55.00
5	2960	436.72	197.80	340.00	0.00	176.94	0.00	0.00	0.00	23.64	0.00
6	3256	470.00	460.00	340.00	0.00	243.00	160.00	0.00	0.00	0.00	55.00
7	3404	248.91	220.61	329.73	262.94	0.00	85.08	0.00	0.00	0.00	55.00
8	3552	470.00	459.81	340.00	283.10	242.53	0.00	0.00	0.00	0.00	0.00
9	3848	451.43	441.85	324.50	286.42	205.69	159.96	0.00	0.00	0.00	55.00
10	4144	470.00	460.00	340.00	300.00	242.43	160.00	129.09	0.00	0.00	0.00
11	4292	470.00	460.00	340.00	300.00	243.00	160.00	130.00	120.00	0.00	0.00
12	4440	438.79	460.00	152.71	276.58	217.22	140.31	130.00	0.00	74.06	55.00
13	4144	470.00	460.00	340.00	300.00	243.00	160.00	130.00	0.00	0.00	0.00
14	3848	408.93	460.00	296.77	300.00	241.67	160.00	0.00	0.00	0.00	55.00
15	3552	470.00	460.00	340.00	300.00	243.00	0.00	0.00	0.00	0.00	0.00
16	3108	470.00	460.00	340.00	0.00	243.00	0.00	130.00	0.00	0.00	0.00
17	2960	470.00	460.00	340.00	0.00	243.00	0.00	0.00	0.00	38.29	0.00
18	3256	470.00	460.00	340.00	0.00	243.00	160.00	0.00	0.00	0.00	55.00
19	3552	469.61	460.00	340.00	300.00	216.38	0.00	0.00	0.00	0.00	0.00
20	4144	470.00	418.47	340.00	299.96	243.00	160.00	130.00	0.00	0.00	0.00
21	3848	470.00	417.54	334.70	269.36	238.71	159.23	0.00	0.00	0.00	55.00
22	3256	470.00	288.57	198.99	0.00	217.07	160.00	0.00	0.00	0.00	55.00
23	2664	470.00	460.00	0.00	0.00	243.00	0.00	0.00	110.05	80.00	55.00
24	2368	470.00	460.00	0.00	0.00	0.00	0.00	0.00	120.00	80.00	55.00

TABLE 2.58

Optimal Dispatch Using CSO for the 20-Unit Test System

Hour	P_D (MW)	FC ($/h)	P_L (MW)	P_T (MW)	CT (s)
1	2072	50,617.22	9.41	2081.41	0.67
2	2220	53,761.74	12.68	2177.13	0.64
3	2516	61,371.82	8.98	2292.04	0.86
4	2812	70,635.87	19.63	2831.63	0.67
5	2960	71,925.94	7.82	2095.80	0.84
6	3256	78,536.43	17.10	3273.10	0.67
7	3404	82,916.77	8.23	2530.10	0.64
8	3552	85,787.34	16.48	3567.48	0.84
9	3848	93,610.44	18.85	3810.00	0.89
10	4144	98,489.26	22.08	4166.08	0.92
11	4292	103,473.28	20.50	4312.50	0.75
12	4440	108,538.75	22.97	3800.11	0.69
13	4144	98,513.54	22.08	4166.08	0.66
14	3848	93,423.72	13.13	3861.13	0.72
15	3552	85,552.66	10.09	3562.09	0.75
16	3108	73,545.25	18.34	3126.34	0.84
17	2960	71,340.12	8.49	2968.49	0.72
18	3256	78,401.68	10.59	3266.59	0.73
19	3552	85,670.34	10.07	3562.07	0.67
20	4144	98,383.30	15.98	4159.98	0.69
21	3848	93,401.40	13.28	3845.98	0.87
22	3256	78,389.65	10.73	3060.43	0.62
23	2664	68,185.93	18.87	2682.87	0.67
24	2368	76,343.92	17.32	2370.00	0.70

TABLE 2.59

Comparative Analysis of Fuel Cost

Test System	Intelligent Technique	Minimum Fuel Cost ($)	Mean Fuel Cost ($)	Maximum Fuel Cost ($)	Standard Deviation
IEEE 30 bus system	GA-FRBFN	12,967.9	12,970.95	12,974.28	3.191024
	GA-EPSO	12,918.56	12,927.78	12,938.45	9.953805
	GA-DE-OBL	12,913.69	12,913.99	12,914.13	0.224796
	GA-IDE-OBL	**12,660.71**	**12,660.92**	**12,661.09**	**0.1264282**
	GA-ABC	12,912.05	12,912.24	12,912.72	0.345302
	GA-CSO	12,919.66	12,935.75	12,942.46	11.71717
Ten-unit system	GA-FRBFN	973,453.4	973,459.58	973,468.14	7.401955
	GA-EPSO	972,282.3	972,286.43	972,291.67	4.695945
	GA-DE-OBL	973,049.1	973,049.54	973,049.91	0.405504
	GA-IDE-OBL	**972,158.58**	**972,158.77**	**972,158.97**	**0.134646**
	GA-ABC	972,323.2	972,323.49	972,323.83	0.315331
	GA-CSO	972,888.5	972,891.43	972,895.37	3.447352
Indian utility 75-bus system	GA-FRBFN	2,149.20	2,151.58	2,154.72	2.76728
	GA-EPSO	**2,094.71**	**2,096.72**	**2,099.54**	**2.428643**
	GA-DE-OBL	2,129.18	2,129.28	2,129.49	0.159954
	GA-IDE-OBL	**2,119.68**	**2,119.89**	**2,120.06**	**0.11827**
	GA-ABC	**2,088.41**	**2,088.72**	**2,088.79**	**0.204512**
	GA-CSO	2,123.33	2,129.87	2,130.73	4.04991
Twenty-unit system	GA-FRBFN	1,961,066	1,961,072.47	1,961,080.44	7.232973
	GA-EPSO	**1,957,971.9**	**1,957,977.58**	**1,957,984.25**	**6.171891**
	GA-DE-OBL	1,960,091.2	1,960,091.57	1,960,091.85	0.326037
	GA-IDE-OBL	**1,958,035.8**	**1,958,036.09**	**1,958,036.49**	**0.216787**
	GA-ABC	1,958,137.5	1,958,137.81	1,958,138.16	0.330202
	GA-CSO	1,960,816.4	1,960,817.09	1,960,821.79	2.940017

Note: Bold signifies optimal fuel cost.

TABLE 2.60

Comparison of Computational Time

Case Study	Computational Time (in Seconds) of Bio-Inspired Algorithms					
	GA-FRBFN	GA-EPSO	GA-DE-OBL	GA-IDE-OBL	GA-ABC	GA-CSO
I	23.02	55.58	13.56	13.20	72.43	15.87
II	88.53	59.14	17.44	14.40	89.93	19.23
III	70.38	55.099	16.93	16.18	95.68	19.69
IV	89.72	80.57	17.55	15.15	100.65	17.72

TABLE 2.61

Comparison of Algorithmic Efficiency

Case Study	Algorithmic Efficiency (in %) of Bio-Inspired Algorithms					
	GA-FRBFN	GA-EPSO	GA-DE-OBL	GA-IDE-OBL	GA-ABC	GA-CSO
I	90.41	89.92	91.07	90.97	91.79	97.49
II	90.87	89.07	90.44	90.13	92.05	97.36
III	90.35	89.47	91.18	90.48	92.19	97.08
IV	90.08	88.98	90.42	90.27	91.45	96.75

TABLE 2.62

Observations of Algorithms for UC-ELD Problem

Advantages	Performance of Algorithms
Optimal fuel cost	GA-IDE-OBL and GA-DE-OBL produced minimum fuel cost for the 6- and 10-unit test systems. GA-ABC resulted in minimum fuel cost for the 15-unit test system. GA-EPSO resulted in minimum fuel cost for the 20-unit test system.
Robustness	GA-IDE-OBL is capable of producing stable solutions over several trials in 6-, 10-, 15-, and 20-unit test systems, thus proving robustness.
Computational time	The computational time of GA-IDE-OBL is less than that of GA-FRBFN, GA-EPSO, GA-DE-OBL, GA-ABC, and GA-CSO for 6-, 10-, 15-, and 20-unit test systems.
Algorithmic efficiency	Algorithmic efficiency is higher in case of GA-CSO algorithm for all the test systems with 6, 10, 15, and 20 generating units.

2.8 Summary

In this chapter, intelligent techniques based on bio-inspired paradigms were used for solving the UC-ELD problems. The UC problem was solved using GA to determine the ON/OFF schedule for a 24-hour time horizon. Based on the GA committed/de-committed schedule, power was dispatched economically for the corresponding load requirement using FRBFN, EPSO, DE-OBL, IDE-OBL, ABC, and CSO algorithms. FCM clustering was adopted as a preprocessing algorithm to the RBFN in order to dimensionally reduce the data, allowing a simpler RBF model for ELD problems. An enhanced PSO was applied for ELD solution, in which a constriction factor was introduced to the velocity update process, thus improving the convergence rate of the algorithm. In the DE-OBL algorithm, the concept of opposition-based learning was applied in the initialization phase to accelerate the SDE algorithm with the motive of achieving optimal solutions with faster convergence characteristics. Likewise, in IDE-OBL, the concept of OBL was applied in the population initialization as well as in the generation phase to ensure stability in convergence. The main advantage of ABC algorithm is its simplicity due to fewer control parameters. This led to the implementation of ABC for solving the ELD problem. In addition to these algorithms, a new heuristic technique, the CSO, was presented for solving ELD problems. Its versatile properties such as randomization, few algorithmic control parameters, and less computational steps helped CSO to solve the ELD problem.

The effectiveness and efficiency of the heuristics were validated on four test systems consisting of 6, 10, 15, and 20 generating units. Each individual test system was solved using GA to determine the committed and de-committed units. Based on the commitment, the ELD problem was solved, and the power generated in the individual units, fuel cost, power loss, total power, and computational time were computed for each hour of the day. The experimental results reported in Table 2.62 in terms of optimal fuel cost for all the four test systems imply that GA-DE-OBL, GA-IDE-OBL, GA-EPSO, GA-ABC, and GA-CSO are capable of producing optimal fuel cost. Though these algorithms showed marginal improvements in fuel costs, it is seen from the SD values in Table 2.62 that GA-IDE-OBL is stable in solving UC-ELD.

It is also observed from Table 2.60 that the computational time of GA-IDE-OBL for all the test systems improved compared to GA-FRBFN, GA-EPSO, GA-DE-OBL, GA-ABC, and GA-CSO algorithms. Similarly, on comparing the algorithmic efficiency (Table 2.61) for the

6-, 10-, 15-, and 20-unit systems, GA-CSO is found to be better due to the smaller number of computational steps in implementing the code. Thus it can be concluded that GA-IDE-OBL shows significant improvements from the perspectives of optimal solution, robustness, and computational efficiency. In future, additional improvements can be included in the techniques by considering the practical constraints of the UC and ELD problems. Besides, new optimization techniques such as stud GA, population-based incremental learning, intelligent water drop algorithm, bio-geography-based algorithm and hybrid combination of these paradigms can also be applied to obtain optimal solutions to the UC-ELD problems.

References

Abookazemi, K., Mustafa, M.W., and Ahmad, H., Structured genetic algorithm technique for unit commitment problem, *International Journal of Recent Trends in Engineering*, 1(3), 135–139, 2009.

Balujia, S. and Davies, S., Fast probabilistic modeling for combinatorial optimization, in *Proceedings of the American Association of Artificial Intelligence: AAAI'98*, 1998, pp. 469–476.

Brown, C., Liebovitch, L.S., Glendon, R., Levy flights in Dobe Ju/'hoansi foraging patterns, *Human Ecology*, 35, 129–138, 2007.

Fogel, D.B., Phenotypes, genotypes, and operators in evolutionary computation, in *Proceedings of the IEEE International Conference on Evolutionary Computation 1995*, Perth, Australia, November 29–December 1, 1995, pp. 193–199.

Gibbons, D.W., Brood parasitism and cooperative nesting in the moorhen, *Gallinula chloropus, Journal on Behavioral Ecology and Sociobiology*, 19, 221–232, 1986.

Goldberg, D.E., *Genetic Algorithms in Search Optimization and Machine Learning*, Addison Wesley Longman Publishing Co., Boston, MA, 1989.

Hathaway, R.J. and Bezdek, J.C., Fuzzy c-means clustering of incomplete data, *IEEE Transactions on Systems, Man, and Cybernetics-Part B: Cybernetics*, 31(5), 735–744, 2001.

Karaboga, D. and Basturk, B., A powerful and efficient algorithm for numerical function optimization: Artificial bee colony (ABC) algorithm, *Journal of Global Optimization*, 39(3), 359–371, 2007.

Meng, K., Dong, Z.Y., Wang, D.H., and Wong, K.P., A self-adaptive RBF neural network classifier for transformer fault analysis, *IEEE Transactions on Power Systems*, 25(3), 1350–1360, 2010.

Orero, S.O. and Irving, M.R., Scheduling of generators with a hybrid genetic algorithm, Presented at *The First International Conference on Genetic Algorithms in Engineering Systems: Innovations and Applications GALESIA*, Sheffield, U.K., September 12–14, 1995, pp. 200–206.

Park, J.-B., Jeong, Y.-W., Shin, J.-R., and Lee, K.Y., An improved particle swarm optimization for nonconvex economic dispatch problems, *IEEE Transactions on Power Systems*, 25(1), 156–166, February 2010.

Pavlyukevich, I., Levy flights, non-local search and simulated annealing, *Journal of Computational Physics*, 226(2), 1830–1844, 2007.

Payne, R.B., Sorenson, M.D., and Klitz, K., *The Cuckoos*. Oxford, U.K.: Oxford University Press, 2005.

Prabhakar, K.S., Palanisamy, K., Jacob, R.I., and Kothari, D.P., Security constrained UCP with operational and power flow constraints, *International Journal of Recent Trends in Engineering*, 1(3), 106–114, 2009.

Rahnamayan, S., Tizhoosh, H.R., and Salama, M.M.A., Opposition based differential evolution, *IEEE Transactions on Evolutionary Computation*, 12(1), 64–79, 2008.

Rajan, C.C.A., Genetic algorithm based simulated annealing method for solving unit commitment problem in utility system, in *Proceedings of the IEEE Transmission and Distribution Conference and Exposition: IEEE PES 2010*, New Orleans, LA, April 19–22, 2010, pp. 1–6.

Reynolds, A.M. and Frye, M.A., Free-flight odor tracking in Drosophila is consistent with an optimal intermittent scale-free search, *PLoS One*, 2(4), 2007.

Ruxton, G.D. and Broom, M., Intraspecific brood parasitism can increase the number of eggs that an individual lays in its own nest, *Proceedings of Biological Science*, 269(1504), 1989–1992, 2002.

Tran, T., Nguyen, T.T., and Nguyen, H.L., Global optimization using Levy flight, in Proceedings of the International Conference on Information and Communication Technology: ICT.rda '04, Hanoi, September 24–25, 2004, pp. 1–12.

Viswanathan, G.M., Buldyrev, S.V., Havlin, S., da Luz, M.G.E., Raposo, E.P., and Stanley, H.E., Optimizing the success of random searches, *Nature*, 401, 911–914, 1999.

Vlachogiannis, J.G. and Lee, K.Y., Quantum-inspired evolutionary algorithm for real and reactive power dispatch, *IEEE Transactions on Power Systems*, 23(4), 1627–1636, 2008.

Wang, Y., Zhou, J., Xiao, W., and Zhang, Y., Economic load dispatch of hydroelectric plant using a hybrid particle swarm optimization combined simulation annealing algorithm, in *Proceedings of the Second WRI Global Congress on Intelligent Systems: GCIS*, Wuhan, China, December 16–17, 2010, pp. 231–234.

Wang, Y.-R., Lin, W.-H., and Yang, L., An intelligent PSO watermarking, in *Proceedings of the International Conference on Machine Learning and Cybernetics: ICMLC*, Qingdao, July 11–14, 2010, pp. 2555–2558.

Yang, X.-S. and Deb, S., Cuckoo search via Levy flights, in *Proceedings of the World Congress on Nature and Biologically Inspired Computing: NaBIC*, 2009, pp. 210–214.

3

Harmonic Reduction in Power Systems

Learning Objectives

On completion of this chapter, the reader will have knowledge on

- Harmonic effects and methods to eliminate harmonics
- VSI drives and switching frequency of drives
- Harmonic measurement at a pulp and paper industry
- Development of MATLAB®/Simulink® models for drives applied for harmonic reduction
- Application of computational intelligent algorithms such as GA and BFO for reducing harmonics

3.1 Harmonic Reduction in Power System

Rapid growth in power electronics technologies has made phenomenal contribution to the human race. The advent of power electronics has opened up new possibility and hopes in various fields ranging from household, commercial, and industrial applications. One of the valuable contributions has been in the field of electrical drives. In India, 75% of electrical power is used to drive the motors, either AC or DC. In the pulp and paper industry, 30%–35% of electrical motors employ variable-speed drives to drive AC or DC motors. Compared to other industries, the pulp and paper industry requires more numbers of electrical drives because of its process and system requirement.

In recent years, induction motor drives have become increasingly employed as variable frequency drives (VFDs) in the pulp and paper industry. The cost-effective VFDs enhance the overall performance, efficiency, and reliability of industrial processes. The most commonly used front-end converter topology in VFD is diode/thyristor rectifier. This converter topology is preferred because of its well-known advantages such as low cost, robustness, and reliability. However, line-commutated diode and thyristor rectifiers exhibit nonlinear load characteristics and draw nonsinusoidal currents from the supply even when fed from sinusoidal supply voltages. These harmonic currents are injected into the supply systems and pollute the power line, causing power quality problems (Rice 1994). The main objective is to systematically analyze, simulate, and verify through experiments various harmonic elimination methods for variable frequency induction motor drives. The following section discusses the harmonic effects in industrial electrical distribution and provides a detailed discussion on various harmonics mitigation methods. A thorough survey of the literature on harmonic elimination methods is also presented in this chapter.

3.2 Harmonic Effects

Harmonic currents are injected into the supply systems and pollute the power distribution system, causing power quality problems. The injected current harmonics cause line voltage distortion and notches, which is a major problem for utilities. The distorted voltage frequently results in malfunctioning or tripping of other linear/nonlinear loads connected to the same point of common coupling (PCC), as shown in Figure 3.1.

The PCC is the point where the consumers are connected together and it is generally defined as the point at which harmonic limits shall be evaluated. Specifically, from the utility side, this point will be in the supply system owned by the utility. From the customer side, it is the point where the end user consumes energy and where other customers are provided with electric service. The effect of current distortion on power distribution systems can be serious, primarily because of the increased current flowing in the system. In other words, because the harmonic current does not deliver any power, its presence simply uses up system capacity and reduces the number of loads that can be powered. Harmonic current occurring in an electrical system can cause equipment malfunction, data distortion, transformer and motor insulation failure, overheating of neutral buses, tripping of circuit breakers, and breakdown of solid-state components. The cost of these problems can be enormous. Harmonic currents also increase heat losses

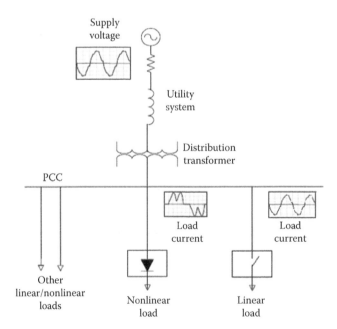

FIGURE 3.1
Point of Common Coupling (PCC). (From Zubi, H.M., Low pass broadband harmonic filter design, MS thesis, Middle East Technical University, Ankara, Turkey, 2005.)

in transformers and windings. Since transformer impedance is frequency-dependent, increasing with the amount of harmonics, the impedance at the fifth harmonic is five times that at the fundamental frequency. So each ampere of the fifth harmonic current causes five times more heating than an ampere of the fundamental current. More specifically, the effects of harmonics can be observed in many sections of electrical equipment and various machines and motors.

The new IEEE Standard 519 requires careful harmonic analysis of the entire power distribution system for specifying proper VFD technology (Keskar 1996). It is not sufficient to look at one location, but requires the combined effect of all VFD drive locations on the PCC. We focus only on the impact of IEEE Standard 519 on the specification of VFDs, which deals with the quality of the input power supply. Sundareswaran et al. (2007) have developed the concept of ACS for the continuous optimization problem of selective harmonic elimination in a pulse width modulation (PWM) inverter. Verma and Singh (2010) presented a genetic algorithm (GA) to design series tuned and second-order bandpass passive filters. The design emphasized the reduction of harmonic current together with minimization of the root-mean squared (rms) source current and reactive requirement. Through analysis of the result and the observed performance, they demonstrated the ability of the proposed designed passive filters to compensate the current harmonics effectively along with the reduction of the net rms source current. Damoun Ahmadi et al. (2011) presented modified four-equation method for selected harmonics elimination for both two-level inverters and multilevel inverters with unbalanced DC sources. For this case with a fairly low number of switching angles and unbalanced multiple voltage levels, the weight-orientated junction point distribution was applied to enhance the performance of the method.

3.3 Harmonics Limits and Standards

The IEEE 519 recommended harmonic standard was introduced as a guide in 1981 and revised in 1992. The IEEE Standard 519-1992 proposes to limit harmonic current injection from end users. The recommended limits are provided for individual harmonic components, and the total distortion indices have been presented (McGranaghan and Mueller 1999). Total harmonic distortion (THD) is commonly used as the index for measuring the harmonic content of a waveform and may be applied to either voltage or current. Restrictions on current and voltage harmonics maintained in many countries, through IEEE 519-1992 (in the United States), AS/NZS 61000.3.6 (in Australia), and IEC 61000-3-2/IEC 61000-3-4 (in Europe), ensure good quality of power in the electrical distribution system. Table 3.1 shows

TABLE 3.1

IEEE 519-1992 Harmonic Current Distortion Limits

I_{SC}/I_{Ls}	<11	$11 \leq h < 17$	$17 \leq h < 23$	$23 \leq h < 35$	$35 \leq h$	TDD/THD
<20[a]	4.0	2.0	1.5	0.6	0.3	5.0
20 < 50	7.0	3.5	2.5	1.0	0.5	8.0
50 < 100	10.0	4.5	4.0	1.5	0.7	12.0
100 < 1000	12.0	5.5	5.5	2.0	1.0	15.0
>1000	15.0	7.0	7.0	2.5	1.4	20.0

[a] All power generation equipment is limited to these values of current distortion, regardless of actual I_{SC}/I_L.

the IEEE 519-1992 current harmonics limit. I_{SC} is the maximum short-circuit current and I_L is maximum demand load current (fundamental frequency component) at the PCC. The IEEE Standard 519-1992 recommended harmonic voltage limits are shown in Table 3.2.

The European standard IEC 61000-3-2, IEC 1000-3-2 CLASS D defines the current distortion limits for equipment connected to the public supply system. Table 3.3 shows the IEC 61000-3-2, IEC 1000-3-2 CLASS D equipment current limits. Its objective is to limit the current distortion, and it addresses public, low-voltage, and household customers. The standard IEC 61000-3-4, IEC 1000-3-4 shown in Table 3.4 is applicable for larger customers (single- and three-phase harmonic limits). It takes into consideration the short-circuit ratio R_{SCC} (Cichowlas 2004).

TABLE 3.2

IEEE 519-1992 Harmonic Voltage Distortion Limits

Bus Voltage at PCC	Maximum Individual Harmonic Component (%)	Maximum THD (%)
69 kV and below	3.0	5.0
69.001 kV through 161 kV	1.5	2.5
161.001 kV and above	1.0	1.5

Note: High-voltage systems can have up to 2.0% THD where the cause is an HVDC terminal that will attenuate by the time it is tapped for a user.

TABLE 3.3

IEC 61000-3-2, IEC 1000-3-2 CLASS D Equipments Current Limits

Harmonic Order	Maximum Permissible Harmonic Current per Watt	Maximum Permissible Harmonic Current
N	mA/W	A
3	3.4	2.3
5	1.9	1.14
7	1.0	0.77
9	0.5	0.40
11	0.35	0.33
$13 \leq n \leq 39$ (odd har. only)	$3.85/n$	Refer to class A

TABLE 3.4

IEC 61000-3-4, IEC 1000-3-4 Three-Phase Equipment Limits

Minimal R_{SCC}	Upper Limits for Harmonics Distortion Factors		Limits for Individual Harmonic in % of I_1			
	THD	PWHD	I_5	I_7	I_{11}	I_{13}
66	17	22	12	10	9	6
120	18	29	15	12	12	8
175	25	33	20	14	12	8
250	35	39	30	18	13	8
350	48	46	40	25	15	10
450	58	51	50	35	20	15
>600	70	57	60	40	25	18

3.4 Method to Eliminate Harmonics

Various harmonic reduction techniques have been developed to meet the requirements imposed by the voltage and current harmonic standards. In general, these techniques can be classified into five broad categories (Zubi 2005):

1. Passive filters
2. Phase multiplication techniques
3. Active filters
4. Hybrid systems
5. PWM rectifiers.

3.4.1 Passive Filters

The passive filter consists of elements such as the inductor (L), capacitor (C), and resistor (R) for filtration of the harmonics. This makes the filter configuration simple and easy to implement. The passive filter can be easily designed for a load power that is predetermined. The filter is connected with the power distribution system and it is tuned to present low impedance to particular harmonics so that these harmonics are diverted from their normal flow path through the filter; or it is tuned to present high impedance to particular harmonics to stop them from affecting the circuit. The tuning depends on the configuration of the filter. The passive filter is a very good choice for low- and medium-rated loads and is a cost-effective solution to harmonic reduction and power factor improvement. For a two-pulse and a six-pulse rectifier drive with low power rating, the passive filter is best suited. In most cases, a passive filter involves an LC combination tuned to serve the purpose. Various passive filters with different configurations of L and C are considered, and their values are obtained using both the conventional methods and the optimization procedure.

3.4.2 Phase Multiplication Technique

The phase multiplication technique is based on increasing the pulse number for the converter. This increases the lowest harmonic order for the converter and reduces the size of the passive filter needed to filter out the current harmonics. A 12-pulse converter ideally has the lowest harmonic order of 11 (fifth and seventh current harmonics are theoretically nonexistent). Similarly, an 18-pulse converter has the lowest harmonic order of 17. However, a 12-pulse converter, shown in Figure 3.2, needs two six-pulse bridges and two sets of 30° phase-shift AC inputs, and an 18-pulse converter needs three six-pulse bridges and three sets of 20° phase-shift AC inputs. Many different topologies exist for achieving the phase shift. In general, the phase multiplication technique is effective in reducing low-order current harmonics as long as there is a balanced load on each of the converters. However, their large size, low efficiency, and high cost are the main drawbacks.

3.4.3 Active Power Filters

The basic principle of active power filters (APFs) is to utilize power electronics technologies to produce specific current components that have the same magnitude but opposite phase as those of the harmonic load current and hence cancel each other so as to make the input current waveform sinusoidal. Using active power filters is the best and most efficient

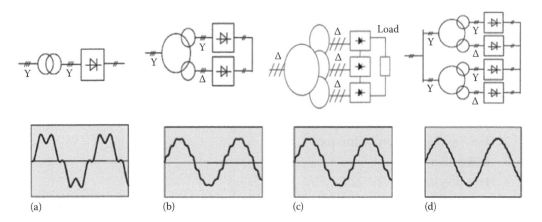

FIGURE 3.2
Phase-shifting transformer-fed drives and current waveforms: (a) 6-pulse, (b) 12-pulse, (c) 18-pulse, and (d) 24-pulse.

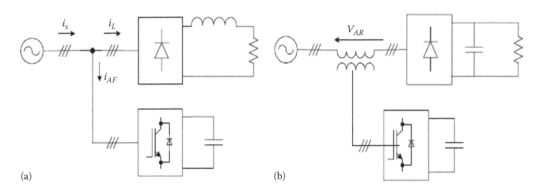

FIGURE 3.3
Active filter topology: (a) parallel active filter; (b) series active filter.

way to achieve unity power factor and to reduce harmonics. Two commonly used active filters are (1) the series active filter and (2) the parallel active filter. Active filters operate on the principle of harmonic cancellation; hence they do not cause any harmful resonance with the power system network. A series active power filter is connected in between the source and the load and offers high impedance to the harmonics blocking any harmonic current from flowing through the source to the load, and vice versa. A parallel active power filter is connected in parallel to the load and produces harmonic currents of the same magnitude but different phase as the load harmonic currents, thereby compensating the harmonics.

The parallel active power filter is basically a PWM inverter. The topologies of both series and parallel active filters are shown in Figure 3.3.

3.4.3.1 Drawbacks of Using APF

Active filtration is a relatively new technology when compared to passive filtration. There is still need for further research and development to make this technology well established. An unfavorable feature of APF is the necessity for fast switching of high currents in the power circuit of the APF. This results in high-frequency noise, which may cause electromagnetic interference (EMI) in the power distribution system. The design procedure is complex, as it needs an additional power converter, and the setup is very expensive. For low-power application, the two-pulse and six-pulse diode/thyristor rectifier front-end drive is always advantageous to add a passive filter circuit, which is less expensive.

3.4.4 Hybrid Active Filters

Hybrid active filters, as shown in Figure 3.4, combines active and passive filters in various configurations. The main purpose of hybrid active filters is to reduce harmonics and to improve efficiency. They are also used to improve the compensation characteristics of passive filters and alleviate any series or parallel resonance due to the supply or load, respectively. Practically, more viable and cost-effective hybrid filter topologies have been developed than stand-alone active filters. They enable the use of significantly small rating active filters, with less than 5% of the load kVA compared to stand-alone parallel (25%–30%) or series active filter solutions.

Usually, with shunt passive filter combinations, the passive filter is tuned up to a specific frequency to suppress the corresponding harmonic and decrease the power rating of the active filter. Another typical combination is a series active filter and a series passive filter. High fundamental component current through the series active filter and the fundamental component voltage across the shunt active filter are problematic. High initial and running costs and complexity are major drawbacks of the active harmonic filtering technique. PWM voltage source rectifiers (PWM-VSR) offer benefits such as power regeneration, low harmonic distortion, unity power factor, and controlled DC link. They are often used in applications where substantial regenerative operating mode occurs. PWM-VSR operation principle is based on direct sinusoidal current generation, whereas the active filter is based on load harmonic compensation. However, the topology with high cost is the main drawback, which makes it unpractical in many applications. To conclude, most of the mentioned filtering techniques have the common drawback of high cost compared to passive filtering techniques. Consequently, passive harmonic filtering techniques, to a large extent, are still the most commonly used techniques for current harmonic mitigation of six-pulse front-end diode/thyristor rectifier applications. In this chapter, we deal with passive harmonic filtering topologies.

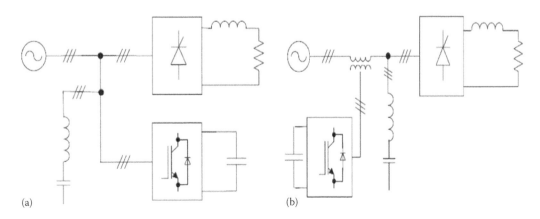

FIGURE 3.4
Hybrid active filter topology: (a) parallel active filter; (b) series active filter.

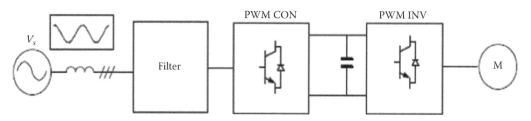

FIGURE 3.5
PWM rectifier topology.

3.4.5 PWM Rectifier

The other possible reduction technique of current harmonics is the application of the PWM rectifier shown in Figure 3.5. Two types of PWM converters, with a voltage source output and a current source output, can be used. The first one is called a boost rectifier (increases the voltage) and operates at fixed DC voltage polarity; the second, called a buck rectifier (reduces the voltage), operates with fixed DC current flow.

The main features of PWM rectifiers are the following:

- Bidirectional power flow
- Nearly sinusoidal input current
- Regulation of input power factor to unity
- Low harmonic distortion of line current (THD below 5%)
- Adjustment and stabilization of DC-link voltage (or current)
- Reduced capacitor (or inductor) size due to the continuous current.

Furthermore, it can be properly operated under line voltage distortion and notching as well as line voltage frequency variations.

3.5 Voltage Source Inverter-Fed Induction Motor Drives

In industries, for adjustable speed operation, various types of drives are used. These drives are categorized into AC drives and DC drives. DC drives were popular because of their overall performance, efficiency, and static and dynamic characteristics. However, they suffer from the major drawback of high initial and maintenance costs. Advancements in semiconductor devices have made significant contributions to AC drives and also made AC drives competitive to DC drives. AC drives are fed from either a voltage source inverter (VSI) or a current source inverter (CSI). VSI-fed drives have proven to be more efficient, with higher reliability and faster dynamic response compared to CSI-fed drives, and are capable of running motors without de-rating. Hence, VSI- fed drives save money with higher efficiencies, minimize install time, eliminate interconnect power cabling costs, and reduce building floor space. The VSI drive PWM inverter along with a rectifier and capacitor filter circuit forms the basic structure of any AC drives. However introduction of rectifier in the front end, introduce current harmonics to supply current. When the harmonics go unnoticed, they cause serious problems such as frequent failures of the drive rectifier diodes, measuring meter failure, production relay malfunctioning, and sudden failure in motor windings. Hence, it has become customary to add a filter circuit and develop harmonic elimination methods to mitigate the problems mentioned. Therefore an understanding of the nature of the harmonics and the source of their generation is critical in achieving the tasks mentioned before.

In this chapter, a detailed analysis 2-pulse, 6-pulse, and 12-pulse rectifier-fed drives is carried out to understand the behaviors of the drives' various modes of operations and explain the comparison of the rectifier, filter, and inverter. Further, all the-drives mentioned have a PWM inverter in common, and hence a discussion on the PWM inverter is also provided. In addition, the various problems in VFDs are elaborated.

3.5.1 Two-Pulse Rectifier Drive

In single-phase drive systems, front-end two-pulse rectifiers are widely employed. The two-pulse drives are suitable for low-power applications in domestic and industrial appliances such as compressors, fans in heat pumps, air conditioners, refrigerators, lab equipments, measuring instruments, and hoists. The schematic of a two-pulse drive is shown in Figure 3.6. In this power structure, four diodes form the rectifier section, a capacitor rated at 500 µF acts as a filter, and the third stage forms the inverter section. The inverter is made up of high-speed transistors or IGBTs. The PWM technique is applied to modulate the voltage and frequency applied to the motor.

All VFD manufacturers commonly provided embedded compact design. The drive power cards, control cards, optional input/output cards (filed bus), digital input cards, and keypad are common to all drives. Normally, drives are controlled remotely from a control room. The ON/OFF and speed reference changes are made by operators via digital/analog input and output cards. The speed reference is fed through an analog input card, with analog voltage 0 to +10 V, R_i = 200 kΩ (–10 V to +10 V joystick control), resolution 0.1%, accuracy ±1% or analog current 0(4) mA–20 mA, R_i = 250 Ω differential.

3.5.1.1 Two-Pulse Rectifier Drive Operation

The two-pulse rectifier front end is made up of a diode rectifier. The power circuit of the diode bridge rectifier is shown in Figure 3.7, and its voltage and current waveforms are shown in Figure 3.8.

It consists of four diodes $D1$, $D2$, $D3$, and $D4$. At the input side, the source impedance is represented by a series RL circuit. A capacitor at the output end maintains constant output voltage and minimizes voltage harmonics. The bridge circuit is characterized by three different modes.

- **Mode 1:** During mode 1, diode $D1$ and $D2$ conduct, and the load resistance is connected to the input voltage. A positive current flows in the circuit through the two diodes of the rectifier, charging up the output DC capacitor. This mode is characterized by the following equations:

$$\begin{bmatrix} i_s \\ v_c \end{bmatrix} = \begin{bmatrix} -\dfrac{R_s}{L_s} & -\dfrac{1}{L_s} \\ \dfrac{1}{C_l} & -\dfrac{1}{R_d C_l} \end{bmatrix} \begin{bmatrix} i_s \\ v_c \end{bmatrix} + \begin{bmatrix} \dfrac{1}{L_s} \\ 0 \end{bmatrix} \begin{bmatrix} V_s \end{bmatrix} \qquad (3.1)$$

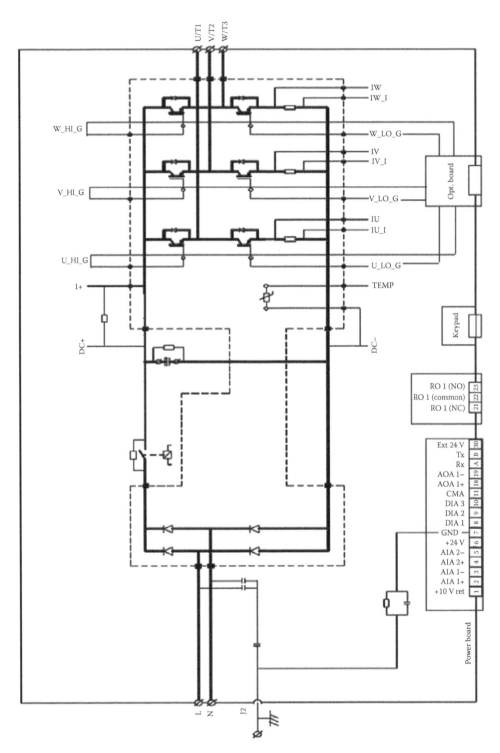

FIGURE 3.6
Power and control structure for a two-pulse drive.

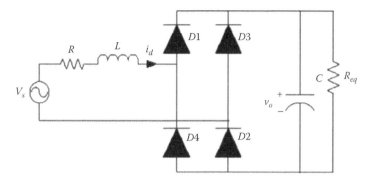

FIGURE 3.7
Two-pulse drive rectifier section.

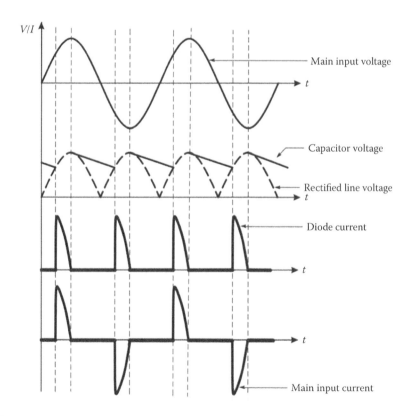

FIGURE 3.8
Voltage and current waveforms in rectifier section.

- **Mode 2:** In this mode, all the rectifier diodes are off, and the current becomes zero, while the DC output voltage is maintained constant by discharging the stored capacitor voltage. The equation representing this mode is given by

$$[v_c] = \left[-\frac{1}{R_d C_l} \right] [v_c] \tag{3.2}$$

- **Mode 3:** A negative current flows in the circuit, charging up the output DC capacitor, while the other set of diodes are conducting.

$$\begin{bmatrix} i_s \\ v_c \end{bmatrix} = \begin{bmatrix} \dfrac{R_s}{L_s} & \dfrac{1}{L_s} \\ -\dfrac{1}{C_l} & -\dfrac{1}{R_d C_l} \end{bmatrix} \begin{bmatrix} i_s \\ v_c \end{bmatrix} + \begin{bmatrix} \dfrac{1}{L_s} \\ 0 \end{bmatrix} [V_s] \tag{3.3}$$

where
V_s is the source voltage
i_s is the input source current
v_c is the output capacitor voltage
R_s is the source resistance
L_s is the source impedance
C_l is the output capacitor
R_d is the diode resistance

The two-pulse drive current waveform is nonsinusoidal in nature, hence a two-pulse rectifier drive generates high THD current, where the dominant harmonics are the third, fifth, seventh, and ninth order harmonics.

3.5.2 Six-Pulse Rectifier Drive

In a three-phase supply system, low and medium power application the six-pulse diode rectifier-fed drives are employed. In industries, 20%–30% drives are employed below base speed applications, such as conveyors, agitators, and pumps. This type of application requires only below base speed and operates in a single quadrant. Hence, in single-quadrant applications a -pulse diode rectifier drive is cost effective and its structure is simple. The schematic layout of six-pulse drive is shown in Figure 3.9. In this power structure, six diodes form the rectifier section, two capacitors rated 500 μF act as the filter, and the third stage is the inverter section. The inverter is made up of high-speed transistors or IGBTs.

The PWM technique is applied to modulate the voltage and frequency applied to the motor. In regenerative mode and four-quadrant applications, the diode rectifier is replaced with a controlled rectifier. In such cases, controlled rectifier is made up of a half-controlled rectifier or a fully controlled rectifier.

3.5.3 Twelve-Pulse Rectifier Drive

In the mid-1960s, when power semiconductors were available only in limited ratings, 12-pulse drives provided a simpler and more cost-effective approach to achieve higher

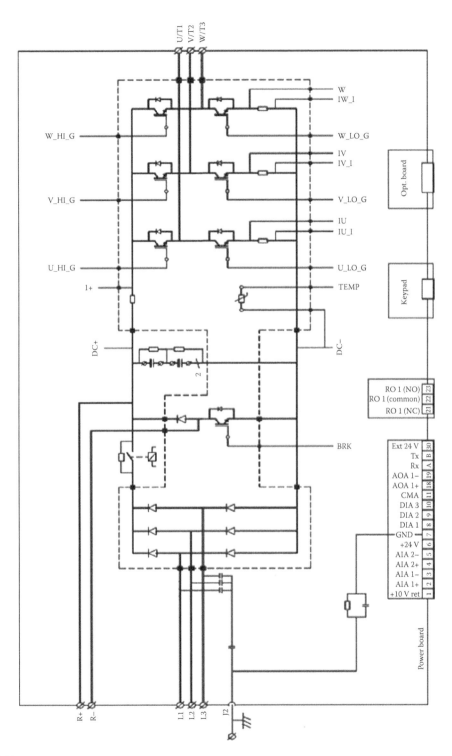

FIGURE 3.9
Power and control structure for a six-pulse drive.

current ratings than direct paralleling of power semiconductors. This technique is still employed today in very large drive applications. A typical power circuit of a 12-pulse drive is shown in Figure 3.10. In this 12-pulse arrangement, two 6-pulse drives are connected in parallel, by which the current rating is improved; for improved voltage rating, the two drives are connected in series, which is shown in Figure 3.11. The drives' input circuit consists of two six-pulse rectifiers, displaced by 30 electrical degrees, operating in parallel. The 30° phase shift is obtained by introducing a phase-shifting transformer. Phase shifting involves separating the electrical supply into two or more outputs, each output being phase-shifted with respect to the others with an appropriate angle for the harmonic pairs to be eliminated.

The concept is to displace the harmonic current pairs in order to bring each to 180° phase shift, so that they cancel each other. Positive sequence currents will act against negative sequence currents, whereas zero sequence currents act against each other in a

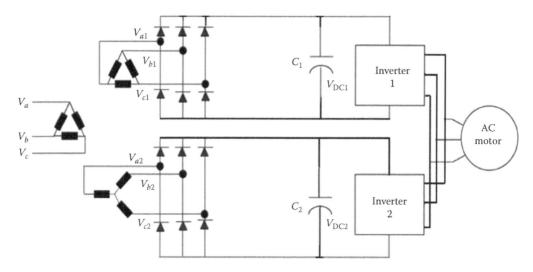

FIGURE 3.10
12-pulse parallel-connected AC drive power structure.

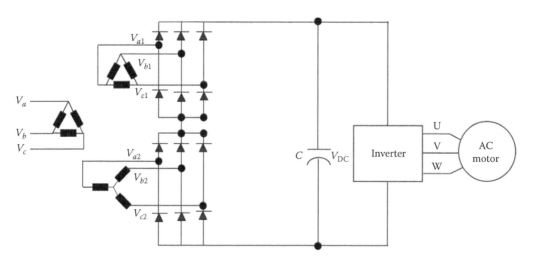

FIGURE 3.11
12-pulse series-connected AC drive power structure.

three-phase system. Recall that triple harmonics are zero-sequence vectors; 5th, 11th, and 17th harmonics are negative sequence vectors; and 7th, 13th, and 19th harmonics are positive sequence vectors. Hence, an angular displacement of 60° is required between two three-phase outputs to cancel the 3rd harmonic currents; 30° is required between two three-phase outputs to cancel the 5th and 7th harmonic currents; and 15° is required between two three-phase outputs to cancel the 11th and 13th harmonic currents.

The circuit in Figures 3.10 and 3.11 simply uses an isolation transformer with a delta primary, a delta-connected secondary, and a second star-connected secondary to obtain the necessary phase shift. The primary current in the transformer is the sum of each 6-pulse rectifier or a 12-pulse waveform. The delta-connected drive phase current I_{a1} waveform, the star connected drive phase current I_{a2} waveform, and total sum of two 6-pulse or 12-pulse phase current I_a waveform are shown in Figure 3.12. The phase current harmonics spectrum is shown in Figure 3.13.

For instance, in the case of two six-pulse variable frequency drives of similar rating, installing a delta-star transformer (30° with respect to the primary) on one drive and delta, delta transformer (0° with respect to the primary) on the other drive gives an angular displacement of 30° between the two outputs, providing the equivalent of a 12-pulse system. On the common supply of both transformers on the primary, phase shifting between the systems will cancel the fifth and seventh harmonic currents. As compared with the 6-pulse

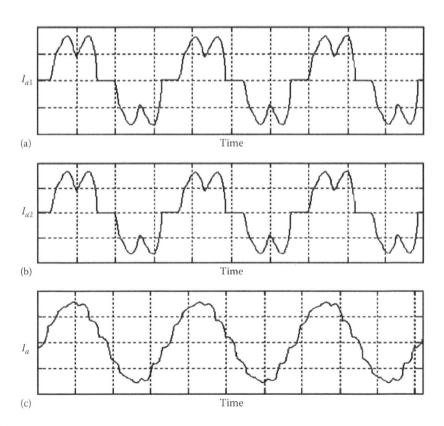

(a)

(b)

(c)

FIGURE 3.12
12-pulse AC drive current waveforms. (a) Delta connect drive phase current I_{a1} waveform, (b) star connect drive phase current I_{a2} waveform, and (c) phase shift transformer input phase current I_{a3} waveform.

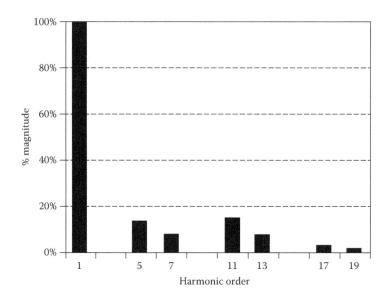

FIGURE 3.13
12-pulse drive harmonics spectrum.

harmonics spectrum and the 12-pulse drive harmonic spectrum (Figure 3.12) the fifth and seventh current harmonics are reduced. An angular displacement 15° between the outputs provides the equivalent of a 24-pulse system, but requires four 6-pulse loads. The earlier approach, that is, phase-shifting nonlinear loads, can thus be used to reduce the effects of selected current harmonics order. Figure 3.13 shows the 12-pulse drive harmonics spectrum.

The real-time 12-pulse drive converter connection arrangements are shown in Figure 3.14a. The two drive input supplies are fed through a phase-shifting transformer. The real-time arrangement with a switch fuse unit for the drive input side is denoted as $K1$ and $K2$ in Figure 3.14a. The two six-pulse converters with the same rating are connected in parallel. The inverter power module and control modules are shown in Figure 3.14b. The power module consists of an IGBT, the IGBT cooling fan board, the measurement board, and the IGBT gate pulse driver board. The control module consist an application-specific integrated circuit (ASIC) board, the input and output interface boards, and the keypad.

3.5.3.1 PWM Inverter

In the voltage source variable-frequency drive, the third stage is made up of an inverter, with the inverter switches being formed by the IGBTs. The switching pattern in PWM technology provides a means to generate motor phase currents of the required shape, permitting digital control usage in modern drives. The voltage pulse injected into the motor winding produces a current that cannot change instantly because of motor having a large inductance. Thus, it is possible to change the current waveform by the width of the voltage pulse.

This is convenient with the digital logic "0 or 1" used in microcontrollers. To generate the pulse of the required width, a microcontroller has to compare the desired (modulating) signal with the reference signal and set 0 or 1 on its output pins. The reference signal often works with a constant frequency, called the carrier frequency, which is equal to the switching frequency. The signals of the microcontroller outputs switch the IGBTs of the inverter (Dzhankhotov 2009). The inverter is the unit that transforms the digital signals

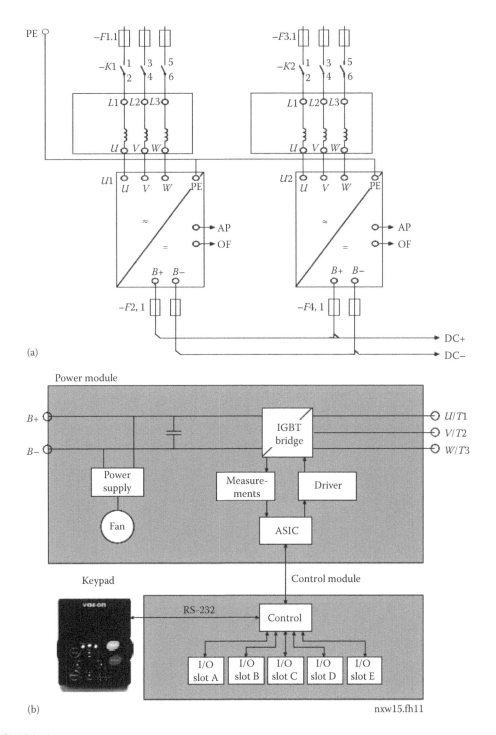

FIGURE 3.14
Power structure for a 12-pulse drive. (a) Converter arrangement for 12-pulse drive and (b) inverter and control module arrangement (Vacon drive).

from the control unit to power voltages that are necessary for motor rotation. It is desirable to provide as high a switching frequency as possible (16 kHz is a typical maximum for modern IGBT transistors in hard switching) at the inverter design stage to prevent audible noise and additional motor losses caused by the nonsinusoidal motor input. In practice, the switching frequency of present-day industrial inverters varies in the range 1–16 kHz. A typical VFD inverter section in all three phases is illustrated in Figure 3.15. Let the motor phases be star-coupled and connected to a PWM inverter. Figure 3.16 shows the possible phase connections to the DC link during one PWM cycle. The connection can be changed 7 times per unit PWM period. In practice, the impedances Z of the phases u, v, w are almost equal, so that we can assume that $Zu = Zv = Zw = Z$.

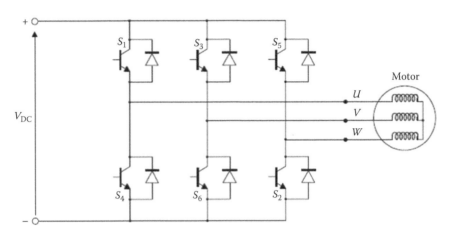

FIGURE 3.15
Variable-frequency drive inverter structure.

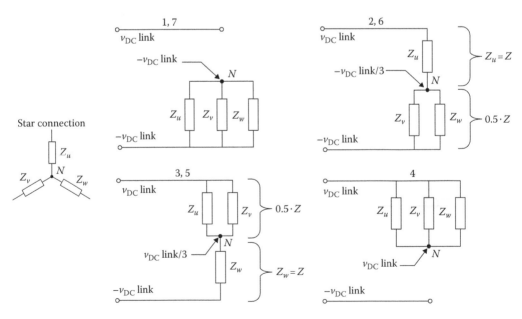

FIGURE 3.16
Equivalent electrical circuits of the phase connections to the DC link for the time ranges from 1 to 7.

Let us consider the possible schemes for phase connections to the DC link for each PWM period. It is evident from Figure 3.16 that the schemes for the time ranges 1 and 7, 2 and 6, and 3 and 5 will be the same. Within 2, 6 and 3, 5, the system can be considered as a simple voltage divider, and the voltage at the motor winding star point N is $-V_{DC}$ link/3 and $+V_{DC}$ link/3, respectively. In the time ranges 1, 4, and 7, all the phases are connected in parallel, and the potential at point N can be assumed to be equal to the full potential of the connected DC link terminal. Therefore, the potential of the star point is nonzero and variable. In the literature, this potential in relation to earth is called common-mode voltage. Now we can obtain the shape of the common-mode voltage, which is presented in Figure 3.17a.

Usually, the aim of the motor control is to obtain sinusoidal currents in the motor phases, but the width of the pulses is not constant in each PWM period. Therefore, the common-mode voltage fundamental also changes with triple frequency of the modulated sinusoidal signal. Now we can state that PWM is characterized by a varying potential at the star point of the motor with amplitude equal to half the DC link voltage, V_{DC} link. Being a nonsinusoidal signal, this potential produces high-frequency harmonics. Thus, high-frequency currents between the neutral point and earth are possible. Therefore, stray capacitances inside the motor have to be taken into account. A star connection is used here just as an example to help us determine the common-mode voltage. Practice shows that the described problem does not depend on the connection of phases or the number of motor phases. Along with common-mode voltages, there are differential-mode signals that can be explained by reflections in the power cable. Such signals do not propagate into the motor but return to the converter protective earth via the cabling system.

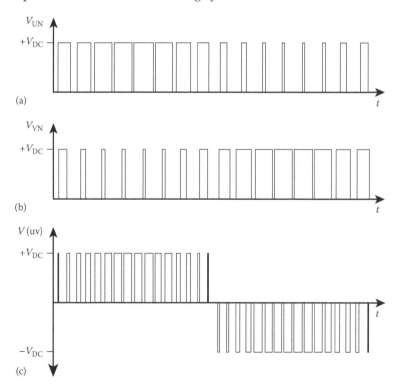

FIGURE 3.17
PWM waveforms: (a) common mode voltage for U phase and neutral point, (b) common mode voltage for V phase and neutral point, and (c) PWM waveform for UV terminal.

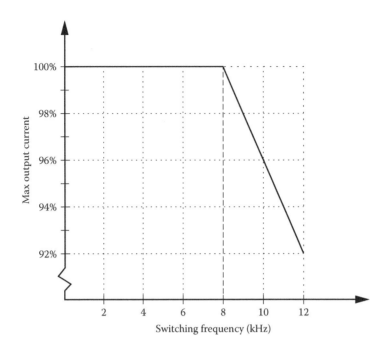

FIGURE 3.18
Switching frequency de-rating graph.

3.5.4 Selection of PWM Switching Frequency

Many modern AC drives have a selectable output switching frequency from 1 to 16 kHz and the tendency is to use the highest output frequency to reduce audible noise. The higher the switching frequency, the higher the leakage current losses. The selection of the PWM switching frequency is a compromise between the losses in the motor and those in the inverter. When the switching frequency is low, the losses in the motor are higher because the current waveform becomes less sinusoidal. When the switching frequency increases, motor losses are reduced but the inverter losses will increase because of the increased number of commutations. Losses in the motor cable also increase due to the leakage current through the shunt capacitance of the cable. Manufacturers of converters usually provide a de-rating table or graph, which would be similar to the typical one shown in Figure 3.18.

3.6 Case Study: Pulp and Paper Industry

Power electronics circuits are associated with the control of power flow from a source to a load in a desired manner. The advent of power electronics opened up new possibilities in many industrial, domestic, and commercial applications. Some of the important and familiar applications include the domestic fan regulator, switched-mode power supplies in personal computers, heating and lighting controls, electronic ballasts in fluorescent lamps, variable speed drives, drives for industrial motion control, induction heating, battery chargers, traction applications, solid-state relays, circuit breakers, off-line DC power supplies, spacecraft power systems, uninterruptible power supplies, conditioning of renewable energy sources, automobile electronics, and electric vehicles. It is estimated that during twenty-first century, 90% of electrical energy generated will be processed by power electronic circuits before its final consumption (Briault et al. 2000). Power quality problems are increasing in industry, particularly with progress in the application of power electronic circuits. In industries, 75% of the load is used to drive the motors, and one-third of the motors are incorporated with variable speed drives. The variable speed drives inject harmonics and create power quality problem in industrial distribution. This drives need proper harmonic management methods and to explore the possibility of harmonics minimization.

Harmonic analysis and study becomes an important and necessary task for maintenance engineers in almost every industrial power distribution system. This is because most industrial plant equipments are controlled by power electronic circuits. This chapter deals with the various possible harmonics that are generated and the power quality issues that rise due to the harmonics generation. For the study of harmonics, a pulp and paper industry power distribution system is taken and step-by-step measurements are done at different locations of the system. Various harmonics measurement schemes have been carried out to characterize the level of harmonics generation for the existing nonlinear loads as a means for verifying the harmonic model. Voltage and current harmonics levels are measured at multiple sites to accomplish this. These measurements are done for 6-pulse and 12-pulse drives. In order to effectively mitigate the harmonics generated in the drives, it is essential to develop mathematical models for the analysis and simulation of the drives. Hence modeling of the drive is done in MATLAB/Simulink, and the Simulink model character is matched with a real-time drive experimental setup. The data are taken in real time and fed to the Simulink model. Further, this simulation model is used for developing harmonics mitigation methods.

3.6.1 Power Distribution

Tamil Nadu Newsprint and Papers Limited (TNPL) was established by the Government of Tami Nadu during early 1980s to produce newsprint, printing, and writing paper using bagasse, a sugarcane residue, as the primary raw material. The company commenced production in the year 1984 with an initial capacity of 90,000 tonnes per annum (tpa). Over the years, the production capacity has been increased to 245,000 tpa, and the company has emerged as the largest bagasse-based paper mill in the world consuming about one million tonnes of bagasse every year.

The company completed a mill expansion plan during December 2010 to increase the mill capacity to 400,000 tpa. The mill consists of the following sections: (1) Paper Machine-1, (2) Paper Machine-2, (3) Paper Machine-3, (4) Finishing House, (5) Pulp Mill, (6) Soda Recovery Plant (SRP), (7) Power and Steam generation Plant, (8) Water Treatment Plant (WTP), (9) Effluents Treatment Plant (ETP), and the (10) Coal Handing Unit. TNPL has six turbogeneratos (TGs) with various capacities and four power transformers. The ratings of TGs and the power transformers are given in Table 3.5. The TGs and power transformers are synchronized with the 110 kV Tamil Nadu Electricity Board (TNEB) grid at the time of power import/export. TNPL's entire power distribution layout is displayed in Figure 3.19.

The distribution of the harmonics measurement points are also indicated in Figure 3.19. The entire distribution has 125 points for measurement of load profiles, power factor, frequency, grid frequency, active power, reactive power, voltage profiles, current profiles, voltage harmonics, and current harmonics. In this chapter, harmonic analysis is restricted to 6-pulse and 12-pulse drives, since they share the major part of the load and are also the pertinent sources of harmonics.

TABLE 3.5

TNPL Power Transformer and TG Ratings

Description	Voltage Rating	Rating
TRF-1 15 MVA	110 kV/12.1 kV[a]	15,000 kVA
TRF-2 15 MVA	110 kV/12.1 kV[a]	15,000 kVA
TRF-3 16.5 MVA	110 kV/12.1 kV[a]	16,500 kVA
TRF-4 20 MVA	110 kV/11 kV	20,000 kVA
TG1	11 kV	8 MW
TG2	11 kV	18 MW
TG3	11 kV	10.5 MW
TG4	11 kV	24.62 MW
TG5	11 kV	20 MW
TG6	11 kV	41 MW

[a] Step down to 11 kV with 12.1 kV/11 KV isolation transformer.

3.6.2 Harmonics Measurement at Grid and Turbo Generators

Harmonic analysis of the distribution system is essential to study the behavior of equipment connected in the nonsinusoidal system environment and for optimal design and location of filters. The TNPL TGs and power transformers are synchronized with the 110 kV TNEB grid at the time of import/export of the power. The harmonics levels are

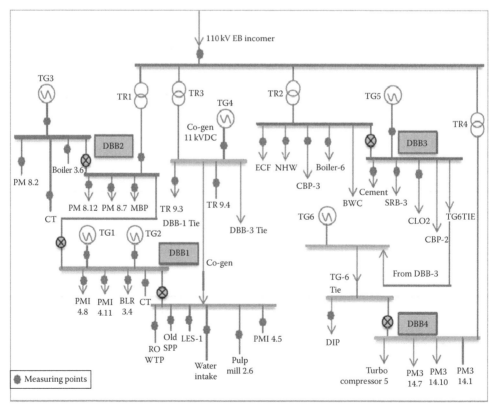

FIGURE 3.19

TNPL power distribution layout.

measured at grid incomer during power export and power import. During power import, the power factor and voltage THD percentage and current THD percentage values are well within limits, and odd-order harmonics such as the fifth-, seventh-, and ninth-order harmonics are strictly within limits. During power export, the power factor is very poor, and voltage and current THD values are above the defined limits.

This is because, when source is exported to the grid, it is supplying to infinite loads, which may be highly nonlinear in nature. For brevity, the grid voltage and current waveforms furring power exponent are shown in Figures 3.20 and 3.21. During export, the current waveform is nonsinousoidal and current THD is 23.4%. The grid current harmonics spectrum is shown in Figure 3.21. In the current harmonics spectrum, the 5th, 7th, and 11th order harmonics are highly dominant in the grid current waveform. Table 3.6 shows the power, power factor, and percent THDs in the power transformers and the grid.

TNPL has six TGs with various ratings. Among these, TG-6 is a newly erected one with capacity of 41 MW; at the time of the case study TG-6 was not running. The trial run and load test were in progress on TG-6, which was to be put to continuous run after the completion of load test. Hence the harmonic study was carried out only for TG-1 to TG-5, and the study results are presented in Table 3.7.

All the five TGs are continuous-running with different loading percentages according to the plant power requirements. In this study, results are taken for all five TGs while running with low harmonic distortion and good power factor. TG-4 was connected to a 4 MW

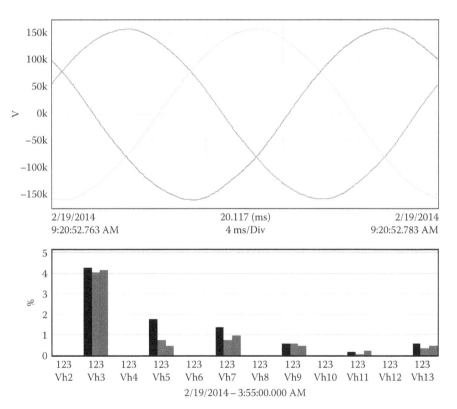

FIGURE 3.20
110 kV grid voltage waveform and its harmonics spectrum.

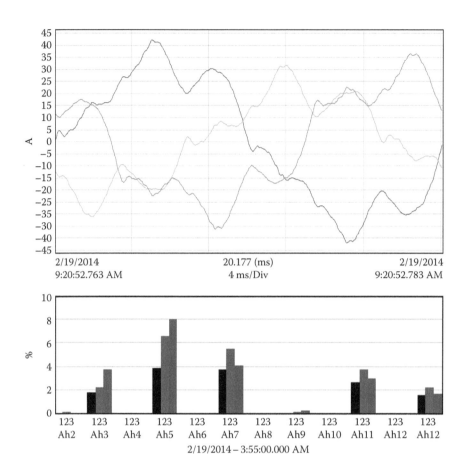

FIGURE 3.21
110 kV grid current waveform and its harmonics spectrum.

TABLE 3.6

Power Transformers and Grid Measurements

Parameters		110 kV I/C Import	110 kV I/C Export	TR1-11 kV	TR2-11 kV	TR3-11 kV	TR4-11 kV
Voltage (V_{rms})		111167	112133	11083	11100	10943	11113
Current (I_{rms})		44	39	351	194	199	579
Power (kW)		8311	−4130	−5618	3064	1487	9952
Power factor		0.974	−0.56	−0.831	0.82	0.396	0.891
% V_{thd}		0.7	1.1	0.4	0.8	1.9	2.1
% I_{thd}		**9.8**	**23.4**	2.4	**6.9**	**9**	4.5
Individual current	3rd	1.6	1.7	3.1	2.4	2.2	3.4
Harmonics (rms)	5th	3.5	5.9	6.5	12.0	15.6	22.6
	7th	1.8	1.8	3.8	4.8	5.6	7.0
	9th	0.1	0.3	0.0	0.0	0.9	0.0
	11th	1.3	1.2	0.4	0.4	4.7	9.2
	13th	0.8	0.7	0.0	0.0	3.5	4.1

TABLE 3.7

Harmonics Measurements in TGs

Parameters		TG1	TG2	TG3	TG4	TG5
Voltage (V_{rms})		11,047	11,077	11,073	11,007	11,140
Current (I_{rms})		349	604	467	1,169	1,106
Power (kW)		6,128	9,255	8,106	18,452	18,990
Power factor		0.915	0.796	0.904	0.827	0.888
% V_{thd}		0.3	0.4	0.4	2	0.7
% I_{thd}		0.7	0.5	0.9	1.9	0.6
Individual current	3rd	0.3	1.8	0.5	4.5	2.2
Harmonics (rms)	5th	2.1	2.4	4.2	18.8	5.6
	7th	1.4	0.6	0.9	7.2	4.4
	9th	0.0	0.6	0.0	1.1	0.0
	11th	0.0	0.0	0.0	8.2	0.0
	13th	0.4	0.0	0.0	4.8	0.0

high-power rectifier for electrolyzer applications. TG-4 voltage and current THDs were higher, and also individual order odd harmonics such as the 5th, 7th, and 11th were high compared to those of the other TGs. Figures 3.22 and 3.23 show TG-4's voltage and current waveforms with the respective harmonics spectra. Since TG-4 was feeding a high power rectifier load, its harmonics levels were high.

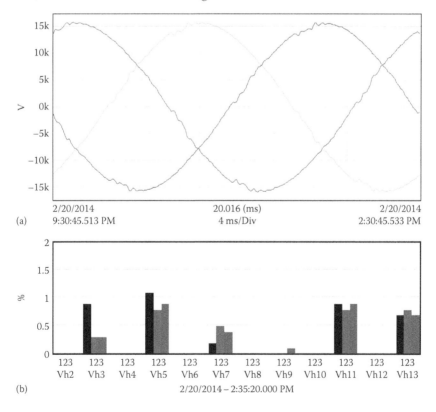

FIGURE 3.22
TG-4 voltage (a) waveform and its (b) harmonics spectrum.

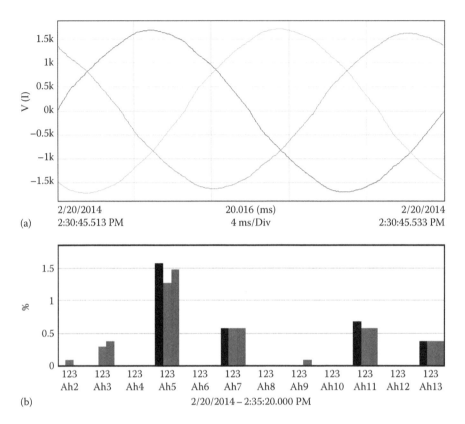

FIGURE 3.23
TG-4 current (a) waveform and its (b) harmonics spectrum.

3.6.3 Harmonic Measurement at Recovery Boiler Distribution

The physical arrangement of the recovery boiler is shown in Figure 3.24. In order to recover the chemicals, the black liquor is first concentrated in the evaporator. The heavy black concentrated liquor is burned in the recovery boiler where lignin becomes the fuel. The recovery boiler burning process is carried out by four burner groups. Each burner mixes the furnace oil/black liquor with a controlled flow of air given by ID fans, SA fan, PA fan, and TA fan, which are driven by 12-pulse and 6-pulse drives. These fans control the furnace oil/black liquor burning process. Each device runs independently of the others but controlled by a closed control loop that maintains the burning process. Therefore, depending on the required burning, the fans run at different speeds and different loadings. Table 3.8 shows the connected load nominal data and drive running data.

The recovery boiler power distribution and the 12-pulse drive installation diagram is shown in Figure 3.25. The recovery boiler plant is supplied from the 24.5 MW TG through distribution transformers of 2 MVA, Dyn11, and 11 kV/415 V. The harmonic study conducted in recovery boiler and the results are presented. The effects of harmonic distortion on the operation of electric drives are discussed. To provide a detailed analysis of the system, one of the 6- and 12-pulse drives was investigated in the recovery boiler.

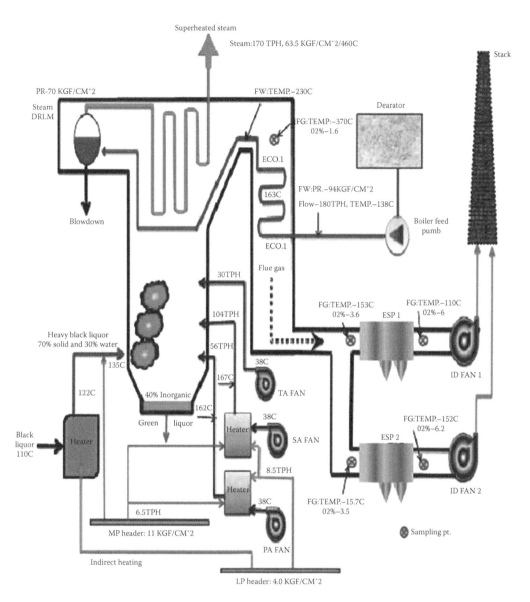

FIGURE 3.24
Recovery power boiler layout.

3.6.3.1 Six-Pulse Drive Harmonic Study

In the recovery boiler power distribution, the primary air fan uses a six-pulse drive for air feeding purpose but same may be changed to a 12-pulse drive because of high harmonic current generation. The three-phase line voltage and current waveform were measured by using a power quality analyzer, and its line voltage and current waveform in PA fan six-pulse drive are shown in Figure 3.26. The drive speed was varied from 1% to 100%, and the harmonics percentages were measured initially and at a current 97.6 A when the drive rpm was the highest. At full speed, the drive line voltage, line current, phase voltage, active

TABLE 3.8

Nominal and Drive Running Data

		Nominal Data				Drive Running Data				Drive
SI. No	Process Equip. Name	kW	V	FLC	RPM	kW (%)	V	RLC	RPM	Type
1.	ID fan-1	585	690	595A	1000	17	335.5	262	486	12-pulse
2.	ID fan-2	585	690	595A	1000	18	345.7	286	493	12-pulse
3.	SA fan	315	690	312A	1500	27.8	428.2	147	910	12-pulse
4.	PA fan	132	415	209A	1500	19.1	243.2	84.5	880	12-pulse
5.	TA fan	90	415	108A	1500	80	380	98	980	6-pulse

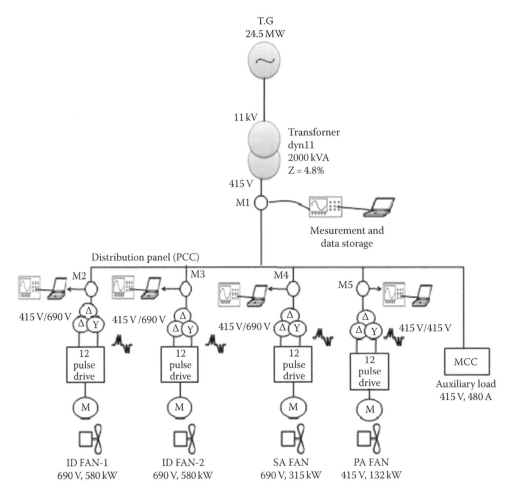

FIGURE 3.25
Boiler power distributions and drives installation.

power, reactive power, power factor, drive frequency, voltage harmonics THD, and current harmonics THD are shown in Figure 3.28. Figure 3.27 shows the three-phase line voltage and line current harmonic spectra. It is clear from this figure that the supply current has very high THD, mostly from the fifth and seventh harmonics. It is clear from this figure that the supply current becomes to nonsinusoidal with 33.87% THD.

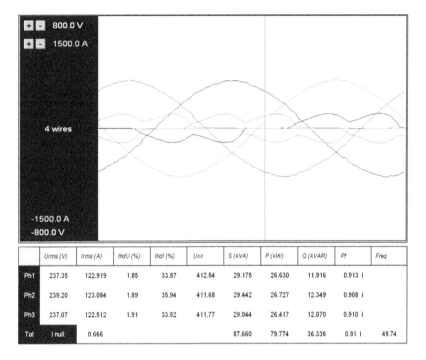

FIGURE 3.26
Six-pulse drive line voltage and current waveform.

Figure 3.28 shows the average current and 5th, 7th, 11th, and 13th harmonic currents from no load to full load. A six-pulse converter would generate harmonic currents of 5th, 7th, 11th, 13th, 17th, 19th, 23rd, 25th, order, and so on.

3.6.3.2 Twelve-Pulse Drive Harmonic Study

The harmonic study was conducted in a 12-pulse drive, and the results are discussed in this section. The effects of harmonic distortion on the operation of electric drives are discussed. To provide a detailed analysis of the system, one of the 12-pulse drives of the 585 kW, 690 V ID fan was investigated in the recovery boiler. The two six-pulse variable frequency drives of similar rating, installed in a delta-star transformer 415/690 V (30° with respect to the primary) on one drive and delta–delta transformer 415/690 V (0° with respect to the primary) on the other drive give an angular displacement of 30° between the two outputs, providing the equivalent of a 12-pulse system for the 700 kVA rating. The three-phase line voltage and current waveforms and its harmonic spectrum are shown in Figures 3.29 and 3.30. A 12-pulse converter would generate harmonic current in the order of 11th, 13th, 23rd, 25th, and so on, at the primary side of the transformer.

The 585 kW, 690 V 12-pulse drive was loaded from 20% to 100%. The I_{THD} varies widely as a function of load current I_L. From this study, a 12-pulse converter would generate a harmonic current in the order of 11th, 13th, 23rd, 25th, and so on, with 7.45% I_{THD} and 1.29% V_{THD}. Hence, 12-pulse converter is nearly in line with IEEE-519 and other standards at the PCC side. The transformer's secondary side voltage and current waveform were measured by using a power quality analyzer and its harmonic spectra are presented. Each six-pulse

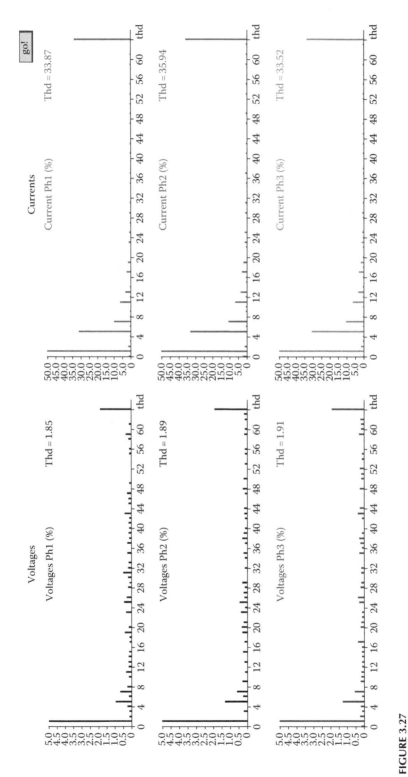

FIGURE 3.27
Six-pulse drive line voltage and current harmonic spectrum.

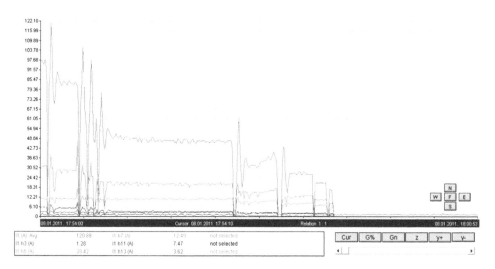

FIGURE 3.28
Average current and 5th, 7th, 11th, and 13th harmonics current.

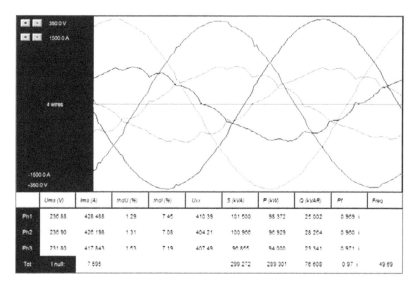

FIGURE 3.29
12-pulse drive line voltage and current waveform.

converter would generate harmonic currents of the 5th, 7th, 11th, 13th, 17th, 19th, 23rd, 25th, and so on, order.

Figures 3.31 and 3.32 show the line voltage and current waveforms and their FFT components with respect to line voltages V_{ab}, V_{bc}, V_{ca} and line currents I_{ab}, I_{bc}, I_{ca} in delta-connected drive and Figures 3.33 and 3.34 show the line voltage and current waveforms and their FFT components with respect to line voltage V_{ab}, V_{bc}, V_{ca} and line current I_{ab}, I_{bc}, I_{ca} in star-connected drive. It is clear from these figures that the secondary current has very high I_{THD}, mostly of 5th and 7th harmonics, and the current becomes nonsinusoidal. Because of the high current harmonics and transformer winding design, the voltage

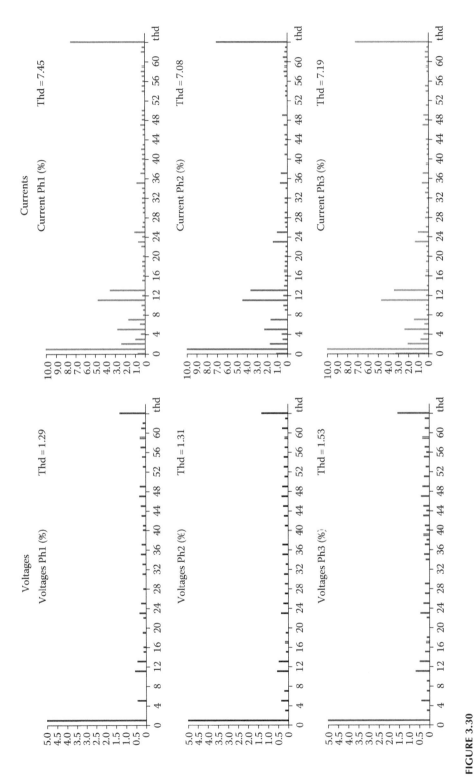

FIGURE 3.30
12-pulse drive line voltage and current harmonic spectrum.

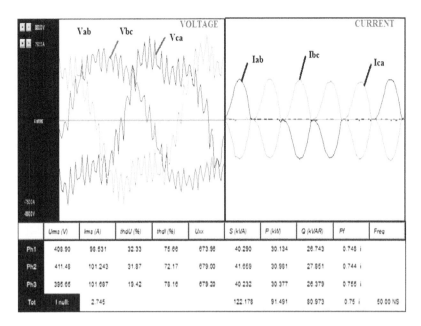

The table in the figure:

	Urms (V)	Irms (A)	thdU (%)	thdI (%)	Uxx	S (kVA)	P (kW)	Q (kVAR)	Pf	Freq
Ph1	408.90	98.531	32.33	75.66	673.96	40.290	30.134	26.743	0.748 i	
Ph2	411.48	101.243	31.87	72.17	679.00	41.659	30.981	27.851	0.744 i	
Ph3	395.65	101.687	19.42	78.16	679.20	40.232	30.377	26.379	0.755 i	
Tot	I null:	2.745				122.178	91.491	80.973	0.75 i	50.00 NS

FIGURE 3.31
Delta winding line voltage and current waveform.

waveforms get distorted and high voltage harmonics would be generated at the second-ary side. Due to the poor quality of the voltage waveform, the drive converter diodes fail frequently. From this case study, a passive filter is required at secondary side of the phase-shifting transformer. The photos of the PCC and the drive side are shown in Figures 3.35 and 3.36.

In the pulp and paper industry, the major nonlinear load is the variable speed drives. These drives would generate harmonics and affect the critical power distribution at the PCC. A six-pulse drive would generate a high harmonic current THD with the order 5th, 7th, 11th, 13th, 17th, 19th, 23rd, 25th, and so on. In 3%–6% line reactor connected drives current THD reduced from 80% to 35%, but lower order harmonics 5th, 7th, and 11th highly dominated. The 12-pulse drive study results also verified the drive supply current, which has very low THD of 7.82% mostly from the 11th and 13th harmonics. But secondary side phase-shifting transformer (star and delta) have high current THD, and this current harmonics affect the secondary side line voltage. The star and delta winding voltage wave-forms get distorted, and this voltage distortion affects the drive rectifier diodes. Based on case study, the 6-pulse drive and the 12-pulse drive inject current harmonics in the sup-ply system. The phase-shifting transformer's secondary side voltage harmonics levels are high, which reduces the drive rectifier diode life time.

3.6.4 MATLAB®/Simulink® Model for 6-Pulse Drive

Figure 3.37 shows the six-pulse drive system model in Simulink. The system consists of a six-pulse rectifier model with a DC link choke and a PWM inverter. The same system is simulated with and without line reactors.

Figures 3.38 and 3.39 shows the simulation results of voltage and current on the 11 kV side and three-phase voltages and line current I_{abc} on the 415 V side of the transformer and

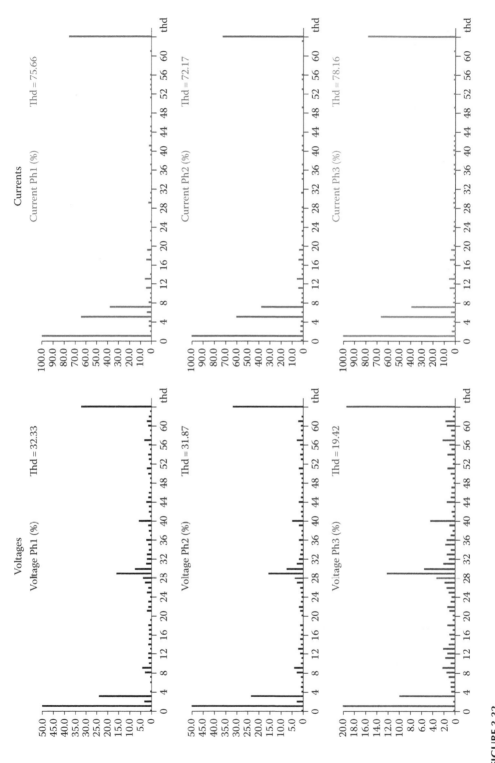

FIGURE 3.32
Delta winding line voltage and current harmonic spectrum.

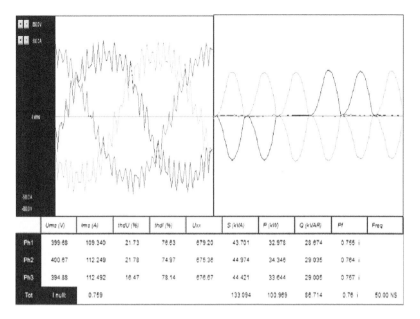

FIGURE 3.33
Star winding line voltage and current waveform.

the simulation waveforms for voltage and current on the load side. (a) V_{DC}, (b) V_{inv}, (c) V_{ab}, (d) I_{ab}, and (e) modulation index.

Figure 3.40 shows the simulation results of the phase current I_a and Figure 3.41 shows the real-time measured waveforms for the phase voltage V_a and phase current I_a without the line reactor. Line reactors commonly used are of two types: one is the AC-side three-phase line reactor, and the other the is DC-side line reactor. The simplest and most economical passive harmonic reduction technique involves the use of AC line reactors (L_{ac}) in front of the VFDs. The series inductance filter or inline reactance is a well-established method. Typically, 1%–5% L_{ac} inductors are used. In the United States, 3% and in Europe 4% are commonly utilized. The effective impedance value in percent is based on the actual loading, given by

$$Z_{eff} = \frac{2\sqrt{3}\pi f L I_{fnd}}{V_{LL}} \times 100 \tag{3.4}$$

The AC inductor's reactance increases proportional to the system frequency. Therefore, the inductance smoothens the line current drawn by the converter. Thereby, a significantly lower current harmonic distortion can be achieved, down to 30%–35% THDI range compared to the basic VFD THDI. This THDI range can be improved when a DC link inductance is combined with the AC line reactors. Unlike the AC line reactors, the DC link inductance does not cause any reactive voltage drop while contributing to shaping the current waveforms. It is known that the effective impedance of the DC link inductance, when referred to the AC side, is approximate the half of its numerical value. A DC link inductor size between 3% and 5% is typically built into some commercial VFD systems.

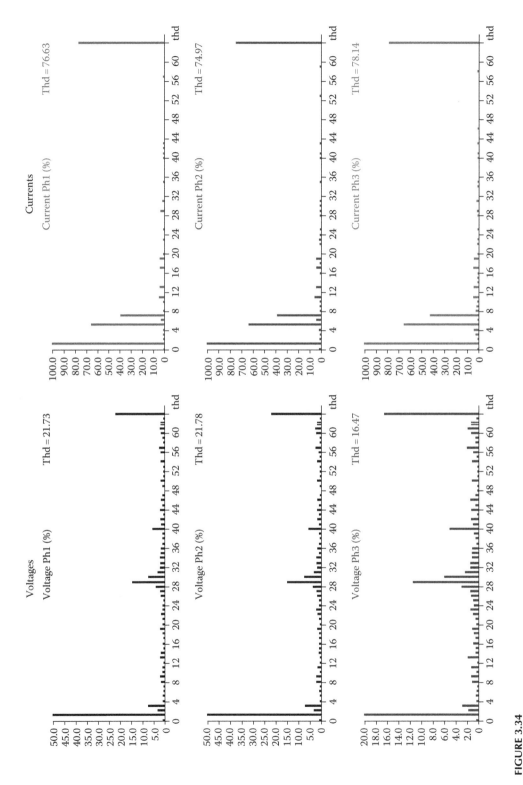

FIGURE 3.34
Star winding line voltage and current harmonic spectrum.

FIGURE 3.35
Photo of the measuring equipment installed at PCC.

FIGURE 3.36
Photo of the harmonic measuring equipment.

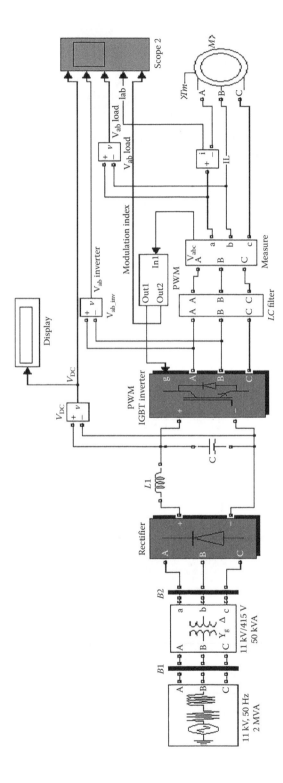

FIGURE 3.37
Simulation model of the six-pulse drive.

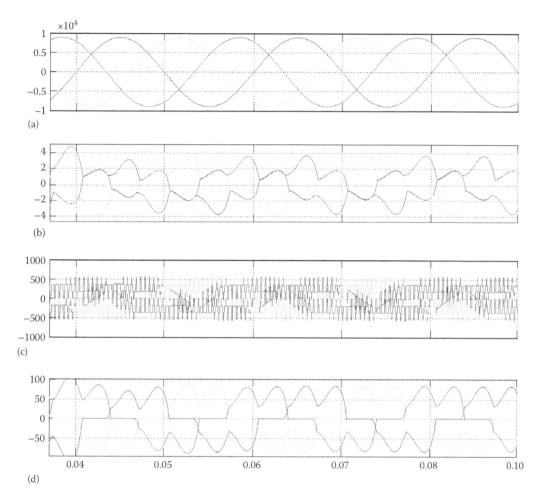

FIGURE 3.38
Simulation waveforms for voltage and current on 11 kV/415 V side of the transformer primary: (a) three-phase voltages, (b) line current, I_{abc}. Transformer secondary: (c) three-phase voltages, and (d) line current, I_{abc}.

Figure 3.42 shows the supply current I_{ab} waveform and its FFT components with respect to phase voltage V_{ab}. It is clear from this figure that the supply current has very high THD, mostly of the fifth and seventh harmonics. It is clear from Figure 3.43 that the supply current becomes to nonsinusoidal with 38.93% THD, and Table 3.9 shows the results of the simulation and case study result comparison for six-pulse drive.

3.6.5 MATLAB®/Simulink® Model for 12-Pulse Drive

The 12-pulse converter is simulated in MATLAB environment along with Simulink toolboxes. Figure 3.44 shows the MATLAB model of the proposed phase-shifting transformer fed by a 12-pulse drive. The simulation results were analyzed to study the effect of load variation on the drive. Figure 3.45 shows the supply voltage and current waveform as well as its FFT components with respect to the line voltage V_{ab} and current I_{ab}. From this figure, it is seen that the supply current has very low THD (7.82%), mostly of 11th and 13th harmonics. Figures 3.46 and 3.47 show the star and delta winding voltage and current

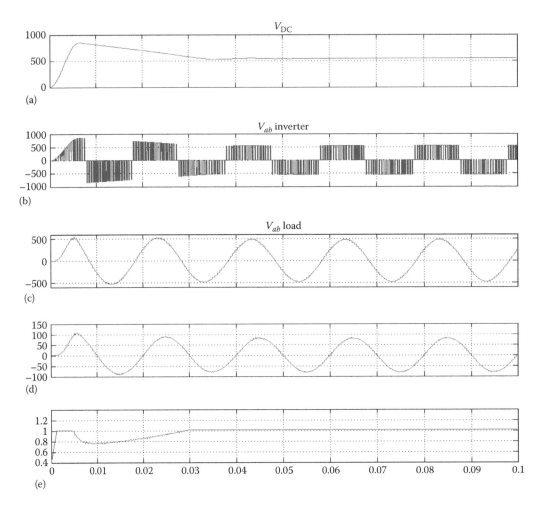

FIGURE 3.39
Simulation waveforms for voltage and current on load side. (a) V_{DC}, (b) V_{in}, (c) V_{ab}, (d) I_{ab}, and (e) modulation index.

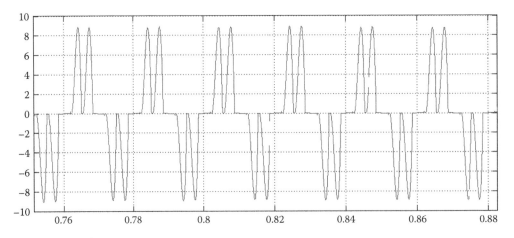

FIGURE 3.40
Simulation waveforms for phase current I_a without line reactor.

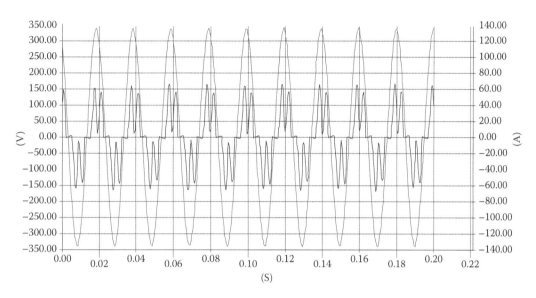

FIGURE 3.41
Real-time measured waveforms for phase voltage V_a and phase current I_a without line reactor.

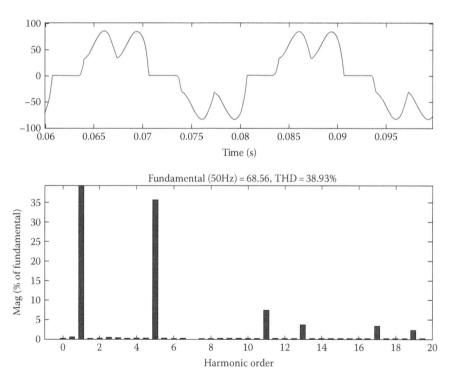

FIGURE 3.42
Simulation FFT result for line current I_{ab}.

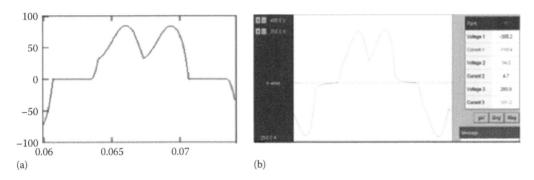

FIGURE 3.43
Simulation and case study phase current waveforms: (a) simulation waveform and (b) real-time case study waveform.

TABLE 3.9

Simulation and Case Study Result Comparison for Six-Pulse Drive

Harmonic Order	Simulation Results		Case Study Result	
	Phase Current	% Distortion	Phase Current	% Distortion
1	117.45	100	116.41	100
5	34.33	29.23	36.45	31.31
7	9.86	8.40	11.62	9.98
11	7.50	6.38	7.33	6.30
13	4.16	3.54	3.72	3.19
17	2.87	2.45	2.88	2.48
19	2.30	1.96	2.01	1.72
%THD	%THD V	%THD I	%THD V	%THD I
%	1.60	33.93	1.83	33.87

waveform along with the FFT results. The simulation waveform and case study measurement waveform FFT results are compared in Table 3.10.

The simulated waveform would generate 11th and 13th harmonics with 7.82% THD, and measured waveform would generate harmonics with 7.45% THD. The simulation results of THD were acceptable with the practical measurements. These models can hence be employed for harmonic analysis of a practical system and to design a suitable passive filter to mitigate harmonics in the 12-pulse drive. The passive filter is a very good choice for drive applications and a cost-effective solution to harmonic reduction and power factor improvement. The passive filter consists of elements such as inductors, capacitors, and resistors for filtration.

In the pulp and paper industry, more variable speed drives are used because of system and process requirement compared to other industries. These variable speed drives create power quality issues such as harmonics in the power distribution system. This chapter clearly explains the harmonics measurement and harmonics levels at various points from TG to 6-pulse and 12-pulse drives. The harmonics study covers TNEB power grid, power transformer, power distribution PCC points, and distribution transformers, and the recovery boiler drive system harmonics are clearly explained. Six-pulse drive

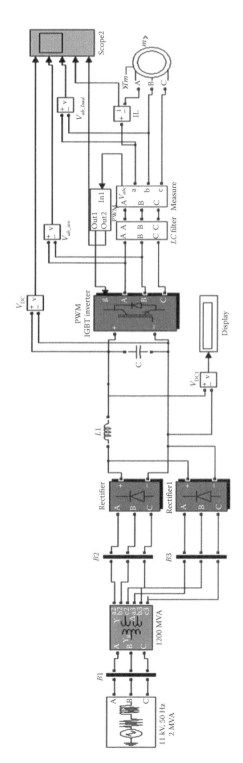

FIGURE 3.44
Simulation model of the 12-pulse drive.

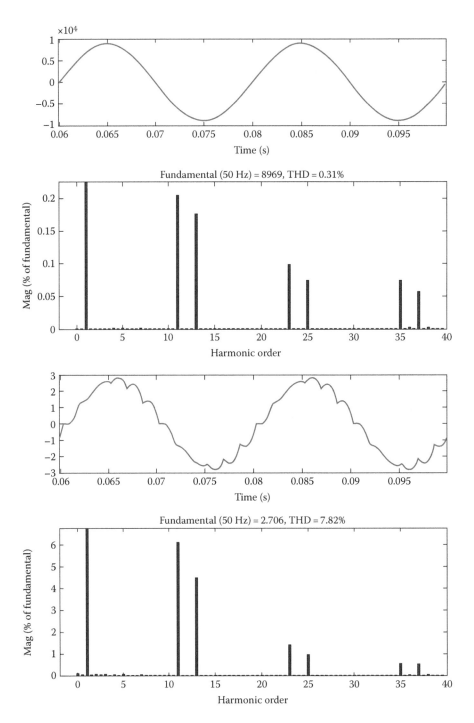

FIGURE 3.45
Simulation and FFT results for input phase voltage and current.

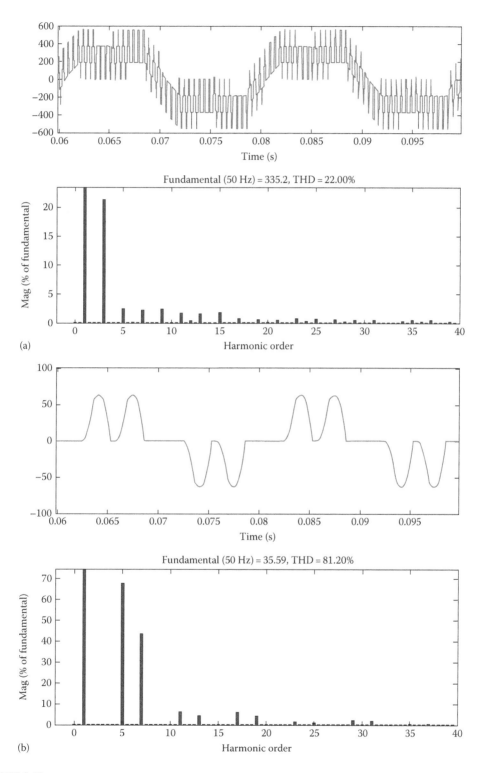

FIGURE 3.46
Simulation and FFT results for star winding (a) voltage and (b) current.

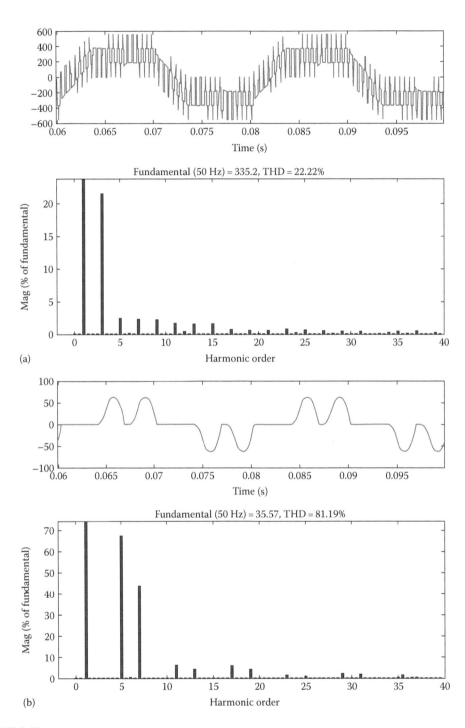

FIGURE 3.47
Simulation and FFT results for delta winding (a) voltage and (b) current.

TABLE 3.10

Case Study and Simulation Result Comparison for 12-Pulse Drive

Transformer	Case Study Results		Simulation Result	
	%THDV	%THD I	%THDV	%THD I
Pry. side Δ	1.29	7.45	0.31	7.82
Sec. side Y	21.73	76.63	22.22	81.19
Sec. side Δ	32.33	75.55	32.00	81.30

and 12-pulse drive harmonics measurements present various load and speed conditions. The cumulative effects of linear and nonlinear loads reduce the harmonics level at the PCC.

Based on case study, MATLAB simulations were carried out, and the results are presented. The case study parameters are input to the MATLAB/Simulink circuit simulator. The simulation results and case study results are matched, and the simulated 6-pulse drive models and 12-pulse drive models are used to design a suitable passive filter to mitigate harmonics in the 6-pulse and 12-pulse drives. The passive filter is a very good choice for drive applications and is a cost-effective solution to harmonics reduction and power factor improvement. The passive filter consists of elements such as inductors, capacitors, and resistors for filtration.

3.7 Genetic Algorithm-Based Filter Design in 2-, 6-, and 12-Pulse Rectifier

3.7.1 Problem Formulation

The mode of operation and experimental analysis of real-time VFDs presented in Sections 1.5 and 1.6 confirm the frequency of distortions in voltage and current waveform attributable to the voltage and current in the system. Further, the existence of harmonics in the system leads to poor supply power factor, motor torque pulsation, and failure of drives. Hence, it becomes mandatory to add a filter to mitigate unwanted noise and distortions in the current and voltage waveforms. The most common practice for harmonic mitigation is the installation of a passive harmonic filter. Such a filter is a simple and cost-effective means of mitigating harmonics and improving the power factor. It supplies reactive power to the system, and at the same time is highly effective in attenuating the harmonic components.

To minimize the computational complexity associated with the filter design, a MATLAB model of a two-pulse drive with different combination of *LC* filters is first developed. The design of the passive *LC* filter is then transformed to as optimization task, and steps of GA optimization are employed to get the optimum values of *L* and *C*. GA is a biologically inspired population-based algorithm. The GA optimization method has the ability to find near-global optimal values with initial random guesses. A quantitative analysis on the performance of the drive is carried out using MATALB simulation. The proposed GA-based *LC* filter, when introduced into the drive, showed considerable improvement in power factor and lower current harmonics. To verify the simulation findings, a hardware prototype was fabricated and added to the two-pulse drive. The experimental results confirm the reduction in harmonics and improvement in the power factor. The objective of power factor improvement is drafted as an optimization task. Maximize

$$F(\phi) = \text{Power factor } (\cos\theta) \tag{3.5}$$

subject to

$$\phi_{min} \leq \phi \leq \phi_{max}$$

where $\phi = \{L, C\}$, and the subscripts indicate the values of the boundary values of the filter components. In GA-based design, emphasis is also given to minimize the size of the filter components.

3.7.2 Genetic Algorithm-Based Filter Design

In this section, an overview of GA and the steps involved in the GA algorithm are explained. The objective of power factor improvement together with harmonic elimination is framed as an optimization task, and the same is solved using GA, which generates solutions to the optimization problem using techniques inspired by natural evolution, such as inheritance, selection, crossover, and mutation. It is a biologically inspired, population-based algorithm and was developed by John Holland, University Of Michigan, to understand the process of natural systems. It is widely used in scientific and engineering fields. The main steps involved are initializing population, evaluation of fitness, selection of survivors based on fitness, and the crossover and mutation operations on the survivors. Here, a population of strings (called chromosomes), which encode candidate solutions (called individuals) to an optimization problem, evolves toward better solutions.

Usually, the algorithm terminates when either a maximum number of generations has been produced or a satisfactory fitness level has been reached for the population. If the algorithm terminates because of a maximum number of generations, a satisfactory solution may or may not have been reached. A typical genetic algorithm requires representation of the solution domain and fitness function to evaluate the solution domain. A flowchart representing the various steps involved is presented in Figure 3.48. Only the fitter individuals in the population are allowed to pass their chromosomes to the next generation. Random individual values are initialized and made to run till the best solution is reached.

3.7.2.1 Steps Involved in GA Based Filter Design

The following section explains how GA is used for design of filter components:

Step 1: Create a population of initial solution of parameters (*L* and *C*)

This step primarily requires the population size. Each variable in the problem is called a gene, and in the present problem there are two (i.e., *L* and *C*) genes. A chromosome consists of the genes, and thus each chromosome represents a solution to the problem. This is illustrated in Figure 3.49. The population consists of a set of chromosomes. It is well articulated in the literature that a population size of 10–30 is an ideal one, and hence the population size is selected as 15 in this example.

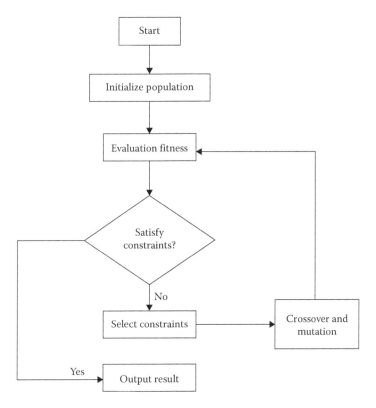

FIGURE 3.48
Flowchart of genetic algorithm.

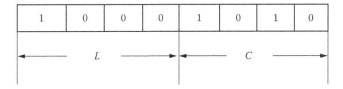

FIGURE 3.49
Chromosome structure.

Step 2: Evaluation of objective function

In the present problem, the input power factor is to be maximized, and the corresponding objective function $F(L, C)$ is computed.

$$F(L, C) = \text{Power factor} \tag{3.6}$$

Step 3: Evaluation of the fitness function

The degree of "goodness" of a solution is qualified by assigning a value to it. This is done by defining a proper fitness function to the problem. Since GA can be used for maximization and minimization problems, the following fitness function is used:

$$\text{Fitness function} = F(\phi) = 0.95 \le \cos\theta \le 1 \tag{3.7}$$

Step 4: Generation of offspring

Offspring is a new chromosome obtained through the steps of selection, crossover, and mutation. After the fitness of each chromosome is computed, parent solutions are selected for reproduction. It emulates the survival of the fittest mechanism in nature. The roulette wheel selection is the most common and easy-to-implement selection mechanism. A virtual wheel is implemented for this selection process. Each chromosome is assigned a sector in this virtual wheel, and the area of the sector is proportional to its fitness value. Thus the chromosome with largest fitness value will occupy the largest area, while the chromosome with a lower value takes the slot of a smaller sector. Let there be five chromosomes labeled A, B, C, D, and E, and let their fitness value increase in the order D, B, A, E, and C.

Figure 3.50a shows a typical allocation of five sectors of chromosomes in the roulette wheel. In roulette wheel selection, an angle is generated randomly, and the chromosome corresponding to this angle is selected. Figure 3.50b shows a randomly generated angle of 4/3 rad. In this case, chromosome C is selected. The chromosomes thus selected are called parent population and are subjected to undergo crossover and mutation to produce offspring for the next generation. Conventional method adopted in GA is the roulette wheel selection, and in this chapter we modify this selection method by combining it with elitism. Using elitism, a finite number of best solutions are retained and are reused in the next generation without undergoing the steps of mutation and crossover.

Following the selection of the parent population, crossover and mutation are performed to generate the offspring population. In crossover, randomly selected subsections of two individual chromosomes are swapped to produce the offspring. Here, multipoint crossover is adopted for increased efficiency since three variables are embedded in one chromosome mutation is another genetic operation by which a bit within a chromosome may toggle to the opposite binary. Figure 3.51 illustrates crossover and mutation. These operations are performed based on the probability of crossover and mutation.

Step 5: Replace the current population with the new population

Step 6: Terminate the program if the termination criterion is reached; else go to step 2.

The GA method offers an advantage in terms of computational burden. The optimal values of L and C using conventional method are obtained after trying out many combinations of L and C, where as they are obtained very easily through the GA method.

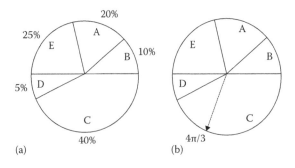

(a)   (b)

FIGURE 3.50
(a, b) Roulette wheel selections.

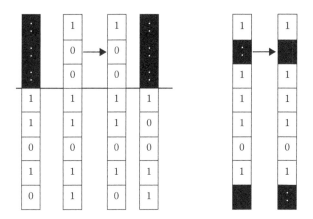

FIGURE 3.51
Crossover and mutation.

3.7.3 MATLAB®/Simulink® Model of Filters

This section presents two pulse rectifier configurations with different *LC* filters. The selection methodology and the design procedure involved are explained. Four different combinations are explained and their GA-0based simulation results are presented.

3.7.4 Series *LC* Filter Configuration

In this section, a series passive *LC* filter configuration is introduced. A two-pulse rectifier with a series *LC* filter connection is shown in Figure 3.52.

The filter impedance varies with the supply frequency. Since the load draws a nonsinusoidal current containing harmonics from the supply, the filter circuit offers infinite impedance to the harmonic components and zero impedance to the fundamental component. The purpose of the series filter is to block the harmonic frequency components and allow the fundamental frequency. The selection methodology of *L* and *C* is explained later.

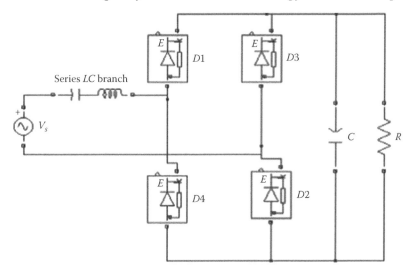

FIGURE 3.52
Diode bridge rectifiers with series *LC* filter.

Determination of L and C values using the conventional method: The filter circuit impedance changes with frequency, and its value is given by the well-known series impedance equation

$$Z = X_L - X_C \tag{3.8}$$

where

$$X_L = 2 * \pi * f * L$$

$$X_C = \frac{1}{2 * \pi * f * C}$$

f_r is the resonant frequency

For the impedance of the filter to become zero at the fundamental frequency ($f_r = 50$ Hz), the inductive reactance should be equal to the capacitive reactance at this frequency according to the formula given.

Hence, $X_L = X_C$

$$2 * \pi * f_r * L = \frac{1}{(2 * \pi * f_r * C)}$$

$$f_r = \frac{1}{2\pi\sqrt{LC}}$$

The thumb rules to find the L and C values are as follows: fix the value of L and determine C. This way, there exist different combinations of L and C and a few combinations. A MATLAB/Simulink model is developed, and the values are substituted. The voltage and current waveforms and the current harmonic spectrum for a typical combination of L and C are shown in Figures 3.53 and 3.54.

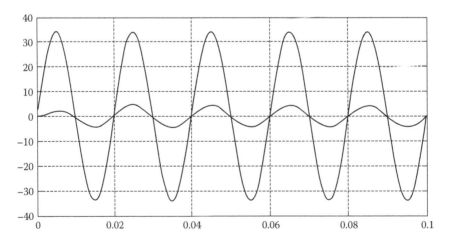

FIGURE 3.53
Simulated input voltage and current waveforms with series *LC* filter.

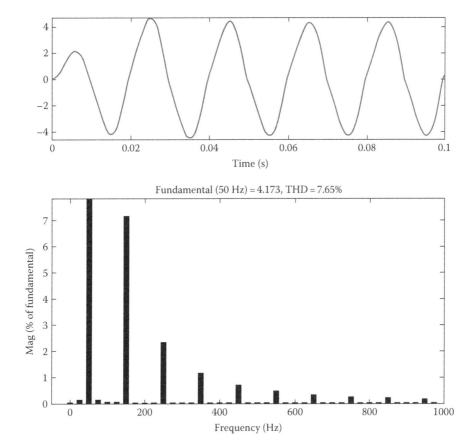

FIGURE 3.54
Simulated harmonic spectra and input current waveform with series *LC* filter.

The different parameters such as the input power factor, output voltage, input rms current, and THD values are computed and tabulated in Table 3.11. From the values, it is clear that the output voltage across the load is not maintained for any combination of *L* and *C*. It is reduced since the filter components are connected is series. This could be attributed to the voltage drop across the series impedance. It can also be inferred that to obtain a low THD value and a good power factor, the value of the inductor should be considerably high (~25 mH), which results in a large inductor size and an increase in the cost of the inductor. And to obtain a very low value of THD, the inductor value has to be further increased, which complicates the circuit with the presence of a huge inductor. Thus this configuration

TABLE 3.11

Series *LC* Filter Simulation Results

S. No.	L (mH)	C (μF)	PF (cos θ)	V_o	I (rms)	THD (%)
1.	5	2026.4	0.9439	18.64	3.566	34.79
2.	10	1013.2	0.9635	18.7	3.362	24.84
3.	15	675.47	0.9758	18.69	3.288	21.29
4.	20	506.6	0.9870	18.67	3.188	13.75
5.	25	405.28	0.9928	18.64	3.059	10.86

has its own drawbacks and limitations, and these results would be compared to the other configurations whose results will be discussed.

3.7.5 Parallel *LC* Filter Configuration

In this section, a parallel passive *LC* filter configuration is introduced. In a parallel *LC* filter, *L* and *C* connected in parallel, as shown in Figure 3.55. The filter impedance varies with the supply frequency. Since the load draws a nonsinusoidal current containing harmonics from the supply, the filter circuit offers infinite impedance to the harmonic components and zero impedance to the fundamental component.

The circuit configuration where the *LC* filter is connected in parallel is shown in Figure 3.55. The basic concept is to block the dominant third-harmonic frequency and to pass the remaining frequencies. This allows the fundamental frequency and the other low-dominant harmonics to pass through.

3.7.5.1 Determination of L and C Values Using Conventional Method

The impedance of a parallel *LC* filter is given by

$$Z = \frac{X_L * X_C}{(X_L - X_C)} \tag{3.9}$$

where
$$X_L = 2 * \pi * f * L$$

$$X_C = \frac{1}{2 * \pi * f * C}$$

f_r is the resonant frequency

For the filter to offer high impedance to the third harmonic, the filter impedance should be assumed to be infinite at $f_r = 150$ Hz. This blocks all the third harmonics, and allows the fundamental and other harmonics to pass through.

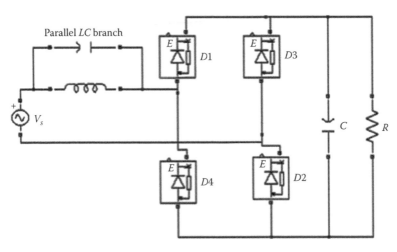

FIGURE 3.55
Parallel *LC* connected in series with the source.

Hence, by equating $Z = \infty$ and letting $X_L = X_C$,

$$2\pi * f_r * L = \frac{1}{(2\pi * f_r * C)}$$

$$f_r = \frac{1}{2\pi\sqrt{LC}}$$

To find the L and C values, fix the value of L and determine C. This way, we arrive at different combinations of L and C. Table 3.12 represents the results obtained with these different combinations of L and C. A MATLAB/Simulink model is developed, and the values are substituted. The voltage and current waveforms and the current harmonic spectrum for a typical combination of L and C are shown in Figures 3.56 and 3.57. For a parallel LC filter, the different parameters such as the input power factor, output voltage, input rms current, and THD values are computed and tabulated in Table 3.12.

TABLE 3.12

Parallel LC Filter Simulation Results

S. No.	L (mH)	C (μF)	PF (cos θ)	Vo	I (rms)	THD (%)
1.	1	1100	0.9651	18.21	3.1290	24.46
2.	2	562.9	0.9714	18.29	3.1202	22.79
3.	3	375.26	0.9771	18.35	3.1109	21.19
4.	4	281.45	0.9810	18.3918	3.0997	19.79
5.	5	225.16	0.9829	18.39	3.0856	18.59

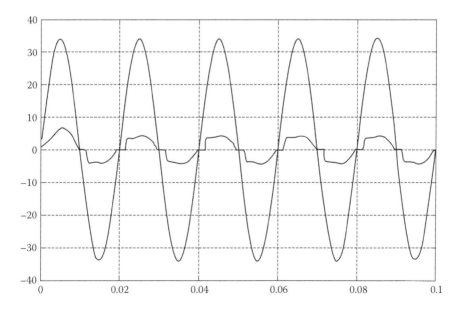

FIGURE 3.56
Simulated voltage and current waveform with parallel filter.

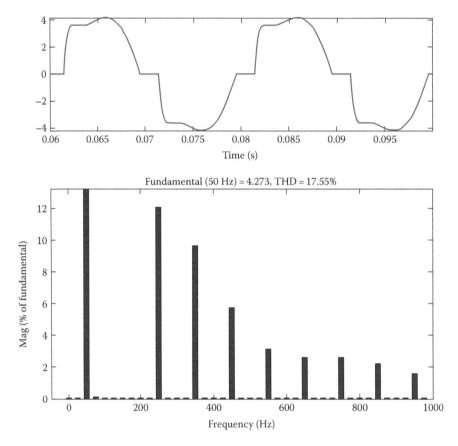

FIGURE 3.57
Simulated harmonic spectra for parallel *LC* filter.

The parallel *LC* filter for the harmonic elimination is designed using the conventional method. From the values obtained, it can be observed that the output voltage across the load is not maintained for any combination of *L* and *C* like in the series configuration. The THD, which is ~16%, is a bit higher than expected. The waveform is still distorted though the power factor is good. When we look at the harmonic spectrum of the line current, we observe that the third harmonic is completely eliminated and the other higher harmonics such as the 5th, 7th, 9th, 11th, and so on, exist, which can also be eliminated using the respectively designed individual parallel *LC* filters connected in series. But this again increases the complexity of the circuit, for which an alternative has to be found. Thus all these factors are to be taken into account when designing a new configuration of *L* and *C* that can overcome these drawbacks.

3.7.6 Shunt *LC* Filter Configuration

The previously designed filter circuit has the drawback of introducing a voltage drop in the circuit. An increase in the voltage drop results in poor efficiency. Hence in this section a shunt passive *LC* filter configuration is designed using GA for harmonics elimination and power factor improvement. The current waveform and the harmonic spectrum are shown, and the other parameters such as the power factor, output voltage, input rms

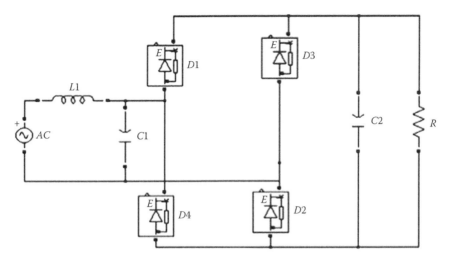

FIGURE 3.58
Full bridge rectifier circuit with shunt *LC* filter.

current, and THD values are tabulated, where a detailed analysis is made based on these values obtained for different values of *L* and *C* produced the optimization method.

The circuit is shown in Figure 3.58. A full bridge rectifier circuit with a shunt *LC* filter was modeled in MATLAB, and a GA program was used to get the optimal values of *L* and *C*, which resulted in a good power factor and low THD. Table 3.13 shows the different parameter values obtained using the GA-based design of the *L* and *C* values for a shunt filter configuration.

The values obtained are very satisfactory, with all the constraints being satisfied. The output voltage is maintained constant for any value of *L* and *C* with less THD and near-unity power factor. The filter component values are also small compared to those of the series filter components.

The simulated input voltage and current are shown in Figure 3.59. The voltage and current waveform coincide and are in phase. The simulated harmonic spectrum of the input current is shown in Figure 3.60. A closer observation reveals that the harmonic content has reduced to a larger extent when compared to series *LC* filter. From the earlier discussion, it is inferred that the GA-based design gives better results and is able to overcome all the limitations.

3.7.7 *LCL* Filter Configuration

In this section, an *LCL* filter configuration is designed using GA for harmonic elimination and power factor improvement. Figure 3.61 shows the full bridge rectifier circuit with the *LCL* filter. The voltage and current waveforms and the harmonic spectrum are shown in Figures 3.62 and 3.63. The other parameters such as power factor, output voltage, input rms current and

TABLE 3.13

Shunt *LC* Filter Simulation Results

S. No.	*L* (mH)	*C* (µF)	PF	*Vo*	*I* (rms)	THD (%)
1.	14.3	220	0.9915	23.13	4.96	6.99
2.	15.5	254.59	0.9966	25.29	4.505	6.95
3.	16.7	266.24	0.9972	25.77	4.686	6.1
4.	17	250.71	0.9966	25.13	4.451	6.35

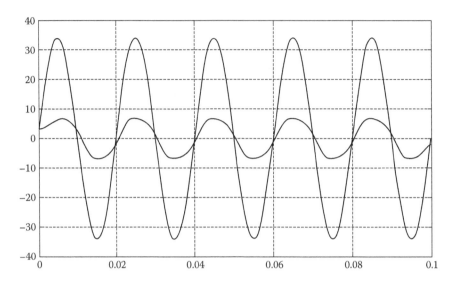

FIGURE 3.59
Simulated voltage and current waveforms for shunt *LC* filter.

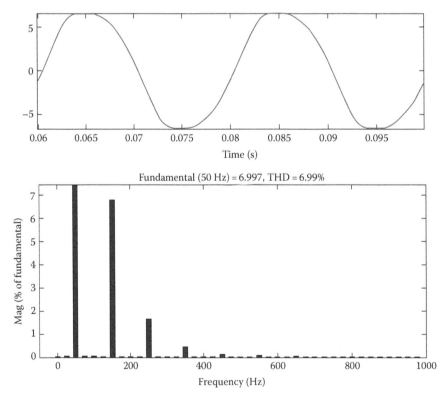

FIGURE 3.60
Simulated harmonic spectra for shunt *LC* filter.

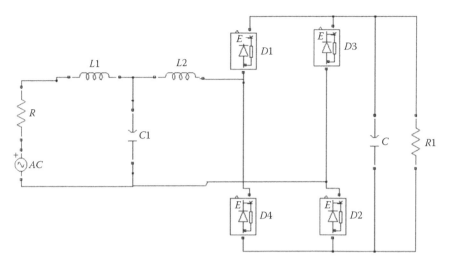

FIGURE 3.61
Full bridge rectifier circuit with *LCL* filter.

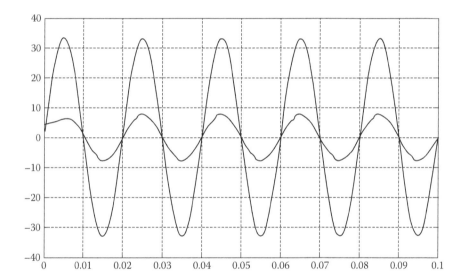

FIGURE 3.62
Simulated voltage and current waveforms for *LCL* filter.

THD values are also tabulated. The circuit mentioned was implemented in MATLAB and GA code was modified accordingly to get the optimal values of *L*1, *L*2, and *C*, which could result in a good power factor and less THD. Table 3.14 presents the different parameter values obtained using the GA-based design of the *L*1, *L*2, and *C* values in the *LCL* filter configuration.

3.7.7.1 Analysis of the Observations

Four *LC* filter combinations were designed using GA, and the simulated voltage and current waveforms and harmonic spectra were presented. When the readings were analyzed, it was observed that the results were more satisfactory than those obtained using a shunt *LC*

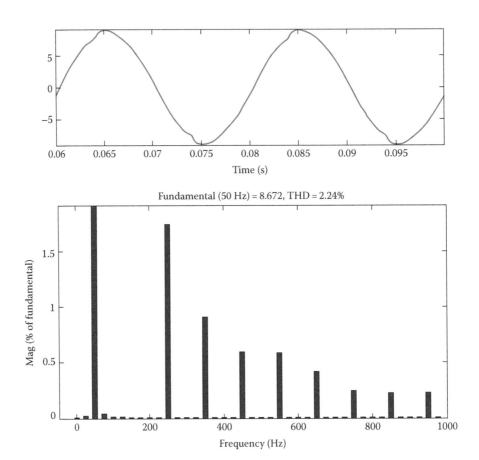

FIGURE 3.63
Simulated harmonic spectra for *LCL* filter.

TABLE 3.14

LCL Filter Simulation Results

S. No.	L1 (mH)	C (µF)	L2 (mH)	PF (cos θ)	Vo	I (rms)	THD (%)
1.	12.5	90	1.2	0.9928	25.07	5.98	2.328
2.	10	105	2.1	0.9954	22.63	4.778	4.036
3.	11	110	2.4	0.9998	23.96	5.4	3.45

filter configuration and an *LCL* filter. The desired power factor was obtained, with the output voltage regulated almost near 24 V, and the THD was very low. The current waveform shape was also nearly sinusoidal. Third harmonic was found to be completely eliminated, and the other harmonics were significantly less. When the *LCL* filter was implemented in hardware for a two-pulse drive, the circuit gave almost similar results to those obtained in simulation. Thus the simulated results were validated by the hardware implementation too.

3.7.7.2 Comparison of Results

To indicate the effectiveness of the GA-based design, a comparative study was made of the four filter configurations. A comparison table was made for the filter performances in

TABLE 3.15

Simulation Results Comparison

Configuration	Power Factor	THD (%)
Without filter	0.6600	105.70
Parallel LC	0.9827	17.79
Series LC	0.9900	7.01
Shunt LC	0.9970	6.11
LCL	0.9998	3.45

terms of the power factor and percent THD values. From Table 3.14, it was clear that the addition of filter not only improved the power factor but also reduced the current harmonics THD to a large extent. The series LC, parallel LC, shunt LC, and LCL filter combinations were designed using GA, and the filter performances were studied for different values, out of which the shunt LC and LCL filters showed improved power factor and also reduced THD. The series and parallel filters work with poor voltage regulation and also optimized power factor improvement, but THD reduction was not achieved. Table 3.15 shows the simulation results comparison of the LCL filter.

3.7.7.3 Input LCL Filter for Two-pulse Drive

Two-pulse drives are used for low-power applications, and in industries such drives are used for paper cutting wheels, measuring equipment, and hoist applications. The experimental analysis is made from a pulp Kappa analyzer drive from Metso. The pump is operated by a capacitor-run single-phase induction motor. Based on the Kappa value of the pulp, the pump has to run for the pre determined time. This will operate in a closed loop system, the kappa value of the pulp will be taken from the Kappa analyser. It pumps hot water for pulp washing at different rpm values. Figure 3.64 shows the two-pulse VFD with an input passive filter. The proposed filter combination is shown in Figure 3.65. The value of $L2$ is the 3% line reactor, and the values of $L1$ and C are designed using the GA method. The two-pulse drive generates low-order harmonics of the 3rd, 5th, 7th, 9th, 11th, and 13th order. These harmonics need to be eliminated for energy-efficient functioning of the drive. Hence in the following input filter, the lower order harmonic elimination and power factor improvement is drafted as an optimization task, which is solved using GA.

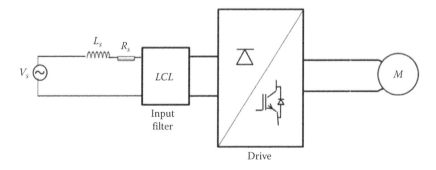

FIGURE 3.64
Two-pulse drive with proposed input filter.

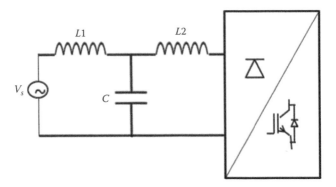

FIGURE 3.65
Proposed input filter connections.

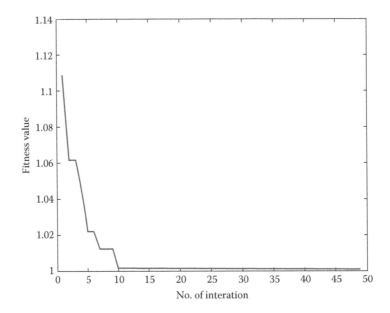

FIGURE 3.66
Input *LCL* filter GA convergence curve.

The optimal *LCL* filter values were identified through the steps of GA procedure, and the performance of the designed filter was analyzed. The best values obtained through the GA design are $L1 = 1.1$ mH, $L2 = 12.22$ mH, and $C = 22.21$ μF. The convergence characteristic curve for GA is shown in Figure 3.66. The power factor improvement is drafted as optimization task, and the same may be achieved through the GA method. The designed filter with the earlier values was connected to a two-pulse drive, and the drive speed and load were varied from 10% to 100%; at the same time, power factor was recorded using a power quality analyzer. The recorded values are shown in Figure 3.67. The experimental results are shown with and without the filter, and from the results it is evident that the power factor value is near 0.99. At the same time, the current harmonics are measured with and without the filter, and the experimental results are shown in Figure 3.68. Form the results, the current THD reduced from 71.9% to 8.9%, and lower order harmonics such as the 3rd, 5th, 7th, and 9th were eliminated in current waveform.

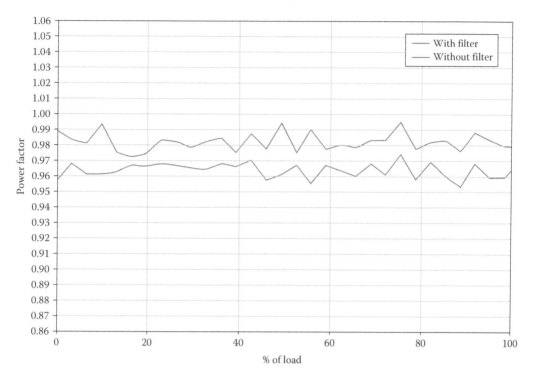

FIGURE 3.67
Experimental power factor results for various loading condition.

In summary, four different filter configurations with L and C were explained with the selection methodology. The design procedure using GA method was explained and the simulation results were presented. From the results, it is clear that conventional method of filter design takes a long time and provides poor filtering. Thus the series LC, parallel LC, shunt LC, and LCL filters were designed using GA, and the comparison results were presented. Next the LCL configuration was designed using GA algorithm for the pulp Kappa analyzer two-pulse drive, and the results both the simulation and hardware results were presented.

3.7.8 Genetic Algorithm-Based Filter Design in 6- and 12-Pulse Rectifier-Fed Drive

In the pulp and paper industry, a number of variable speed drives are used because of system and process requirements. We present now the harmonics study results from six-pulse and 12-pulse variable frequency drives. The front-end topology for the VFDs is still the six-pulse diode rectifier because of its well-known advantages such as high efficiency, low cost, robustness, and reliability. Nevertheless, the major drawback of this type of VFD is the generation of harmonic currents. Many choices exist for harmonic minimization such as adding line reactors, passive harmonic filters, harmonic traps, and active filters.

The most common practice for harmonic mitigation is the installation of passive harmonic filters. Passive filters show the best cost–benefit relationship among all mitigation techniques when dealing with low- and medium-voltage rectifier systems. Filter banks installed in medium-voltage systems are able to provide satisfactory reduction in voltage and current distortions. The input filter has four primary functions. One is to prevent electromagnetic interference generated by the switching source from reaching the power

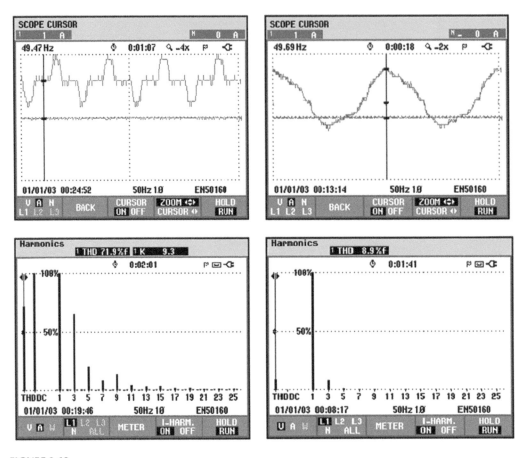

FIGURE 3.68
Experimental harmonics spectra for input *LCL* filter.

line and affecting other equipment. The second is to prevent high-frequency voltage on the power line from passing through the output of the power supply. Third is to improve the power factor, and the fourth is to eliminate the harmonics. A passive filter consists of elements such as inductors, capacitors, and resistors for filtration. This makes the filter configuration simple and easy to implement. The passive filter is a very good choice for low- and medium-load applications and is a cost-effective solution to harmonic reduction and power factor improvement. All these advantages can be lost if the input filter is not properly designed. An oversized input filter unnecessarily adds cost and volume to the design and compromises system performance.

In this chapter, the optimal filter components, that is, the *L* and *C* values, are identified by applying GA. In a 6-pulse drive, the input side and output side passive filters are designed, whereas in case of a12-pulse drive, the phase-shifting transformer secondary side filters are designed for harmonic minimization. The case study based simulation circuits are used to design a passive filter with different problem formulations. The lower order harmonics elimination, sine wave output voltage, and phase-shifting transformer secondary side voltage harmonics elimination are drafted as the optimization task, and the same is solved by using GA. The designed filter is tested in both simulation and experiment. The simulation and experimental results are presented.

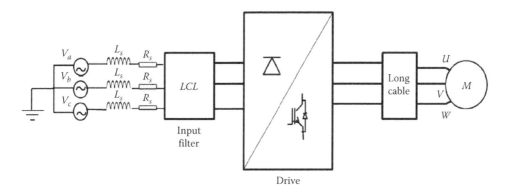

FIGURE 3.69
Six-pulse drive with proposed input filter.

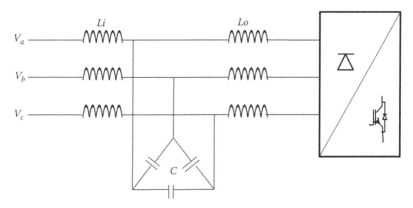

FIGURE 3.70
Proposed input filter connections.

Input LCL filter: Figure 3.69 shows the six-pulse VFD with an input passive filter. Some of the manufacturers use a DC choke inside the drive after rectifier section, but in small-rated drives 3%–6% AC line reactors are added in the input side. The proposed filter combination is shown in Figure 3.70. The value of *Lo* is the 3% of the line reactor, and the values of *Li* and *C* are designed using the GA method. From the real-time study presented in Section 3.6, the rectifier in a six-pulse drive generates low-order harmonics of the 5th, 7th, 11th, and 13th order, and these harmonics need to be eliminated for energy-efficient performance of the drive. Hence, in the following input filter, the lower order harmonic elimination with reduced total harmonics distortion is drafted as an optimization task and is solved using GA.

As mentioned earlier, the objective here is the elimination of lower order harmonics (5th, 7th, 11th, and 13th) and the minimization of THD. Hence the objective function is framed to minimize the harmonics, and is given here:

Minimize

$$F(\phi) = \{(5\text{th}, 7\text{th}, 11\text{th}, \text{and } 13\text{th})\% \leq 5\%\} \tag{3.10}$$

and

$$\text{Min}\%I_{THD} = \frac{\sqrt{\sum_{h=2}^{h=\infty} I_h^2}}{I_1} \times 100\% \tag{3.11}$$

where $\phi = \{L, C\}$, and the subscript indicates the boundary values of the filter components. In the GA-based design, emphasis is also given to minimizing the size of the filter components. The following are the steps involved in the implementation of GA for the framed optimization problem.

3.7.8.1 Steps for Genetic Algorithm

Step 1: Create a population of initial solution of parameters (*L* and *C*)
This step primarily requires the population size. Each variable in the problem is called a gene, and in the present problem formulation there are two genes (i.e., *L* and *C*). A chromosome consists of the genes, and thus each chromosome represents a solution to the problem. This is illustrated in Figure 3.71. The population consists of a set of chromosomes.

Step 2: Evaluation of the objective function
In the present problem, lower order current harmonics of the drive input are to be minimized and the corresponding objective function $F(L, C)$ is computed using

$$F(LC) = \{\%5\text{th}, 7\text{th}, 11\text{th}, \text{and } 13\text{th} \leq 5\%\} \tag{3.12}$$

Step 3: Evaluation of the fitness function
The degree of "goodness" of a solution is qualified by assigning a value to it. This is done by defining a proper fitness function to the problem, since GA can be used for maximization and minimization problems.

Step 4: Generation of offspring
Offspring is a new chromosome obtained through the steps of selection, crossover, and mutation. In this present problem, the crossover probability is 0.7, and the mutation probability is 0.01. The roulette wheel selection mechanism is used to select the reproduction chromosome from the parent.

Step 5: Replace the current population with the new population

Step 6: Terminate the program if the termination criterion is reached; else go to step 2.

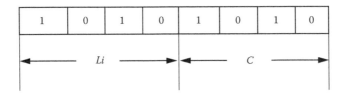

FIGURE 3.71
Chromosome structure.

3.7.8.2 Simulation Results

A dedicated software program was written to implement GA in MATLAB. The simulation is done by two parts: one is creating the MATLAB/Simulink circuit, and another is MATLAB/M-file programming coding. GA is a biologically inspired, population-based algorithm, and this algorithm coding is written to present problem of passive *LC* filter design. In the Simulink circuit, the *L* and *C* parameters values are considered as chromosome genes. The following are the GA parameters used for the GA-based filter design in present problem.

- *Population size*: 15
- *Coding*: Binary
- *Number of generations*: 800
- *Selection scheme*: Combination of roulette wheel selection with elitism
- *Crossover operator*: Multipoint crossover
- *Crossover probability*: 0.7
- *Mutation probability*: 0.01

The convergence characteristic of GA for the typical case of an input *LCL* passive filter design is shown in Figure 3.72. From the characteristics, it is clear that the GA converges to lower objective function, minimizing the harmonics. It converges to the best solution at 15th iteration. The convergence obtained is seen to be satisfactory, as the objective of lower order harmonics elimination is achieved.

To verify the effectiveness of the GA-based filter design, the designed *L* and *C* values were incorporated in the MATLAB/Simulink model of the six-pulse drive, and simulations were carried out. For verification, the simulation results are presented for without filter,

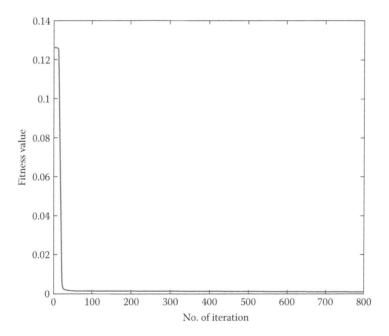

FIGURE 3.72
Input *LC* filter GA convergence curve.

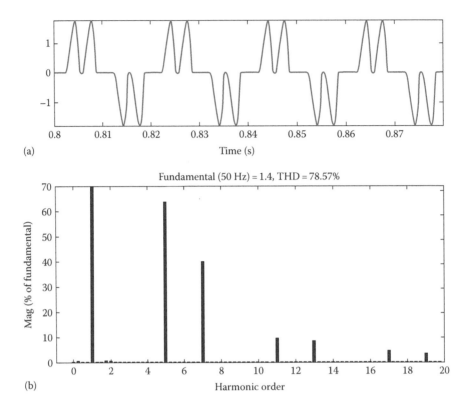

FIGURE 3.73
Simulated phase current (a) waveform and (b) spectrum.

with 3% line reactors, and with GA filter included. For every case, the simulated drive input phase current I_a waveform and harmonics spectrum were taken, and they are shown in Figures 3.73 through 3.75, respectively. From these figures, the drive current waveform is discontinuous in nature and has 78.57% current THD with 5th, 7th, 11th, and 13th order harmonics in case of a drive without any filtering components. The six-pulse drive Simulink circuit module was simulated with 3% line reactor with a value of 12 mH. The simulated drive input phase current I_a waveform and harmonic spectrum show a reduction in harmonics from 78% to 39% as well as an improvement in the current waveform nature.

The GA-designed L and C values are 5.2 mH and 20.22 µF, respectively. The designed filter was incorporated with a line reactor and simulation results were noted. The simulated phase current waveform and harmonic spectrum with filter are shown in Figure 3.75. Introduction of designed filter in the system reduces the harmonics by a large amount and percentage current THD value reduces to the allowed 5%; further, the current waveform is also sinusoidal, indicating minimized harmonics.

3.7.8.3 Experimental Results

In order to validate the theoretical findings, a hardware prototype was fabricated and used to drive a 415 V, 0.37 kW, 1500 rpm, three-phase induction motor. The filter Li, Lo, and C values were 4.1 mH, 12 mH, and 18 µF, respectively, in order to match standard values. The voltage and current harmonics were measured, and the measured harmonic spectra are shown in Figure 3.76b without and with the filter. The experimental results are on par with

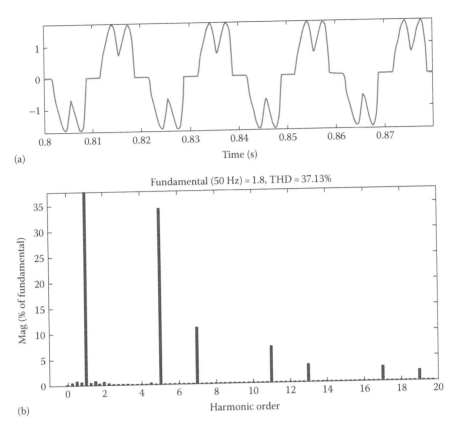

FIGURE 3.74
Simulated phase current (a) waveform and (b) spectrum with line reactor.

the simulated values. Further, both experimental and simulated values confirm the lower order harmonic minimization. From experimental results, it is evident that the first harmonic distortion in the current is reduced from 82.6% to 18.1% and 5th, 7th, 11th, and 13th order harmonics are also reduced. The objective of lower order harmonics reduction was drafted as an optimization, which was also satisfied from simulation and experimental results. Also, compared with 3% line reactor, the harmonics THD was reduced from 39.05% to 18.1%.

To confirm with the lower order harmonics, the 5th, 7th, 11th, and 13th were noted, and all the values were below 5%. Fifth-order harmonics reduced from 50.8% to 3.0%, 7th order harmonics from 46% to 1.3%, 11th order harmonics from 32% to 3.6% and 13th order reduced from 28% to 3.8%.

3.7.8.4 Output Sine Wave Filter Design in 6-Pulse Drive

Figure 3.77 shows the six-pulse VFD with the output sine wave filter. Drive manufactures usually specify the cable length from the drive to the motor location to minimize the *dv/dt* effect and the cable de-rating factor for small-rated drives. However, the plant requirement and space availability might warrant a longer cable length than the specified value. In such cases, the increase in cable length results in a higher receiving-end voltage compared to sending-end voltage. This is due to the PWM switching waveform and cable capacitance effect. Hence it is mandatory to incorporate filters between the drive and the motor load to smoothen the PWM output voltage waveform. Hence this problem of sine wave filter design with reduction in harmonic distortion is framed as an optimization problem, and

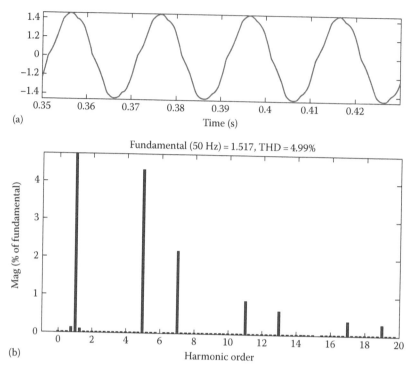

FIGURE 3.75
Simulated phase current (a) waveform and (b) spectrum with *LC* filter.

it is solved using the GA method. The proposed filter combination along with drive structure is shown in Figure 3.78. Detailed hardware and simulation study is performed to verify the design proportion.

Problem formulation: The objective of voltage THD (V_{THD}) minimization and synthesis of sine wave from the PWM output voltage is drafted as an optimization task, which is given by

$$F(\phi) = \text{Sine wave output voltage} \qquad (3.13)$$

and

$$\text{Min}\%V_{THD} = \frac{\sqrt{\sum_{h=2}^{h=\infty} V_h^2}}{V_1} \times 100\% \qquad (3.14)$$

where $\phi = \{L, C\}$, and the subscripts indicate the values of boundary values of the filter components. In the GA-based design, emphasis is also given to minimizing the size of the filter components.

3.7.8.4.1 Simulation Results of Drive Output Filter

The simulation is done in two parts. The first is to create a MATLAB/Simulink circuit, and the second is MATLAB/M-file programming coding. The six-pulse drive Simulink circuit module is simulated without the *LC* filter. The simulated drive output "UV" voltage waveform and its harmonic spectrum are shown in Figure 3.79a and b.

From this figure, it is seen that without the filter the drive waveform is PWM in nature and has 22.94% THD in voltage. The high switching frequency is to improve the current waveform due to the high inductor value in motor winding. But same time dv/dt

effect, cable de-rating factor, and inverter switching losses are the side effects of the PWM. Hence sine wave output filter design is drafted as problem formulation, and a dedicated software program is written to implement GA in MATLAB.

GA is a biologically inspired, population-based algorithm, and this algorithm coding is written to present the problem of a passive sine wave *LC* filter design. In the Simulink circuit, the *L* and *C* parameter values are considered as chromosome genes. The following GA parameters are considered for the simulation.

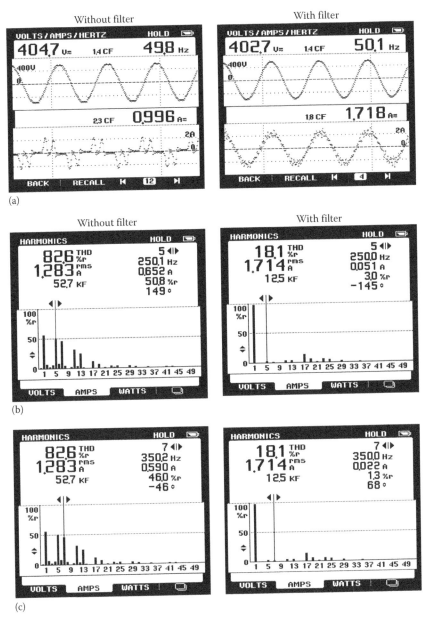

FIGURE 3.76
Experimental harmonics spectra for input *LCL* filter. (a) Voltage and current waveforms, (b) 5th order harmonics, (c) 7th order harmonics.
(Continued)

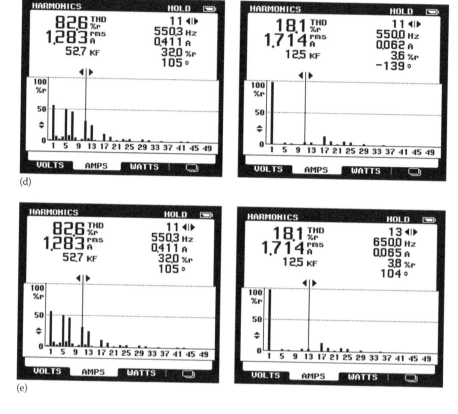

FIGURE 3.76 (Continued)
Experimental harmonics spectra for input *LCL* filter. (d) 7th order harmonics and (e) 13th order harmonics.

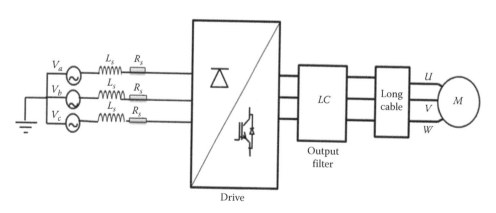

FIGURE 3.77
Six-pulse drive with proposed output filter.

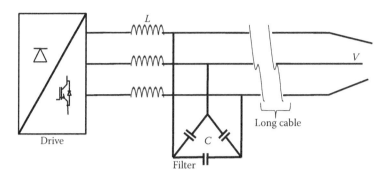

FIGURE 3.78
Proposed output filter connections.

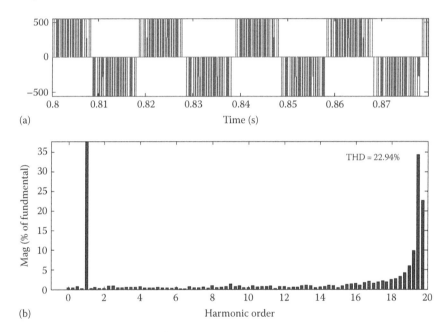

(a)

(b)

FIGURE 3.79
Simulated PWM voltage (a) waveform and (b) spectrum without filter.

- *Population size*: 15
- *Coding*: Binary
- *Number of generations*: 800
- *Selection scheme*: Combination of roulette wheel selection with elitism
- *Crossover operator*: Multipoint crossover
- *Crossover probability*: 0.7
- *Mutation probability*: 0.01.

The convergence characteristics of GA for a typical case of a sine wave output passive filter is shown in Figure 3.80. From the characteristics, it is clear that the GA converges to objective function of sine wave output with lower voltage THD at 20th iteration. The convergence obtained is seen to be satisfactory.

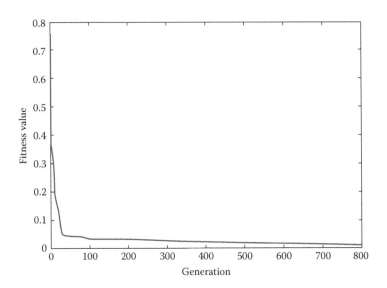

FIGURE 3.80
Output *LC* filter GA convergence curve.

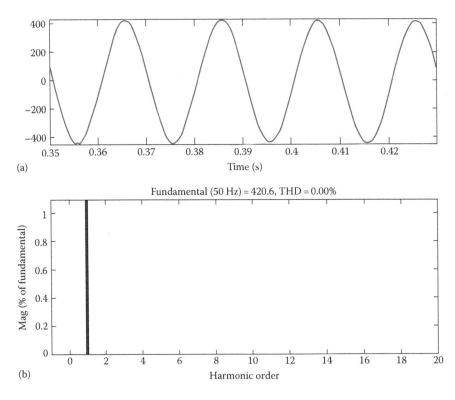

FIGURE 3.81
Simulated PWM voltage (a) waveform with filter and (b) spectrum without filter.

The simulated phase voltage waveform and harmonic spectrum with the filter are shown in Figure 3.81. The sine wave LC filter formation was made, and same was used to find the L and C values by the used GA program. The program gave the L value as 26.31 mH and C value as 2.22 µF at 20th iteration, the convergence curve was already illustrated in Figure 3.80.

3.7.8.4.2 Experimental Results

In order to validate the theoretical findings, a hardware prototype was fabricated, which was used to drive a 415 V, 0.37 kW, 1500 rpm, 1.4 A, three-phase induction motor. The drive switching frequency and drive to motor cable length were maintained as 3.5 kHz and 100 m, respectively. The filter L and C values were 29 mH, and 2.2 µF, because the simulated values were not standard. The drive output reference was maintained with three different set points to validate the filter performance at variable operating speed conditions.

The set point was maintained at 80%, and the voltage and current harmonics were measured. The measured voltage waveform, current waveform, and voltage harmonic spectra are shown in Figure 3.82 without and with the filter. The 80% set point drive will give 40 Hz frequency; Hence motor will run at 1.31 A load current and 1150 rpm. As compared with experimental results, the voltage harmonics Vthd% were reduced from 22.3% to 8.4%, and voltage waveform changed from PWM to sine wave.

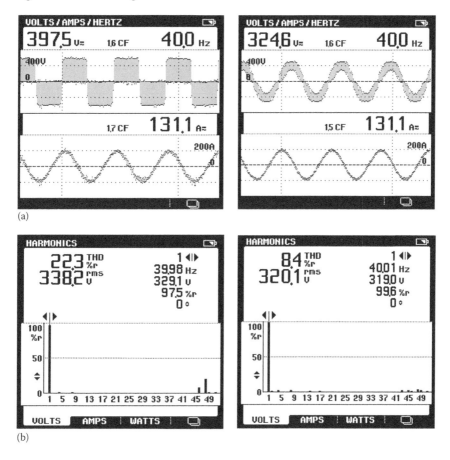

FIGURE 3.82
Sine wave output filter experimental result for 80% set point. (a) Voltage and current waveforms and (b) voltage harmonics spectrum.

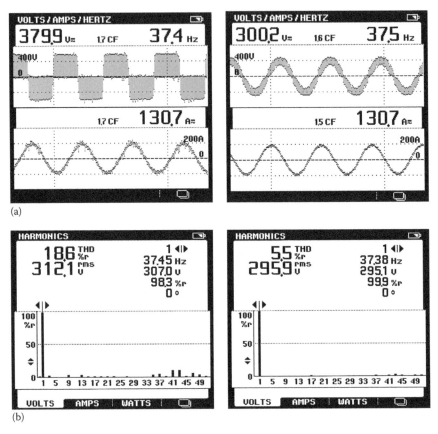

(a)

(b)

FIGURE 3.83
Sine wave output filter experimental result for 75% set point. (a) Voltage and current waveforms and (b) voltage harmonics spectrum.

Figure 3.83 shows the experimental output at 75% set point. At this set point, the drive maintained 37.5 Hz frequency, and motor would run at 1080 rpm with 1.30 A load current. Figure 3.84 shows the experimental output at the 50% set point. At this set point, the drive maintained 25 Hz frequency, and motor would run at 720 rpm and 1.27 A load current. Figure 3.85 shows a photo of the experimental setup. The sine wave output filter performed well in all three set points, and gave sinusoidal output voltage waveform with reduced voltage THD.

3.7.8.5 Input LC Filter Design in 12-Pulse Drive

The objectives of voltage THD (V_{THD}) minimization in a phase-shifting transformer secondary side is framed as an optimization task, and the same is solved by using GA. The 12-pulse drive secondary side voltage harmonics are generated by two six-pulse drives operated in parallel with 30 degree phase shift by using the phase-shifting transformer. The high current harmonics affect the nature of the phase-shifting transformer voltage. The poor voltage wave leads to failure of the drive rectifiers. Hence, an *LC* filter is proposed to improve the voltage wave. The simulation and experimental results are presented here.

The objective of voltage THD (V_{THD}) minimization is drafted as an optimization task, and is given by minimizing

$$F(\phi_1, \phi_2) = \{\%\Delta V_{THD} \leq 3\%, \%Y V_{THD} \leq 3\%\} \tag{3.15}$$

50% reference wave form without filter   50% reference wave form with *LCL* filter

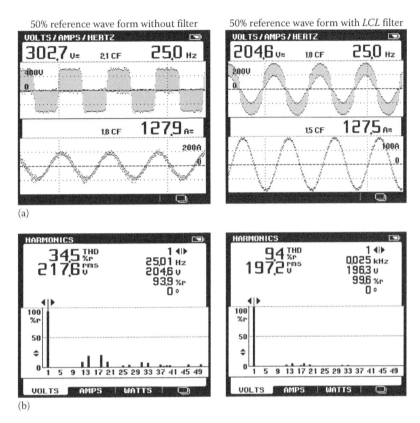

(a)

(b)

FIGURE 3.84
Sine wave output filter experimental result for 50% set point. (a) Voltage and current waveforms and (b) voltage harmonics spectrum.

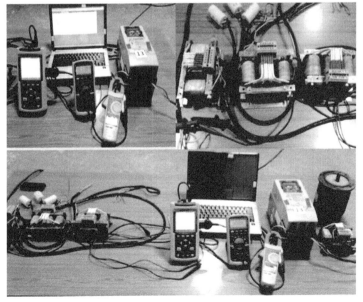

FIGURE 3.85
Photo of the experimental setup.

Delta winding

$$\text{Min}\%\Delta V_{THD} = \frac{\sqrt{\sum_{h=2}^{h=\infty} V_h^2}}{V_1} \times 100\% \qquad (3.16)$$

Star winding

$$\text{Min}\%\text{Y}V_{THD} = \frac{\sqrt{\sum_{h=2}^{h=\infty} V_h^2}}{V_1} \times 100\% \qquad (3.17)$$

where

$\phi_1 = \{L_1, C_1\}$,

$\phi_2 = \{L_2, C_2\}$, and the subscripts indicate the boundary values of the filter components

In the GA-based design, emphasis is also given to minimizing the size of the filter components.

3.7.8.5.1 *Steps Involved in GA-Based Filter Design*

The following section explains how GA is used for design the filter components:

Step 1: Creation of a population of initial solution of parameters (*L* and *C*)
This step primarily requires the population size. Each variable in the problem is called as gene, and in the present problem there are four (i.e., L1, C1, L2, and C2) genes. A chromosome consists of the genes, and thus each chromosome represents a solution to the problem. This is illustrated in Figure 3.86. The population consists of a set of chromosomes. It is well articulated in the literature that a population size of 10–30 is an ideal one, and hence the population size is selected as 15 in this example.

Step 2: Evaluation of objective function
In the present problem, $\%\Delta V_{THD}$ and $\%\text{Y}V_{THD}$ are used to minimize and the corresponding objective functions *F* (L1, C1) and *F* (L2, C2) are computed using

$$F(\phi_1) = \{\%\Delta V_{THD} \le 3\%\} \qquad (3.18)$$

where, $\phi_1 = \{L_1, C_1\}$

$$F(\phi_2) = \{\%\text{Y}V_{THD} \le 3\%\} \qquad (3.19)$$

where, $\phi_2 = \{L_2, C_2\}$.

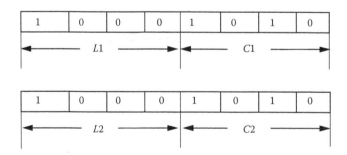

FIGURE 3.86
Chromosome structure.

Step 3: Evaluation of the fitness function

The degree of "goodness" of a solution is qualified by assigning a value to it. This is done by defining a proper fitness function to the problem. Since GA can be used only for maximization or minimization problems, the following fitness function is used:

$$\text{Fitness function} = F(\phi_1, \phi_2) \tag{3.20}$$

Step 4: Generation of off spring

Step 5: Replacing the current population with the new population

Step 6: Terminate the program if the termination criterion is reached; else go to step 2.

3.7.8.5.2 Simulation Results

A dedicated software program was written to implement GA in MATLAB. The following GA parameters were considered for the simulation:

- *Population size*: 15
- *Coding*: Binary
- *Number of generations*: 800
- *Selection scheme*: Combination of roulette wheel selection with elitism
- *Crossover operator*: Multipoint crossover
- *Crossover probability*: 0.7
- *Mutation probability*: 0.01.

The convergence characteristics of GA for the typical case of a phase-shifting transformer secondary side sine wave output passive filter is shown in Figure 3.87. The convergence characteristic of the GA is obtained at the 20th iteration. The convergence obtained is seems satisfactory, as the objective functions in our case $\%\Delta V_{THD} \leq 3\%$, and $\%Y V_{THD} \leq 3\%$. After obtaining convergence, the phase-shifting transformer secondary side voltage waveforms and its harmonics spectrums are presented.

Figure 3.88 shows the star winding voltage waveform and the voltage harmonic spectrum of the star winding before the filter connection. As compared with Figures 3.31 and 3.46, the phase shifting transformer's star winding voltage waveform has been improved. The proposed *LC* filter approach is able to improve the line voltage THD ($\%Y V_{THD}$) from 22.22% to 2.50%. Also, Figure 3.89 shows the delta winding voltage waveform and voltage harmonic spectrum of the delta winding before the filter connection.

As compared to Figures 3.33 and 3.47, the phase shifting transformer delta winding voltage waveform also has been improved. The proposed LC filter approach is able to improve the line voltage THD ($\%\Delta V_{THD}$) from 32.00% to 2.47%. Figure 3.90 shows the star winding voltage waveform and voltage harmonic spectrum of the star winding after the filter connection. Figure 3.91 shows the voltage waveform and voltage harmonic spectrum of the delta winding after the filter connection.

From this simulation results, the proposed GA-based filter design is seen to reduce the phase-shifting transformer secondary side voltage harmonics distortion. The case study results show that the phase-shifting transformer fed by the 12-pulse drive injects less current and voltage harmonics at the primary side/PCC side. But the secondary side generates more current and voltage harmonics, and due to this the drive rectifier diode lifetime gets reduced. Hence, the phase-shifting transformer design is an important criterion for poly-pulse AC/DC

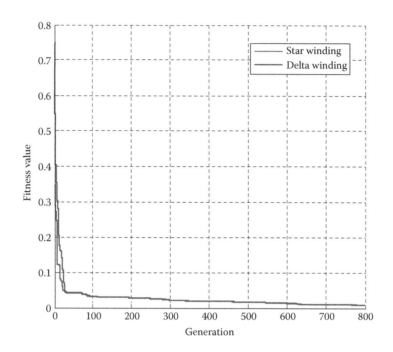

FIGURE 3.87
Input *LC* filter and GA convergence curve.

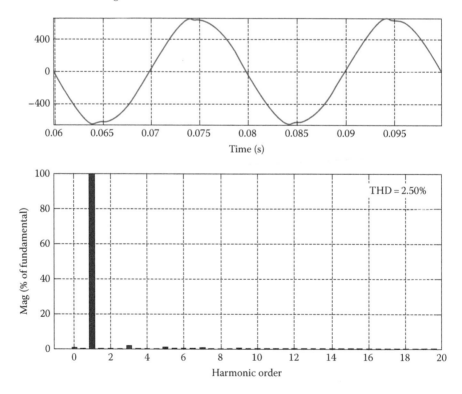

FIGURE 3.88
Voltage waveform and harmonic spectrum of star winding before filter connection.

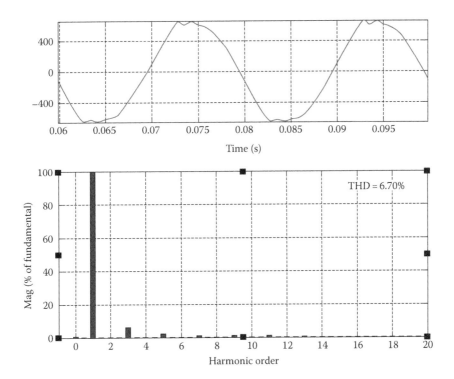

FIGURE 3.89
Voltage harmonic spectrum of delta winding before filter connection.

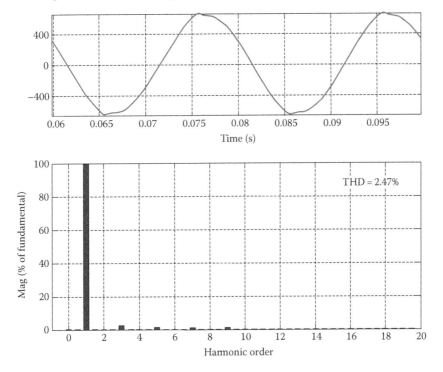

FIGURE 3.90
Voltage harmonic spectrum of star winding after filter connection.

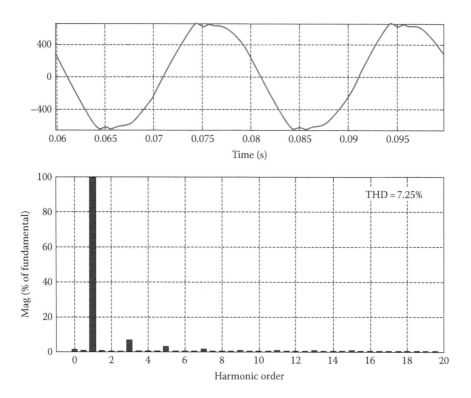

FIGURE 3.91
Voltage harmonic spectrum of delta winding after filter connection.

converter-fed drives. At the design stage, some transformer manufactures concentrate only on the primary voltage, primary current, secondary voltages, secondary currents, and kVA ratings. This case study and simulation results clearly show that additional attention should be paid to required actual load conditions and harmonic generation in poly-phase transformers.

3.7.8.5.3 Experimental Results

The proposed passive LC filter for harmonics reduction was connected to a phase-shifting transformer secondary side at the star and delta windings on a 12-pulse PA fan drive. Figures 3.92 and 3.93 show the experimental results with the filter. The two six-pulse rectifier operates under high current harmonics.

The measured THD of star current is 55.69%. Figure 3.93 shows the experimental waveform and the current harmonics spectrum at the star winding side. The measured THD of the delta current is 59.77%. Figure 3.94 shows the experimental waveform and the current harmonics spectrums at the delta winding side. Because of improper design of the phase-shifting transformer, voltage harmonics are generated at the secondary side. Hence it is proposed to add an LC filter at the secondary side of the phase-shifting transformer. This LC filter reduces the current and voltage harmonics. The main task is to reduce the voltage harmonics below 3% at the secondary side. Figures 3.92 and 3.95 show the voltage waveform and spectrum at the star and delta windings. From these results, the voltage harmonics are seen to be reduced below 3% and current harmonics from 75% to 55% as compared to the figures without filters.

The low current rated six-pulse drives and the higher current rated 12-pulse drives are used in industries for many applications. The uncontrolled rectifier-fed variable frequency

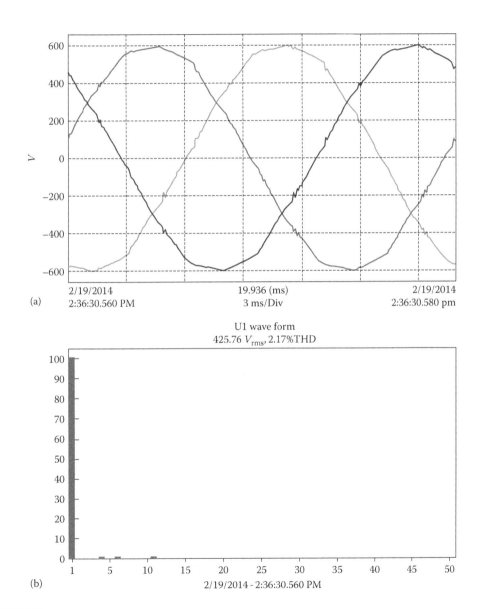

FIGURE 3.92
Experimental result for star winding voltage (a) waveforms and (b) harmonic spectrum.

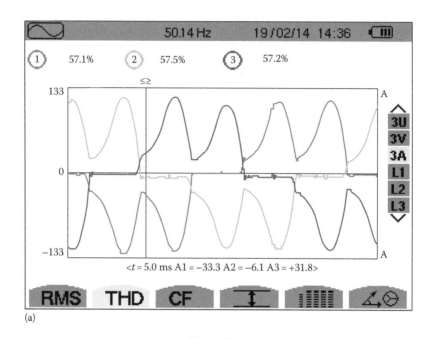

(a)

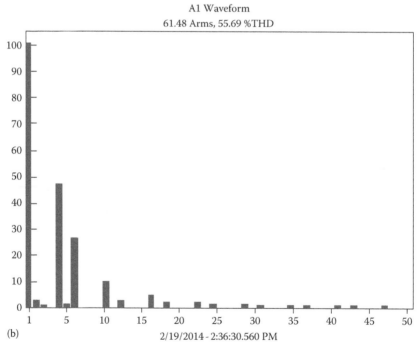

(b)

FIGURE 3.93
Experimental result for star winding current (a) waveforms and (b) harmonic spectrum.

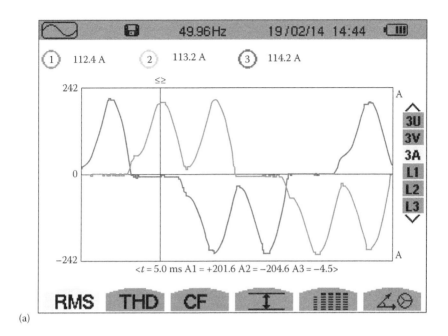

(a)

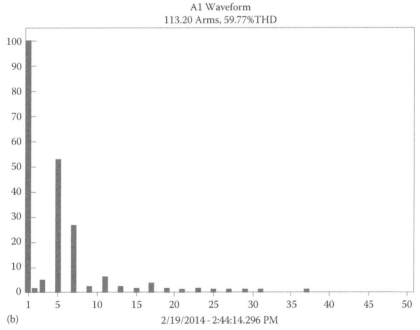

(b)

FIGURE 3.94
Delta winding current harmonic spectrum. (a) Experimental result for delta winding voltage waveforms and (b) star winding voltage harmonic spectrum.

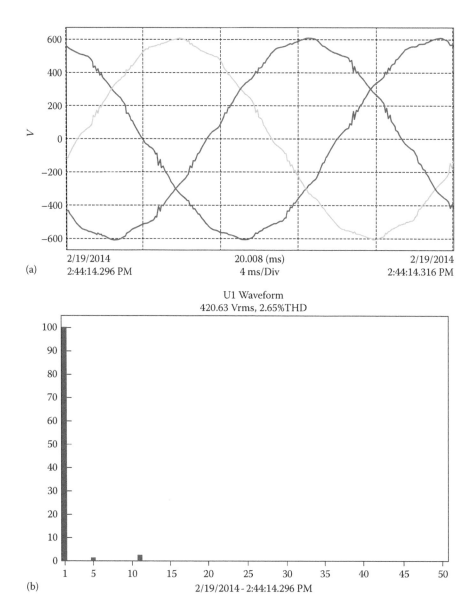

FIGURE 3.95
Delta winding voltage (a) waveform and (b) harmonic spectrum.

drive always generates current harmonics at the input side and the PWM voltage generates voltage harmonics at the output side. The PWM wave creates oscillating torque, noise, voltage doubling effect, cable de-rating, and inverter switching losses. These problems are eliminated by using a passive filter. The passive filter is a very good choice for low current rating. The drives, phase shifting transformer, and passive filters are procured from different supplier at the required voltage, current, and power rating. But in harmonic elimination point of view, it does not work effectively in real-time operation. Hence, real-time harmonic analysis is required under various load conditions and various power distribution points for effectively eliminating harmonics.

In this section, a GA based passive filter design was proposed based on real-time case study. Three different problems—drive input current harmonics, drive output voltage harmonics, and phase-shifting transformer voltage harmonics—were solved using the GA optimizing algorithm. The MATLAB/Simulink circuits based on the case study and GA optimizing M-file program codes were used to find the passive filter values. The simulation and experimental results presented show the effectiveness of GA.

3.8 Bacterial Foraging Algorithm for Harmonic Elimination

With the recent developments in low-cost semiconductor devices and high-speed digital processors, PWM inverters have become cost effective (Tutkun 2010). The most commonly favored switching scheme for inverter output voltage regulation is the PWM technique. Among the several PWM methods used, selective harmonic elimination PWM (SHEPWM) method has been the main focus of research since it offers several advantages over traditional methods, such as acceptable performance with low switching frequency to fundamental frequency ratio, direct control over output waveform harmonics, and the ability to leave triplen harmonics uncontrolled to take advantage of circuit topology in three-phase systems (Holmes 2003). This method has been proposed by Passino (2002) and has been successfully applied to solve various engineering optimization problems (Hossein Nouri and Tang Sai Hong 2013, Mishra 2005, 2007). The BFA method is based on the foraging (methods for locating, handling, and ingesting food) behavior of *Escherichia coli* (*E. coli*) bacteria present in our intestines. The SHEPWM method was first examined by Turnbull (1964) and later developed into matured form by Patel and Hoft (1973). The common characteristic of these methods is that waveform analysis is done in the Fourier domain from which a set of nonlinear transcendental equations are derived, and the solution is obtained via iterative procedure mostly by the Newton–Raphson method. This method is derivative-dependent, and can end in local optima; further, a judicial choice of initial values alone will guarantee faithful convergence.

Various meta-heuristic algorithms have been proposed for selective harmonic elimination in the PWM converter and a few worth mentioning are GA (Shi and Li 2005), PSO (Ray et al. 2010), ACO (Sundareswaran et al.2007), and artificial bee colony algorithm (Kavousi et al. 2012). GA, which is inspired by the laws of natural selection and genetics, has been extensively studied in the literature for harmonic elimination problem in PWM inverter as well as in multilevel inverters (Dahidah et al. 2008, Jegathesan and Jerome 2011). GA offers simple structure, makes easy computation, converges to near-optimum solution, and finds promising regions of the search space quickly (Al-Othman et al. 2013). However, this method has the inherent drawback of convergence to local optima and premature and slow convergence. On the other hand, in comparison with GA, PSO is easy to implement, as there are very few parameters to be adjusted. Two types of switching for PWM inverter were tested using the PSO method (Ray et al. 2009). PSO performance depends strongly on its parameters; so it might easily lose the diversity and may be influenced by premature convergence, especially when the best solution is local (Askarzadeh and Rezazadeh 2013). Further, ACO and others metahuristic algorithms have also been applied for the harmonic elimination problem. However, the following are the shortcomings of the previous works in harmonics elimination: (1) converges to local optima, (2) premature convergence, (3) the use of either global or local search, and (4) heavy computational burden. Further, the method used for harmonic elimination problem must have the capability to handle a large solution space having multiple solutions. Hence, in order to address the earlier problems, an attempt is made in this example to apply BFA for the harmonics elimination problem.

The foraging strategy is governed by four main processes: chemotaxis, swarming, reproduction, and elimination–dispersal. Among the many optimization techniques proposed, BFA has been reported to have better performance than GA and PSO (Biswas et al. 2007, 2007, 2010, Tripathy and Mishra 2007) in terms of convergence speed and solution quality. Recently, to increase the probability of obtaining a global solution in a reasonable time, optimization technique employing population-based search was found to be a viable alternative for solving nonlinear transcendental equations. These methods offer various advantages such as starting with random initial guess, convergence to the global/near-global optimal value, and the provision of multiple optimal solutions compared to a single solution in conventional search algorithms. The objective of harmonic elimination, together with output voltage regulation, is suitably framed as an optimization task, and the switching instances are identified through the steps of BFA. The performance of BFA is estimated, and the results are compared with GA and PSO. The results reveal that the proposed method is successful in achieving selective harmonic elimination with output voltage regulation.

3.8.1 Inverter Operation

The circuit diagram of the single-phase PWM inverter-fed induction motor is shown in Figure 3.96. The power circuit consists of four IGBTs $S1$, $S2$, $S3$, and $S4$. The input is a robust DC voltage source of magnitude V_{DC}. In the positive half-cycle, $S1$ is fired at an angle $0°$ and is kept on for the entire half-cycle. $S3$ is fired at various angles to obtain a PWM output voltage. The firing of $S3$ is done as many times as the number of pulses required in the output waveform. Diode $D2$ is used to free-wheel the motor current when $S3$ is switched off. Similarly, in the negative half-cycle, $S2$ is fired at an angle of $180°$ and kept on for the entire half-cycle. $S4$ is fired for the required number of pulses. Diode $D1$ is used for free-wheeling the motor current when $S4$ is switched off; for illustration, IGBT firing pulse together with typical output voltage is depicted in Figure 3.97 for a typical case of five pulses per half-cycle.

3.8.1.1 Mode of Operations

There are two distinct modes of operation of the drive with the proposed inverter circuit.

- **Mode I:** When IGBTs $S1$ and $S3$ are switched on during positive half-cycle

$$V_m = V_a = V_{DC} \tag{3.21}$$

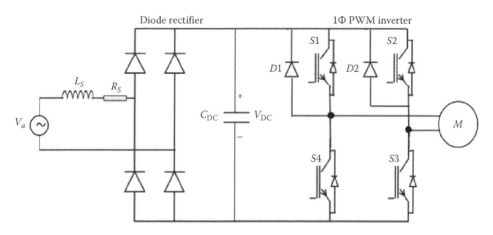

FIGURE 3.96
Single-phase drive power structure.

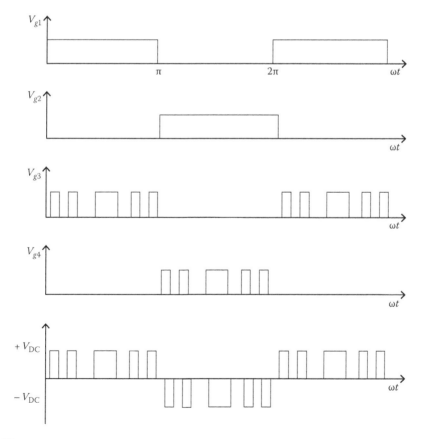

FIGURE 3.97
Output voltage waveform of a single-phase PWM inverter.

- **Mode II:** When IGBT $S3$ is turned off, load current freewheels through $S1$ and $D2$. Hence

$$V_m = V_a = 0 \tag{3.22}$$

During negative half-cycle, $S2$ and $S4$ are turned on in mode-1 such that

$$V_m = V_a = -V_{DC} \tag{3.23}$$

When IGBT $S4$ is turned off in mode-II, the load current passes through $S2$ and $D1$. Therefore,

$$V_m = V_a = 0 \tag{3.24}$$

where
V_m is the main winding voltage
V_a is the auxiliary winding voltage

The typical output voltage of single-phase PWM inverter with k pulses per-half cycle is shown in Figure 3.98. For generalization, the output voltage is assumed to have k pulses per half-cycle with switching angles symmetrical with respect to $\pi/2$. The value of k is an odd number and it varies from 3, 5, 7, 9, and so on. The PWM inverter output voltage can be expressed using the Fourier series as

$$V_0 = a_0 + \sum_{n=1}^{\infty} A_n \cos(n\omega t) + \sum_{n=1}^{\infty} B_n \sin(n\omega t) \tag{3.25}$$

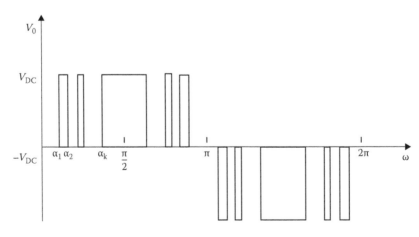

FIGURE 3.98
PWM inverter output voltage waveforms.

Due to the quarter wave symmetry of the output voltage, even harmonics are absent (Ray et al. 2008). Further, the values of the Fourier coefficients A_n and a_0 are equal to zero due to symmetry. Thus the earlier equation reduces to the following form:

$$V_0 = \sum_{n=1}^{\infty} B_n \sin(n\omega t), \tag{3.26}$$

The values of Fourier coefficients B_n are computed using the following equation:

$$B_n = \frac{4V_{DC}}{n\pi} \left[\cos(n\omega t)\right]_{\alpha_2,\alpha_4,\ldots,\pi/2}^{\alpha_1,\alpha_3,\ldots,\alpha_k} \tag{3.27}$$

The fundamental component value is given by

$$B_1 = \frac{4V_{dc}}{\pi} \left[\cos(n\omega t)\right]_{\alpha_2,\alpha_4,\ldots,\pi/2}^{\alpha_1,\alpha_3,\ldots,\alpha_k} \tag{3.28}$$

The principle of selective harmonic elimination technique is to identify the switching instants $\alpha_1, \alpha_2, \ldots, \alpha_k$ of the inverter output voltage waveform for k pulses per half-cycle in such a way that the values of the harmonic components $B_3, B_5, \ldots, B_{k-1}$ present in output voltage will be made zero, in addition to making the fundamental component value B_1 equal to the desired output voltage. Therefore, the objective of SHEPWM method is to regulate the output voltage and eliminate the output voltage harmonics. This task can be suitably framed as an optimization problem, and the solution can be derived via an iterative procedure. One such objective function along with the constraints is presented in Equation 3.25. The earlier problem can be mathematically written as Minimize

$$F(\alpha) = F(\alpha_1, \alpha_2, \ldots, \alpha_k) = e_r + h_c \tag{3.29}$$

subject to constraints:

$$0 \le \alpha_1 \le \alpha_2 \ldots \le \alpha_{k-1} \le \alpha_k \le \frac{\pi}{2}$$

$$\text{Where,} \ e_r = \left|V_0^* - B_1\right| \quad \text{and} \quad h_c = |B_3| + |B_5| + |B_7| + |B_9| \tag{3.30}$$

The objective is twofold: (1) maintaining the output voltage at the desired level: V_0, B_1, and (2) eliminating the harmonic content that is, h_c. Depending upon the number of pulses per half-cycle (odd number), the desired harmonics (even number) can be eliminated. For instance, if there are k pulses in the half-cycle in the output voltage, then $k-1$ harmonics will be eliminated.

3.8.2 Basics of Bacterial Foraging Algorithm

BFA was proposed by Passino (2002) and is a new addition to the family of nature-inspired optimization algorithms. The algorithm is based upon the fact that the genes of the fitter species, with a successful foraging strategy, are likely to survive and get propagated in the evolution chain. *E. coli* bacteria present in the human intestine also undergo a similar foraging strategy. Application of group foraging strategy for a swarm of *E. coli* bacteria in multioptimal function optimization is the key idea of this new algorithm (Das et al. 2009). BFA is an effective and flexible optimization tool that can solve a large class of problems by exploring all regions of the state space and exponentially exploit promising areas through processes such as chemotaxis, swarming, reproduction, and elimination–dispersal operations (Biswas et al. 2007).

3.8.2.1 Chemotaxis

During chemotaxis, a bacterium undergoes two processes: swimming and tumbling. Both refer to the movement of *E. coli* cell in search of nutrients (i.e., food). It can either swim for a period in the same direction, it may tumble altogether in a different directions, or it may alternate between these two modes of operation for the entire lifetime. For instance, if $\theta^i(j,k,l)$ represents *i*th bacterium at *j*th chemotactic, *k*th reproductive, and *l*th elimination–dispersal step, then the bacterium tumbles a unit step size of $C(i)$ in a random direction $\phi(i)$ to find a new direction of movement. The new direction of movement after a tumble is represented by

$$\theta^i(j+1,k,l) = \theta^i(j,k,l) + C(i) \times \phi(i) \tag{3.31}$$

The bacterium continues to swim after the tumble if the present direction is rich in nutrients but only up to a maximum number of steps N_s. This chemotaxis process is a major step in the BFA. Chemotaxis is a foraging strategy that implements a type of local optimization, where the bacteria try to climb up the nutrient concentration to avoid noxious substances and search for ways out of neutral media (Dasgupta and Das 2009).

3.8.2.2 Swarming

During swarming, the bacterium that has already followed the optimum path tries to attract other bacteria so that they move together to the desired location. This behavior is called swarming. The bacteria congregate into groups, and hence move as groups in concentric patterns with high bacterial density. The cell-to-cell signaling between bacteria can be mathematically represented by the following equation:

$$
\begin{aligned}
J_{cc} &= \sum_{i=1}^{S} J_{cc}(\theta, \theta^i(j,k,l)) \\
&= \sum_{i=1}^{S} \left[-d_{attract} \exp\left(-w_{attract} \sum_{m=1}^{P} (\theta_m - \theta_m^i)^2 \right) \right] \\
&\quad + \sum_{i=1}^{S} \left[-h_{repelent} \exp\left(-w_{repelent} \sum_{m=1}^{P} (\theta_m - \theta_m^i)^2 \right) \right]
\end{aligned}
\tag{3.32}
$$

where J_{cc} represents a time-varying objective function whose values depend on cell-to-cell signaling via attractant–repellent profile.

3.8.2.3 Reproduction

After the chemotactic step, the health of each bacterium during its lifetime is accumulated and calculated using the following equation:

$$J_{health}^{i} = \sum_{j=1}^{j=N_c+1} J_{sw}(i,j,k,l) \tag{3.33}$$

The health values are sorted in ascending order, and the least healthy bacteria are removed, whereas each of the healthier bacteria (those yielding lower value of the objective function) is split into two, maintaining the population of the bacteria constant. This process allows healthier bacteria to stay and the deletion of the bad ones. Further, it increases the speed of searching for a Pareto front (Ben Niu et al. 2011, Biswas et al. 2010).

3.8.2.4 Elimination and Dispersal

Gradual or sudden changes in the local environment, due to various reasons, may kill a group of bacteria or disperse a group of bacteria into a new location. Sometimes, this process places the bacteria near good nutrients. Furthermore, the elimination–dispersion event reduces the chance of convergence to local optimum position, that is, premature convergence. BFA, due to its unique dispersal and elimination technique, can find favorable regions when the population involved is small. This unique feature of the algorithm avoids the premature convergence problem and hence increases the search capability (Abd-Elazim and Ali 2012).

The aforementioned BFA processes—chemotaxis, swarming, reproduction, elimination and dispersal—are implemented by applying the following steps.

Step 1: Initialize the parameters p, S, N_c, N_s, N_{re}, N_{ed}, P_{ed}, $C(i)$, θ^i, $i = 1, 2, 3, \ldots, S$

Step 2: Elimination–dispersal loop $l = l + 1$

Step 3: Reproduction loop $k = k + 1$

Step 4: Chemotaxis loop $j = j + 1$

   a. For, $i = 1,2,3,\ldots, S$, take the chemotactic step for each bacterium.
   b. Compute objective function $J(i,j,k,l)$. Let $J(i,j,k,l) = J(i,j,k,l) + J_{cc}(\theta^i(j,k,l))$
   c. Let $J_{last} = J(i,j,k,l)$, to save this value since the objective is a minimization function.
   d. *Tumble*: Generate a random vector with each element $\Delta(i)$, a random number in the range of [−1, 1].
      Move: Let

$$\theta^i(j+1,k,l) = \theta^i(j,k,l) + C(i)\frac{\Delta(i)}{\sqrt{\Delta^T(i)\Delta(i)}}$$

This results in a step size of $C(i)$ in the direction of the tumble for bacterium i.

Compute $J(i, j + 1, k, l)$ and then let

$$J(i, j+1, k, l) = J(i, j+1, k, l) + J_{cc}\left(\theta^i(j+1, k, l), P(j+1, k, l)\right) \qquad (3.34)$$

e. Swim:
   i. Let $m = 0$ (counter for swim length)
   ii. While $m < N_S$
   Let $m = m + 1$
   If $J(i, j + 1, k, l) < J_{last}$ (if doing better), let $J_{last} = J(i, j + 1, k, l)$ and let

$$\theta^i(j+1, k, l) = \theta^i(j+1, k, l) + C(i)\frac{\Delta(i)}{\sqrt{\Delta^T(i)\Delta(i)}}$$

and use the $\theta^i(j + 1, k, l)$ to compute new $J(i, j + 1, k, l)$ (as was done in step f).
Else, let $m = N_S$ (This is the end of while statement.)
Go to the next bacterium $(i + 1)$
If $j < N_c$, (go to step 4b) till all the bacteria undergo chemotaxis.

Step 5: Reproduction:

a. For the given k and l, for each $i = 1, 2, 3, ..., S$, let $J_{health}^i = \sum_{j=1}^{j=N_c+1} J(i, j, k, l)$ be the health of the ith bacterium and sort J_{health} in ascending order

b. The S_r bacteria with the highest J_{health} value die, and other S_r bacteria with the best values split (and the copies made are replaced at the same location of their parent).

Step 6: If $k < N_{re}$, go to step 3. In this case, when the specified reproduction steps are not reached, we start the next generation in the chemotactic loop.

Step 7: Elimination–dispersal: For $i = 1, 2, 3, ..., S$, a random number is generated, and if it is less than or equal to P_{ed}, then that bacterium is dispersed to a new random location; else it remains at its original location.

Step 8: If $l < N_{ed}$, then go to step 2; otherwise end. Stop and print the result. The flow-chart corresponding to the previous implementation steps is given in Figure 3.99.

3.8.3 Application of BFA for Selective Harmonic Elimination Problem

The following section discusses how BFA is applied for selective harmonic elimination in a PWM inverter. In case of SHEPWM, a set of optimum switching instants that is, $(\alpha_1, \alpha_2, ..., \alpha_k)$, have to be identified in order to minimize the objective function defined by Equation 3.25. The objective function framed addresses two main issues.

1. Elimination of lower order harmonics
2. Improvization of fundamental components.

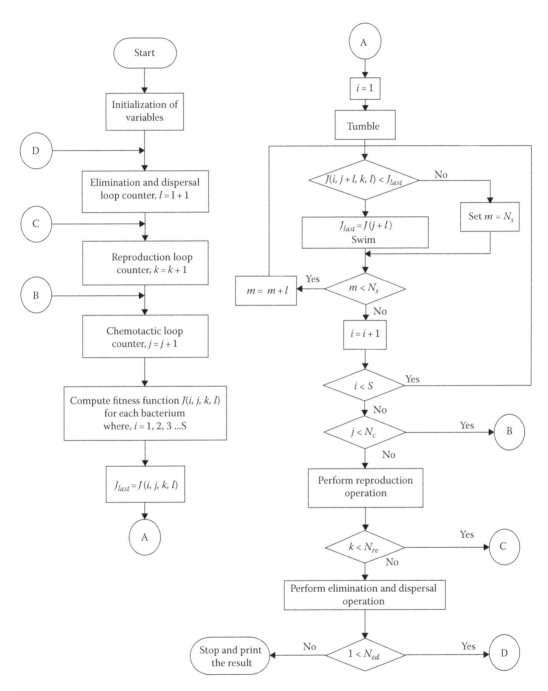

FIGURE 3.99
Flowcharts for bacterial foraging algorithm.

The optimization techniques applied to the earlier problem must have the capacity to explore and exploit the search space. BFA is best suited to solve the earlier problem objective since it possess the property of exploring all regions of the solution space and exponentially exploit promising areas via the steps of BFA. In this sense, BFA is the most appropriate method for switching angle selection for the SHEPWM method. In this application, $\theta^i(j, k, l)$ and $J(i, j, k, l)$ represent the switching instants and the objective function, respectively. The following are the steps involved in BFA implementation:

Step 1: Initialize the BFA parameters and specify the lower and upper boundaries of the switching angles.

Step 2: Generate the positions of the switching angles parameters randomly for a population of bacteria.

Step 3: Determine the objective value of each bacterium in the population using Equation 3.34.

Step 4: Modify the positions of the switching angles for all the bacteria by applying tumbling/swimming and the swarming process.

Step 5: Perform reproduction and elimination–dispersal operation.

Step 6: If the maximum number of chemotactic, reproduction, and elimination–dispersal steps are reached, then go to Step 7. Otherwise, go to Step 4.

Step 7: Obtain the switching angles corresponding to the overall best bacterium.

Step 8: Compute the harmonic spectra for different pulses per half-cycle.

3.8.4 Performance Analysis of BFA for Selective Harmonic Elimination

To demonstrate the superiority of the BFA method, extensive simulations were carried out at various operating points with different pulses per half-cycle and the results were compared with the GA and PSO method. For simulation, a dedicated software program was developed in MATLAB for all the three optimization techniques for eliminating the voltage harmonics. Simulations were carried out using a 2.4 GHz INTEL i3 processor with 2.0 GB RAM.

For fairness of comparison, all the methods were initialized with the same population size and tested for various operating points. Further, for better convergence, the GA parameters crossover probability and mutation probability were set to 0.7 and 0.3, respectively. In addition, the PSO parameters were selected based on the guidelines presented by Barkat et al. (2009) and Kennedy and Russell Eberhart (1995). The PSO parameter values of the inertia weight factor $w1$ and acceleration factors $c1$ and $c2$ were set to 0.4, 1.6, and 1.43, respectively. The efficacy of the BFA method largely relies on parameter selection because the speed of convergence and convergence to the optimal value are greatly influenced by the parameter selection. Therefore, to identify the best parameter values, the BFA algorithm was tested with different compositions for 15 independent trials, and based on the maximum, minimum, and standard deviation of objective function, the best combination of parameters was arrived at. Further, it is important to mention that the parameter selection is made by applying a thumb rule, that is, $N_c > N_{re} > N_{ed}$, given in Sakthivel et al. (2011) for better results. The results are tabulated in Table 3.16. When the chosen values of N_c, N_s, N_{ed} are low, then their corresponding objective function value is large, indicating that it converges to the local optimum. The objective function value is improved in case B, when N_c is increased,

TABLE 3.16

Maximum, Minimum and Standard Deviation Values of Objective Function for Different Cases

Case	N_c	N_s	N_{re}	N_{ed}	Minimum Value	Maximum Value	Standard Deviation
A	5	5	4	2	0.1251	0.1464	0.0059
B	10	5	4	4	0.0385	0.0733	0.0098
C	15	8	4	2	0.0011	0.0429	0.013
D	20	10	6	4	0.0004	0.0071	0.0015
E	100	4	4	2	0.00017	0.0018	0.0003

but it is not sufficient to reach the global optimum. When the value of N_c is increased further, as in case E, the bacterium reaches a better fitness function value but at the cost of a large convergence time.

Hence, to achieve better fitness value with faster convergence, the other parameters were adjusted, and the best solution was arrived at for case D. In addition, the remaining swarm parameters $d_{attract}$, $W_{attract}$, $h_{repellent}$, and $W_{repellent}$ were set to 0.01, 0.04, 0.01, and 10, respectively. To determine the performance of proposed BFA method, the results obtained were compared with GA and PSO methods in terms of their convergence characteristics, solution quality, and computational efficiency. Further, case studies have also been done with various operating points for different pulses per half-cycle. For validation, the computed harmonic spectra for different pulses per half-cycle were compared.

3.8.4.1 Convergence Characteristics

To estimate the quality of the solution, the convergence characteristics of GA, PSO, and BFA methods were taken for different pulses per half-cycle. The best computed convergence characteristics of GA, PSO, and BFA methods corresponding to 3, 5, 7, and 9 pulses per half-cycle are shown in Figure 3.100. From the graphs, it is evident that, among the curves, the fitness function value of the BFA method is the lowest for 3, 5, and 7 pulses per half-cycle. Further, in this method the convergence occurs steadily with a lower fitness value.

In case of three pulses per half-cycle, the fitness value starts at 0.08426 and converges steadily toward the global optimum value of 0.0003845 at ~40th iteration. Further, similar type of characteristics is seen for the BFA method even with higher pulses per half-cycle also. GA and PSO methods show comparatively good performance in case of fewer pulses per half-cycle, since the distance between the α values are larger. However, their performance deteriorates with a larger number of pulses per half cycle, and this could be attributed to the fact that both methods do not have a mechanism like elimination and dispersion to avoid convergence to the local optima.

Hence the methods settle at the local minima. Moreover, these methods take a longer time to converge. Hence, from the convergence characteristics, it can be concluded that BFA performs well compared to GA and PSO, irrespective of the number of pulses per half-cycle.

Further, to demonstrate the effectiveness of the proposed approach, the results obtained with BFA were compared with the Newton–Raphson (NR) and GA methods since both methods are extensively employed in the literature for harmonics elimination.

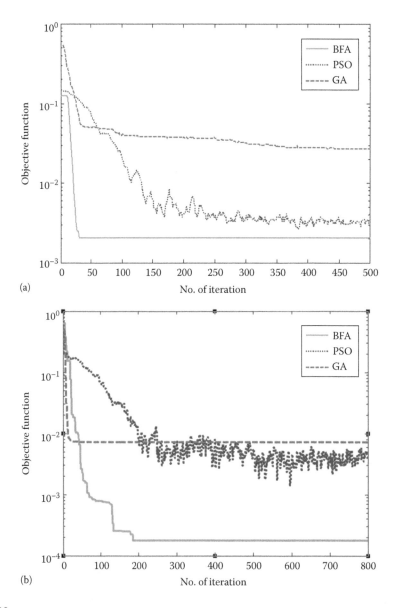

FIGURE 3.100
Comparison of convergence characteristics of BFA, PSO, and GA methods. (a) 3 pulses per half-cycle and (b) 5 pulses per half-cycle. *(Continued)*

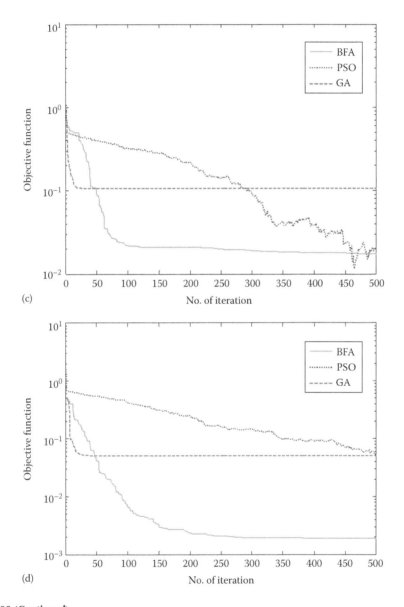

(c)

(d)

FIGURE 3.100 (*Continued*)
Comparison of convergence characteristics of BFA, PSO, and GA methods. (c) 7 pulses per half-cycle and (d) 9 pulses per half-cycle.

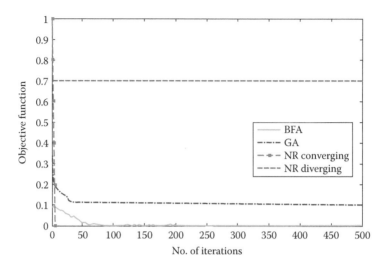

FIGURE 3.101
Comparison of convergence characteristics of BFA, GA, and NR methods.

The convergence characteristics of all methods are plotted in Figure 3.101. When the NR method was applied for harmonic elimination, it was found that the convergence strongly depended on the initial value selected and it failed to converge with random initial guess. However, this shortcoming is absent with GA and BFA, since both the methods share the basic property of population-based methods. It is interesting to compare the convergence characteristics of both methods. Both the methods start with a random initial guess, and try to converge to optimum value.

3.8.4.2 Solution Quality

To analyze the consistency of the solution obtained, the objective function values are noted for GA, PSO, and BFA for different runs. From the results, a comparison table is made by taking the minimum and maximum objective function values as well as their standard deviation, which are given in Table 3.17. From the table, it is apparent that BFA solution yields the best fitness value compared to the other two methods. Further, the BFA method has the lowest minimum and maximum values. Based on the standard deviation values, it can be inferred that BFA holds the lowest standard deviation value, indicating that this method is capable of producing quality solutions with any number of runs.

TABLE 3.17

Comparison of Solution Quality with GA, PSO, BFA

Algorithm	Minimum Value	Maximum Value	Standard Deviation
GA	0.2576	0.0124	0.008795
PSO	0.001976	0.0741	0.0183
BFA	0.000428	0.00621	0.000944

Note: Computational efficiency.

TABLE 3.18

Comparison of Computational Time (s) for Various Methods

Computational Time (s)	GA	PSO	BFA
	6.6002	5.43	3.84

In many applications, computational burden imposed on the optimization technique restricts its usage. Further, the computational burden increases the computational time for any optimization algorithm. To analyze the computational burden associated with GA, PSO, and BFA methods, the computation times were recorded for best run, and they are presented in Table 3.18. BFA takes a moderate amount of computation time compared to the other two methods. The PSO and GA methods take longer time to find the optimum value.

To test the effectiveness of all the three methods in solving the objective function framed in Equation 3.30, the harmonic spectra were taken for GA, PSO and BFA methods, and they are compared for four different cases with 3, 5, 7, and 9 pulses per half-cycle. The computed harmonic spectra taken at various operating points with BFA, GA, and PSO are shown in Figure 3.102. From the spectra, it is evident that BFA works successfully for the present problem by completely eliminating the undesired voltage harmonics with proper selection of the α value. In addition, the algorithm performs very well with higher number of switching pulses per half-cycle.

3.8.4.3 Experimental Validation of BFA Results

In order to substantiate the simulation findings, a hardware prototype was fabricated in the laboratory, and the same was used to drive a 220 V, 55 W, 1250 rpm, capacitor-run induction motor driving a domestic fan load. The Aurdino processor was used to generate the switching pulses to the IGBTs.

The switching angles were calculated offline for different operating points of the inverter, and they were subsequently stored in the processor. The harmonic spectra obtained with the experiments for 3, 5, 7, and 9 pulses per half-cycle for GA and BFA methods are presented in Figure 3.103. From the figure, it is apparent that there is good agreement between the computed harmonic spectra and the measured one. The small discrepancies could be attributed to the fact that, while simulation results are obtained for 50 Hz, the same frequency could not be maintained in the hardware implementation. Further, the measured motor terminal voltage waveform taken for three pulses per-half cycle is presented in Figure 3.104.

The problem of selective harmonic elimination in a PWM inverter was drafted as an optimization task, and the solution was sought through proposed bacterial foraging algorithm. Extensive simulations were carried out using MATLAB/Simulink and the results were taken for different operating points. For brevity, the results obtained with BFA were compared with the other optimization techniques GA, NR, and PSO. Based on the results, it was concluded that BFA works efficiently for selective harmonic elimination together with output voltage regulation. This method showed superior convergence, reduced computational burden, and better solution quality. In addition, hardware results were presented to validate the computed results.

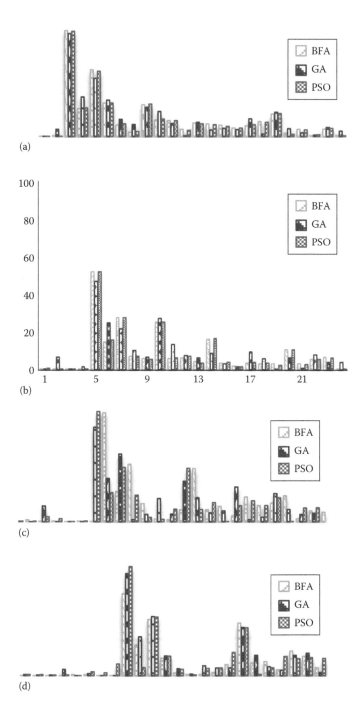

FIGURE 3.102
Simulated harmonic spectra for (a) $k = 3$ pulses per half-cycle, (b) $k = 5$ pulses per half-cycle, (c) $k = 7$ pulses per half-cycle, and (d) $k = 9$ pulses per half-cycle.

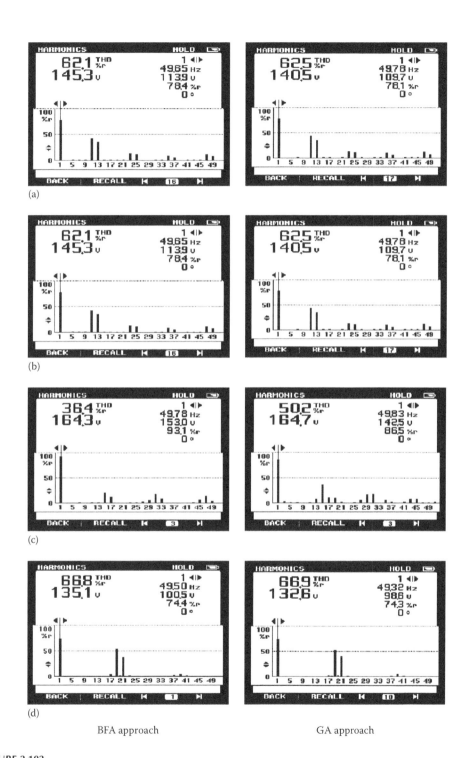

BFA approach　　　　　　　　　　GA approach

FIGURE 3.103
Experimental harmonic spectra: (a) $k = 3$ pulses per half-cycle, (b) $k = 5$ pulses per half-cycle, (c) $k = 7$ pulses per half-cycle, and (d) $k = 9$ pulses per half-cycle.

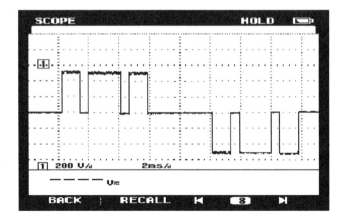

FIGURE 3.104
Experimental motor terminal voltage waveform for $k = 3$ pulses per half-cycle.

3.9 Summary

The problem of harmonic elimination in VFD input current and voltage harmonics and PWM output selective voltage harmonics was drafted as an optimization task, and the same was solved using GA and BFA. From the results, it was observed that both algorithms worked efficiently for output voltage regulation together with harmonic elimination. The proposed algorithms GA and BFA completely eliminated lower order current harmonics, improved the power factor at input side of the VFD, and reduced inverter output selective voltage harmonics, together with voltage regulation. To test the feasibility and quality of the solution, a MATLAB program was developed and successfully implemented for different operating conditions and compared with other optimization techniques. Hardware prototypes of a two-pulse rectifier drive, a 6-pulse rectifier drive, and a 12-pulse rectifier drive filter were also built to compare and validate the computed results. The VFDs are most commonly preferred in induction and permanent magnet motors for speed variation due to their high static and dynamic performance. They find numerous applications such as in pumps, fans, conveyors, agitators, pulping machines, paper machines, paper packing machines, and so on. VFDs are switched periodically in order to maintain the output voltage with a *V/F* ratio at the motor end. However, these VFDs inject current harmonics at source side and voltage harmonics at motor side. The front-end converter topology in VFD is diode/thyristor converters, which are the best choice for low power rating applications because of their low cost, robustness, and reliability.

The problem of selective harmonics elimination in PWM inverter was drafted as an optimization task, and it was solved through proposed BFA. Extensive simulations were carried out using MATLAB/Simulink, and the results were taken for different operating points. For brevity, the results obtained with BFA were compared with for three other optimization techniques: GA, NR, and PSO. Based on the results, it was concluded that BFA works efficiently for selective harmonic elimination together with output voltage regulation. This method showed superior convergence, reduced computational burden, and better solution quality. In addition, hardware results were presented to validate the computed results.

References

Asiminoaei, L., Hansen, S., and Blaabjerg, F., Predicting harmonics by simulations. A case study for high power adjustable speed drive, *Electrical Power Quality and Utilisation, Magazine*, 2(1), 65–75, 2006.

Abraham, A., Biswas, A., Dasgupta, S., and Das, S., Analysis of reproduction operator in bacterial foraging optimization algorithm, *IEEE Congress on Evolutionary Computation*, 1476–1483, 2008.

Avirneni, V. and Obulesh, Y.P., Harmonic analysis of electrical drives in DELTA Paper Mill, in *Proceedings of IEEE-International Conference on Emerging Trends in Electrical and Computer Technology (ICETECT)*, 2011, pp. 226–231.

Ahmadi, D., Zou, K., Li, C., Huang, Y., and Wang, J., A universal selective harmonic elimination method for high-power inverters, *IEEE Transactions on Power Electronics*, 26(10), 2743–2752, 2011.

Abd-Elazim, S.M. and Ali, E.S., Bacteria foraging optimization algorithm based SVC damping controller design for power system stability enhancement, *Electrical Power and Energy Systems*, 43(1), 933–940, 2012.

Al-Othman, A.K., Ahmed, N.A., AlSharidah, M.E., and AlMekhaizim, H.A., A hybrid real coded genetic algorithm—Pattern search approach for selective harmonic elimination of PWM AC/AC voltage controller, *Electrical Power and Energy Systems*, 44, 123–133, 2013.

Askarzadeh, A. and Rezazadeh, A., Artificial bee swarm optimization algorithm for parameters identification of solar cell models, *Applied Energy*, 102, 943–949, 2013.

Briault, F., Hélier, M., Lecointe, D., Bolomey, J.C., and Chotard, R., Broad-band modeling of a realistic power converter shield for electric vehicle applications, *IEEE Transactions on Electromagnetic Compatibility*, 42(4), 477–486, 2000.

Biswas, A., Dasgupta, S., Das, S., and Abraham, A., Synergy of PSO and bacterial foraging optimization: a comparative study on numerical benchmarks, *Second International Symposium on Hybrid Artificial Intelligent Systems*, 44, 255–263, 2007.

Biswas, A., Dasgupta, S., Das, S., and Abraham, A., A synergy of differential evolution and bacterial foraging optimization for faster global search, *International Journal on Neural and Mass-Parallel Computing and Information Systems—Neural Network World*, 17(6), 607, 2007.

Barkat, S., Berkouk, E.M., and Boucherit, M.S., Particle swarm optimization for harmonic elimination in multilevel inverters, *Electrical Engineering*, 91, 221–228, 2009.

Biswas, A., Das, S., Abraham, A., and Dasgupta, S., Analysis of the reproduction operator in an artificial bacterial foraging system, *Applied Mathematics and Computation*, 215(9), 3343–3355, 2010.

Biswas, A., Das, S., Abraham, A., and Dasgupta, S., Stability analysis of the reproduction operator in bacterial foraging optimization, *Theoretical Computer Science*, 411, 2127–2139, 2010.

Bindu, J., Selvaperumal, S., Muralidharan, S., and Muhaidheen, M., Genetic algorithm based selective harmonic elimination in PWM AC-AC converter, in *Proceedings of IEEE International Conference on in Recent Advancements in Electrical, Electronics and Control Engineering*, 2011, pp. 393–397.

Chen, Y.M., Passive filter design using genetic algorithms, *IEEE Transactions on Industrial Electronics*, 50(1), 202–207, 2003.

Cichowlas, M., PWM rectifier with active filtering. PhD thesis, Warsaw University of Technology, Warsaw, Poland, 2004.

Dahidah, M.S., Agelidis, V.G., and Rao, M.V., Hybrid genetic algorithm approach for selective harmonic control, *Energy Conversion and Management*, 49(2), 131–142, 2008.

Das, S., Biswas, A., Dasgupta, S., and Abraham, A., Bacterial foraging optimization algorithm: theoretical foundations, analysis, and applications, *Foundations of Computational Intelligence in Computational Intelligence*, 3, 23–55, 2009.

Dasgupta, S., Das, S., Abraham, A., and Biswas, A., Adaptive computational chemotaxis in bacterial foraging optimization an analysis, *IEEE Transactions on Evolutionary Computation*, 13(4), 919–941, 2009.

Dzhankhotov, V., Hybrid LC filter for power electronic drives: theory and implementation. PhD thesis, Lappeenranta University of Technology, Lappeenranta, Finland, 2009.

El-Naggar, K. and Abdelhamid, T.H., Selective harmonic elimination of new family of multilevel inverters using genetic algorithms, *Energy Conversion and Management*, 49(1), 89–95, 2008.

Fukuda, S. and Ohta, M., An auxiliary-supply-assisted twelve-pulse diode rectifier with reduced input current harmonics, *Proceedings of IEEE Conference on Industry Applications*, 1, 2004.

Fukuda, S., Ohta, M., and Iwaji, Y., An auxiliary-supply-assisted harmonic reduction scheme for 12-pulse diode rectifiers, *IEEE Transactions on Power Electronics*, 23(3), 1270–1277, 2008.

Holmes, D.G. and Lipo, T.A., *Pulse Width Modulation for Power Converters: Principles and Practice* (vol. 18). Hoboken, NJ: John Wiley & Sons, 2003.

Huang, L., He, N., and Xu, D., Optimal design for passive power filters in hybrid power filter based on particle swarm optimization, in *Proceedings of the IEEE International Conference on Automation and Logistics*, Jinan, China, 2007, pp.1468–1472.

Jegathesan, V. and Jerome, J., Elimination of lower order harmonics in voltage source inverter feeding an induction motor drive using evolutionary algorithms, *Expert Systems with Applications*, 38(1), 692–699, 2011.

Kennedy, J. and Eberhart, R., Particle swarm optimization, *IEEE International Conference on Neural Networks*, 4, 1942–1948, 1995.

Keskar, P.Y., Specification of variable frequency drive systems to meet the new IEEE 519 standard, *IEEE Transaction on Industry Application*, 32(2), 393–402, 1996.

Keypour, R., Seifi, H., and Yazdian-Varjani, A., Genetic based algorithm for active power filter allocation and sizing, *Electric Power Systems Research*, 71(1), 41–49, 2004.

Kavousi, A., Vahidi, B., Salehi, R., Bakhshizadeh, M., Farokhnia, N., and Fathi, S.S., Application of the bee algorithm for selective harmonic elimination strategy in multilevel inverters, *IEEE Transactions on Power Electronics*, 27(4), 1689–1696, 2012.

Liserre, M., Aquila, A.D., and Blaabjerg, F., Genetic algorithm-based design of the active damping for an LCL-filter three-phase active rectifier, *IEEE Transactions on Power Electronics*, 19(1), 76–86, 2004.

Lian, K.L., Perkins, B.K., and Lehn, P.W., Harmonic analysis of a three-phase diode bridge rectifier based on sampled-data model, *IEEE Transactions on Power Delivery*, 23(2), 1088–1096, 2008.

Li, M.S., Ji, T.Y., Tang, W.J., Wu, Q.H., and Saunders, J.R., Bacterial foraging algorithm with varying population, *Bio Systems*, 100(3), 185–197, 2010.

Li, Y., Luo, L., Rehtanz, C., Yang, D., Rüberg, S., and Liu, F., Harmonic transfer characteristics of a new HVDC system based on an inductive filtering method, *IEEE Transactions on Power Electronics*, 27(5), 2273–2283, 2012.

Liu, C., Chu, Y., Wang, L., and Zhang, Y., Application and the parameter tuning of ADRC based on BFO-PSO algorithm, in *Proceedings of IEEE in Control and Decision Conference*, Firenze, Italy, 2013, pp. 3099–3102.

Maheswaran, D., Straight talk about harmonics problems and case study for using AC drives in pulp and paper industries, *International Journal of the Indian Pulp and Paper Technical Association (IPPTA), ISSN (0379–5462)*, 24(2), 113–118, 2012.

Maheswaran, D., Kandasamy, R., and Rajasekar, N., Installation and operating experience of 12 pulse AC/DC converter fed AC drives in recovery boiler, *International Journal of the Indian Pulp and Paper Technical Association (IPPTA), ISSN(0379-5462)*, 24(4), 105–110, 2012.

Maheswaran, D., Rajasekar, N., and Ashok Kumar, L., Design of passive filters for reducing harmonic distortion and correcting power factor in two pulse rectifier systems using optimization, *Journal of Theoretical and Applied Information Technology(JATIT)*, 62(3), 720–728, 2014.

Maheswaran, D., Rajasekar, N., and Ashok Kumar, L., A detailed current harmonics analysis and GA based passive LC filter design in six-pulse variable frequency drive, *International Journal of the Indian Pulp and Paper technical Association (IPPTA)*, 26(2), 80–88, 2014.

McGranaghan, M.F. and Mueller, D.R., Designing harmonic filters for adjustable-speed drives to comply with IEEE-519 harmonic limits, *IEEE Transactions on Industry Applications*, 35(2), 312–318, 1999.

Mishra, S., A hybrid least square-fuzzy bacterial foraging strategy for harmonic estimation, *IEEE Transactions on Evolutionary Computation*, 9(1), 61–73, 2005.

Mishra, S. and Bhende, C.N., Bacterial foraging technique-based optimized active power filter for load compensation, *IEEE Transactions on Power Delivery*, 22(1), 457–465, 2007.

Mingwei, R., Yukun, S., and Xiang, R., A study on optimization of passive filter design, in *Proceedings of the 29th Chinese Control Conference*, Beijing, China, 2010, pp. 4981–4985.

Nazarzadeh, J., Razzaghi, M., and Nikravesh, K.Y., Harmonic elimination in pulse-width modulated inverters using piecewise constant orthogonal functions, *Electric Power Systems Research*, 40(1), 45–49, 1997.

Nouri, H. and Hong, T.S., Development of bacteria foraging optimization algorithm for cell formation in cellular manufacturing system considering cell load variations, *Journal of Manufacturing Systems*, 32(1), 20–31, 2013.

Niu, B., Wang, H., Wang, J., and Tan, L., Multi-objective bacterial foraging optimization, *Neuro Computing*, 116, 336–345, 2013.

Okaeme, N.A. and Zanchetta, P., Hybrid bacterial foraging optimization strategy for automated experimental control design in electrical drives, *IEEE Transactions on Industrial Informatics*, 9(2), 668–678, 2013.

Patel, H.S. and Hoft, R.G., Generalized techniques of harmonic elimination and voltage control in thyristor inverters: Part 1—harmonic elimination, *IEEE Transactions on Industry Applications*, IA-9, 310–317, 1973.

Passino, K.M., Biomimicry of bacterial foraging for distributed optimization and control, *IEEE Control System Magazine*, 22(3), 52–67, 2002.

Patnaik, S.S. and Panda, A.K., Comparative evaluation of harmonic compensation capability of active power filter with conventional and bacterial foraging based control, in *Proceedings of IEEE Ninth International Conference on Power Electronics and Drive Systems*, Singapore, 2011, pp. 95–99.

Rice, D.E., A detailed analysis of six-pulse converter harmonic currents, *IEEE Transaction on Industry Application*, 30(2), 294–304, 1994.

Ray, R.N., Chatterjee, D., and Goswami, S.K., A modified reference approach for harmonic elimination in pulse-width modulation inverter suitable for distributed generations, *Electric Power Components and Systems*, 36(8), 815–827, 2008.

Ray, R.N., Chatterjee, D., and Goswami, S.K., An application of PSO technique for harmonic elimination in a PWM inverters, *Applied Soft Computing*, 9(4), 1315–1320, 2009.

Ray, R.N., Chatterjee, D., and Goswami, S.K., A PSO based optimal switching technique for voltage harmonic reduction of multilevel inverter, *Expert Systems with Applications*, 37(12), 7796–7801, 2010.

Sundareswaran, K. and Kumar, A.P., Voltage harmonic elimination in PWM A.C. Chopper using genetic algorithm, *IEEE Proceedings on Electron Power*, 151(1), 26–31, 2004.

Shi, K.L. and Li, H., Optimized PWM strategy based on genetic algorithms, *IEEE Transactions on Industrial Electronics*, 52(5), 1458–1461, 2005.

Suriadi, S., Analysis of harmonics current minimization on power distribution system using voltage phase shifting concept. Doctoral dissertation, Universiti Sains Malaysia, Penang, Malaysia, 2006.

Singh, B., Bhuvaneswari, G., and Garg, V., Harmonic mitigation using 12-pulse AC–DC converter in vector-controlled induction motor drives, *IEEE Transactions on Power Delivery*, 21(3), 1483–1495, 2006.

Singh, B., Bhuvaneswari, G., Garg, V., and Gairola, S., Pulse multiplication in AC-DC converters for harmonic mitigation in vector-controlled induction motor drives, *IEEE Transactions on Energy Conversion*, 21(2), 342–352, 2006.

Singh, B., Bhuvaneswari, G., Garg, V., and Chandra, A., Star connected autotransformer based 30-pulse AC-DC converter for power quality improvement in vector controlled induction motor drives, in *Proceedings of IEEE Conference on Power India*, New Delhi, India, 2006, p. 6.

Sanglikar, A. and John, V., Novel approach to develop behavioral model of 12-pulse converter, in *Proceedings of IEEE International Conference on Power Electronics, Drives and Energy Systems*, New Delhi, India, 2006, pp. 1–5.

Sundareswaran, K., Krishna, J., and Shanavas, T.N., Inverter harmonic elimination through a colony of continuously exploring ants, *IEEE Transactions on Industrial Electronics*, 54(5), 2558–2565, 2007.

Salehi, R., Vahidi, B., Farokhnia, N., and Abedi, M., Harmonic elimination and optimization of stepped voltage of multilevel inverter by bacterial foraging algorithm, *Journal of Electrical Engineering & Technology*, 5(4), 545–551, 2010.

Subudhi, B. and Ray, P.K., A hybrid adeline and bacterial foraging approach to power system harmonics estimation, in *Proceedings of IEEE International Conference on Industrial Electronics, Control & Robotics*, Orissa, India, 2010, pp. 236–242.

Salam, Z. and Bahari, N., Selective harmonics elimination PWM (SHEPWM) using differential evolution approach, in *Proceedings of IEEE International Conference on Power Electronics, Drives and Energy Systems*, New Delhi, India, 2010, pp. 1–5.

Sakthivel, V.P., Bhuvaneswari, R., and Subramanian, S., An accurate and economical approach for induction motor field efficiency estimation using bacterial foraging algorithm, *Measurement*, 44, 674–684, 2011.

Shirabe, K., Swamy, M., Kang, J., Hisatsune, M., Wu, Y., Kebort, D., and Honea, J., Efficiency comparison between Si-IGBT-based drive and GaN-based drive, *IEEE Transaction on Industry Applications*, 50(1), 566–572, 2014.

Shankar, M., Monisha, S., Shesna, H., Vignesh, T., Sikkandar, N., Sundaramoorthi, S., and Venkatesh, S., Implementation of space vector pulse width modulation technique with genetic algorithm to optimize unified power quality conditioner, *American Journal of Applied Sciences*, 11(1), 152–159, 2013.

Sousa Santos, V., Quispe, E.C., Sarduy, J.R.G., Viego, P.R., Lemozy, N., Jurado, A., and Brugnoni, M., Bacterial foraging algorithm application for induction motor field efficiency estimation under harmonics and unbalanced voltages, in *Proceedings of IEEE International Conference in Electric Machines & Drives*, Chicago, IL, 2013, pp. 1108–1111.

Turnbull, F.G., Selected harmonic reduction in static dc-ac Inverters, *IEEE Transactions on Communication and Electronics*, 83(73), 374–378, 1964.

Twining, E. and Cochrane, I., Modelling variable speed ac drives and harmonic distortion, in *Proceedings of Australasian Universities Power Engineering Conference, AUPEC*, Darwin, Australia, vol. 99, 1999, pp. 26–29.

Tripathy, M. and Mishra, S., Bacteria foraging-based to optimize both real power loss and voltage stability limit, *IEEE Transactions on Power Systems*, 22(1), 240–248, 2007.

Tutkun, N., Improved power quality in a single-phase PWM inverter voltage with bipolar notches through the hybrid genetic algorithms, *Expert Systems with Applications*, 3(7), 5614–5620, 2010.

Villablanca, M.E., Nadal, J.I, and Bravo, M.A., A 12-pulse AC–DC rectifier with high-quality input/output waveforms, *IEEE Transactions on Power Electronics*, 22(5), 1875–1881, 2007.

Vishal, V. and Singh, B., Genetic-algorithm-based design of passive filters for offshore applications, *IEEE Transactions on Industry Applications*, 46(4), 1295–1303, 2010.

Variable Speed Drives and Power Electronics. http://www.idc-online.com.

Vacon Drive Manuals. http://www.vacon.com.

Wu, W., He, Y., and Blaabjerg, F., An LLCL power filter for single-phase grid-tied inverter, *IEEE Transactions on Power Electronics*, 27(2), 782–789, 2012.

Wang, Y., Wen, X., Zhao, F., and Xinhua, G., Selective harmonic elimination PWM technology applied in PMSMs, in *Proceedings of IEEE Vehicle Power and Propulsion Conference*, Seoul, South Korea, 2012, pp. 92–97.

You-hua, J. and Dai-fa, L., Multi-objective optimal design for hybrid active power filter based on composite method of genetic algorithm and particle swarm optimization, in *Proceedings of IEEE International Conference on Artificial Intelligence and Computational Intelligence*, Shanghai, China, vol. 2, pp. 549–553, 2009.

Zobaa, A.F., Harmonic problems produced from the use of adjustable speed drives in industrial plants: case study, in *International Conference on Harmonics and Quality of Power*, Giza, Egypt, 2004, pp. 6–10.

Zubi, H.M., Low pass broadband harmonic filter design. M.S. Thesis, Middle East Technical University, Ankara, Turkey, 2005.

Zhou, H., Li, Y.W., Zargari, N.R., Cheng, Z., Ruoshui, N.I., and Zhang, Y., Selective Harmonic Compensation (SHC) PWM for grid-interfacing high-power converters, *IEEE Transactions on Power*, 29(3), 1118–1127, 2014.

4

Voltage and Frequency Control in Power Systems

Learning Objectives

On completion of this chapter, the reader will have knowledge on

- Dynamics of LFC and AVR systems and their modeling and control
- Design of Intelligent Fuzzy Logic Controller using MATLAB®/Simulink®
- Controllers for Single-Area Power System and Multi-Area Power System
- Design of PID controller based on GA, PSO, and ACO
- Hybrid algorithms: EPSO, MO-PSO, SPSO, FPSO, BF-PSO, and HGA-based PID controller design

4.1 Introduction

Power is an important infrastructure that decides the economic growth of a country. Both electrical utilities and end users of electricity have become more concerned about the quality and reliability of electric power. Power quality is defined as "any power problem manifested in voltage, current or frequency deviations that result in failure or misoperation of customer equipment" (Ali 2012). Throughout the world, with the increase in generation and demand, the electric power industry has gone under various changes in improving power quality. Problems of power quality arise from a variety of different reasons, and different solutions can be used to improve quality and equipment performance. It is a term used to broadly encompass the entire scope of interaction among electric suppliers, environment, and users of systems and products (Bollen 2000). Figure 4.1 gives the general steps that are often required in a power quality investigation, along with the major considerations that must be addressed at each step.

The basic steps in investigating the problem category and the general procedure of finding the optimum solution must consider the economics and technical limitations. During transmission and distribution, both the active power balance and the reactive power balance must be maintained between the generation and utilization of electric power. The active and reactive power balances correspond to two equilibrium points: voltage and frequency. Both frequency and voltage are required to be maintained at nominal values during operation in order to obtain a good quality electric power system. Voltage fluctuations are systematic variations of the voltage envelope or a series of random voltage

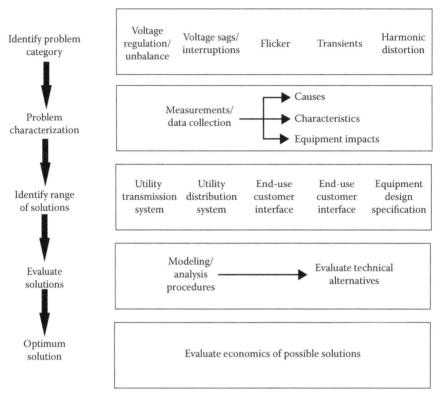

FIGURE 4.1
Basic steps involved in power quality evaluation.

changes, the magnitude of which does not exceed the voltage range specified as 0.9–1.1 p.u. The purpose of voltage control in distribution networks is to compensate for load variations and events in the transmission system, such that a supply voltage to customers is within certain bounds. The power system frequency is directly related to the rotational speed of the generators supplying the system. Due to change in dynamic load and generation, slight variation in the frequency will be noticed. The change in frequency and its duration depend on the characteristics of the load and generation control system responses. Although the individual suppliers on the power market usually have independent goals, such as survival and maximization of profits, the overall goal in the operation of power systems is a reliable energy supply to consumers at a low cost. Seen from the point of view of the engineer, some additional constraints have to be met; for example, the voltage and frequency in all parts of the network should be acceptable and operational limits of the individual power system components are not exceeded. Load frequency control (LFC) and automatic voltage regulator (AVR) are the two control loops necessary to regulate the output frequency and voltage of the generator. LFC is a system that tries to keep the frequency of the supply constant by increasing or decreasing the real power output from generators. AVR, as the name suggests, tries to keep the voltage constant, or at least within limits for all consumers. AVR acts on the exciter of a synchronous machine, which supplies the field voltage and consequently the current in the field winding of the machine, and can thereby regulate its terminal voltage. A properly designed and operated power system should be able to meet the continually changing load demand for active and reactive power.

A typical large-scale power system is composed of several areas of generating units. In order to enhance the fault tolerance of the entire power system, these generating units are connected via tie-lines. The usage of tie-line power imports a new error into the control problem, that is, tie-line power exchange error. When a sudden active power load change occurs to an area, the area will obtain energy via tie-lines from other areas. But eventually, the area that is subject to the load change should balance it without external support. Hence, each area requires a separate load frequency controller to regulate the tie-line power exchange error so that all the areas in an interconnected power system can set their set points differently. A power system control is required to control the frequency deviation and the voltage magnitude of the generating system. The proportional-integral-derivative (PID) controller contributes to this satisfactory operation of the power system by upholding these parameters within the acceptable limits, and so the balance between the generations against system loads and losses is also achieved. Conventional PID controllers are employed in most engineering applications because of their simplicity, easy implementation, and robustness. The PID controller has the ability to cope up with various load changes and profound effect on the dynamic performance of the power system. The gains of PID controllers must be properly tuned to guarantee security, dynamic performance, and sustainable utilization of the plants. The conventional methods for improving the performance in tuning K_P, K_I, and K_D gain parameters such as, Ziegler–Nichols, simplex method, orthogonal test, Newton method, quadratic method, etc., need a complete set of information related to the plant behavior and prerequisite knowledge of the problem. The conventional controllers are designed for particular operating condition, and the gain values are tuned manually. Hence, they fail to provide better results, and when the operating conditions change, they may not provide suitable control. In this chapter, considerable advancements and improvements to mitigate the requirements of ends users in terms of quality power supply have been discussed.

The most recent advancement in all engineering control domains is the application of intelligent computing concepts such as neural network (NN), fuzzy logic (FL), genetic algorithm (GA), and evolutionary algorithm (EA) to tackle the difficulties associated with the design of controllers for the efficient control of voltage and frequency. The significant advantage of using evolutionary search lies in the gain of flexibility and adaptability to the task at hand, in combination with robust performance and global search characteristics (Back et al. 1997). The artificial intelligence paradigms are appropriate to model the uncertainties found in the power demand and improve the flexible nature of controllers. The current trend focuses on the application of intelligent computing techniques for optimum tuning of PID gain parameters to improve the efficiency of the controller. The optimized values of gains K_P, K_I, and K_D are applied in the PID controller in the secondary control loop of LFC and AVR models. The fundamental features justifying the strong emphasis on the application of evolutionary computing include ease of implementation, computational efficiency, and provision for fine tuning to obtain predetermined results. EAs often yield excellent results when applied to complex optimization problems where other conventional methods are either not applicable or turn out to be unsatisfactory (Back et al. 1997). In an interconnected system, all the connected generator units of the system contribute to the overall change in generation, irrespective of the location of the load change. Hence, the focus is also extended to regulate the system frequency between interconnected areas through tie-lines.

The models are simulated for different loads and regulation parameters to validate the robustness of the proposed algorithms. The performance measures used to study the effectiveness of the proposed controllers include settling time, overshoot, and oscillations. The computational complexity of the proposed algorithms is compared based on the program execution time taken by the CPU in generating optimum results.

4.2 Scope of Intelligent Algorithms in Voltage and Frequency Control

Electric power converts energy from one of the naturally available forms to the electric form and transports it to the point of consumption. Energy is seldom consumed in the electric from, but it is converted into other forms by the user. The advantage of the electrical form of energy is that it can be transported and controlled with relative ease at high degree of efficiency and reliability. Users of electric power change the loads randomly and momentarily. It will be impossible to maintain the balances of both the active and reactive powers without control. Pursuant to the imbalance, the frequency and voltage levels will vary with the change of the loads. Thus, a control system is essential to cancel the effects of the random load changes and to keep the frequency and voltage at the standard values. The scope of this chapter is to identify the need for a dispatching system that can control large numbers of electrical generators without being excessively complicated. In order to have a better idea of the inherent requirements of a system, it would be wise to review the control engineering aspects of power systems to get a good impression of the related problems. Due to their noncontinuous, nondifferentiable, and highly nonlinear nature, these problems are difficult or impossible to solve using the conventional optimization techniques. Computational intelligence algorithms are applied to solve the complex power system problems relating to the frequency and voltage profile of generating systems.

The different stages of solving this problem are as follows:

1. Analyze the problems related to LFC and AVR in maintaining optimum frequency and voltage of the power generating system and recommend intelligent computing techniques to improve the quality of generating systems.
2. Develop an efficient methodology for solving these complex problems by using different evolutionary optimization algorithms.
3. Estimate the potential of EAs for obtaining optimal solutions for different loads and regulation parameters related to convergence rate of the algorithm and computational complexity.
4. Test the validity of hybrid algorithms for two-area LFC systems with identical generating system to investigate the performance of the algorithm by applying different load and regulation in both areas.

Due to such high occurrence of voltage and frequency instability events, there is a serious concern for remedial measures. The major contribution is the application of intelligent computing techniques for solving power system optimization problems, which were previously addressed by conventional problem-solving methods. This chapter elaborates on the different intelligent computing techniques for online voltage and frequency stability monitoring. Online stability monitoring is the process of obtaining voltage and frequency stability information for a given operating scenario. The prediction should be fast and accurate such that control signals can be sent to appropriate locations quickly and effectively.

The performance of LFC and AVR is enhanced with the application of FL, GA, particle swarm optimization (PSO), ant colony optimization (ACO), and hybrid EA optimization techniques, described as follows:

- An intelligent fuzzy logic controller (FLC) is designed to improve the control performance and robustness of isolated and interconnected power generating systems. The conventional PID controller is replaced by fuzzy control in the control loop to enhance the performance of LFC and AVR. Simulation results obtained from applying the FL control technique to the decentralized LFC and AVR problem have been compared to conventional PID control.

- Genetic algorithm is one of the most important search algorithms based on the mechanics of natural selection and natural genetics. It has been verified to be effective to solve a complex optimization problem of selecting optimum gains of PID controller in a secondary loop of LFC and AVR of the power generating system. Since GA has the high potential toward global optimization, the GA-PID controller increases the dynamic performance of the isolated and interconnected generating system.

- Evolutionary algorithm is a useful paradigm in finding an optimum solution to the real-time optimization problem. The evolutionary paradigms such as PSO and ACO are proposed to tune the gain parameters for the PID controller in LFC and AVR. This method improves the computational efficiency and performance of the controller in satisfying the requirements related to settling time, oscillations, and overshoot. The simulation results are compared with conventional PID controller for its validity and stability. The effectiveness of the proposed PSO and ACO algorithms is tested on two-area LFC system, and the results indicate the robust behavior when compared to traditional controllers. In order to truly identify the changing dynamics of the power grid at all times and provide the appropriate control actions, the high computational power for fast dynamic modeling capability is needed. Hence, to overcome the computational challenges for real-time online modeling, hybrid EA is proposed to increase computational capabilities and speed.

The conventional PSO algorithm is combined with fuzzy, GA, and BF to obtain hybrid algorithms to optimize the PID gain parameters. An advanced hybrid EA is applied to increase the searching speed, reliability, and efficiency of the original PSO algorithm. The proposed LFC and AVR with hybrid EA-based PID controller contribute to the satisfactory operation of the power system by maintaining system voltages and frequency for different load and regulation parameters.

Currently, researchers are focusing on exploring the possibilities of applying knowledge-based systems to increase the efficiency and computational complexity of conventional controllers. This, in turn, has prompted the development of novel and intelligent algorithms for power system control with greater reliability and less computational time. The performance measures used to quantify the advantages of the proposed algorithm include settling time, oscillations, overshoot, and computational time as compared to conventional PID controllers.

4.3 Dynamics of Power Generating System

Electrical engineers are concerned with every step in the process of generation, transmission, distribution, and utilization of electrical energy. The electric utility industry is probably the largest and most complex industry in the world. Electrical engineers working in that industry will encounter challenging problems in designing future power systems to deliver quality and reliable electrical energy (Glover et al. 2007). As electric utilities have grown in size and the number of interconnections has increased, planning for future expansion has become increasingly complex. The increasing cost of additions and modifications has made it imperative that utilities consider a range of design options and perform detailed studies of the effects on the system. To assist engineers in this power system control, digital computers and highly developed computer programs are used. Power flow programs compute the voltage magnitudes, phase angles, generator reactive power output, and transmission line power flows for a network under steady-state operating conditions. Today's computers have sufficient speed and storage capacity to efficiently compute solutions to complex power systems with 100,000 buses and 150,000 transmission lines. The large storage capacity of computers and high-speed capabilities allow engineers to run the necessary cases to analyze and design transmission and generation expansion options (Glover et al. 2007). The range of voltage and frequency excursions depends on the type of disturbances in the system. In practice, the value of frequency excursions is less than 5% and the voltage excursions will be more than 10%. Accidental loss of load often leads to high-frequency and high-voltage conditions; loss of generation and system collapse due to excessive rise of load usually lead to low-frequency and low-voltage conditions. Immediately, after a major disturbance, the power system's frequency rise and decay are arrested automatically by load rejection, load shedding, controlled separation, and isolation mechanisms. The success rate of these automatic restoration mechanisms has been about 50%. The challenge is to coordinate the control and protective mechanisms with the operation of the generating plants and the electrical system. During the subsequent restoration, plant operators, in coordination with system operators, attempt manually to maintain a balance between load and generation. The duration of these manual procedures has been invariably longer than limitations that equipment can accommodate.

Hence there is a danger that the operation of power plants and the power system may not maintain the necessary coordination resulting in greater impacts on the quality of power supply (Adibi and Fink 2006). The control of frequency is achieved primarily through a speed governor mechanism aided by LFC for precise control. The speed of the shaft is sensed and compared with a reference value, and the feedback signal is utilized to increase or decrease the power generated by controlling the inlet valve. In the case of interconnected operation, the tie-line power flows must be maintained at the specified values. It is fulfilled by deriving an error signal from the deviations in the specified tie-line power flows to the neighboring utilities and adding this to the control signal of the LFC (Murty 2005). The AVR senses the terminal voltage and adjusts the excitation to maintain a constant terminal voltage. The process involves continuous sensing of the terminal voltage, its rectification, smoothening, and comparison with reference value. The "error voltage" generated after comparison of results is used to control the field excitation of the alternator. The AVR loop maintains reactive power balance at a generator bus by indirectly maintaining a constant voltage level. Control strategies such as PI controllers, decentralized controllers, optimal controllers, and adaptive controllers are adopted for LFC and AVR to regulate the real and reactive power output (Cam and Kocaarslan 2005, Elgerd 1982, Shanmugham et al. 2001). In recent years, intellectual computing methods such as FL, GA, PSO, ACO, and hybrid evolutionary algorithm (HEA) are applied for voltage and frequency control in synchronous generators (Eberhart and Shi 2000, Gaing 2004, Juang and Lu 2004, Karnavas and Papadopoulos 2002, Mukherjee and Ghosal 2008, Pratima Singh and Jha 2001, Shigenori Naka et al. 2004, Yildiz et al. 2006). Extensive efforts are made by applying these algorithms to create a more competitive environment for electricity markets and to promote greater efficiency.

In order to implement computer control of a power system, it is imperative to gain a clear understanding of the representation of the power system components. Studies of electrical energy systems are based on the simulation of actual phenomena using models behaving exactly in the identical way as the elements in the physical system (Chakrabarti and Halder 2006). In this application, the synchronous generator with non-reheat turbine is modeled by including the prime mover and excitation control loops. Intelligent computing techniques such as fuzzy, GA, ACO, PSO, and Hybrid PSO are used to limit the transient oscillation due to instantaneous load perturbations and to maintain the voltage and frequency at its nominal value. Hence, an efficient and economical means to tune the PID controller gains used in the secondary control loop of LFC and AVR is presented in this chapter. The effectiveness and usefulness of the proposed methods are tested for their efficacy by simulating the LFC and AVR models for different loads and regulations. The simulation was also carried out in two-area interconnected power system for controlling the frequency of the generators working in parallel. The investigations on simulation results show that the proposed controllers have profound effect on the dynamic performance of the power system and on its ability to cope with disturbances. In a nutshell, the primary motivation of the work is to explore the behavior of the plant and to analyze the effect of the controllers based on the simulation results. The chapter provides strong theoretical support for implementation due to the robustness with sufficient tolerances to the model and parameter uncertainties. This chapter deals with the modeling of an LFC and AVR for single-area and two-area interconnected power system using the Simulink® available in MATLAB® 7.1.

4.3.1 Control of Active and Reactive Power

Sustained operation of power systems is impossible, unless generator frequencies and bus voltages are kept within strict limits. The control of active and reactive power flows in a transmission network is a very important task in power system control. Active power control is closely related to frequency control, and reactive power control is closely related to voltage control. As constancy of frequency and voltage is an important factor in determining the quality of power supply, the control of active power and reactive power is vital for the satisfactory performance of the power system.

The controllers are set for a particular operating condition and take care of small changes in load demand to maintain the frequency and voltage magnitude within the specified limits. Small changes in real power are mainly dependent on change in rotor angle δ and thus the frequency. The reactive power is mainly dependent on voltage magnitude (i.e., on generator excitation). A properly designed and operated power system should have the following requirements (Wood and Wollenberg 1984):

- The system must be able to meet the continually changing load demand for active and reactive power.
- The system should supply energy at minimum cost and with minimum ecological impact.
- The "quality" of power supply must meet certain minimum standards with regard to the following factors:
  - Constancy of frequency
  - Constancy of voltage
  - Level of reliability

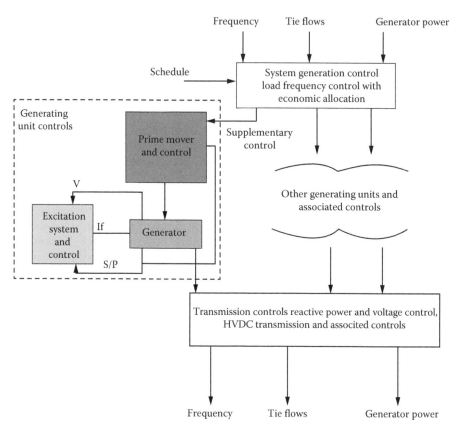

FIGURE 4.2
Power system and associated controls.

Several levels of controls involving a complex array of devices are used to meet the aforementioned requirements. The various subsystems of a power system and the associated controls are depicted in Figure 4.2. In this overall structure, there are controllers operating directly on individual system elements. The generating unit consists of two types of controls: prime mover controls and excitation controls. The prime mover controls are concerned with speed regulation and control of energy supply system variables such as boiler pressures, temperatures, and flows. The function of the excitation control is to regulate generator voltage and reactive power output. The desired output (in megawatts) of the individual generating units is determined by the system-generation control. The transmission control includes power and voltage control devices, such as static VAR compensators, synchronous condensers, switched capacitors and reactors, tap-changing transformers, phase shifting transformers, and HVDC transmission controls. The controls described earlier contribute to the satisfactory operation of the power system by maintaining system voltage and frequency within their acceptable limits.

The primary purpose of the system-generation control is to balance the total system generation against system load and losses so that the desired frequency and power interchange with the neighboring systems (tie-line flows) are maintained (Kundur 1994).

4.3.2 Modeling of Synchronous Generator

Power system engineering is a branch where practically all the results of modern control theory can be applied. A well-designed and well-operated power system should cope with changes in the load and with system disturbances, and it should provide an acceptable high level of power quality while maintaining the voltage and frequency within tolerable limits (Shayegi 2008). Control system design, in general, for its analytical treatment, requires the determination of a mathematical model from which the control strategy can be derived. The first step in the analysis and design of a control system is mathematical modeling of the system. Models of essential components used in the power system can be expressed in state–space form or block-diagram form. Block diagrams for standard generator voltage and frequency controllers have been developed by IEEE working group (IEEE Committee Report 1968, 1973). The state–space form requires the component models to be expressed as a set of first-order differential equations. In the block-diagram form, the Laplace operator "s" replaces the differential operator d/dt of the state–space form. The excitation system time constant is much smaller than the prime mover time constant, and its transient decay is much faster and does not affect the LFC dynamics. Thus, the cross-coupling between the LFC loop and the AVR loop is negligible, and the load frequency and excitation voltage control are analyzed independently.

4.3.3 Modeling of LFC

The regulation of the frequency in the power system requires the speed control of the prime mover, using the governor. Based on the IEEE proposed standards, suitable mathematical models are developed to integrate the prime mover, governor, and the network. This section presents a brief analysis on the mathematical modeling of the generator, load, prime mover, and governor.

4.3.3.1 Generator Model

The swing equation forms the basis for modeling the LFC loop of the power system. A change in the rotor angle δ results in the change in real power, which ultimately affects the frequency. Under normal operating conditions, the relative positions of the rotor axis and the resultant magnetic field axis are fixed. The angle between these axes is known as the power angle or torque angle. During sudden load disturbance, rotor will decelerate or accelerate with respect to synchronously rotating air gap MMF, and a relative motion begins. The equation describing this relative motion is known as the swing equation. The swing equation of a synchronous machine is given by

$$\frac{2H}{\omega_s} \frac{d^2 \Delta\delta}{dt^2} = \Delta P_m - \Delta P_e \qquad (4.1)$$

where
$\Delta P_m - \Delta P_e$ is the increment in power input to the generator
H is the inertia constant
ΔP_m is the mechanical power output
$\Delta\delta$ is the measure of change in frequency
ω_s is the synchronous speed

Expressing speed deviation in per unit, Equation 4.1 can be rewritten as

$$\frac{d\Delta(\omega/\omega_s)}{dt^2} = \frac{1}{2H}(\Delta P_m - \Delta P_e) \tag{4.2}$$

With speed expressed in per unit, without explicit per unit notation, we have

$$\frac{d\Delta\omega}{dt} = \frac{1}{2H}(\Delta P_m - \Delta P_e) \tag{4.3}$$

Taking Laplace transform of Equations 4.3 and 4.4, we have

$$\Delta\Omega(s) = \frac{1}{2H(s)}[\Delta P_m(s) - \Delta P_e(s)] \tag{4.4}$$

The block-diagram representation of Equation 4.4 is shown in Figure 4.3.

4.3.3.2 Load Model

The load model on a power system consists of a variety of electrical devices. For resistive loads, such as lighting and heating loads, the electric power is independent of frequency. Motor loads are sensitive to changes in frequency. The speed–load characteristics of a composite load are approximated by

$$\Delta P_e = \Delta P_L + D\Delta\omega \tag{4.5}$$

where
ΔP_L is the nonfrequency-sensitive load change
$D\Delta\omega$ is the frequency-sensitive load change
D is expressed as the percent change in load divided by the percent change in frequency

Including the load model in the generator block diagram results in the block diagram of Figure 4.4.

Eliminating the simple feedback loop in Figure 4.4 results in the block diagram as shown in Figure 4.5.

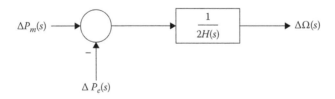

FIGURE 4.3
Generator block diagram.

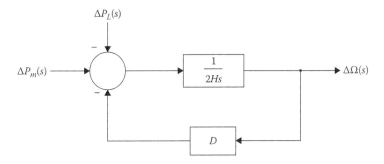

FIGURE 4.4
Generator and load block diagram.

FIGURE 4.5
Simplified generator and load block diagram.

4.3.3.3 Prime Mover Model

The source of the mechanical power, commonly known as the prime mover, may be hydraulic turbines at waterfalls or steam turbines whose energy comes from the burning of coal, gas, nuclear, and gas turbines. The models for the prime mover must take account of the steam supply and boiler control system characteristics in the case of a steam turbine. The turbine model relates to changes in mechanical power output ΔP_m to changes in a steam valve position ΔP_v. The simplest prime mover model for the non-reheat steam turbine can be approximated with a single time constant τ_T, resulting in the following transfer function:

$$G_T(s) = \frac{\Delta P_m(s)}{\Delta P_v(s)} = \frac{1}{1 + \tau_T s} \tag{4.6}$$

The block diagram for a simple turbine is shown in Figure 4.6.

The time constant τ_T is in the range of 0.2–2.0 s.

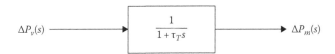

FIGURE 4.6
Block diagram for a simple non-reheat steam turbine.

4.3.3.4 Governor Model

When the generator electrical load is suddenly increased, the electric power exceeds the mechanical power input. This power deficiency is supplied by the kinetic energy stored in the rotating system. The reduction in kinetic energy causes the turbine speed and, consequently, the generator frequency to fall. The change in speed is sensed by the turbine governor, which acts to adjust the turbine input valve to change the mechanical power output to bring the speed to a new steady state. The earliest governors were the watt governors, which sense the speed by rotating fly balls and provide mechanical motion in response to speed changes. However, most modern governors use electronic means to sense speed changes. The major parts of a conventional watt governor are discussed as follows.

4.3.3.4.1 Speed Governor

The essential parts are centrifugal fly balls driven directly or through gearing by the turbine shaft. The mechanism provides upward and downward vertical movements proportional to the change in speed.

4.3.3.4.2 Linkage Mechanism

These are links for transforming the fly balls movement to the turbine valve through a hydraulic amplifier and providing a feedback from the turbine valve movement.

4.3.3.4.3 Hydraulic Amplifier

Very large mechanical forces are needed to operate the steam valve. Therefore, the governor movements are transformed into high-power forces via several stages of hydraulic amplifiers.

4.3.3.4.4 Speed Changer

The speed changer consists of a servomotor, which can be operated manually or automatically for scheduling load at a nominal frequency. By adjusting this set point, a desired load dispatches can be scheduled at a nominal frequency. For stable operation, the governors are designed to permit the speed to drop as the load is increased. The speed governor mechanism acts as a comparator whose output ΔP_g is the difference between the reference set power ΔP_{ref} and the power $(1/R)\Delta\omega$ in Equation 4.7:

$$\Delta P_g = \Delta P_{ref} - \frac{1}{R}\Delta\omega \qquad (4.7)$$

Taking Laplace transform of Equation 4.7, we get

$$\Delta P_g(s) = \Delta P_{ref}(s) - \frac{1}{R}\Delta\Omega(s) \qquad (4.8)$$

The command ΔP_g is transformed through the hydraulic amplifier to the steam valve position command ΔP_{ref} assuming a linear relationship and considering a simple time constant τ_g, as follows:

$$\Delta P_v(s) = \frac{1}{1+\tau_g s}\Delta P_g(s) \qquad (4.9)$$

Equations 4.8 and 4.9 are represented by the block diagram shown in Figure 4.7.

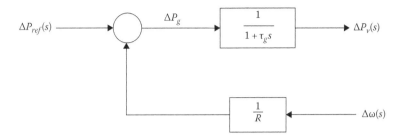

FIGURE 4.7
Block diagram for speed-governing system of steam turbine.

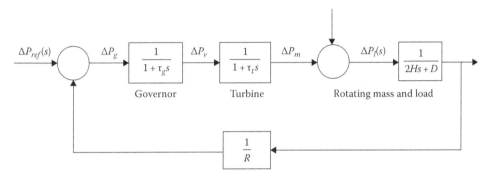

FIGURE 4.8
LFC block diagram of an isolated power system.

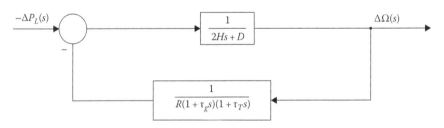

FIGURE 4.9
LFC block diagram with input $\Delta P_L(s)$ and output $\Delta P_L(s)$.

Combining the block diagrams of Figures 4.5 through 4.7 results in the complete block diagram of the LFC of an isolated power station, as shown in Figure 4.8.

Redrawing the block diagram of Figure 4.8 with the load change $\Delta P_L(s)$ as the input and the frequency deviation $\Delta P_L(s)$ as the output results in the block diagram shown in Figure 4.9.

The open-loop transfer function of the block diagram in Figure 4.9 is

$$KG(s)H(s) = \frac{1}{R} \frac{1}{(2Hs+D)(1+\tau_g s)(1+\tau_T s)} \tag{4.10}$$

and the closed-loop transfer function relating the load change $\Delta P_L(s)$ to the frequency deviation is

$$\frac{\Delta\Omega(s)}{-\Delta P_L(s)} = \frac{(1+\tau_g s)(1+\tau_T s)}{(2Hs+D)(1+\tau_g s)(1+\tau_T s)+(1/R)} \tag{4.11}$$

or

$$\Delta\Omega(s) = -\Delta P_L(s)T(s) \tag{4.12}$$

The load change is a step input, $\Delta P_L(s) = \Delta P_L/s$. Utilizing the final value theorem, the steady-state value of $\Delta\omega$ is

$$\Delta\omega_{ss} = \lim_{s\to 0} s\Delta\Omega(s) = (-\Delta P_L)\frac{1}{D+(1/R)} \tag{4.13}$$

It is clear that for the case of no frequency-sensitive load (i.e., with $D = 0$), the steady-state deviation in frequency is determined by the governor speed regulation, as given by Equation 4.14:

$$\Delta\omega_{ss} = (-\Delta P_L)R \tag{4.14}$$

When several generators with governor speed regulations $R1, R2, \ldots, Rn$ are connected to the system, the steady-state deviation in frequency is given by

$$\Delta\omega_{ss} = (-\Delta P_L)\frac{1}{D+(1/R_1)+(1/R_2)+(1/R_3)+\cdots+(1/R_n)} \tag{4.15}$$

The steady-state speed deviation depends on the governor speed regulation. When generators operate in parallel, the composite frequency power characteristics depend on the combined effect of the droops of all the generator speed governors.

4.3.4 Modeling of AVR

The analysis of transmission of active and reactive power gives useful insight into the characteristics of the transmission system. As the constancy of frequency is an important factor in determining the quality of supply, the control of active power is vital to the satisfactory performance of the power system. The LFC and AVR are fairly independent of each other and influenced by different control strategies; hence the analysis of the characteristics and responses can be done separately (Kundur 1994). In a normal state, the system variables are within the normal range, and no equipment is being overloaded. The system operates in a secured manner and can withstand a contingency without violating any of the constraints. During restoration, when individual generators are being brought up to speed and large blocks of load are being reconnected, perturbations outside the range of automatic controls are inevitable; hence hands-on controls by system operators are necessary. The knowledge of optimization techniques and optimal control methods is essential to understand the multilevel approach that has been successfully utilized. System frequency is the mean frequency of all the machines that are online, and deviations by individual machines must be strictly minimized to avoid mechanical damage to the generator and disruption of the entire system. This is generally accomplished by picking up loads in increments that can be accommodated by the inertia and response of the restored and synchronized generators (Adibi and Fink 2006).

The operating objectives of the LFC are to maintain reasonably uniform frequency, to divide the load between generators, and to control the tie-line interchange schedules (Mohammad et al. 2002). The change in frequency and tie-line power is sensed, which is a measure of the change in the rotor angle δ, that is, the error $\Delta\delta$ is to be corrected. The error signals (i.e., Δf and ΔP_{tie}) are amplified, mixed, and transformed into a real power command signal ΔP_v, which is sent to the prime mover to call for an increment in the torque. The prime mover, therefore, brings change in the generator output by an amount ΔP_g, which will change the values of Δf and ΔP_{tie} within the specified tolerance. The turbine governor models are designed to give representations of the effects of the power plants in the power system stability (PSS/E-26 1998). A functional diagram of the speed-governing system and its relationship to the generator is given in Figure 4.10.

The generator excitation of conventional machines may be provided through slip rings and brushes by means of DC generators mounted on the same shaft as the rotor of the synchronous machine. However, modern excitation systems usually use AC generators with rotating rectifiers and are known as brushless excitation. The sources of reactive power are generators, capacitors, and reactors, and they are controlled by field excitation. The voltage profile on electric transmission systems can be improved by transformer load tap changers, switched capacitors, step-voltage regulators, and static vary control equipment.

The primary means of a generator reactive power control is the generator excitation control using an AVR. A synchronous generator or alternator is equipped with an AVR, which is responsible for keeping the output voltage constant under normal operating conditions at various load levels (Abul and Sadrul 1994). The schematic diagram of a simplified AVR is shown in Figure 4.11. An increase in the reactive power load of the generator is accompanied by a drop in the terminal voltage magnitude. The voltage magnitude is sensed

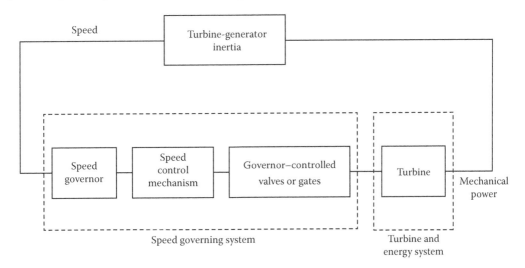

FIGURE 4.10
Speed governor and turbine in relation to the generator.

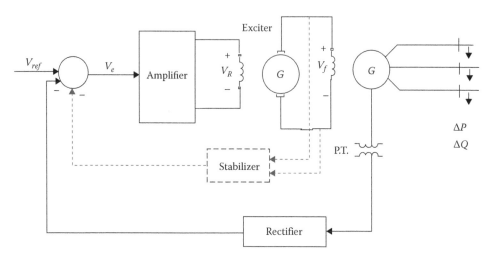

FIGURE 4.11
A typical arrangement of a simple AVR.

through a potential transformer on one phase. This voltage is rectified and compared to a DC set point signal. The amplified error signal controls the exciter field and increases the exciter terminal voltage. The controller will make an intelligent decision on the amount of field currents that should be applied to the generator in order to keep the output voltage at its rated value (LaMeres 1999). The reactive power generation is increased to a new equilibrium, raising the terminal voltage to the desired value.

The voltage regulator controls the exciter output, such that the terminal voltage of the AC generator equals the desired voltage, often called the reference voltage. The IEEE has suggested standard computer models for excitation system studies. The excitation control is one of the important factors in the transient study of power system analysis. It is the normal requirement to have a high gain for excitation controller as an effective means of providing transient stability (ANSI/IEEE Std 421.1–1986). When a disturbance occurs, the excitation controller can moderate the control signal quickly and provide good damping of oscillations in the system. A typical relationship between the excitation control system and the generator is illustrated in Figure 4.12. The voltage regulator controls the exciter output, such that the terminal voltage of the AC generator equals

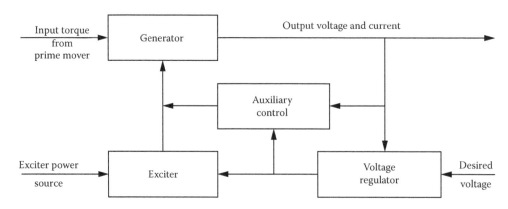

FIGURE 4.12
Typical arrangements of excitation components.

the desired voltage or reference voltage. Auxiliary controls are desirable for feedback of speed, frequency, and acceleration signals to improve system conditions such as stability, damping to overshoot, etc. (Uma Rao 2007). The mathematical models of amplifier, exciter, generator, and sensor are derived in this section. All the loops are designed to operate around a normal state with small variable excursions. Thus, the loops may be modeled with linear, constant coefficient differential equations and represented with linear transfer functions.

4.3.4.1 Amplifier Model

The excitation system amplifier may be a magnetic amplifier, rotating amplifier, or modern electronic amplifier. The amplifier is represented by a gain K_A and a time constant τ_A as in transfer function given in the following:

$$\frac{V_R(s)}{V_e(s)} = \frac{K_A}{1 + \tau_A s} \tag{4.16}$$

Typical values of K_A are in the range of 1–400. The amplifier time constant is very small in the range of 0.02–0.1 s and is often neglected.

4.3.4.2 Exciter Model

The exciter of the alternator is the main component in the AVR loop. The output voltage of the exciter is a nonlinear function of the field voltage because of the saturation effects in the magnetic circuit. Thus, there is no simple relationship between the terminal voltage and the field voltage of the exciter. A reasonable model of a modern exciter is a linearized model, which takes into account the major time constant and ignores the saturation or other nonlinearities. In the simplest form, the transfer function of a modern exciter is represented in Equation 4.17 by a single time constant τ_E and a gain K_E, that is, the time constant of modern exciters is very small.

$$\frac{V_F(s)}{V_R(s)} = \frac{K_E}{1 + \tau_E s} \tag{4.17}$$

4.3.4.3 Generator Model

The synchronous machine generated emf is a function of the machine magnetization curve, and its terminal voltage is dependent on the generator load. In the linearized model, the transfer function relating the generator terminal voltage to its field voltage can be represented by a gain K_G and a time constant τ_G is represented as

$$\frac{V_t(s)}{V_F(s)} = \frac{K_G}{1 + \tau_G s} \tag{4.18}$$

where K_G may vary between 0.7 and 1, and τ_G between 1.0 and 2.0 s from a full load to no load.

4.3.4.4 Sensor Model

The voltage is sensed through a potential transformer, and it is rectified through a bridge rectifier. The sensor is modeled by a simple first-order transfer function given by

$$\frac{V_S(s)}{V_t(s)} = \frac{K_R}{1 + \tau_R s} \tag{4.19}$$

where τ_R varies from 0.01–0.7 s. Combining the mathematical models in Equations 4.16 through 4.19 results in the AVR block diagram shown in Figure 4.13.

The open-loop transfer function of the block diagram in Figure 4.13 is

$$KG(s)H(s) = \frac{K_A K_E K_G K_R}{(1 + \tau_A s)(1 + \tau_E s)(1 + \tau_G s)(1 + \tau_R s)} \tag{4.20}$$

and the closed-loop transfer function relating the generator terminal voltage $V_t(s)$ to the reference voltage is

$$\frac{V_t(s)}{V_{ref}(s)} = \frac{K_A K_E K_G K_R (1 + \tau_R s)}{(1 + \tau_A s)(1 + \tau_E s)(1 + \tau_G s)(1 + \tau_R s) + K_A K_E K_G K_R} \tag{4.21}$$

or

$$V_t(s) = V_{ref}(s)T(s) \tag{4.22}$$

For a step input, $V_{ref}(s) = 1/s$, the steady-state output of an AVR is

$$V_{tss} = \lim_{s \to 0} V_t(s) = \frac{K_A}{1 + K_A} \tag{4.23}$$

4.3.4.5 Excitation System Stabilizer

Excitation systems, especially using DC and AC exciters, comprise elements with significant time delays. The relative stability of the excitation system can be increased by introducing a controller, which would add a zero to the AVR open-loop transfer function as shown in Figure 4.14.

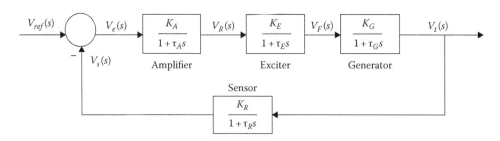

FIGURE 4.13
A simplified AVR block diagram.

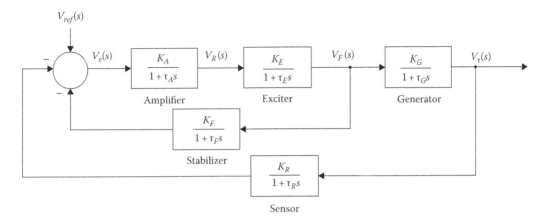

FIGURE 4.14
Block diagram of a compensated AVR system.

Feedback parameters K_F and τ_F are tuned properly to obtain a satisfactory response and to improve dynamic performance of the system. For small value of amplifier gains, the step response of AVR is not satisfactory, and for higher values, the response is unbounded. Hence, the relative stability is increased by introducing a controller in the feedback path of the system.

4.3.5 Interconnection of Power Systems

Interconnections between power utilities or areas form an important part of the overall power system, and they contribute greatly to increase the reliability of the supply and economy of power production (Chowdhury et al. 1999). The purpose of interconnections in the power system is to form larger and more robust systems and to exploit the diversity of power generation forms. Interconnection of the power system provides the ability to balance the load between different areas with differing consumption patterns. If the load on the system is increased, the turbine speed drops before the governor can adjust the input of the steam to the new load. As the change in the value of speed diminishes, the error signal becomes smaller and the position of the governor fly balls gets closer to the point required to maintain a constant speed. However, a small offset from the set speed of the governor is unavoidable and needs proper controller action to reach the desired value. One way to restore the speed or frequency to its nominal value is to add an integrator. The integral unit monitors the average error over a period of time and will overcome the offset. Because of its ability to return a system to its set point, integral action is also known as rest action. Thus, as the system load changes continuously, the generation is adjusted automatically to restore the frequency to the nominal value, and this scheme is known as the automatic generation control (AGC).

In an interconnected system consisting of several pools, the role of AGC is to divide the loads among the system, stations, and generators so as to achieve maximum economy and also to control the scheduled interchanges of tie-line power while maintaining a reasonably uniform frequency. During large transient disturbances and emergencies, AGC is bypassed and emergency control relays are triggered to restore the system. The large-scale power systems are normally divided into control areas based on the principle of coherency.

The coherent areas are interconnected through the tie-lines, which are used for contractual energy exchange between areas and provide inter-area support during abnormal operations (Mathur and Manjunath 2007).

Each area supplies its user pool, and tie-lines allow electric power to flow between areas. Therefore, each area affects others, that is, a load perturbation in one of the areas affects the output frequencies of other areas as well as power flows on tie-lines. Due to this, the control system of each area needs information about the transient situation in all areas to bring the local frequency to its steady-state value. While the information about each area is found in its frequency, the information about the other areas is in the perturbations of tie-lines power flows (Demiroren and Yesil 2004). A vast power system can be decomposed into a number of LFC areas and tied together through tie-line. The problem of real power control can be studied by first studying the problem for a single area and extending for a multi-area system (Gupta 2003).

Frequency control in an interconnected system is accomplished by two different methods: primary control to make initial adjustments of the frequency among the various generators in the control area and the secondary control loop reduces the frequency error to zero through the controller action (Mathur and Manjunath 2007). With the primary LFC loop, a change in the system load will result in steady-state frequency deviation, depending on the governor speed regulation. A reset action must be provided to reduce the frequency deviation to zero. The rest action can be achieved by introducing an integral controller to act on the load reference setting to change the speed set point. The integral controller increases the system type by 1, which forces the final frequency deviation to zero. The LFC system, with the addition of the secondary loop, is shown in Figure 4.15. The integral controller gain K_I must be adjusted for a satisfactory transient response. Combining the parallel branches of Figure 4.15 results in the equivalent block diagram shown in Figure 4.16. The closed-loop transfer function of the control system shown in Figure 4.16 with only $-\Delta P_L$ as input becomes

$$\frac{\Delta\Omega(s)}{-\Delta P_L(s)} = \frac{s(1+\tau_g s)(1+\tau_T s)}{s(2Hs+D)(1+\tau_g s)(1+\tau_T s) + K_I + s/R} \qquad (4.24)$$

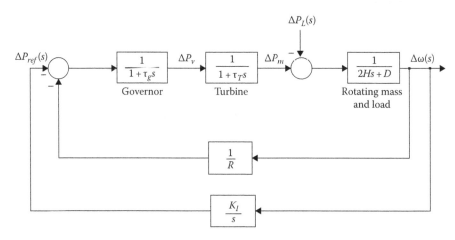

FIGURE 4.15
AGC for an isolated power system.

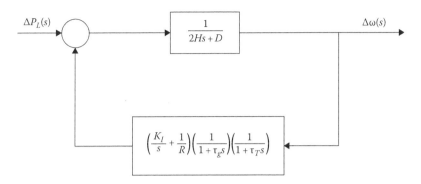

FIGURE 4.16
Equivalent block diagram of AGC for an isolated power system.

where R is the regulation and may be expressed in per unit as well as in Hz/MW. When both frequency and power output are expressed in p.u., regulation is simply the magnitude of the slope of the speed versus power output characteristics of the alternator.

4.3.5.1 AGC in Multiarea System

Planning, operation, and control of an isolated or interconnected power system exhibit challenging problems that require mathematical techniques from various branches for solution. In recent years, power generating units have been located at large distances from urban areas due to technical, economic, and environmental constraints, and the power generating capacity has been greatly increasing to meet the excess load demand (Kumar and Ibrahim 1998). For the generator turbines having the same response characteristics, the LFC loop can be referred to as control area to represent the whole system. Yang et al. (2002) proposed a design of decentralized robust LFC for interconnected multi-area system and obtained satisfactory performance of the system. Consider two areas represented by an equivalent generating unit interconnected by a lossless tie-line with reactance X_{tie}. Each area is represented by a voltage source behind an equivalent reactance as shown in Figure 4.17.

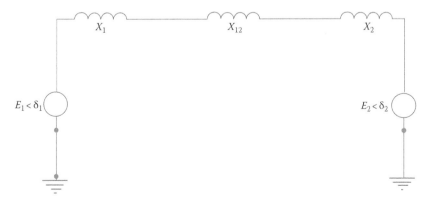

FIGURE 4.17
Equivalent network for a two-area power system.

During normal operation, the real power transferred over the tie-line is given by

$$P_{12} = \frac{|E_1||E_2|}{X_{12}} \sin \delta_{12} \tag{4.25}$$

where $X_{12} = X_1 + X_{tie} + X_2$, and $\delta_{12} = \delta_1 - \delta_2$. Equation 4.25 can be linearized for a small deviation in the tie-line flow ΔP_{12} from the nominal value, that is,

$$\Delta P_{12} = \frac{dP_{12}}{d\delta_{12}}\bigg|_{\delta_{120}} \Delta\delta_{12} = P_s \Delta\delta_{12} \tag{4.26}$$

The quantity P_s is the slope of the power angle curve at the initial operating angle $\delta_{120} = \delta_{10} - \delta_{20}$. This is known as synchronizing power coefficient. The synchronizing power coefficient P_s is given by Equation 4.27.

$$P_s = \frac{dP_{12}}{d\delta_{12}}\bigg|_{\delta_{120}} = \frac{|E_1||E_2|}{X_{12}} \cos \Delta\delta_{120} \tag{4.27}$$

The tie-line power deviation then takes the form

$$\Delta P_{12} = P_s(\Delta\delta_1 - \Delta\delta_2) \tag{4.28}$$

The tie-line power flow appears as a load increase in one area and a load decrease in the other area, depending on the direction of the flow. The direction of flow is indicated by the phase angle difference; if $\Delta\delta_1 > \Delta\delta_2$, m area 1 to area 2. A block diagram representation for the two-area system with LFC containing only the primary loop is shown in Figure 4.18.

Consider a load change ΔP_{L1} in area 1. In the steady state, both areas will have the same steady-state frequency deviation, that is,

$$\Delta\omega = \Delta\omega_1 = \Delta\omega_2 \tag{4.29}$$

$$\Delta P_{m1} - \Delta P_{12} - \Delta P_{l1} = \Delta\omega D_1$$
$$\Delta P_{m2} + \Delta P_{12} = \Delta\omega D_2 \tag{4.30}$$

The change in mechanical power is determined by the governor speed characteristics given by

$$\Delta P_{m1} = \frac{-\Delta\omega}{R_1}$$
$$\Delta P_{m2} = \frac{-\Delta\omega}{R_2} \tag{4.31}$$

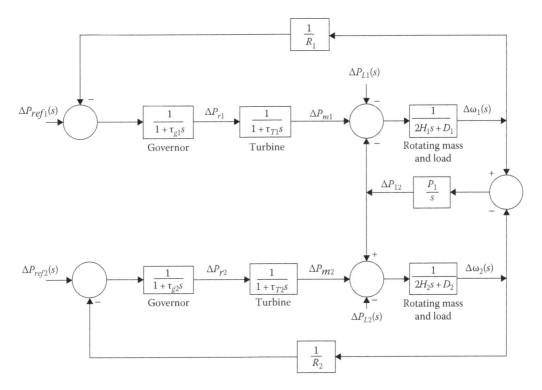

FIGURE 4.18
Two-area interconnected system with only primary LFC loop.

Substituting from Equation 4.31 into 4.30 and solving for $\Delta\omega$ result in Equation 4.32:

$$\Delta\omega = \frac{-\Delta P_{L1}}{\left((1/R_1)+D_1\right)+\left((1/R_2)+D_2\right)} = \frac{-\Delta P_{L1}}{B_1+B_2} \tag{4.32}$$

where

$$B_1 = ((1/R_1) + D_1); \; B_2 = ((1/R_2) + D_2) \tag{4.33}$$

B_1 and B_2 are known as the frequency bias factors. The change in the tie-line power is

$$\Delta P_{12} = \frac{-\left((1/R_2)+D_2\right)\Delta P_{L1}}{\left((1/R_1)+D_1\right)+\left((1/R_2)+D_2\right)} = \frac{B_2}{B_1+B_2}(-\Delta P_{L1}) \tag{4.34}$$

4.3.5.2 Tie-Line Bias Control

In LFCs equipped with only the primary control loop, a change of power in area 1 is met by the increase in generation in both areas associated with a change in the tie-line power and a reduction in frequency. In the normal operating state, the power system is operated so that the demands of areas are satisfied at the nominal frequency. A simple control strategy for the normal mode is to keep the frequency approximately at the nominal value (50 Hz), maintain the tie-line flow, and each area should absorb its own load changes.

Conventional LFC is based on tie-line bias control, where each area tends to reduce the area control error (ACE) to zero. The ACE in Equation 4.35 is a linear combination of frequency and tie-line error.

$$ACE_i = \sum_{j=1}^{n} \Delta P_{ij} + K_i \Delta \omega \tag{4.35}$$

The area bias K_i determines the amount of interactions during a disturbance in the neighboring areas. An overall satisfactory performance is achieved when K_i is selected equal to the frequency bias factor of that area:

$$B_i = \left(\frac{1}{R_i} + D_i \right) \tag{4.36}$$

Thus, the ACEs for a two-area system are

$$ACE_1 = \Delta P_{12} + B_1 \Delta \omega_1$$

$$ACE_2 = \Delta P_{21} + B_2 \Delta \omega_2 \tag{4.37}$$

where ΔP_{12} and ΔP_{21} are departures from scheduled interchanges. ACEs are used as actuating signals to activate changes in the reference power set points, and when the steady state is reached, ΔP_{12} and $\Delta \omega$ will be zero. The integrator gain constant must be chosen small enough so as not to cause the area to go into a chase mode. The block diagram of a simple AGC for a two-area system is shown in Figure 4.19.

The model of the two-area LFC system is developed in Simulink by considering the similar parameters used for single area, as shown in Table 4.1, for governor, turbine, and load. In this chapter, LFC is implemented for a two-area interconnected power system, since the governor will take care of the power output of both generator and tie-line power. The problem of maintaining voltages within the required limits is complicated by the fact that power is supplied to a vast number of loads and is fed from many generating units. As load varies, the reactive power requirements of the transmission system vary. Since reactive power cannot be transmitted over long distance, voltage control has to be dispersed throughout the system. This contrasts with the control of frequency, which depends on the overall system active power balance. Hence, the AVR loop is confined for a particular machine only.

Synchronous generators are normally equipped with AVRs, which continually adjust the excitation to control the armature voltage (Fang and Chen 2009). In short, the chapter focuses only on regulating the output of the AVR for a single-area isolated power generating system.

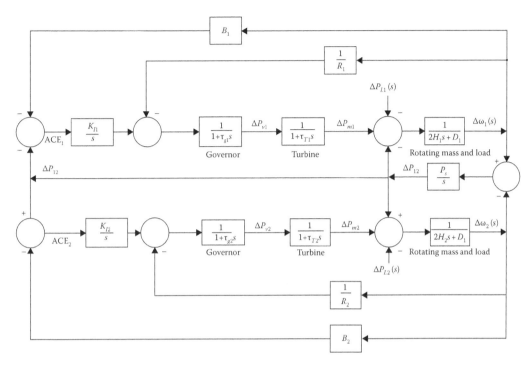

FIGURE 4.19
AGC for a two-area system.

TABLE 4.1

Parameters used in LFC system with PID controller

Symbol	Parameters
ΔP_{ref}	Change in reference power
K_P	Proportional gain
K_I	Integral gain
K_D	Differential gain
τ_g	Governor time constant
τ_t	Turbine time constant
R	Regulation parameter
ΔP_g	Change in power generated
ΔP_v	Change in steam valve position
ΔP_m	Change in mechanical power output
ΔP_l	Change in nonfrequency-sensitive load
H and D	Inertia constants of the load

4.4 Fuzzy Logic Controller for LFC and AVR

Electric power is an essential ingredient for the industrial and economic development of a country. The control of a power system involves many elements and is one of the major responsibilities of system operators. The parameters to be controlled are system frequency, tie-line flows, line currents, equipment loading, and voltage. All these quantities must be kept within limits in order to provide satisfactory service to power system customers (Devaraj 2008). The automatic control system detects these changes and initiates in real time a set of control actions, which will eliminate as effectively and quickly as possible the state deviations. Synchronous generators are equipped with two major control loops: AVR loop and LFC loop. The AVR loop keeps track of output voltage of the generator and initiates the control action under varying loads. The function of LFC is to keep the system frequency and the inter-area tie-line power near to the scheduled values as possible through control action. This section demonstrates the principle of operation of speed-governing and excitation systems of a generator.

4.4.1 Basic Generator Control Loops

Both utility and consumer equipment are designed to operate at a certain frequency and voltage rating. Prolonged use of the equipment outside the range of frequency and voltage could adversely affect their performance. The proper selection and coordination of equipment for controlling real and reactive power are an important challenge in power system engineering. The primary controllers installed in the synchronous generator take care of small changes in load demand and maintain the frequency and voltage magnitude within the specified limits. Changes in rotor angle δ are caused by a momentary change in generator speed, which affects real power and thus the frequency f. Voltage magnitude (i.e., on the generator excitation) is mainly dependent on the reactive power. Therefore, load frequency and excitation voltage controls are noninteractive for small changes and can be modeled and analyzed independently. Furthermore, excitation control is fast acting in which the major time constant encountered is that of the generator field and its transient decay much faster, while the power–frequency control is slow acting in which the major time constant contributed by the turbine and generator moment of inertia-time constant is much larger than that of the generator field (Hadi 2002). Since the AVR loop is much faster than the LFC loop, cross-coupling between the controls can be neglected. Hence, the LFC and AVR are implemented and analyzed separately (Kundur 2006, Uma Rao 2007).

4.4.1.1 LFC Loop

LFC takes important part in the reliable operation of electric power systems. It aims to maintain real power balance in the system through the control of system frequency. The control of generation and frequency is commonly referred to as LFC; whenever the real power demand changes, a frequency change occurs. This frequency error is amplified, mixed, and changed to a command signal, which is sent to the turbine governor. The governor operates to restore the balance between the input and output by changing the turbine output. This method is also referred to as megawatt frequency or power–frequency (*P-f*) control. The regulation of the frequency in the power system requires the speed control of the prime mover using the governor. This would require sensing of the speed and translating it to suitable control action. The prime mover control must have drooping characteristics to ensure proper division of load, when generators are operating in parallel. It is also necessary for the prime mover control to adjust the generation according to economic dispatch schedule. The functional block diagram in Figure 4.20 indicates the speed-governing system, which includes AGC, speed governor, speed controller, turbine, and load. Speed sensing and conditioning are done by electronic circuits in electrohydraulic systems and by mechanical components in mechanical hydraulic systems.

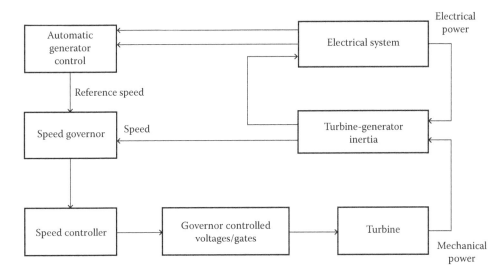

FIGURE 4.20
Speed-governing system for steam turbines.

4.4.1.2 AVR Loop

The aim of this control is to maintain the system voltage between limits by adjusting the excitation of the machines. The AVR senses the difference between a rectified voltage derived from the stator voltage and a reference voltage. This error signal is amplified and fed to the excitation circuit. The change of excitation maintains the VAR balance in the network. This method is also referred to as megawatt volt amp reactive (MVAR) control or reactive-voltage (QV) control (Wood 2006). The field winding of the AC generators needs a DC excitation. The excitation system consists of the device supplying the DC, called the exciter, and a voltage regulator that controls the output of the exciter so that the required voltage is generated. The voltage regulator is normally a continuously acting system, which takes corrective action for any deviation in the AC terminal voltage.

The general arrangement of excitation system components is shown in Figure 4.21. The voltage regulator controls the exciter output, such that the terminal voltage of the AC generator equals the desired voltage, often called the reference voltage. According to the error signal and the value of load current, the field current is increased or decreased in order to make the actual voltage closer to the desired voltage (Awadallah and Morcos 2001). The auxiliary controls for feedback of speed, frequency, and accelerations are required to improve the stability, damping, and overshoot of the excitation system (Uma Rao 2007).

4.4.2 Design of Intelligent Controller Using MATLAB®/Simulink®

With increasing demands for high-precision autonomous and intelligent controllers with wide operating regions, conventional PID control approaches are unable to adequately deal with system complexity, nonlinearities, spatial and temporal parameter variations, and with uncertainty. Intelligent control or self-organizing/learning control is a new emerging discipline that is designed to deal with real-time problems. Rather than being a model-based intelligent controller, it is experiential based and is an amalgam of the disciplines of artificial intelligence, systems theory, and operations research. For practical implementation,

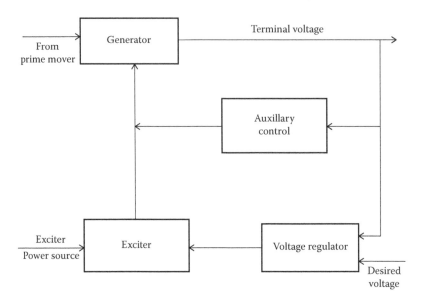

FIGURE 4.21
Excitation system.

intelligent controllers must demonstrate rapid learning convergence, be temporally stable, be robust to parameter changes and internal and external disturbances. Conventional PID controllers generally do not work well for nonlinear systems, higher-order and time-delayed linear systems, as well as complex and vague systems that have no precise mathematical models. In a conventional controller, what is modeled is the system or process being controlled, whereas in an FLC, the focus is on the human operator's behavior. Conventionally, the system is modeled analytically by a set of differential equations from which the control parameters are adjusted to satisfy the controller specification. Hence, to overcome this drawback, alternative methods using improved dynamic models or adaptive and intelligent controllers are required (Barjeev 2003, Mathur 2006, Moon et al. 2002). A fuzzy-logic-based design can resolve the weaknesses of conventional approaches cited earlier. In the FLC, the adjustments of the control parameters are handled by a fuzzy-rule-based expert system.

The use of FL control is motivated by the need to deal with high complex and performance robustness problems. It is well known that FL is much closer to human decision making than traditional logical systems. Fuzzy control provides a new design paradigm such that a controller can be designed for complex, ill-defined processes without the knowledge of quantitative data regarding the input–output relations. In this chapter, the conventional controllers in the secondary control loop are replaced with FLC, and their performance is analyzed for various operating conditions by detailed digital computer simulations. The efficacy of the proposed method was validated and compared against the conventional controller for different load and regulation parameters (Khodabakshian and Goldbon 2005, Mathur and Manjunath 2007, Yukita 2000).

4.4.3 Fuzzy Controller for Single-Area Power System

A fuzzy control system is a real-time expert system that enhances conventional system design with engineering expertise. It implements a part of the human operator's or process engineer's expertise, which does not lend itself to be easily expressed in PID

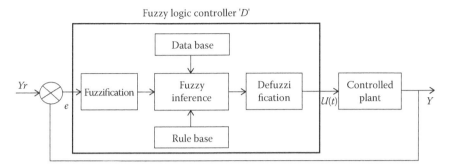

FIGURE 4.22
Basic structure of the fuzzy control systems.

parameters or differential equations but rather in situations/action rules (Driankov et al. 2001). Furthermore, it provides a mathematical morphology to emulate certain perceptual and linguistic attributes associated with human cognition. The basic structure of the fuzzy control systems is shown in Figure 4.22, where controlled plant represents the LFC and AVR model developed using transfer function approach. The purpose of designed fuzzy controller "D" is to guarantee the desired response of the plant output "Y."

The main objective of the controller design is to develop an intelligent FLC "D" that takes care of the response characteristics of the plant output "Y." The model shown in Figure 4.22 is implemented using Simulink, an interactive environment in MATLAB for modeling, analyzing, and simulating a wide variety of dynamic systems. The fuzzy file (.FIS) created in the fuzzy toolbox is linked to the model through the FLC block included in the control loop.

The FLC with a rule viewer is selected from the library of the fuzzy toolbox in MATLAB and placed in the feedback loop, as shown in Figure 4.23. It is regarded as a nonlinear static function that maps controller inputs into controller outputs. The inputs to the system can change the state of the system, which causes variations in the response characteristics. The task of the controller is then to take corrective action by providing a set of inputs that ensures the desired response. The steps in designing a fuzzy control system are as follows:

1. Identify the input and output variables of the plant.
2. Divide the universe of discourse into a number of fuzzy subsets and assign a linguistic label to each subset.
3. Assign membership function to each fuzzy subset.
4. Form a rule base to relate the input and output variables.
5. Choose appropriate scaling factors for input and output variables in order to normalize the variables to (0, 1) or (−1, 1) interval.
6. Fuzzify the inputs to the controller.
7. Verify the output contributed from each rule.
8. Aggregate the fuzzy outputs recommended by each rule.
9. De-fuzzify the output for converting into crisp values.

The design steps are followed in sequence by creating fuzzy inference system (FIS) editor, as shown in Figure 4.24. The inputs and outputs are selected from the edit window in a pop-up menu. The diagram indicates the names of each input variable on the left and output variable in the right.

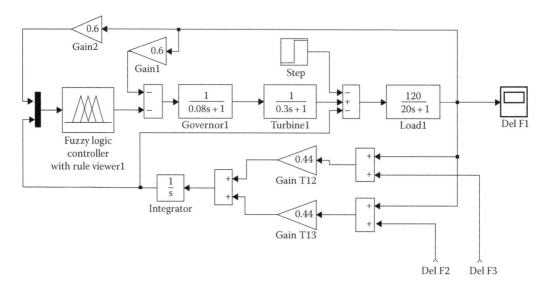

FIGURE 4.23
Area 1 of a three-area power system using FLC.

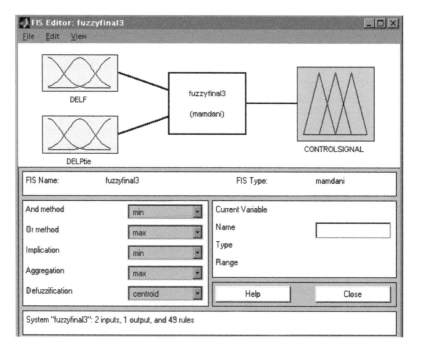

FIGURE 4.24
FIS editor.

The FIS editor displays the detailed information about input/output, input/output, file-name, FIS type, name, fuzzification and de-fuzzification methods. The input and output of the system are labeled, and the membership function is edited by double clicking the input variable icon. The rules are edited by selecting "Edit rules" in the "Edit" command in the drop-down menu. Below the FIS name in the left is the pop-up menu that allows to modify the inference process variables. To its right, the name of the current variable, its associated membership function type, and its range are displayed. The basic structure of FLC consists of three main sections: fuzzification, knowledge base, and defuzzification.

4.4.3.1 Fuzzification

The fuzzy-based control system is designed to control the voltage and frequency of the synchronous generator. The controller uses two input state variables and one output control variable. The difference between the set value and the actual value (error) is termed the first input variable. The change in error, that is, the difference between the errors in consecutive steps of simulation, is assigned as another input variable. That is,

Input 1: Error $\Delta f = f_{nom} - f_1 = e_t$
Input 2: Change in Error $\Delta e_t = e_t - e_{t-1} = ce_t$

The fuzzy output control variable is the change in the control signal. This control signal acts as the input signal to the speed governor and excitation system of LFC and AVR of the generator. The input and output variables in the proposed controller are represented as a set of nine linguistic variables:

NVL	Negative very large
NH	Negative high
NM	Negative medium
NL	Negative low
ZE	Zero error
PL	Positive low
PM	Positive medium
PL	Positive large
PVL	Positive very large

In fuzzification, the precise numerical values obtained by measurements are converted to membership values of the various linguistic variables. The degree to which a fuzzy number belongs to a set or not is known as membership function (MF). In LFC, the universe of discourse of the input state variable MF is −0.8 to +0.8, as shown in Figure 4.25.

The degree of membership plays an important role in designing a fuzzy controller. In LFC design, the linguistic variables are represented by triangular MF except for NVL and PVL by trapezoidal membership. The shape of the fuzzy set is not a concern. However, in practice symmetric triangles (or) trapezoids centered about representative values are used to represent fuzzy sets.

In AVR, the universe of discourse of the input state variable MF is −1.5 to +1.5, as shown in Figure 4.26. In this design, triangular MF is selected because of its simplicity and more efficient for this application. The MF is symmetrical in shape and overlaps the adjacent MF by 50%. Overlapping in MF is important because it allows for a good interpolation of input values, that is, the entire input space is accommodated. The fuzzification module converts these crisp values of the control inputs into fuzzy values so that they are compatible with the fuzzy set representation in the rule base.

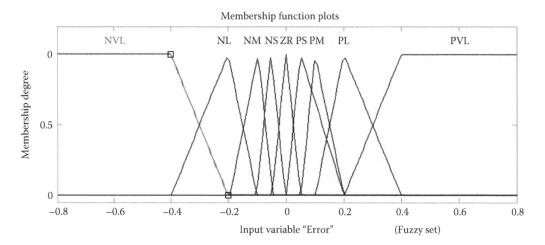

FIGURE 4.25
Membership function for error in LFC.

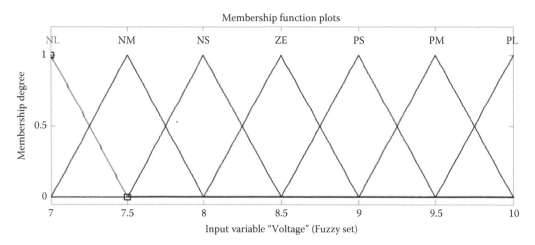

FIGURE 4.26
Membership function for voltage in AVR.

4.4.3.2 Knowledge Base

The knowledge base consists of a database of the LFC and AVR models. It provides all the necessary definitions for the fuzzification process such as membership functions, fuzzy set representation of the input–output variables, and the mapping functions between the physical and fuzzy domain. The rule base should cover all the possible combinations of input value, but such coverage is typically neither practical nor necessary. Rule conditions are joined by using minimum intersection operator so that the resulting MF for a rule is given by $\mu(e, ce) = \min(\mu_{Ai}(e), \mu_{Bi}(ce))$. The rules of an FLC give the controller its intelligence, since the rules are developed based on the expert knowledge obtained from the experienced operator. In FLC design, the desired effect is to keep the output voltage and frequency of the generator at its rated value under varying loads. From this desired objective, the rules are derived for every combination of input state variables in order to obtain the desired output variable.

As the number of rule increases, the computational efficiency and robustness of the system will also be improved. Fuzzy rule bases are developed using a conjunctive relationship of the antecedents in the rules. This is termed an intersection rule configuration (IRC) by Combs and Andrews (1998) because the inference process maps the intersection of antecedent fuzzy sets to output consequent fuzzy sets. IRC is a general exhaustive search of solutions that utilize every possible combination of rules in determining an outcome.

The combinatorial explosion in rules is given by $R = l^n$, where R is the number of rules, l is the number of linguistic labels for each input variable, and n is the number of input variables. In the present work, the value of "l" is 9 and "n" is 2, hence the number of rules R for LFC is 81, as mentioned in Table 4.2. Similarly for AVR, the value of "l" is 7 and "n" is 2, hence R is 49.

In the IRC approach, the intersection of the input values that are related to the output is achieved with an "AND" operation and are of IF-THEN type. The basic function of the rule base is to represent in a structured way the control policy of an experienced human operator in the form of a set of production rules such as

IF (process state) THEN (control output)

For example

*If the change in error is **NM** and error is **NS** then the output is **PM***

*If error voltage is **NS** and change in error voltage is **PL**, then the field current is **NM**.*

This IF-THEN rule is a fuzzy description of the control logic representing the human expert's qualitative knowledge. Since the controller selected from the fuzzy toolbox is with a rule viewer, the effect of control signal for change in rules can be viewed through the graphical user interface (GUI), as shown in Figure 4.27. The first two columns of plots (the six yellow plots) show the membership functions referenced by the antecedent, or the "if" part of each rule. The third column of plots (the three blue plots) shows the membership functions referenced by the consequent, or the "then" part of each rule. Seven linguistic variables are used to represent each input variable; hence 49 rules are developed in the rule base. The rule viewer is a MATLAB technical computing environment based display

TABLE 4.2

Rule Base for LFC

		Error								
		NVL	NL	NM	NS	ZR	PS	PM	PL	PVL
	NVL	PVL	PVL	PL	PL	PM	PM	PS	PS	ZR
	NL	PVL	PL	PL	PM	PM	PS	PS	ZR	NS
	NM	PL	PL	PM	PM	PS	PS	ZR	NS	NS
Change in error	**NS**	PL	PM	PM	PS	PS	ZR	NS	NS	NM
	ZR	PM	PM	PS	PS	ZR	NS	NS	NM	NM
	PS	PM	PS	PS	ZR	NS	NS	NM	NM	NL
	PM	PS	PS	ZR	NS	NS	NM	NM	NL	NL
	PL	PS	ZR	NS	NS	NM	NM	NL	NL	NVL
	PVL	ZR	NS	NS	NM	NM	NL	NL	NVL	NVL

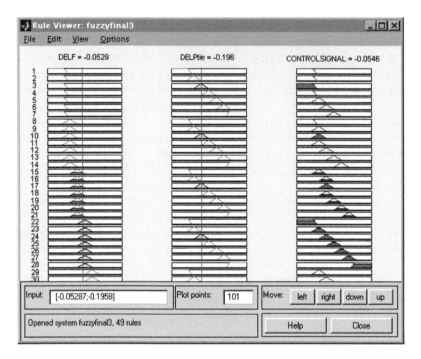

FIGURE 4.27
Fuzzy rule viewer.

of the FIS. It is used as diagnostic test to find which rules are active and how individual membership function shapes influence the results. The rule viewer is a road map for the whole fuzzy inference process. The three plots across the top of Figure 4.27 represent the antecedent and consequent of the first rule. Each rule is a row of plots, and each column is a variable. The rule numbers are displayed on the left of each row (Table 4.3).

From the fuzzy rule viewer screen shot in Figure 4.27, it is inferred that **If** DELF is −0.0529 and DEL Ptie is −0.196, **then** control signal is −0.0546. At this particular instance, if the change in frequency is 0.0529 and the tie-line power change is −0.0196, the control signal generated to initiate the turbine governor action is −0.0546. The rule viewer also shows how the shape of certain membership functions influences the overall result. After rules are evaluated, crisp values are produced by defuzzification of the corresponding membership function.

TABLE 4.3

Rule Base for AVR

		Error						
		NL	**NM**	**NS**	**ZR**	**PS**	**PM**	**PL**
	NL	PL	PL	PL	PL	PM	PS	ZE
	NM	PL	PL	PM	PM	PS	ZE	NS
Change in error	**NS**	PL	PM	PS	PS	NS	NM	NL
	ZR	PL	PM	PS	ZE	NS	NM	NL
	PS	PL	PM	PS	NS	NS	NM	NL
	PM	PM	ZE	NS	NM	NM	NL	NL
	PL	ZE	NS	NM	NL	NL	NL	NL

4.4.3.3 Defuzzification

The mathematical procedure of converting fuzzy values into crisp values is known as defuzzification. It plays a great role in a fuzzy-logic-based control system design, since it converts fuzzy set into a numeric value without loosing any information. Different defuzzification methods exist to accomplish the task and naturally there exist trade-offs to each method. The selection of right strategy depends on the application and the type of MF used. The performance of FLC depends on the defuzzification process, since the system under control is determined by the defuzzified output. The center of gravity method is used because of its computational speed and accuracy in real-time control (Cirstea et al. 2002). Figure 4.28 shows the graphical representation of the center of gravity method and an output condition with two significant linguistic values. The output fuzzy variable is converted into a crisp value by the centroid method as given in Equation 4.38:

$$\Delta\mu = \frac{\sum_{j=1}^{nr}\mu_j\mu_i}{\sum_{j=1}^{nr}\mu_j} \tag{4.38}$$

where
  μ_j is the membership value of linguistic variable recommending the fuzzy controller action
  μ_i is the precise numerical value corresponding to that fuzzy controller action

Since the final output is a combination of recommended actions of many rules, the controller is more robust to accommodate the changes in power system parameters.

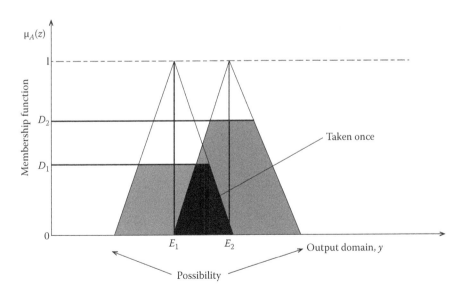

FIGURE 4.28
Graphical representation of the center of gravity method.

4.4.4 Fuzzy Controller for Single-Area Power System

In order to demonstrate the effectiveness of the fuzzy controller, the Simulink model for a single-area power system is simulated and the frequency response is plotted for a period of 100 s. The change in frequency for different loads and regulation parameter "R" is obtained, and transient responses are found to be stable. It is clear from the simulation results that FLC can bring down the frequency to its rated value immediately after the disturbance and without any oscillations. Figure 4.29 shows the response of frequency obtained for a sudden increase in load of 0.1 p.u. and for regulation value of 75. The FLC makes an intelligent decision on the amount of steam input to the turbine, and hence the speed of the generator is altered to maintain the desired frequency response. From Figure 4.29, it is clear that the frequency deviation ranges from −0.0066 to 0 and the settling time is 14 s, which is less by a factor of 61% when compared to PID controller. In short the proposed fuzzy controller reduces the settling time, thereby making the system to respond immediately for the sudden increase in load.

The voltage characteristics of fuzzy-based AVR are obtained with change in load of 0.1 p.u., as shown in Figure 4.30. The simulation is conducted, and the response of the proposed controller is plotted for a duration of 100 s. The AVR reduces the settling time by 73%, and the desired voltage is reached in 10.5 s without any transient oscillations.

The success of the proposed excitation control lies in increasing the dynamic performance of the system under varied operating conditions. The performance of the controller cannot be judged by the single result; hence, the models are simulated for various load changes and regulations to validate the efficiency of the proposed algorithms. Load disturbances of 0.3 p.u. and 0.8 p.u. are applied along with change in regulation values of 100 and 125, each at a time, and the results are tabulated in Table 4.4. The value of R determines the slope of the governor characteristics, and it determines the change in the output for a given change in frequency. In practice, R is set on each generating unit so that change in load on a system

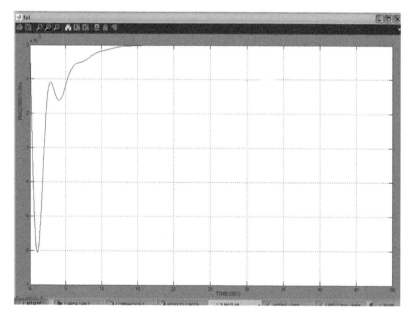

FIGURE 4.29
LFC with fuzzy controller for $R = 75$ and $\Delta P_L = 0.10$ p.u.

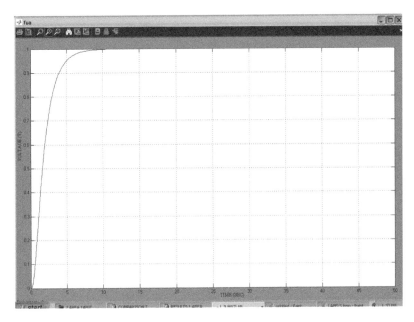

FIGURE 4.30
AVR with fuzzy controller for $\Delta P_L = 0.10$ p.u.

TABLE 4.4

Performance Analysis of Fuzzy Controller for Single Area Network

| | R = 100 | | | | R = 125 | | | |
| | $\Delta P_L = 0.30$ | | $\Delta P_L = 0.80$ | | $\Delta P_L = 0.30$ | | $\Delta P_L = 0.80$ | |
Parameters	LFC	AVR	LFC	AVR	LFC	AVR	LFC	AVR
Settling time (s)	21.5	13.1	28.3	20	27.5	16	29.2	20.8
Overshoot	0.015	0	0.032	0	−0.0046	0	−0.0213	0
Oscillation	0 to −0.015	0 to 1	0 to −0.032	0 to 1	0 to −0.0046	0 to 1	0 to −0.0137	0 to 1

will be compensated by the generated output. The mesh and contour plots of the peak's surface for the change in frequency are plotted in Figure 4.31. The plots clearly indicate how the control parameters evolve with change in load with respect to time.

Controllers based on the FL give the linguistic strategies control conversion from expert knowledge in automatic control strategies. In addition, the application of FL provides desirable small and large control signals to provide dynamic performance, which is not possible with linear control technique. Therefore, FLC has the ability to improve the robustness of the synchronous generator by bringing its output frequency and voltage to pre-established values. Further, the scope of this chapter is extended for evaluating the performance of proposed controller in multi-area power system.

4.4.5 Fuzzy Controller for Two-Area Power System

A multi-area interconnection consists of regions or areas that are interconnected by tie-lines. Tie-lines have the benefit of providing inter-area support for abnormal conditions as

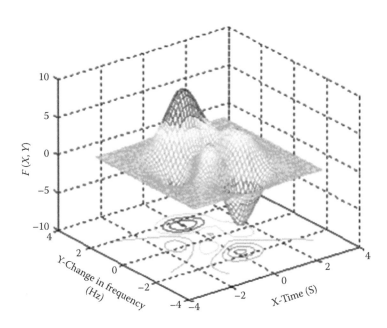

FIGURE 4.31
Contour and mesh plots of change in frequency.

well as transmission paths for contractual energy exchange between the areas. A two-area system consists of two single areas connected through a tie-line. Each area feeds its user pool, and the tie-line allows electric power to flow between areas. In each control area, all generators are assumed to form a coherent group. Since both areas are interconnected, a load disturbance in one area affects the output frequencies of both areas as well as the power flow on the tie-lines (Sabahi et al. 2008).

The most important aspect of electrical system reliability is to keep the synchronous generators working in parallel and with adequate capacity to satisfy the load demand. The trend of frequency measured in any area is an indicator of the trend of mismatch power in the interconnection and for that particular area alone. Information about the other areas found in the output frequency fluctuation of that area and in the tie-line power fluctuations.

Hence, the tie-line power is sensed and the resulting tie-line power signal is fed back into both areas (Ertugrul and Kocaarslan 2005). The dynamic model in the state–space variable form obtained from the associated transfer function is as shown in Equation 4.39:

$$\dot{x}(t) = Ax(t) + Bu(t) + Ld(t) \tag{4.39}$$

where
  A is the system matrix
  B and L are the input and disturbance distribution matrices

A two-area power system is considered for simulation where state vector $x(t)$, control vector $u(t)$, and disturbance vector $d(t)$ are defined as in Equation 4.40:

$$x(t) = (\Delta f_1, \Delta P_{G1}, \Delta P_{v1}, \Delta P_{tie12}, \Delta f_2, \Delta P_{G2}, \Delta P_{v2}) \tag{4.40}$$

$$u(t) = (u_1, u_2)$$

$$d(t) = (\Delta P_{D1}, \Delta P_{D2})$$

where Δf_1, Δf_2, ΔP_{G1}, ΔP_{G2}, ΔP_{v1}, and ΔP_{v2} denote deviation from nominal values of frequency, governor power, and valve powers of area 1 and area 2, respectively. The tie-line power between area 1 and area 2 is given by ΔP_{tie12} (Mathur and Manjunath 2007).

In the design of fuzzy controller, the ACEs of area 1 (ACE-1) and area 2 (ACE-2) and changes in errors of ACE-1 and ACE-2 are taken as the inputs, and the control outputs are ΔP_{c1} and ΔP_{c2}. The comparative study of system dynamic performance is carried out by investigating the frequency response for a change in load of 0.20 p.u. and speed regulation of 75 in both areas. The response of LFC in Figure 4.30 shows that the controller is successful in stabilizing the frequency of the generating system for interconnected areas. Any load change in one of the control areas affects the tie-line power flow causing other control areas to generate the required power to damp the frequency oscillations. The response time of LFC is very important to have the power system to gain control with increased stability margins.

The performance characteristics of change in frequency in Figure 4.32 reveal that the oscillations for area 1 and area 2 vary between −0.0063 and 0, and the settling time is 10 s, which is reduced by 80% when compared to PID controller. Table 4.5 indicates the simulation results obtained for two-area network with fuzzy controller for speed regulations of 100 and 125 for sudden load perturbations of 0.2 and 0.3 p.u.

From the investigations of results in Table 4.5, the influence of FLC for interconnected power system outperforms the behavior of conventional PID controller. Irrespective of the load in either area or both areas, the controller exhibits efficient control to guarantee zero steady-state error. It is observed that the proposed controller provides better dynamic response with reduced settling time, oscillations, and overshoot in both areas.

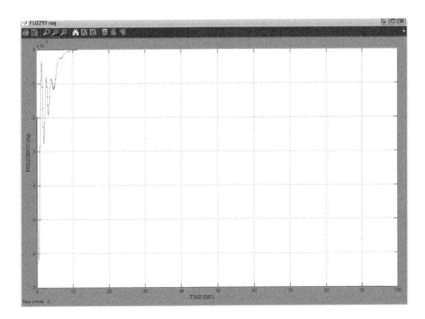

FIGURE 4.32
LFC with fuzzy controller for $R = 75$, $\Delta P_{L1} = 0.20$ p.u., and $\Delta P_{L2} = 0.20$ p.u.

TABLE 4.5

Performance Analysis of Fuzzy Controller for Two Area Network

Parameters				Settling Time (s)	Overshoot	Oscillation
$R = 100$	LFC	$\Delta P_{L1} = 0.10$	Area 1	11.32	−0.0042	0 to −0.0042
		$\Delta P_{L2} = 0.20$	Area 2	13.18	−0.0056	0 to −0.0056
$R = 100$	LFC	$\Delta P_{L1} = 0.20$	Area 1	14.5	−0.0057	0 to −0.0057
		$\Delta P_{L2} = 0.30$	Area 2	13.2	−0.0086	0 to −0.0086
$R = 125$	LFC	$\Delta P_{L1} = 0.20$	Area 1	17.4	−0.0052	0 to −0.0052
		$\Delta P_{L2} = 0.30$	Area 2	17.5	−0.0079	0 to −0.0079
$R = 125$	LFC	$\Delta P_{L1} = 0.30$	Area 1	18.3	−0.0084	0 to −0.0084
		$\Delta P_{L2} = 0.40$	Area 2	20.2	−0.0096	0 to −0.0096

4.5 Genetic Algorithm for LFC and AVR

In this section, the design is extended in developing GA-based PID control for LFC in a two-area interconnected power system. To demonstrate the effectiveness of the proposed method, the step responses of a closed-loop system were compared with that of the conventional controllers. The model of the plant has been developed using the MATLAB Simulink package and simulated for different load and regulation parameters. Simulation results are presented to show the effectiveness of the proposed method in handling processes of plant under different operating characteristics.

4.5.1 Design of GA-Based PID Controller

Designing and tuning a PID controller for an application that needs multiple objectives are difficult for a design engineer. The conventional PID controller with fixed parameters usually derives poor control performance. When gain and time constants change with operating conditions, conventional controllers result in suboptimal corrective action and hence fine tuning is required. This necessitates the development of tools that can assist engineers to achieve the best PID control for the entire operating envelope of a given process. A GA belongs to the family of evolutionary computational algorithms that have been widely used in control engineering applications. It is a powerful optimization algorithm that works on a set of potential solutions, which is called population. GA finds the optimal solution through cooperation and competition among the potential solutions. The genetic search space is thus extended to all possible sets of controller parameter values to minimize the objective function, which in this case is the error criterion. GA is used to minimize the error criteria of the PID controller in each iteration. The integral square error is used to define the PID controller's error criteria.

4.5.1.1 GA Design Procedure

GA-based controllers have the ability to adapt to a time-varying environment and may be able to maintain good closed-loop system performance. In the design of PID controller, the performance criterion or objective function is first defined based on the desired specification such as time domain specifications. As a mathematical means

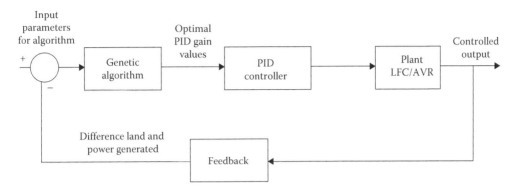

FIGURE 4.33
GA-based PID controller.

for optimization, GAs can be naturally applied to the optimal tuning of classical PID controllers. The PID controller parameters are optimized offline by GA at all possible operating conditions. The optimized PID gains are used in LFC and AVR control loops for efficient control of frequency and voltage. The frequency and voltage profile of the generator validates the design of the GA-based tuning algorithm. The closed-loop system consisting of proposed GA-based PID controller and plant depicting LFC/AVR is illustrated in Figure 4.33.

The transfer function of a standard PID controller structure is given in Equation 4.41:

$$G(s) = K_P(1 + K_I/s + K_D s) \tag{4.41}$$

where
K_P is the proportional gain providing overall control action
K_I is the integral gain reducing the steady-state error
K_D is the differential gain improving the transient response of the system

The fitness function $f(x)$ depends on the problem type, that is, maximization or minimization. The fitness function used for tuning PID controller is the minimization of the integral of time multiplied by the absolute value of error (ITAE) as shown in Equation 4.42:

$$F(x) = \int e(t)dt \tag{4.42}$$

The role of PID controller is to drive the output response of LFC and AVR within the tolerance limit set in the under frequency and under voltage relays. Here, GA is used to optimize the proportional, integral, and derivative gains of the PID controller such that the system will have a better performance in terms of settling time, peak overshoot, and oscillations. For each operating condition, GA is used to optimize the PID controller parameters in order to minimize the performance index. The control system is given a step input, and the error is assessed using the appropriate error performance criteria, that is, ITAE. Each chromosome is assigned an overall fitness value according to the magnitude of error; the smaller the error, the larger the fitness value.

The performance index is calculated over a time interval "*T*," normally in the region of $0 \le T \le t_s$, where t_s is the settling time.

Hence, the optimum values of K_P, K_I, and K_D are obtained for single- and two-area power systems for efficient control of voltage and frequency. For a two-area LFC system, considering ACE as the system output, the control vector for a PID controller in a continuous form can be given as in Equation 4.43:

$$U_i = -(K_P ACE_i + K_D ACE_i + K_I \int ACE_i) \tag{4.43}$$

where K_P, K_I, and K_D are the proportional, derivative, and integral gains, respectively. Since the performance index is related to time and error, the optimum gain values are suitable for controlling under different operating load and regulation parameters.

The GA steps to tune gain values for the PID controller (Ismail 2006) are as follows:

1. Randomly choose the genetic pool of parameters K_P, K_I, and K_D.

2. Compute the fitness of all population.

3. Choose the best subset of the population of the parameters K_P, K_I, and K_D.

4. Generate new strings using the subset chosen in step 3 as parents and the "single point crossover" and "uniform mutation" as operators.

5. Verify the fitness of the new population members.

6. Repeat steps 3 to 5 until the fixed amount of fitness is attained.

A stopping criterion terminates the algorithm after a specified number of iterations have been performed or until the solution is encountered with specified accuracy. To simplify the analysis, the optimal parameter values for two interconnected areas are considered:

$$K_{P1} = K_{P2} = K_P, \quad K_{I1} = K_{I2} = K_I, \quad K_{D1} = K_{D2} = K_D \tag{4.44}$$

The best fitness function value in a population is defined as the smallest value of the objective in the current population. The fitness function is a measure of how well the candidate solution fits the data, and it is the entity that focuses the GA toward the solution. Minimization of the fitness function by GA results in reduced settling time and overshoots.

A flow chart of the GA optimization procedure is given in Figure 4.34. Selecting an objective function is the most difficult part of designing a GA. An objective function is required to evaluate the best PID controller for LFC and AVR. An objective function could be created to find the optimum gains of a PID controller that gives the smallest overshoot, faster settling time, and reduced oscillations. By combining all these objectives, an error minimization function is selected that will reduce the deviation in frequency and voltage of the generating system. The convergence criterion of a GA is a user-specified condition, for example, the maximum number of generations when the string fitness value exceeds the threshold limit.

4.5.2 Simulink® Model of Single-Area Power System

The transfer function approach is followed in modeling the LFC and AVR for a power system with GA-PID controller. The optimum gains of the PID controller are obtained by running the M-file and the values are stored in the workspace. Series of experiments

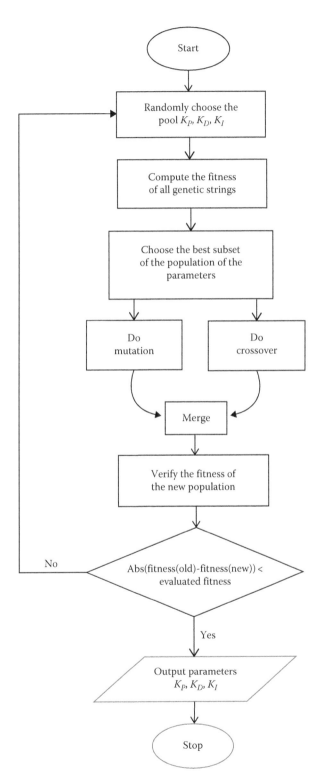

FIGURE 4.34
Genetic algorithm for PID controller.

TABLE 4.6

GA Parameters

Parameter	Value
Population size	25
Number of generations	500
Crossover rate	4
Mutation rate	18
Probability of selection	0.18

are conducted with different parameter settings before actual runs to collect the results. The parameters used in the simulations are summarized in Table 4.6. The computational efficiency is extremely important in optimization techniques such as GA, where the generation of solution takes more time with a large amount of data. The mean CPU time taken for completing the fixed number of iterations is measured. The GA is initialized with a population of 25 and was iterated for 500 generations. The crossover and mutation rates are set to 4 and 18, respectively. For the given parameter values, the computational time taken by the algorithm to generate optimum value of PID gains is 54 s. The optimized PID gains are transferred to the Simulink models of LFC and AVR of the isolated power generating system. Investigation is extended to two-area LFC to validate the performance of the proposed controller.

The dynamic model of the power system given by Elgerd (1983) is used for implementation of the proposed controller. In this chapter, the Simulink model of LFC and AVR with PID controller is designed with nominal values for each parameter. The nominal parameters for AVR model are $K_a = 10$, $K_e = 1$, $K_g = 1$, $K_r = 1$, $T_a = 0.1$, $T_e = 0.4$, $T_g = 1$, and $Tr = 0.05$. Since the generator is driven by steam turbine, the governor and turbine time constants are selected as $T_g = 0.2$ and $T_t = 0.5$, respectively. In the absence of a speed governor, the system response to a load change is determined by the inertia constant and the damping constant. The inertia constant H and damping constant D are maintained as 10 and 0.8, respectively. The steady-state speed deviation is such that the change in load is exactly compensated by the variation in load due to frequency sensitivity. The Simulink model of LFC and AVR is shown in Figures 4.35 and 4.36, respectively. The AVR model consists of a step input, PID

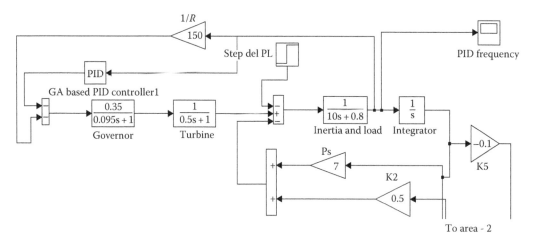

FIGURE 4.35
Simulink® model of LFC with GA-based PID controller.

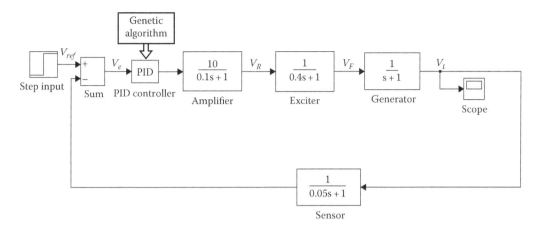

FIGURE 4.36
Simulink® model of AVR with GA-based PID controller.

controller based on GA, an amplifier that amplifies the exciter signal, which in turn controls the voltage of the generator, and a scope to display the terminal voltage. It also contains a sensor that determines the difference between load demand and power generated and feeds it to the controller based on the load changes. The LFC model in Figure 4.3 consists of a step input, PID controller based on GA, a governor to control the speed of the turbine that drives the generator, and the scope that shows the frequency deviation.

The optimum values of K_P, K_I, and K_D are obtained by executing the GA code in M-file. The optimum gain values are included in the PID controller in a secondary control loop of LFC and AVR. The simulation results are plotted for a period of 100 s with GA-based PID controller. The performance of GA-PID was tested and verified for speed regulations 75, 100, and 125 and for change in load of 0.10–0.90 p.u. A step increase in load of 10% is applied and the dynamic response of GA-based LFC is analyzed. The frequency deviation and terminal voltage responses obtained for a change in load of 0.10 p.u. and speed regulation of 75 are shown in Figures 4.37 and 4.38, respectively.

The response of LFC for load perturbation of 10% is plotted in Figure 4.38. It is observed that the frequency (ΔF) oscillates between −0.0030 and 2.4629e−006 and settles at 8.7 s.

It can be interpreted from Figure 4.38 that voltage regulator senses the changes in the terminal voltage and initiates corrective action to drive the exciter to maintain the rated output within 10.5 s. The AVR is a continuously acting type so that corrective action is

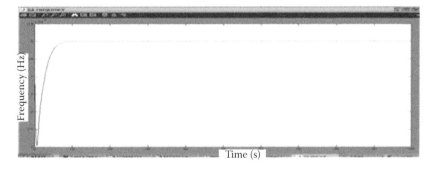

FIGURE 4.37
LFC with GA-based PID controller for $R = 75$ and $\Delta P_L = 0.10$ p.u.

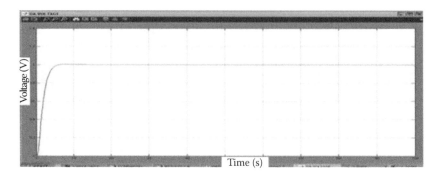

FIGURE 4.38
AVR with GA-based PID controller for $\Delta P_L = 0.10$ p.u.

TABLE 4.7

Performance Analysis of GA Based PID Controller for Single Area Network

	R = 100				R = 125			
	$\Delta P_L = 0.30$		$\Delta P_L = 0.80$		$\Delta P_L = 0.30$		$\Delta P_L = 0.80$	
	Computational Time: 53.6 s		Computational Time: 54.2 s		Computational Time: 54.8 s		Computational Time: 55.3 s	
Parameters	LFC	AVR	LFC	AVR	LFC	AVR	LFC	AVR
Settling time (s)	8.88	11.2	14.8	15.3	11.4	11.25	14.8	15.7
Overshoot	−0.0073	0	−0.054	0	−0.0065	0	−0.051	0
Oscillation	2.1736e−006 to −0.0073	0–1	0 to −0.054	0–1	4.606e−007 to−0.0065	0–1	0 to −0.051	0 to 1

taken proportional to deviations in terminal voltage. Hence, a fast-acting exciter with high-gain AVR reduces the oscillations and provides stable operation of the power system. The behavior of changing frequency and voltage for various loads and regulations is presented in Table 4.7. The transient response of LFC and AVR can be analyzed and compared with conventional PID controllers. During restoration, the power system should be operated at nominal voltage. The voltage deviations should be maintained at a minimum possible level to reduce charging currents. Hence, enough care must be taken to ensure that a fast-acting AVR is installed on each generator. The stability of the system depends on the restoration progress of the generating system. Hence, alternative controls on generators should be placed in a position to ensure instantaneous responses to changes in voltage and frequency. The feasibility of an optimal control scheme is based on the improvement in performance characteristics with respect to the parameters tabulated, for example, settling time, overshoot, and oscillations. The settling time for LFC with a change in load of 0.30 and R value of 100 is reduced by 80.48% when compared to conventional PID and 58.69% with respect to an FLC. The overshoot and oscillations are reduced by 52.9% and 51.4% with respect to conventional PID and FLCs, respectively.

A GA-based PID controller achieves better performance with respect to AVR by achieving 69.44% reduction in settling time in comparison to conventional PID and 15% when compared with fuzzy-based controller. The investigation carried out for a single-area network under different load perturbation and regulations reveals the reliability of the proposed GA-based controller. In order to emphasize the advantages of the proposed controller, the computational time taken by GA to generate optimum gain values for different load regulation and change in loads are mentioned in Table 4.7. For conventional PID-controller-based LFC and AVR, the optimum gain values are selected by a trial-and-error method or based on the experience of an operator. It is hard to manually tune the optimal gains of PID controller because of the higher order, time delays, and nonlinearities. Even though the computational time is less when compared to GA, the conventional controller does not exhibit desired performance characteristics. For these reasons, it is proposed to increase the capability of PID controller by enhancing its features. The LFC and AVR design using modern optimization theory enables power engineers to design an optimal control system with respect to the given performance criterion.

4.5.3 Simulink® Model of Interconnected Power System

In interconnected systems, a group of generators are connected internally to form a coherent group. The LFC loop can be represented for the whole area, referred to as the control area. Consider two areas in Figure 4.39 interconnected by lossless tie-line of reactance X_{tie}, with the power flow P_{12} from area 1 to area 2. For LFC studies, each area may be represented by an equivalent generating unit exhibiting its overall performance. The Simulink model of LFC for a two-area power system is shown in Figure 4.40, with each area represented by an equivalent inertia M, load-damping constant D, turbine and governing system with speed droop R. This model depicts the interconnection of two power systems with LFC, and the results are analyzed from the scope that displays the combined output of the frequencies of the two systems.

If the areas are equipped only with primary control of the LFC, a change in load in one area is met with a change in generation in both areas, a change in tie-line power, and a change in the frequency. Hence, a supplementary control is necessary to maintain the frequency at nominal value. A two-area interconnected power system of non-reheat type turbine with GA-based PID controllers is shown in Figure 4.39. The optimum gain values for the PID controller are tuned using GA and transferred to the Simulink model for obtaining the desired frequency response.

The two areas are loaded simultaneously and independently with load perturbations of 20% and 30%. The performance characteristics are plotted for a period of 100 s to monitor the wave front of tie-line frequency. The response plots for frequency deviation of area 1 and area 2 with change in load of 0.20 p.u. in both areas and speed regulation of 75 are given in Figure 4.41.

In an interconnected power system, if a large generator trips, it results in a shortage of generation. The remaining unit gives up some of their rotating kinetic energy to increase the power output and then the frequency declines. In response to the frequency decline, the governor opens the control valve to increase the power input to the turbines. Hence, the designed controllers must be fast enough to respond to the frequency changes and initiate

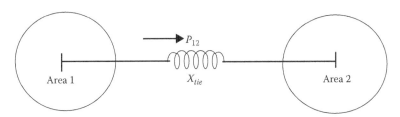

FIGURE 4.39
Two-area system.

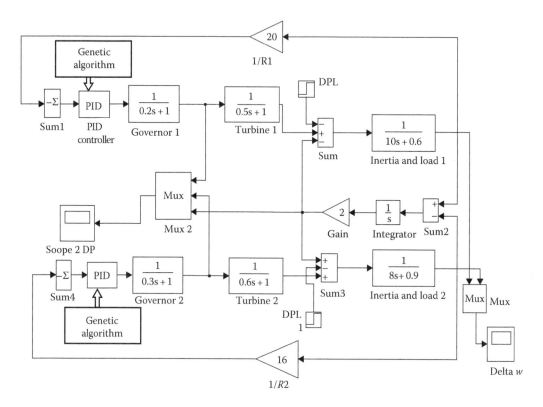

FIGURE 4.40
Simulink® model of LFC for a two-area power system with GA-based PID controller.

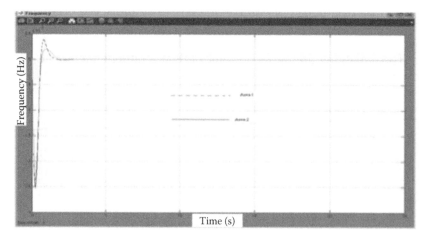

FIGURE 4.41
LFC with GA-based PID controller for $R = 75$, $\Delta P_{L1} = 0.20$ p.u., and $\Delta P_{L2} = 0.20$ p.u.

TABLE 4.8

Performance Analysis of GA Based PID Controller for Two Area Network

Parameters			Settling Time (s)	Overshoot (Hz)	Oscillation (Hz)
$R = 100$	Area 1 $\Delta P_{L1} = 0.20$	Comp. time 55.7 s	4	−0.0019	$1.4454e - 004$ to −0.0019
	Area 2 $\Delta P_{L2} = 0.20$		3.17	−0.0027	$8.7985e - 004$ to −0.0027
$R = 125$	Area 1 $\Delta P_{L1} = 0.20$	Comp. time 56 s	2.25	−0.0025	$5.3295e - 004$ to −0.0025
	Area 2 $\Delta P_{L2} = 0.20$		1.75	−0.0023	$1.2619e - 004$ to −0.0023
$R = 100$	Area 1 $\Delta P_{L1} = 0.20$	Comp. time 56.8 s	1.5	−0.0025	$4.2800e - 004$ to −0.0025
	Area 2 $\Delta P_{L2} = 0.30$		5.5	−0.0045	0.0023 to −0.0045
$R = 125$	Area 1 $\Delta P_{L1} = 0.20$	Comp. time 57.1 s	1.7	−0.0024	$3.5490e - 004$ to −0.0024
	Area 2 $\Delta P_{L2} = 0.30$		1.85	−0.0035	$1.9075e - 004$ to −0.0035

controller action to vary the prime mover input. The simulation results in Figure 4.41 reveal that the proposed GA-based PID controller for a two-area LFC is faster in restoring the frequency to its optimum value. Frequency deviation returns to zero and remains at the same value until a new load is applied. Table 4.8 illustrates the performance comparison of area 1 and area 2 with speed regulation "R" of 75 and change in load of 0.20 p.u. in both areas. The simulation result shows that the proposed GA-based PID controller yields a reduction of 95.45% in settling time, 79.19% in peak overshoot, and reduction of 77% in oscillation when compared to conventional PID controller.

4.6 PSO and ACO for LFC and AVR

A PID controller improves the transient response of a system by reducing the overshoot and settling time of a system. The main reason to develop better methods to design PID controllers is because of the significant impact on performance improvement. The performance index adopted for problem formulation is settling time, overshoot, and oscillations. The primary design goal is to obtain a good load disturbance response by optimally selecting the PID controller parameters. Traditionally, the control parameters have been obtained by a trial-and-error approach, which consumes more amounts of time in optimizing the choice of gains. To reduce the complexity in tuning PID parameters, evolutionary computation techniques can be used to solve a wide range of practical problems, including optimization and design of PID gains. It can obtain suboptimal solutions for very difficult problems, which conventional methods fail to produce in reasonable time. EAs can be a useful paradigm and provide promising results for solving complex optimization functions. Evolutionary computation refers to the study of computational systems that use ideas to draw inspirations from natural evolution.

4.6.1 Evolutionary Algorithms for Power System Control

In general, an electric power network is a large and complex system consisting of synchronous generators, transformers, transmission lines, relays, and switches. Various control objectives such as operating conditions, actions, and design decisions require solving one or more linear or nonlinear optimization problems. EA is considered a useful promising

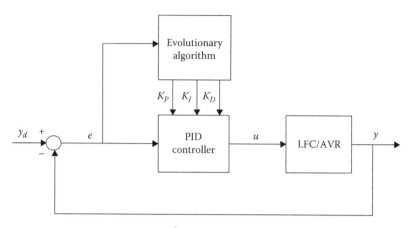

FIGURE 4.42
PID control system with evolutionary algorithm.

technique for deriving the global optimization solution for complex problems. Since the loads are switched on and off, the power system is prone to sudden changes to its configuration. Under these circumstances, keeping voltage and frequency within the allowable range is one of the important tasks for power system control. An online control strategy to achieve this is referred to as real and reactive power control using LFC and AVR. Essentially, LFC takes care of frequency, and AVR ensures voltage of the generating system. The PID control system with plant indicating LFC/AVR and EA-based PID is shown in Figure 4.42. K_P, K_I, and K_D are, respectively, the proportional, integral, and derivative gains of the PID controller that are tuned by EA. In the proposed system, PSO and ACO algorithms are used to optimize the set of PID parameters in the system to achieve the desired output "yd." The control output "u" from EA-PID is based on the error signal "e," which is the difference between the actual output "y" and the desired output "yd." The objective of PSO- and ACO-based optimization is to seek a set of PID parameters such that the feedback control system has a minimum performance index. A set of optimal PID parameters can yield good frequency and voltage characteristics of LFC and AVR. EA is considered a useful and promising technique for deriving the global optimum solution of complex functions. Hence, the application of these algorithms yields improved performance characteristics in terms of settling time, oscillations, and frequency.

The LFC/AVR is subjected to different operating characteristics such as varying load and regulation parameters to verify the validity of the proposed algorithm. Stochastic techniques such as PSO and ACO are applied to tune the controller gains to ensure optimal performance at nominal operating conditions. PSO and ACO are used in offline to tune the gain parameters and applied to PID controller in the secondary control loop of the plant.

4.6.2 ACO-Based PID Controller

The conventional fixed-gain PID controller is a well-known technique for industrial control process. The design of this controller requires the three main parameters: proportional gain (K_P), integral time constant (K_I), and derivative time constant (K_D). The gains of the controller are tuned by a trial-and-error method based on the experience and plant behavior. This process will consume more time and will be suitable only for particular operating conditions. Here ACO algorithm is used to optimize the gains, and the values are transferred to the PID controller of the plant representing LFC and AVR of the power generating system as shown in Figure 4.43.

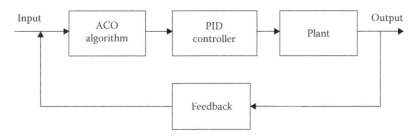

FIGURE 4.43
ACO-PID controller.

The proportional gain makes the controller respond to the error, while the integral gain helps to eliminate steady-state error and the derivative gain prevents overshoot. The plant is replaced by LFC and AVR models developed using Simulink in MATLAB. With the optimum gains generated by the proposed algorithm, the models are simulated for various operating conditions to validate the performance.

The design steps of ACO-based PID controller for AVR are as follows:

1. Initialize the algorithm parameters such as number of iterations, number of ants, strength of pheromone, and decay rate.

2. Initialize the ranges of PID controller gain values.

3. For each ant, the transition probability is calculated using
$$P_{ij} = (\tau_{ij})^\alpha (\eta_{ij})^\beta / \sum_{h \in S} (\tau_{ij})^\alpha (\eta_{ij})^\beta$$
The heuristic factor is computed as $\eta_{ij} = 1/F(X_j)$, $j \in S$, where $F(X)$ represents the cost function of X.

4. Incrementally build a solution and update local pheromone by using $\tau_{ij}(t) \leftarrow (1 - \rho)$ $\tau_{ij}(t) + \rho\tau_0$, where $\tau_{ij}(t)$ is the amount of pheromone on the edge (i, j) at time t, ρ is a parameter governing pheromone decay such that $0 < \rho < 1$, and τ_0 is the initial value of pheromone on all edges.

5. Record the best solution found so far.

6. Do a global pheromone update by using $\tau_{ij}(t) \leftarrow (1 - \rho)\tau_{ij}(t) + \rho\Delta\tau_{ij}(t)$, where $\Delta\tau_{ij}(t) = L^{-1}$, if this is the best tour, otherwise $\Delta\tau_{ij}(t) = 0$, and L indicates the length of the global best tour.

7. Repeat steps 3 through 6 until the maximum iteration is reached.

The algorithm is tested for different values of parameters by simulating the model for different operating conditions. According to the trials, the optimum parameters used for verifying the performance of the ACO-PID controller are listed in Table 4.9.

TABLE 4.9

ACO Parameters

Parameters	LFC	AVR
Number of ants	500	400
Number of nodes	120	130
Number of generations	10	25
Pheromone strength	0.01	0.02
Decay rate	0.99	0.84

The ACO algorithm design steps for LFC are as follows:

1. Initialize the population size, the initial search steps of all variables and number of ants, $t = 0$, and count $t = 0$.
2. Initialize the PID parameters.
3. For each ant ($j = 1, 2, ..., n$), select the jth solution component with a probability P_{ij}.
4. Evaluate the candidate solution and get the best individual ant and the path.
5. Evaluate the local and global pheromone. If no improvement occurs, adjust the current searching step scheme according to the path.
6. Repeat the process until the best searching step is reached or the maximum iteration is performed.

Since the model parameters of LFC are identical, the optimized parameters are used in the PID controller for single- and two-area interconnected LFC systems. The system is stable, and the control task is to minimize the system frequency deviation Δf_1 in area 1, Δf_2 in area 2, and tie-line power deviation ΔP_{tie}. The performance of the system can be tested by applying load disturbance ΔP_{D1} to the system and observing the change in frequency in both areas. To assess the effectiveness of the optimized parameters, the models are tested for different load and regulation parameters.

4.6.3 PSO-Based PID Controller

With the advancement of computational methods in the recent times, optimization techniques are often proposed to tune the control parameters. Stochastic algorithm can be applied for tuning PID controller gains to ensure optimal control performance at nominal operating conditions. In conventional PID controller, the gains are randomly selected and the results are verified for every set of random gain values. PSO algorithm finds the proportional, integral, and derivative gains of the PID controller, and the values are passed to the PID controller of single area LFC and AVR, as shown in Figure 4.44. The gain values are tested for two-area LFC to optimize the change in frequency in both areas.

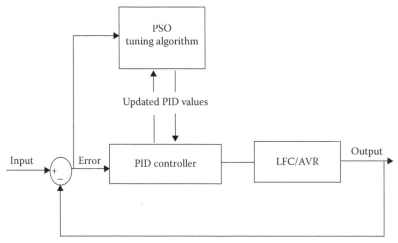

FIGURE 4.44
PSO algorithm-based PID controller.

The design steps of PSO-based PID controller for the LFC of a power generating system are as follows:

1. Initialize the algorithm parameters such as number of generations, population, inertia weight, cognitive and social coefficients.
2. Initialize the values of the parameters K_P, K_I, and K_D randomly.
3. Calculate the fitness function of each particle in each generation.
4. Calculate the local best of each particle and the global best of the particles.
5. Update the position, velocity, local best, and global best in each generation.
6. Repeat steps 3 through 5 until the maximum iteration is reached or the best solution is found.

The objective function represents the function that measures the performance of the system. The fitness (objective) function for PSO is defined as the integral of time multiplied by the absolute value of error (ITAE) of the corresponding system. Therefore, it becomes an unconstrained optimization problem to find a set of decision variables by minimizing the objective function. The AGC performance of a two-area test system has been tested with a PSO-tuned optimal PID controller. The main objectives of the AGC in multi-area power system are maintaining zero steady-state errors for frequency deviation and accurate tracking of load demands. Hence, the optimal parameters obtained by the proposed algorithm guarantee both stability and desired performance in both areas of interconnected system. Each area consists of three first-order transfer functions, modeling the turbine, governor, and power system. In addition, all generators in each area are assumed to form a coherent group. For PID controller, the objective function is defined as

$$f = \sum_{j=1}^{N} \sum_{i=1}^{N} \left(\int_{0}^{\infty} t \, | \Delta f_i^j | \, dt \right) \tag{4.45}$$

where
N is the number of areas in the power system
Δf_i^j is the frequency deviation in area i for step load changes in area j

The flowchart for PSO-based PID controller is shown in Figure 4.45. To design the LFC for two areas, the change in load in both areas must be taken into account along with the parameters of the governor, turbine, and load. Two identical areas with non-reheat type turbine with similar parameters are considered for implementation. Furthermore, the generators tend to have the same response characteristics and are said to be coherent. Then it is possible to let the LFC loop represent the whole system, which is referred to as a common area. The prime mover control must have drooping characteristics to ensure proper division of load, when generators are operating in parallel. In many cases, a group of generators are closely coupled internally and swing in unison. The AGC of a multi-area system can be realized by analyzing AGC for a two-area system. Tie-line power appears as a load increase in one area and a load decrease in the other area, depending on the

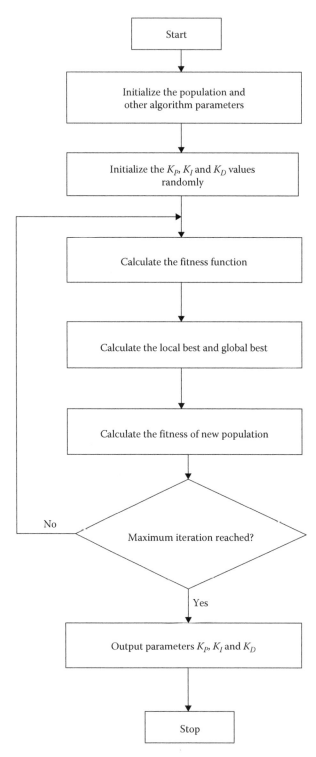

FIGURE 4.45
PSO algorithm for PID controller.

TABLE 4.10

PSO Parameters

Parameters	LFC	AVR
Population size	5	50
Number of generations	10	50
Inertia weight	0.8	0.9
cognitive coefficient	2.05	2.0
social coefficient	2.05	2.0

direction of the flow. The optimum values used for various parameters in PSO implementation are listed in Table 4.10.

The following procedure is used for implementing the PSO algorithm for AVR:

1. Initialize the swarm by assigning a random position in the problem hyperspace to each particle.

2. Evaluate the fitness function for each particle.

3. For each individual particle, compare the particle's fitness value with its P_{best}. If the current value is better than the P_{best} value, then set this value as the P_{best} and the current particle's position, x_i and p_i.

4. Identify the particle that has the best fitness value. The value of its fitness function is identified as g_{best} and its best position as p_g.

5. Update the velocities and positions of all the particles using Equations 5.6 and 5.7.

6. Repeat steps 2 through 5 until the stopping criterion is reached: when the maximum number of iterations or the optimum solution is reached.

4.6.4 Simulink® Model of an AVR

The AVR model consists of a step input, PID controller based on PSO, an amplifier that amplifies the signal to the exciter, which in turn controls the voltage of the generator, and a scope to display the terminal voltage. It also contains a sensor that senses the voltage rise or fall due to the difference between load demand and power generated and feeds it to the controller based on the load changes. The AVR model shown in Figure 4.46 is simulated with system parameter values indicated in Table 4.11.

4.6.5 Simulink® Model of LFC

PID controllers are parametric controllers that affect the behavior of the LFC system, if the parameters are not optimized. Designing an optimum controller ensures improved performance by minimizing the performance index. To illustrate the importance of proposed PSO and ACO algorithms, the LFC model designed using Simulink in MATLAB is considered. It consists of a step input, PID controller based on PSO, a governor that controls the speed of the turbine that drives the generator, and the scope that shows the frequency deviation. The optimum parameters used in LFC model in Figure 4.47 are indicated in Table 4.12.

4.6.6 Effect of PID Controller Using ACO

The Simulink model for LFC and AVR with ACO-based PID controller was simulated. The optimum gain values obtained by the M-file are transferred to the Simulink model and tested for different loads and regulation parameters. The frequency deviation and

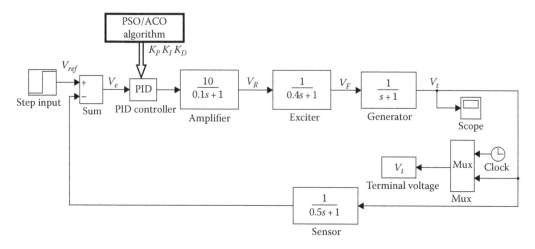

FIGURE 4.46
Simulink® model of automatic voltage regulator with PID controller.

TABLE 4.11

Values for Constants in AVR model

Symbol	Parameters	Optimum Values
K_A	Amplifier gain	10
τ_A	Amplifier time constant	0.1
K_G	Generator gain	1
τ_G	Generator time constant	1
K_R	Sensor gain	1
τ_R	Sensor time constant	0.05

terminal voltage response for a change in load of 0.1 p.u. and regulation of 10 are shown in Figures 4.48 and 4.49, respectively.

From Figure 4.48, it is observed that the frequency deviation and the peak overshoot are minimum. The settling time for frequency deviation is 9 s, and the oscillation varies between −0.0080 and +0.0030, which is very less compared to PID controllers.

From Figure 4.49, it is found that the settling time of AVR with ACO-based integral controller is 5.2 s, and there is a transient overshoot of about 0.16. The LFC and AVR models are simulated for different load conditions in order to replicate the daily load curve of the power system. The computational time taken by the proposed algorithm in generating the optimum values of PID gains is obtained and tabulated. To show the effectiveness of the proposed algorithm, the settling time for different operating conditions of LFC is presented in Table 4.13. As witnessed from the table, the merits of ACO are the response characteristics and computational efficiency. The computational time is reduced by 21.6% when compared to GA-based PID controller. Since the population of ants is operated simultaneously, the computational efficiency is improved. It is achieved because of the parallel search and optimization capabilities inspired by the behavior of ant colonies.

Owing to the randomness of heuristic algorithms, their performance cannot be judged by a single run. Many trials with different initialization should be made to acquire useful conclusion about the performance. An algorithm is robust if it gives a consistent result during all the trials. The simulation results for LFC and AVR with ACO-based PID controller

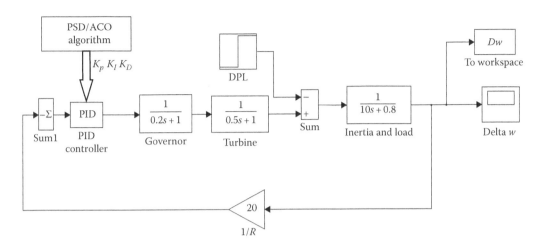

FIGURE 4.47
Simulink® model of LFC with PID controller.

TABLE 4.12

Values for constants in LFC model

Symbol	Parameters	Optimum Values
τ_G	Governor time constant	0.2
τ_T	turbine time constant	0.5
R	Regulation parameter	20, 30
H and D	Inertia constants of the load	10 and 0.8

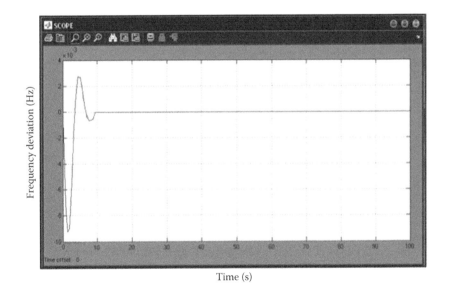

FIGURE 4.48
LFC with ACO-PID controller for $R = 10$ and $\Delta P_L = 0.1$ p.u.

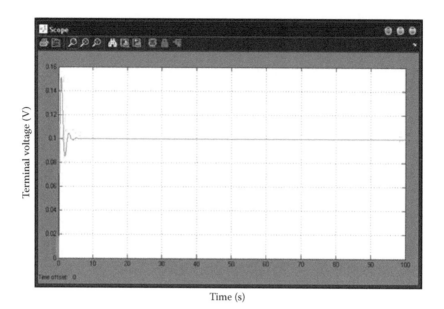

Time (s)

FIGURE 4.49

AVR with ACO-PID controller for $\Delta P_L = 0.1$ p.u.

TABLE 4.13

Performance Analysis of ACO Based PID Controller for LFC

	$R_1 = 10$		$R_2 = 20$	
Parameters	$\Delta P_L = 0.2$	$\Delta P_L = 0.7$	$\Delta P_L = 0.2$	$\Delta P_L = 0.7$
Computational time (s)	42	44.5	41.8	44.2
Settling time (s)	9	8.6	10.3	9.8
Overshoot (Hz)	0.0030	0.0053	−0.00058	−0.00034
Oscillation (Hz)	−0.0080 to +0.0030	−0.018 to +0.0053	−0.0071 to −0.00058	−0.138 to −0.00034

TABLE 4.14

Performance Analysis of ACO based PID Controller for AVR

	Change in Load			
	$\Delta P_L = 0.2$	$\Delta P_L = 0.4$	$\Delta P_L = 0.6$	$\Delta P_L = 0.8$
Computational time (s)	42	44.5	41.8	44.2
Settling time (s)	5.2	5.5	5.86	6.6
Overshoot (V)	0.15	0.301	0.142	0.28
Oscillation (V)	0–0.15	0–0.301	0–0.142	0–0.28

under various load changes and regulations are tabulated in Tables 4.13 and 4.14, respectively. From the tables, it is observed that the settling time, peak overshoots, and oscillations of LFC are reduced by 37%, 21%, and 50%, respectively. The settling time of AVR is reduced by 53% when compared to the conventional controller for a sudden increase in load of 0.2 p.u. The fitness function of the algorithm to generate optimum gain value is same; hence, the execution time for LFC and AVR is similar as tabulated. From the results,

TABLE 4.15

Performance Comparison of ACO based AVR

Methods	Fixed Parameters: $K_A = 10$, $\tau_A = 0.1$, $K_E = 1$, $\tau_E = 0.1$, $K_G = 1$, $\tau_G = 1$, $K_R = 1$, $\tau_R = 0.05$		
	Settling Time (s)	Overshoot (V)	Oscillations (V)
Conventional PID	37.5	0.1	0–0.1
Fuzzy controller	16	0.1	0–0.1
GA-PID	11.38	0.1	0–0.1
ACO-PID	5.2	0.15	0–0.15

TABLE 4.16

Performance Comparison of ACO Based LFC

Methods	Fixed Parameters: $\tau_G = 0.2$, $\tau_T = 0.5$, $K_G = 1$, $H = 5$, $D = 0.8$		
	Settling Time (s)	Overshoot (Hz)	Oscillations (Hz)
Conventional PID	51	−0.0083	0 to 0.0083
Fuzzy controller	20	−0.0052	0 to −0.0052
GA-PID	10.25	−0.0026	0 to −0.0026
ACO-PID	8.6	0.003	0 to 0.005

it is revealed that the ACO method is a potential alternative to be developed in solving LFC and AVR problems.

To assess the effectiveness of the ACO-PID controller, the simulation results are compared with the conventional PID-, fuzzy-, and GA-based controllers in Tables 4.15 and 4.16. The settling time of ACO-based AVR is reduced by 54.3% when compared to GA-PID and 67.5% when compared to the fuzzy controller. The simulation results demonstrate the adaptability of the ACO algorithm and its advantage in solving power system optimization problems.

The settling time, oscillations, and overshoot of the proposed LFC with ACO-based controller are reduced by 63.6%, 88.9%, and 66.3%, respectively, when compared to conventional PID controller. The settling time of AVR with ACO-based controller is decreased by a factor of 86.2%. Hence, the ACO-based controller gives improved performance characteristics when compared to conventional controllers. When compared to the fuzzy controller, the settling time, overshoot, and oscillations of the proposed ACO-PID controller are reduced by 57%, 42.3%, and 3.8%, respectively. When compared to GA-based controller, the settling times of LFC and AVR are reduced by 16% and 54.3%, respectively. With respect to oscillations and overshoot, the performance of the ACO-based controller is found to be very close with GA-PID controller and can be varied by optimum tuning of regulation.

4.6.7 Impact of PSO-Based PID Controller

The model for LFC and AVR with PSO-based PID controller is designed in the Simulink. The K_P, K_I, and K_D values for the PID controller were obtained by running the M-file. The simulation was performed for different regulations and loads to validate the robustness of the proposed controller. The terminal voltage response for a change in load of 0.1 p.u. and regulation of 10 is shown in Figure 4.50.

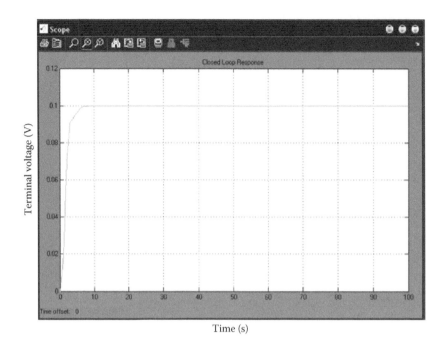

FIGURE 4.50
AVR with PSO-based PID controller for $\Delta P_L = 0.1$ p.u.

From Figure 4.50, it is observed that the settling time of AVR with PSO-based PID controller is 9.03 s and there is no transient peak overshoot.

From Figure 4.51, it is inferred that the settling time of LFC with PSO-based PID controller is 8.2 s, and the peak overshoot is −0.0114. The simulation results for AVR and LFC with PSO-based PID controller under various load changes and regulations are given in

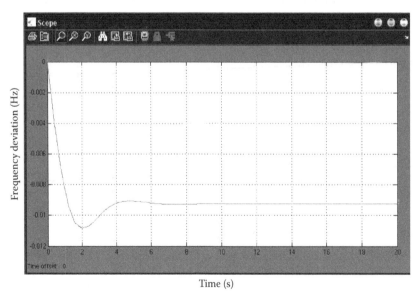

FIGURE 4.51
LFC with PSO-based PID controller for $R = 10$ and $\Delta P_L = 0.1$ p.u.

TABLE 4.17

Performance Analysis of PSO Based PID Controller for AVR

Parameters	Change in Load			
	$\Delta P_L = 0.2$	$\Delta P_L = 0.4$	$\Delta P_L = 0.6$	$\Delta P_L = 0.8$
Computational time (s)	26.8	27.2	28.4	29
Settling time (s)	9.03	10.2	11.2	11.8
Overshoot (V)	0	0.22	0	0.204
Oscillation (V)	0–0.1	0–0.22	0–0.1	0–0.204

TABLE 4.18

Performance Analysis of PSO Based PID Controller for LFC

Parameters	$R_1 = 10$		$R_2 = 20$	
	$\Delta P_L = 0.2$	$\Delta P_L = 0.7$	$\Delta P_L = 0.2$	$\Delta P_L = 0.7$
Computational time (s)	26.8	27.2	28.4	29
Settling time (s)	8.2	8.34	10.3	10.42
Overshoot (Hz)	−0.0014	−0.0213	−0.0147	−0.0076
Oscillation (Hz)	0–0.0014	0–0.0213	0 to −0.0147	0–0.0076

Tables 4.17 and 4.18, respectively. These results show that the proposed algorithm can search optimal PID controller parameters quickly and efficiently. The PSO method does not perform the selection and crossover operation in evolutionary processes; the computation time is reduced by 47% when compared with the GA method.

It is observed from the results that when compared to the conventional controller, the settling time, peak overshoots, and oscillations of LFC are reduced by 73%, 77%, and 77%, respectively. The settling time of AVR is reduced by 66% as compared to the conventional controller for a change in load of 20%. The objective function (ITAE) used for the PSO algorithm is same for AVR and LFC; hence the computational time is similar as mentioned in Table 4.17.

The results in the comparison tables (Tables 4.19 and 4.20) show that for a load of 0.1 p.u. and regulation of 10, the settling time of LFC is reduced by a factor of 59.5%, the oscillations are decreased by 75%, the overshoots are reduced by 75%, and the settling time of AVR is reduced by 44.87% as compared to fuzzy controllers. When compared to GA-based controller, the settling time of LFC is reduced by 20.9%, the oscillations are decreased by

TABLE 4.19

Performance Comparison of PSO Based AVR

Methods	Fixed Parameters: $K_A = 10$, $\tau_A = 0.1$, $K_E = 1$, $\tau_E = 0.1$, $K_G = 1$, $\tau_G = 1$, $K_R = 1$, $\tau_6 = 0.05$		
	Settling Time (s)	Overshoot (V)	Oscillations (V)
Conventional PID	37.5	0	0–0.1
Fuzzy controller	16	0	0–0.1
GA-PID	11.38	0	0–0.1
PSO-PID	8.82	0	0–0.1

TABLE 4.20

Performance Comparison of PSO Based LFC

| Methods | Fixed Parameters: $\tau_G = 0.2$, $\tau_T = 0.5$, $K_G = 1$, $H = 5$, $D = 0.8$ | | |
	Settling Time (s)	Overshoot (Hz)	Oscillations (Hz)
Conventional PID	51	−0.0083	0–0.0083
Fuzzy controller	20	−0.0052	0 to −0.0052
GA-PID	10.25	−0.0026	$1.5919e - 006$ to −0.0026
PSO-PID	8.1	−0.0013	0 to −0.0013

50%, the overshoots are reduced by 50%, and the settling time of AVR is reduced by a factor of 22.49%. It is clear from the results that the proposed PSO method can avoid the drawbacks of the premature convergence problem in GA and obtain a highly reliable solution with reduced computational time.

4.6.8 Simulation Model for LFC in a Two-Area Power System

The normal operation of the multi-area interconnected power system requires that each area maintains the load and generation balance. This system experiences deviations in nominal system frequency and schedules power exchanges to other areas with change in load. AGC tries to achieve this balance by maintaining the system frequency and the tie-line flows at their scheduled values. The AGC action is guided by the ACE, which is a function of system frequency and tie-line flows. The ACE represents a mismatch between area load and generation taking into account any interchange agreement with the neighboring areas (Ibraheem et al. 2005, Kothari and Nagrath 2007). Since both areas are connected together, a load perturbation in one area affects the output frequencies of both areas. The controller employed in each area needs the status about the transient behavior of both areas in order to maintain the frequency to optimal value. The tie-line power fluctuations and frequency fluctuations are sensed, and the signal is fed back into both areas (Ertugrul and Kocaarslan 2005, Yesil and Eksin 2004). The primary speed controller employed makes initial course of adjustment, but it is limited by the time lags of the turbine and the system. Hence, an intelligent and efficient secondary controller is required to adjust the system frequency by reducing the error. The model of LFC for a two-area interconnected system is represented in Figure 4.52 with PSO- and ACO-based PID controller. This model depicts the interconnection of two power systems with LFC, and the results are analyzed from the scope that displays the combined output of the frequencies of the two systems.

In order to emphasize the advantages of the proposed controller, the two-area LFC has been implemented and compared with conventional controllers. In multi-area power networks, the active power generation within each area should be controlled to maintain scheduled power interchanges. Control and balance of power flows at tie-line are required for supplementary frequency control. For successful control of frequency and active power generation, the damping of oscillation at tie-line is important. The simulation result is plotted in Figure 4.53 for a change in load of 20% in area 1 and 60% in area 2.

It can be shown from Figure 4.53 that the proposed secondary controller damps the frequency oscillations in both areas by achieving power balance between them and increasing the tie-line power flow. Initial oscillation is due to time delay in governor control, but then the proposed secondary controller starts acting and decreases the

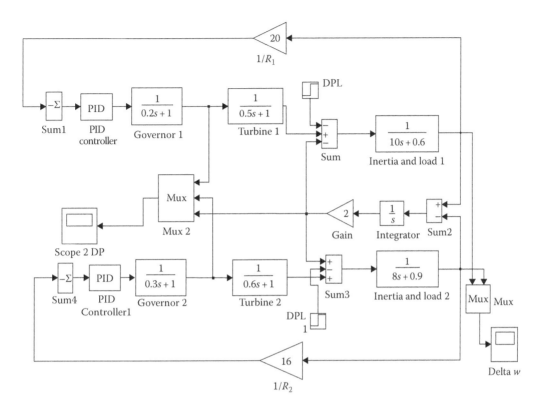

FIGURE 4.52
Simulink® model of LFC for a two-area power system with PID controller.

oscillations. The deviation in frequency is further investigated due to change in load from 20% to 80% in both areas and the results are tabulated in Table 4.21. For comparing the performance of the algorithm, the computational time for different operating conditions is specified in Table 4.21. This approach can be a useful alternative when compared to GA, since the computational time taken for convergence of particles is reduced by 46.5%.

As can be seen from the simulation result, the PSO method has prompt convergence and good evaluation value. The results indicate that the PSO-PID controller is efficient in arresting the frequency oscillation of both areas. The settling time, oscillations, and overshoot are reduced by 76%, 70.8%, and 63.6%, respectively, when compared to conventional PID controller for change in load of 0.2 and 0.6 p.u. In the application of ACO algorithm for a two-area LFC system, the initial population of the colony is randomly generated within the search space. Then the fitness of ants is individually assessed based on their corresponding objective function. In order to examine the dynamic behavior and convergence characteristics of the proposed method, simulation is carried out for the different load and regulation parameters. Figure 4.54 shows the frequency response of the two-area interconnected system for change in load of 20% in area 1 and 40% in area 2.

The low-frequency oscillations, if not damped immediately after a sudden load in a power system, will drive the system to instability. Hence, the secondary controller employed in LFC has to manage efficiently for the increase in load and act dynamically to reduce the

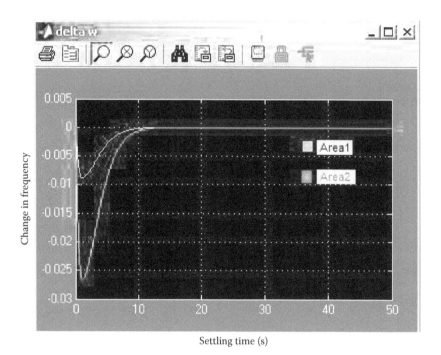

FIGURE 4.53
PSO-PID-based LFC for area 1 loaded by 0.2 p.u. and area 2 loaded by 0.6 p.u.

TABLE 4.21

Performance Analysis of PSO-PID for Two Area LFC

	$R_1 = 20, R_2 = 15$					
	Computational Time = 29.6 s		Computational Time = 30.4 s		Computational Time = 31 s	
	$\Delta P_{L1} = 0.2$	$\Delta P_{L2} = 0.6$	$\Delta P_{L1} = 0.3$	$\Delta P_{l2} = 0.7$	$\Delta P_{L1} = 0.4$	$\Delta P_{l2} = 0.8$
Parameters	Area 1	Area 2	Area 1	Area 2	Area 1	Area 2
Settling time (s)	12.5	13.1	12.8	13.5	13.0	14.0
Overshoot (Hz)	−0.0089	−0.026	−0.008	−0.029	−0.0045	−0.035
Oscillation (Hz)	0–0.0089	0–0.026	0–0.008	0–0.029	0–0.0045	0–0.035

frequency oscillations. Table 4.22 shows the simulation results of a two-area system for loads varying from 0.02 to 0.08 p.u. with R value of 20 and 15. The effectiveness of the algorithm is evaluated by comparing it with conventional PID, and it is found that settling time, oscillations, and overshoot are reduced by 75%, 82.9%, and 61.8%, respectively, for change in load of 0.2 and 0.4 in both areas. The computational efficiency of the proposed ACO-PID controller is found to be improved since the execution time is reduced by 20.8% when compared to GA-PID controller. The computational time taken by the algorithm in generating optimum gain values is given in Table 4.22 for different load and regulation parameters.

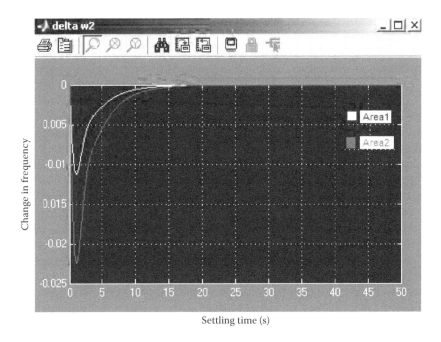

FIGURE 4.54
ACO-PID-based LFC for area 1 loaded by 0.2 p.u. and area 2 loaded by 0.4 p.u.

TABLE 4.22

Performance Analysis of ACO-PID for Two Area LFC

	$R_1 = 20, R_2 = 15$					
	Computational Time = 43.6 s		Computational Time = 44.9 s		Computational Time = 45.2 s	
	$\Delta P_{L1} = 0.2$	$\Delta P_{L2} = 0.4$	$\Delta P_{L1} = 0.3$	$\Delta P_{L2} = 0.6$	$\Delta P_{L1} = 0.2$	$\Delta P_{L2} = 0.8$
Parameters	Area 1	Area 2	Area 1	Area 2	Area 1	Area 2
Settling time (s)	13.1	14.8	13.8	14.1	13.2	15.1
Overshoot (Hz)	−0.011	−0.023	−0.014	−0.028	−0.015	−0.03
Oscillation (Hz)	0–0.011	0–0.023	0–0.014	0–0.028	0–0.015	0–0.03

4.7 Hybrid Evolutionary Algorithms for LFC and AVR

In many engineering disciplines, a large spectrum of optimization problems has grown in size and complexity. In some instances, the solution to complex multidimensional problems by using classical optimization techniques is sometimes difficult and/or computationally expensive. Hence, EAs have been applied successfully to many complex problems in the field of industrial and operational engineering. In power systems, EA is applied to well-known applications, including generation planning, network planning, unit commitment, economic dispatch, load forecasting, power quality, and reliability studies. However, as a consequence of the structural changes in the electric power industry and need for more advanced controllers, the incorporation of hybrid optimization methods in the decision-making process has become inevitable. Moreover, the industry restructuring introduces a wide range of new optimization tasks characterized by their complexity and the amount of variables involved in the optimization process. In some instances, the solution to these multidimensional problems by classical optimization techniques is difficult or even impossible. Hence, to deal with these types of optimization problems, a special class of hybrid algorithms has received increased attention regarding their potential in solving complex problems. In this chapter, hybridization is achieved by combining basic PSO algorithm with fuzzy, GA, and BF for optimal tuning of PID gains. This improves the local and global search ability of PSO and overcomes the premature convergence problem associated with large-scale and complex applications. The hybrid intelligent paradigms exhibit an ability to adapt and learn to new applications or situations under changing environment. This technique puts the adaptively changing terms in original constant terms, so that the parameters of the original PSO algorithm change with convergence rate, which is presented by the objective function. In order to identify the changing dynamics of the power system and to provide appropriate control actions, fast dynamic models are needed. The model of the LFC and AVR of a single-area power system is designed using Simulink in MATLAB. The objective of this work is to design and implement hybrid EA-based PID controller to search the optimal PID gain parameters for efficient control of voltage and frequency. The EAs used in this application are enhanced particle swarm optimization (EPSO), multiobjective particle swarm optimization (MO-PSO), stochastic particle swarm optimization (SPSO), fuzzy particle swarm optimization (FPSO), bacterial foraging particle swarm optimization (BF-PSO), and hybrid genetic algorithm (HGA). The algorithms are designed to generate the optimum proportional, integral, and derivative gains of the controller. These values are sent to workspace and shared with the Simulink model for simulation under various loads and regulation parameters. The proposed LFC and AVR contribute to the satisfactory operation of the power system by maintaining system voltages and frequency for different load and regulation parameters.

4.7.1 Design of EA-Based Controller Using MATLAB®/Simulink®

In a conventional PID controller, the gains are randomly selected by a trial-and-error method. In this application, EA finds the proportional, integral, and derivative gains of the PID controller, and the values are transferred to the PID controller. In these algorithms, the gains of PID controller are searched in the feasible region of response until a determined objective function is minimized. In the design of EA-based controllers, it is desirable that the controlled system includes a suitable transient and steady-state response. Hence, the important characteristics of the system such as overshoot, settling time, and oscillations are improved. Based on the input parameters and on the feedback signal, the proposed EA generates optimal PID gain parameters for efficient control of LFC and AVR, as shown in Figure 4.55.

The Simulink model for LFC and AVR with PID controller is designed based on the transfer function approach with the different values for each parameter. The nominal parameters for AVR model are $K_a = 10$, $K_e = 1$, $K_g = 1$, $K_r = 1$, $T_a = 0.1$, $T_e = 0.4$, $T_g = 1$, and $T_r = 0.05$. The governor and turbine time constants are selected as $T_g = 0.2$ and $T_t = 0.5$, respectively. Since the generator is driven by steam turbine, the inertia constant H and damping constant D are maintained as 10 and 0.8, respectively. This model depicts a plant that encloses LFC loop within it and the PID controller getting a step input and the regulated output is

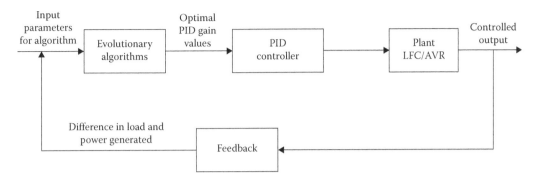

FIGURE 4.55
EA-based PID controller.

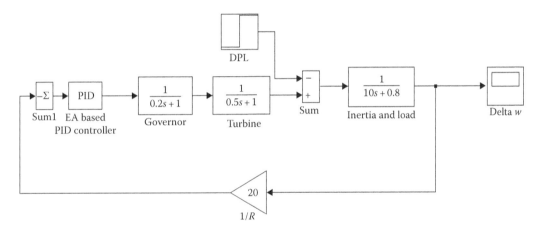

FIGURE 4.56
Simulink® model of load frequency control with EA-based PID controller.

displayed in the scope. The Simulink model of the isolated and interconnected two-area LFC system is shown in Figures 4.56 and 4.57, respectively. The LFC model in Figure 4.56 shows a step input, PID controller based on EA, a governor that controls the speed of the turbine that drives the generator, and the scope that shows the frequency deviation.

The power system normally contains several areas interconnected together by tie-lines. With increased size and enlargement of capacity, interconnections between the power generating systems are the vital solution to meet the demand. This necessitates the need for more advanced and sophisticated control strategies to be incorporated for efficient control. It is very important to keep the system frequency and inter-area tie-line power close to optimum value through appropriate control action. Since the PID parameters are optimized using hybrid EA, the controller action is fast and provides an attempt to increase the efficiency of the power generating system. The Simulink model of the two-area LFC system in Figure 4.57 represents two identical systems with similar machine parameters.

The AVR model in Figure 4.58 consists of a step input, EA-based PID controller, an amplifier that amplifies the signal to the exciter, which in turn controls the voltage of the

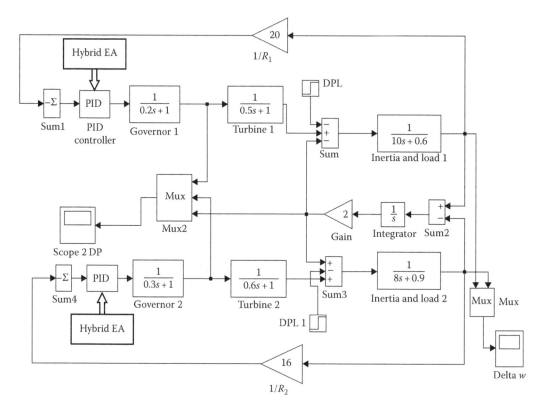

FIGURE 4.57
Two-area LFC with hybrid EA-based PID controller.

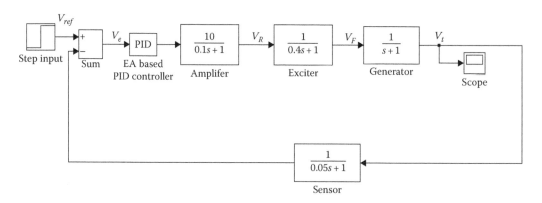

FIGURE 4.58
Simulink® model of automatic voltage regulator with EA-based PID controller.

generator, and a scope to display the terminal voltage. It also contains a sensor that determines the difference between load demand and power generated and feeds it to the controller based on the load changes.

The Simulink model for LFC and AVR with EA-based PID controller is constructed based on the transfer function model of the turbo alternator with non-reheat type turbine.

The K_P, K_I, and K_D values for the PID controller are obtained by running the M-file that calls the fitness function to evaluate the fitness of the solution.

4.7.2 EPSO-Based PID Controller

The EPSO is an improved version of the conventional PSO, being inspired by the study of birds and fish flocking. In EPSO, the constriction factor approach is introduced in the velocity update formula to ensure faster convergence. In EPSO algorithm, each particle in the swarm represents a solution to the problem, and it is defined with its position and velocity. In D-dimensional search space, the position of the ith particle can be represented by a D-dimensional vector, $X_i = (X_{i1}, ..., X_{id}, ..., X_{iD})$. The velocity of the particle V_i can be represented by another D-dimensional vector $V_i = (V_{i1}, ..., V_{id}, ..., V_{iD})$. The best position visited by the ith particle is denoted by $P_i = (P_{i1}, ..., P_{id}, ..., P_{iD})$, and P_g as the index of the particle visited the best position in the swarm; then P_g becomes the best solution found so far. The working of EPSO algorithm is explained in the following steps:

Step 1: The algorithm parameters such as number of generation, population size, inertia weight minimum, maximum (W_{min}, W_{max}), and maximum iterations are initialized.

Step 2: The values of K_P, K_I, and K_D are initialized randomly within the optimal range of values for each gains.

Step 3: The constriction factor (K) is evaluated from the given values of C_1 and C_2 by the Equation 4.46:

$$K = \left| \frac{2}{2 - C - \sqrt{C^2 - 4C}} \right| \tag{4.46}$$

where $C = C_1 + C_2$, C_1 and C_2 are the cognitive and social coefficients of the particles in the search space. The C_1 and C_2 values are selected so that the sum "C" is always greater than 4.

Step 4: The weight of the particle is linearly decreased with each iteration according to the Equation 4.47:

$$W = W_{max} - iter * \frac{W_{min} - W_{max}}{iter_{max}} \tag{4.47}$$

where
$iter_{max}$ is the maximum of iteration in evolution process
W_{max} is the maximum value of inertia weight
W_{min} is the minimum value of inertia weight, $iter$ is the current value of iteration, and the weight W is updated in every iteration

Step 5: The fitness of each particle is evaluated using the integral of time multiplied by the absolute value of error (ITAE) fitness function as in Equation 4.48:

$$F = \int_0^\infty t.e(t)dt \tag{4.48}$$

Step 6: The local best position (P_i) and the global best position (P_g) of particles are found based on the fitness value of the particles calculated from step 5.

Step 7: The velocity and position of the particle are updated using Equations 4.49 and 4.50, respectively:

$$V_{id} = WKV_{id} + C_1R(P_{id} - X_{id}) + C_2r(P_{gd} - X_{id}) \tag{4.49}$$

$$X_{id} = X_{id} + V_{id} \tag{4.50}$$

where R and r are random numbers selected between 0 and 1.

Step 8: Steps 4 through 7 are repeated until the maximum iterations are reached or the best solution is found. The variables used in EPSO algorithm and their definitions are tabulated in Table 4.23.

The fitness function is possibly the most important component of an EA, and its purpose is to map a chromosome representation into a scalar value. Since each chromosome represents a potential solution, the evaluation of the fitness function quantifies the quality of chromosome, that is, how close the solution is to the optimal solution. Selection, crossover, mutation, and elitism operators usually make use of the fitness evaluation of chromosomes (Eberhert and Shi 2000). Also the probability of an individual to be mutated can be a function of its fitness. It is, therefore, extremely important that the fitness function accurately models the optimization problem. The fitness of each particle is evaluated using the ITAE fitness function as in Equation 4.48 (Fang and Chen 2009, Shigenori et al. 2004).

The objective function provides a means for evaluating the performance of the PID controller by determining the gain parameters in the process of search, so that an optimized controller would be developed by the best individual. The proposed EPSO algorithm provides stable and faster convergence toward the global best solution in a minimal computational time.

TABLE 4.23

Variables and their Definitions Used in EPSO Algorithm

Variable	Definition
Iter_max	Maximum number of iteration
X	Position of the particle
X_i	Position of ith particle
V	Velocity of the particle
V_i	Velocity of ith particle
P	Best position of the particle
P_i	Best position previously visited by ith particle
P_g	Best position visited by a particle
K	Constriction Factor
W	Inertia weight
W_{max}	Maximum value of inertia weight
W_{min}	Minimum value of inertia weight
C_1	Cognitive coefficient
C_2	Social coefficient
R and r	Random number between 0 and 1

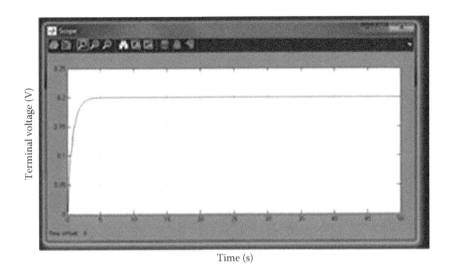

FIGURE 4.59
AVR with EPSO-based PID controller for $\Delta P_L = 0.2$ p.u.

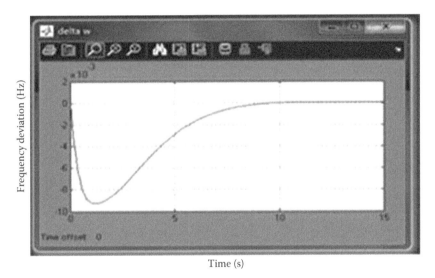

FIGURE 4.60
LFC with EPSO-based PID controller for $R = 20$ and $\Delta P_L = 0.2$ p.u.

The simulation results for AVR and LFC with EPSO-based controllers are presented in Figures 4.59 and 4.60 to validate the efficiency of the proposed algorithm. The algorithm is simulated by keeping the population size and number of generations as 50. The inertia weight minimum is kept at 0.4, and the maximum inertia weight is kept at 0.9. The cognitive and social coefficients are maintained at 2.05 and 2, respectively. The step load change (ΔP_L) of 20% (0.2 p.u.) disturbance is considered for the single-area power system. It can be found that EPSO generates relatively better results with faster convergence rate and higher precision. The computational time taken for generating the optimal PID gains by using this algorithm is 10.2 s. It is observed that the settling time of AVR with EPSO-based PID controller is 4.9 s, and there is no transient peak overshoot. Furthermore, the settling time

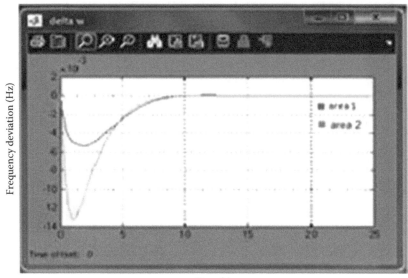

Time (s)

FIGURE 4.61
Two-area LFC with EPSO-based PID controller.

of LFC is 9.5 s, and the peak overshoot is −0.0093, for the same disturbance and at a speed regulation value of 20.

When compared to conventional PID controller, the settling time of AVR is reduced by 88% and no oscillation with overshoot is noticed due to load fluctuations. The results also indicate that the proposed controller could create a very perfect step response of the terminal voltage in the AVR system. The settling time, peak overshoot, and oscillations of LFC are reduced by 83%, 33%, and 33%, respectively. The response of EPSO-based PID controller for the LFC of a two-area system is shown in Figure 4.61. The response is generated for a load of 0.2 p.u. on area 1 and 0.4 p.u. for area 2. From these results, it is observed that when compared to the conventional controller, the settling time, peak overshoot, and oscillations of LFC are reduced by 81.6%, 28.7%, and 28.1%, respectively.

Tables 4.24 and 4.25 show the simulation results for an EPSO-based PID controller under various load changes and regulations for AVR and LFC, respectively. For comparing the performance of the algorithm, the computational time for different operating conditions is also indicated in Tables 4.24 and 4.25.

It is observed from the results in Table 4.24 that the controller provides satisfactory performance for loads varying from 20% to 80%. To demonstrate the superiority of the EPSO over conventional PID controller, the settling time is compared for the different loads as tabulated. When compared to the conventional controller, the settling time is reduced by 88% for change in load of 0.2 p.u. The performance of EPSO algorithm for LFC system is analyzed for regulation values 20 and 75 with a change in load (ΔP_L) of 0.2 and 0.6 p.u. The results reveal that the EPSO algorithm is best suited for PID gain tuning and provide consistent results with increased reliability.

The proposed algorithm provides consistently near-optimum solution for loads varying from 0.2 to 0.6 p.u. It is observed that when compared to the conventional controller, the settling time, peak overshoot, and oscillations of LFC are reduced by 83%, 33%, and 33%, respectively, for regulation R as 75 with 20% increase in load.

TABLE 4.24

Performance Analysis of EPSO Based PID Controller for AVR

Parameter	$\Delta P_L = 0.1$	$\Delta P_L = 0.2$	$\Delta P_L = 0.6$	$\Delta P_L = 0.8$
Computational time (s)	9.84	10.2	10.7	10.9
Settling time (s)	4.5	4.9	5.2	5.5
Overshoot (V)	0	0	0	0
Oscillation (V)	0–0.1	0–0.2	0–0.6	0–0.8

TABLE 4.25

Performance Analysis of EPSO Based PID Controller for LFC

	$R_1 = 20$		$R_2 = 75$	
Parameter	$\Delta P_L = 0.2$	$\Delta P_L = 0.6$	$\Delta P_L = 0.2$	$\Delta P_L = 0.6$
Computational time (s)	10.2	10.7	10.4	10.9
Settling time (s)	9.1	11.6	11.5	12.5
Overshoot (Hz)	−0.0093	−0.028	−0.0066	−0.020
Oscillation (Hz)	0–0.0093	0–0.028	0–0.0066	0–0.020

TABLE 4.26

Performance Analysis of EPSO Based PID Controller for Two Area LFC

	$R_1 = 20, R_2 = 75$			
	Computational Time = 10.9 s		Computational Time = 11.1 s	
Parameter	$\Delta P_L = 0.1$	$\Delta P_L = 0.2$	$\Delta P_L = 0.3$	$\Delta P_L = 0.4$
Settling time (s)	9.5	9.8	10.1	10.3
Overshoot (Hz)	−0.005	−0.013	−0.014	−0.028
Oscillation (Hz)	0–0.005	0–0.013	0–0.014	0–0.028

Table 4.26 shows the simulations' results of EPSO-based PID control for a two-area LFC system for regulation values of 20 and 75. It is observed that when the system frequency is on schedule, each area automatically adjusts to its generation to maintain its net transfers with other areas on schedule, thereby absorbing its own load variations. The peak overshoot, settling time, and oscillation are compared for regulation values of 20 and 75 with increased load of 0.1 p.u. in area 1 and 0.2 p.u. in area 2. When compared to conventional PID controller, the settling time, oscillations, and overshoot are reduced by 83.66%, 50.4%, and 51.2%, respectively.

4.7.3 MO-PSO-Based PID Controller

Real-world problems often have multiple conflicting objectives. In certain problems, no single solution is the best when measured from all objectives. In this method, the acceleration factor (K) is introduced in the velocity update formula and the inertia weight of each particle is made to decrease linearly in all iterations. By adjusting the different objectives, the MO-PSO seeks to discover possible combinations of the available objectives and then the best solutions can be found for the PID controller (Popov et al. 2005). In this algorithm,

TABLE 4.27

Variables and Their Definitions Used in MO-PSO Algorithm

Variable	Definition
iter_max	Maximum number of iteration
X	Position of the particle
X_i	Position of ith particle
V	Velocity of the particle
$V_{j,t+1}^i$	Velocity of ith particle
P	Best position of the particle
$p_{j,t}^i$	Best position previously visited by ith particle
$p_{j,t}^{i,g}$	Best position visited by a particle
α	Acceleration factor
W	Inertia weight
C_1	Cognitive coefficient
C_2	Social coefficient
R_1 and R_2	Random number between 0 and 1

the original PSO operating in continuous space is extended to operate on discrete binary variables. The extended version of PSO has been proven to be very effective for static and dynamic optimization problems. The MO-PSO technique is based on the idea of combining several objective functions that are needed to be satisfied by solution. The variables used in the MO-PSO algorithm and their definitions are tabulated in Table 4.27.

The following steps explain the procedure of MO-PSO algorithm:

Step 1: First, population p_t is initialized, which contains the initial particles, their positions x_t^i, and their initial velocities v_t.

Step 2: The values of K_P, K_I, and K_D are initialized randomly within the optimal range of values for each gains.

Step 3: The acceleration factor (α) is evaluated from Equation 4.51:

$$\alpha = \alpha_0 + \left(\frac{t}{T} \right) \tag{4.51}$$

where
 t denotes the current generation
 T denotes the total number of generation
 α_0 is selected within the range 0.5 and 1

Step 4: After evaluating population p_t, the initial archive A_t is generated with non-dominated solutions in p_t. The weight of the particle is linearly decreased with each iteration according to Equation 4.52:

$$w = w_0 + r * (1 - w_0) \tag{4.52}$$

where
 w_0 is the initial weight
 r is a random number varying between 0 and 1

Step 5: The local guide for each particle is found using the function $findlb(A_{t+1}, x_t^i)$ from the set of nondominated solutions stored in the archive A_t for each particle in the search space.

Step 6: Determine the velocity and position of the ith particle $V_{j,t}^i$ and $x_{j,t}^i$ to direct the swarm toward the optimum solution.

Step 7: The velocity and position of the particle is updated using Equations 4.53 and 4.54, respectively:

$$V_{j,t+1}^i = wV_{j,t}^i + \alpha * (c_1 R_1(p_{j,t}^i - x_{j,t}^i) + c_2 R_2(p_{j,t}^{i,g} - x_{j,t}^i)) \tag{4.53}$$

$$x_{j,t+1}^i = x_{j,t}^i + V_{j,t+1}^i \tag{4.54}$$

The local and global best positions are updated after each iteration based on the fitness values of particles.

Step 8: The fitness value is calculated considering the objective function using the relation in Equation 4.55:

$$E_{val}(k) = \sum_{i=1}^{n} w_i f_i(k) \tag{4.55}$$

Step 9: Steps 2 through 8 are repeated until the maximum iteration is reached or the best solution is attained. Each particle has to change its position x_t i toward the position of a local guide p_t i.g and its best personal position stored in p_t. The particles in the population p_t will be evaluated by the function chosen for the particular application in this case, the objective function. An ideal fitness function correlates closely with the algorithm's goal and yet may be computed quickly. The speed of execution is very important since a typical EA must be iterated many times in order to produce a usable result for a nontrivial problem. The fitness functions for overshoot/undershoot minimization and settling time minimization are given by Equations 4.56 and 4.57, respectively:

$$F_1(K_i K_p K_d) = \max\left(\frac{1}{(1+OU)}\right) \tag{4.56}$$

where OU is the overshoot in single area.

$$F_2(K_i K_p K_d) = \max\left(\frac{1}{(1+T_N)}\right) \tag{4.57}$$

where $T_N = T_{\text{settling time}} / T_{\text{total}}$.

Thus, the fitness of each particle is evaluated, and the particle with minimum fitness for overshoot and settling time is selected as the best. The original PSO sometimes takes time to get into the current effective area in the solution space. On the contrary, MO-PSO moves the evaluated agents to the current effective area directly using the selection method and search is concentrated especially in the current effective area. In this algorithm, the optimal gains of PID controller are searched in the feasible region of response until the determined objective function is minimized. The K_P, K_I, and K_D values for the PID controller are obtained by running the MO-PSO code developed as an M-file in MATLAB R2008b. The optimal parameter values for the population and number of generations are maintained at 50 and 25 for both LFC and AVR. The cognitive and social coefficients are maintained at 2.05 and 3, respectively. The PID gain values are transferred to the AVR and LFC Simulink model for simulating with different load and regulation values. In the proposed hybrid MO-PSO algorithm, the objective functions are collectively minimized by assigning weight for different objective functions. The computational time for the particle convergence to the optimum values of PID gains in MO-PSO is 14.35 s for a change in load of $\Delta P_L = 0.1$. The simulation results for AVR and LFC with MO-PSO-based PID controller under various load changes and regulations are tabulated in Tables 4.28 and 4.29, respectively.

As shown in Table 4.28, when compared to the conventional controller, the settling time, peak overshoot, and oscillations are reduced by 81%, 34%, and 34%, respectively for a change in load of 0.2 p.u. with regulation value of 20. The settling time of AVR for a change in load of $\Delta P_L = 0.1$ is reduced by 87% as compared to the conventional controller (Table 4.29). The results in Table 4.30 are obtained for a two-area model under different regulations $R_1 = 20$ and $R_2 = 75$ for changes in load of ΔP_{L1} and ΔP_{L2} varying from 0.1 to 0.4 p.u.

Here the two control areas are interconnected; hence, the power flow between the areas and tie-line frequency are together responsible for the LFC. The performance of MO-PSO-based PID controller is measured on applying 10% load in area 1 and 20% load in area 2 with regulation values of 20 and 75. Improvement is achieved with respect to reduction in settling time by 86.6%, oscillations by 83.1%, and overshoot by 82.4% when compared to conventional PID controller.

TABLE 4.28

Performance Analysis of MO-PSO Based PID Controller for AVR

Parameter	$\Delta P_L = 0.1$	$\Delta P_L = 0.2$	$\Delta P_L = 0.6$	$\Delta P_L = 0.8$
Computational time (s)	14.35	14.52	14.9	15.4
Settling time (s)	4.8	5.0	5.3	5.7
Overshoot (V)	0	0	0	0
Oscillation (V)	0–0.11	0–0.21	0–0.6	0–0.8

TABLE 4.29

Performance Analysis of MO-PSO Based PID Controller for LFC

Parameter	$R_1 = 20$		$R_2 = 75$	
	$\Delta P_L = 0.2$	$\Delta P_L = 0.6$	$\Delta P_L = 0.2$	$\Delta P_L = 0.6$
Computational time (s)	14.52	14.9	15.52	15.9
Settling time (s)	9.7	11.8	11.8	12.8
Overshoot (Hz)	−0.0091	−0.031	−0.0068	−0.024
Oscillation (Hz)	0–0.0091	0–0.031	0–0.0068	0–0.024

TABLE 4.30

Performance Analysis of MOPSO Based PID Controller for Two Area LFC

| | $R_1 = 20$, $R_2 = 75$ | | | |
| | Computational Time = 14.6 s | | Computational Time = 15.4 s | |
Parameter	$\Delta P_{L1} = 0.1$	$\Delta P_{L2} = 0.2$	$\Delta P_{L1} = 0.3$	$\Delta P_{L2} = 0.4$
Settling time (s)	9.9	10.7	11.1	11.3
Overshoot (Hz)	−0.0045	−0.015	−0.016	−0.029
Oscillation (Hz)	0–0.0045	0–0.015	0–0.016	0–0.029

4.7.4 SPSO-Based PID Controller

In this method, the time-varying acceleration coefficients (TVACs) are introduced for cognitive and social coefficients. The implementation of these TVACs reduces the cognitive component (C_1) meanwhile increasing the social component (C_2) acceleration coefficient with time. Here, the inertia weight and acceleration coefficients are neither set to a constant value nor set as a linearly decreasing time-varying function. Instead, these values are updated nonlinearly in each generation, and so better convergence rate is obtained toward the optimal PID gains in minimal iterations (Reddy and Kumar 2007). The TVACs, that is, cognitive and social coefficients are initialized as in Equations 4.58 and 4.59:

$$c_1 = (c_{i1} - c_{f1}) * \left(\frac{\text{max\_iter} - \text{iter}}{\text{max\_iter}} \right) + c_{f1} \tag{4.58}$$

$$c_2 = (c_{i2} - c_{f2}) * \left(\frac{\text{max\_iter} - \text{iter}}{\text{max\_iter}} \right) + c_{f2} \tag{4.59}$$

where the initial cognitive factor $c_{1i} = 2.05$, the initial social factor $c_{2i} = 2.05$, the final cognitive factor $c_{f1} = 3$, and the final social factor $c_{f2} = 3$.

Now the weight of each particle is updated nonlinearly by using the formula in Equation 4.60:

$$w = (w_{\max} - w_{\min}) * \left(\frac{\text{max\_iter} - \text{iter}}{\text{max\_iter}} \right) + w_{\min} \tag{4.60}$$

where
$w_{\max}$ is the maximum inertia weight
$w_{\min}$ is the minimum value of inertia weight
max_iter is the maximum number of iteration
iter is the current iteration value

In this application, the fitness function or cost function is a minimization function, that is, each particle in the search space should approach the optimal solution available in that space. Let S be the number of particles in the swarm. Each particle will be in its own position, that is, $x_i \in R^n$ in the N-dimensional search space and its velocity $V_i \in R^n$. Let P_i be the current best position of ith particle, and let g be the global best known position in the entire swarm. The steps involved in the SPSO algorithm are as follows:

Step 1: The algorithm parameters such as number of generations, number of dimensions, inertia weight minimum (w_{min}), inertia weight maximum (w_{max}), initial and final values for cognitive and social coefficients, and maximum iterations are initialized.

Step 2: The PID controller gain values K_P, K_I, and K_D are initialized randomly within the optimal range.

Step 3: The TVACs, that is, cognitive (c_1) and social (c_2) coefficients are initialized as in Equations 4.58 and 4.59.

Step 4: The weight of each particle is updated nonlinearly by using Equation 4.60.

Step 5: The fitness function is applied for each particle, and it is evaluated in each iteration for updating the particles toward the best solution in every step.

Step 6: Determine the local best position P_i and the global best position P_g among the particles based on the fitness function.

Step 7: Now the velocity and the position of the particle are updated by using the velocity and position update formula, that is,

$$V_{id} = wV_{id} + c_1R_1(P_{id} - X_{id}) + c_2R_2(P_{gd} - X_{id}) \tag{4.61}$$

$$X_{id} = X_{id} + V_{id} \tag{4.62}$$

Thus, the local and global best positions are updated for each iteration using the fitness function, and the solution moves toward the optimal value in every step.

Step 8: Steps 4 through 7 are repeated until the best solution is obtained or the maximum iteration is reached. The variables used in the SPSO algorithm are tabulated in Table 4.31.

The particles are moved toward the best solution on each iteration based on the fitness function. The fitness function used in this algorithm is the integral time absolute error (ITAE) function given in Equation 4.63, that is, the integral of time is multiplied by the absolute value of error (Moorthy and Arumugam 2009):

$$F = \int_0^\infty t.e(t)dt \tag{4.63}$$

In steady state, the performance criteria must be minimized by the SPSO algorithm. A set of good PID gains can yield a good step response that will result in minimization of performance criteria in time domain. The performance criteria in time domain include overshoot, settling time, and oscillations. The standard PSO algorithm determined by the non-negative real parameter tuple $\{w, c_1, c_2\}$ is analyzed using the stochastic process theory. The K_P, K_I, and K_D values for the PID controller are obtained by running the SPSO source code as an M-file. The optimal parameter values for population size and number

TABLE 4.31

Definitions of the Variables Used in SPSO Algorithm

Variable	Definition
Iter	Value of current iteration
iter_max	Maximum number of iteration
X	Position of the particle
X_{id}	Position of ith particle
V	Velocity of the particle
V_{id}	Velocity of ith particle
P	Best position of the particle
P_i	Best position previously visited by ith particle
G	Global Best position.
W	Inertia weight
W_{min}	Minimum value of inertia weight
W_{max}	Maximum value of inertia weight
C_1	Cognitive coefficient
C_2	Social coefficient
R_1 and R_2	Random number between 0 and 1

of iterations are maintained as 50 and 25, respectively. The inertia weight is linearly varied between 0.35 and 0.4. The minimum and maximum values for C_1 and C_2 are selected between the ranges 2 and 3. To access the computational intelligence of the SPSO, the times taken by the algorithm in generating optimum gain values are tabulated along with simulation results. The simulation results for AVR and LFC with SPSO-based PID controller under various load changes and regulations are tabulated in Tables 4.32 and 4.33, respectively.

TABLE 4.32

Performance Analysis of SPSO Based PID Controller for AVR

Parameters	$\Delta P_L = 0.1$	$\Delta P_L = 0.2$	$\Delta P_L = 0.6$	$\Delta P_L = 0.8$
Computational time (s)	17.2	17.9	18.3	18.8
Settling time (s)	4.7	5.1	5.4	5.7
Overshoot (V)	0	0	0	0
Oscillation (V)	0–0.1	0–0.2	0–0.6	0–0.8

TABLE 4.33

Performance Analysis of SPSO Based PID Controller for LFC

Parameters	$R_1 = 20$ $\Delta P_L = 0.2$	$\Delta P_L = 0.6$	$R_2 = 75$ $\Delta P_L = 0.2$	$\Delta P_L = 0.6$
Computational time (s)	17.9	18.3	18.52	18.94
Settling time (s)	9.9	11.9	11.9	12.8
Overshoot (Hz)	−0.0097	−0.035	−0.0071	−0.027
Oscillation (Hz)	0–0.0097	0–0.035	0–0.0071	0–0.027

TABLE 4.34

Performance Analysis of SPSO Based PID Controller for Two Area LFC

| | $R_1 = 2, R_2 = 75$ | | | |
| | Computational Time = 17.6 s | | Computational Time = 18.2 s | |
Parameters	$\Delta P_{L1} = 0.1$	$\Delta P_{L2} = 0.2$	$\Delta P_{L1} = 0.3$	$\Delta P_{L2} = 0.4$
Settling time (s)	9.8	10.1	11.1	11.3
Overshoot (V)	−0.0053	−0.013	−0.017	−0.027
Oscillation (Hz)	0–0.0053	0–0.013	0–0.017	0–0.027

From Tables 4.32 and 4.33, it is observed that the settling time, peak overshoot, and oscillations of LFC are reduced by 80%, 30%, and 30%, respectively, with 20% increase in load. The settling time of AVR is reduced by 86% when compared to the conventional controller for 10% increase in load. The simulation response of a two-area LFC model obtained for loads changing from $\Delta P_{L1} = 0.1$ p.u. to $\Delta P_{L2} = 0.4$ p.u. and the results are shown in Table 4.34.

As can be witnessed from the results for $\Delta P_{L1} = 0.1$ and $\Delta P_{L2} = 0.2$, the performance indexes of the proposed controller—settling time, oscillations, and overshoot—are compared. The performance parameters are reduced by 83.16%, 85.39%, and 85.1%, respectively, when compared to conventional PID controller.

4.7.5 FPSO-Based PID Controller

Fuzzy particle swarm optimization is a hybrid evolutionary computation-based search algorithm, which can be used to solve optimization problems. In this technique, FL is used to dynamically adapt the inertia weight of the PSO. By linearly decreasing the inertia weight from a relatively large value to a small value through the course of a PSO run, the PSO tends to have more global search ability at the beginning of the run while local search ability at the end of the run (Shi and Eberhert 2001). As a result, a deterministic approach toward the optimal gain value of the PID Controller is obtained. In the proposed FPSO method, the inertia weight factor W is varied according to the mathematical model representing the application. Modify the member velocity of each individual according to the inertia weight factor w, which is obtained from FL. In order to fuzzify the variation of factor W, two fuzzy inputs are used. The first input is called current best performance evaluation (CBFE) and describes the point that has the best performance. In this technique, the normalized form of the current best performance evaluation (NCPBE) is calculated using Equation 4.64. $CBPE_{min}$ is the best acceptable performance of FPSO, and $CBPE_{max}$ is the worst acceptable performance of FPSO. The second fuzzy input is the current value of the inertia weight factor (Shi and Eberhert 2000).

$$NCBPE = \left(\frac{CBPE - CBPE_{min}}{CBPE_{max} - CBPE_{min}} \right) \tag{4.64}$$

To design a fuzzy system to dynamically adapt the inertia weight, normally the inputs to the system are variables that measure the performance of the PSO and the output is the inertia weight or the variation of the inertia weight. The membership functions of normalized CBPE, inertia weight factor w, and variation of inertia weight factor w are depicted in Figures 4.62 through 4.64. In these figures, S, M, and L stand for small, medium, and large, respectively (Banejad and Housman 2001).

The fuzzy quantization is obtained by representing the linguistic states by triangular shape membership function for all linguistic variables, except for S and L, which are represented by trapezoidal membership function. The rule base of the fuzzy system in Table 4.35 contains general knowledge pertaining to the problem domain. In fuzzy expert systems, the knowledge is usually represented by a set of fuzzy production rules, which connect antecedents with consequences or conditions with actions.

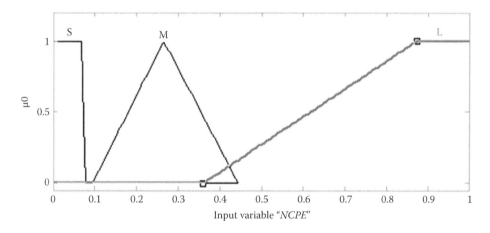

FIGURE 4.62
The membership function of the normalized CBPE.

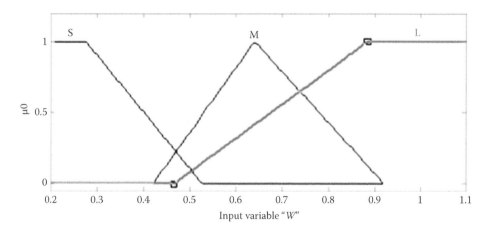

FIGURE 4.63
The membership function of inertia weight factor w.

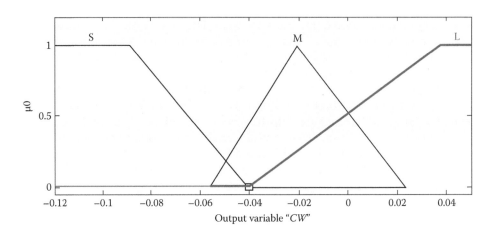

FIGURE 4.64
The membership function of variation of inertia weight factor w.

TABLE 4.35

Rule Base for Fuzzy System

NCBFE	Inertia Weight w		
	Small	Medium	Large
Small	L	S	S
Medium	L	M	S
Large	L	M	S

The input and output variables take their values in their respective universe of discourse or domains. The output obtained from the fuzzy rule base is a variation of factor w. The design steps of FPSO-based PID controller are as follows:

Step 1: Specify the lower and upper bounds of the three controller parameters and initialize randomly the individuals of the population, including searching points, velocities, $pbest^k$, and $gbest^k$.

Step 2: For each initial individual K of the population, employ the Routh–Hurwitz criterion to test the closed-loop system stability.

Step 3: Generate the position of the particle randomly.

Step 4: Calculate the fitness function value of each individual in the population using the evaluation function f given by Equation 4.65:

$$\min_{k\,stabilizing} W(k) = (1 - e^{-\beta})(M_p + E_{ss}) + e^{-\beta}(t_s - t_r) \tag{4.65}$$

where $k = [K_P, K_I, K_D]$.

Step 5: Compare each individual's evaluation value with its $pbest^k$. The best fitness value is denoted by $gbest^k$.

Step 6: Update the velocity of the particle in each individual k with the inertia weight factor W obtained from FL.

$$V_{j,g}^{k+1} = wV_j^k + c_1 rand_1(pbest_{i,g} - x_{j,g}^j) + c_1 rand_2(gbest_{i,g} - x_{j,g}^k)) \tag{4.66}$$

$$x_{j,g}^{k+1} = x_{j,g}^k + V_j^{k+1} \tag{4.67}$$

Step 7: Update the local best P_i and the global best P_g by using the fitness function in Equation 4.65.

Step 8: If the number of iterations reaches the maximum, then go to step 9. Otherwise, go to step 2.

Step 9: The individual that generates the latest $gbest^k$ is an optimal controller parameter. The variables used in the FPSO algorithm are tabulated in Table 4.36.

A fitness function is a particular type of objective function that prescribes the optimality of a solution. The performance criteria in the time domain include the overshoot M_p, rise time t_r, settling time t_s, steady-state error, and vector coefficients of PID controller (Gaing 2004). Therefore, a new performance criterion is shown in Equation 4.65.

Since the performance of PSO is nonlinear, the use of linear equations for inertia weight W is not suitable. Also the algorithm searches the global solution first and then looks for the local solution. Hence, this creates a linear relationship between the local and global searches (Sabahi et al. 2008). In order to improve the optimal performance of PSO and to overcome the problem of trapping in local optimum, FPSO is proposed. The new method of fuzzy tuning improves the convergence speed and provides improved performance characteristics when compared to traditional PSO. The fuzzy adaptive inertia weight is

TABLE 4.36

Definitions of the Variables Used in FPSO Algorithm

Variable	Definition
itermax	Maximum number of iteration
X	Position of the particle
X_i	Position of ith particle
V	Velocity of the particle
V_i	Velocity of ith particle
P	Best position of the particle
P_i	Best position previously visited by ith particle
P_g	Best position visited by a particle
W	Inertia weight
C_1	Cognitive coefficient
C_2	Social coefficient
R_1 and R_2	Random number between 0 and 1
NCBPE	Normalized form of current best performance evaluation
M_p	Overshoot
E_{ss}	Steady state error
t_s	Settling time
t_r	Rise time

used to improve the performance of PSO. The simulation results prove the validity of the algorithm for its application in voltage and frequency control. The population size and number of generations are maintained as 50 for LFC and AVR. The minimum current best performance ($CBPE_{min}$) and the maximum current best performance ($CBPE_{max}$) values are assumed as 0.1 and 1, respectively. The cognitive and social coefficients are selected as 1.2 and 1, respectively. The proposed FPSO algorithm for the autotuning of PID gain generates the optimum values of K_P, K_I, and K_D. The LFC and AVR models with optimum gain values are simulated for different regulations and loads. The time taken for the computation of the PID gains using this algorithm is 10.2 s for a change in load of 0.1 p.u. The computational burden is reduced on the application of this algorithm, and it is true for different operating conditions of the plant, as shown in Tables 4.37 and 4.38.

For a change in load of 0.1 p.u., it is inferred that when compared to the conventional controller, the settling time of AVR is reduced by 87.7%, without any overshoot and initial oscillations due to load fluctuations.

The simulation result for $\Delta P_L = 0.2$ with regulation value 20, the settling time, peak overshoot, and oscillations of LFC are reduced by 78.87%, 39.47%, and 39.47% when compared to the conventional controller. When the frequency is constant, it is assumed that the generator automatically adjusts its output according to its own load and net transfers to other areas on schedule. The result also indicates that the change in load parameters does not affect the quality of the suggested FPSO algorithm. Furthermore, the settling time, peak overshoot, and oscillations of LFC are reduced by 78.87%, 39.47%, and 39.47%, respectively. The Simulink model for the two-area power system with FPSO-based PID controllers simulated for a change in load of 0.1, 0.2, 0.3, and 0.8 for both area 1 and area 2. Table 4.39 shows the simulation results for FPSO-based PID control for a two-area LFC system.

Improvement is achieved in terms of reduction in 82.5%, 69.2%, and 68.6%, with respect to settling time, oscillations, and overshoot for changes in load of 0.1 and 0.2 p.u. It is evident from Table 4.37 that the FPSO algorithm needs less time in the range of 10.2–11.7 s for computation and achieves better performance.

TABLE 4.37

Performance Analysis of FPSO Based PID Controller for AVR

Parameter	$\Delta P_L = 0.1$	$\Delta P_L = 0.2$	$\Delta P_L = 0.8$
Computational time (s)	10.2	10.73	11.2
Settling time (s)	4.5	4.7	5.3
Overshoot (Hz)	0	0	0
Oscillation (Hz)	0–0.1	0–0.2	0–0.8

TABLE 4.38

Performance Analysis of FPSO Based PID Controller for LFC

Parameter	$R_1 = 20$		$R_2 = 75$	
	$\Delta P_L = 0.2$	$\Delta P_L = 0.8$	$\Delta P_L = 0.2$	$\Delta P_L = 0.8$
Computational time (s)	10.73	10.9	11.52	11.94
Settling time (s)	9.4	10.1	11.3	12.4
Overshoot (Hz)	−0.0089	−0.037	−0.0079	−0.0312
Oscillation (Hz)	0 to −0.0089	0 to −0.037	0 to −0.0089	0 to −0.0312

TABLE 4.39

Performance Analysis of FPSO Based PID Controller for Two Area LFC

	$R_1 = 20, R_2 = 75$			
	Computational Time = 10.9 s		Computational Time = 11.6 s	
Parameter	$\Delta P_{L1} = 0.1$	$\Delta P_{L2} = 0.22$	$\Delta P_{L1} = 0.3$	$\Delta P_{L2} = 0.8$
Settling time (s)	10.2	10.5	10.9	11.7
Overshoot (Hz)	−0.013	−0.027	−0.026	−0.037
Oscillation (Hz)	0–0.015	0–0.028	0–0.014	0–0.037

4.7.6 BF-PSO-Based PID Controller

The BF-PSO is based on the idea of combining bacterial foraging oriented PSO. The selection behavior of bacteria tends to eliminate poor foraging strategies and improve successful foraging strategies. Foraging theory is based on the assumption of animals searching for and obtaining nutrients to maximize their energy intake E per unit time T spent on foraging. A group of bacteria moving in search of food and away from noxious elements is known as foraging (Passino 2002). The optimal foraging theory formulates the foraging problem and via computational or analytical methods that specify formulations for dynamic programming. In the proposed BF-PSO approach, after undergoing a chemotactic step, each bacterium is mutated by a PSO operator (Biswas et al. 2000). In this phase, the bacterium is stochastically attracted toward the globally best position found in the entire population at the current time and also toward its previous heading direction. In order to accelerate the convergence speed of the bacterial colony near the global optima, the PSO algorithm has been combined to BFO, which resulted in a significant improvement in the performance of the traditional PSO algorithm in terms of convergence speed, accuracy, and robustness. The PSO model consists of a swarm of particles, which are initialized with a population of random candidate solutions. Each particle has a position represented by a position vector X_{ik} (where i is the particle representing the index of bacteria trying to move toward the food concentration gradient individually to optimize the PID controller gains) and a velocity represented by a velocity vector V_{ik}. Each particle remembers its own best position P_{iLbest}. The best position vector among the swarm is then stored in a vector $P_{iglobal}$ (Kim and Cho 2005). During the iteration time k, the new velocity is updated based on the previous velocity as mentioned in Equation 4.68:

$$V_{id} = wV_{id} + c_1R_1(P_{id} - X_{id}) + c_2R_2(P_{gd} - X_{id})$$ (4.68)

$$X_{id} = X_{id} + V_{id}$$

The new position is then determined by the sum of the previous position and the new velocity, as shown in Equation 4.68. The movement of the particle is decided by the memory of its best past position, and the experience of the most successful particle in the swarm.

In this algorithm, N denotes the number of bacteria, k denotes the number of reproduction loops, and *ell* denotes the number of elimination dispersal.

Step 1: First initialize the parameters such as number of bacteria S, maximum number of swim length N_s, chemotactic steps N_c, the number of reproduction steps N_{re}, elimination and dispersal events N_{ed} and P_{ed}, cognitive coefficient c_1 and social coefficient c_2, and random numbers R_1 and R_2.

Step 2: The values of K_P, K_I, and K_D are initialized randomly within the optimal range of values for each gains.

Step 3: Generate the random direction (n, i) and position $P(i, j)$, dimension of search space n, elimination and dispersal limit l, and k is the reproduction.

Step 4: l is incremented by one for every cycle until it reaches the elimination and dispersal limit $l = l + 1$.

Step 5: k is incremented by one for every cycle until it reaches the reproduction limit.

Step 6: Chemotaxis loop: $j = j + 1$ for $i = 1, 2, \ldots, S$, compute the fitness of $J(i, j, k, l)$,

$$J(i,j,k,l) = Fitness(P(i,j,k,l)) \tag{4.69}$$

Store the best fitness function in J_{last} as in Equation 4.70 and the best fitness function for each bacteria will be selected to be the local best J_{local} as in Equation 4.71:

$$J_{last}(i,j,k,l) = J_{last}(i,j,k,l) \tag{4.70}$$

$$J_{local}(i,j,k,l) = J_{last}(i,j,k,l) \tag{4.71}$$

Update the position and fitness function of the bacteria and allow the bacteria to swim in the right direction and store the bacteria into J_{last} using Equation 4.72:

$$P(i,j+1) = p(i,j) + c(i) * \Delta(n,i)$$

$$J_{last} = (i,j+1, k,l) \tag{4.72}$$

Store the J_{last} and update the position of the bacteria using fitness function:

$$P(i,j+1,k,l) = p(i,j,k,l) + c(i) * \Delta(n,i) \tag{4.73}$$

Step 7: Evaluate the local best position $P_{l(best)}$ in Equation 4.74 and the global best position $P_{g(best)}$ in Equation 4.75 for each bacterium:

$$P_{current}(i,j+1) = p(i,j+1) \tag{4.74}$$

$$J_{local}(i,j+1) j_{last}(i,j+1) \tag{4.75}$$

Step 8: Update the position and velocity of the dth coordinate of the ith bacterium according to Equations 4.76 and 4.77:

$$V_{id} = wV_{id} + c_1 R_i(P_{lbest} - P_{current}) + c_2 R_2(P_{gbest} - P_{current}) \tag{4.76}$$

$$d^{new}(i,j+1,k) = \theta_d^{old}(I,j+1,k) + V_{id} \tag{4.77}$$

Step 9: In reproduction, for the given k and l and for each i = 1, 2, ..., N, the health of the bacterium i is obtained as in Equation 4.78:

$$J_{health} = sum\left(J\left(i,j,k,l\right)\right) \tag{4.78}$$

Sort bacteria and chemotactic parameters $C(i)$ in the order of ascending cost J_{health} (higher cost means lower health).

Substep: The Sr bacteria with the highest J_{health} values die and the remaining Sr bacteria with the best values split (this process is performed by the copies that are made are placed at the same location as their parent) into i and $i + Sr$ as given by Equation 4.79:

$$P\left(i+Sr,j,k+1,l\right) = p\left(i,j,k+1,l\right) \tag{4.79}$$

Step 10: If $k < N_{re}$, go to step 5. (In case if the specified number of reproduction is not reached, start the next generation in chemotaxis until the best solution is obtained.)

Step 11 Elimination–dispersal: For i = 1, 2, ..., N, with probability P_{ed}, eliminate and disperse each bacterium, which results in keeping the number of bacteria in the population constant. To do this, if a bacterium is eliminated, simply disperse one to a random location on the optimization domain. If $l < N_{ed}$, then go to step 4; otherwise end. The variables used in the BF-PSO and their definition are tabulated in Table 4.40.

TABLE 4.40

Variables and Their Definitions Used in BF-PSO Algorithm

Variable	Definition
itermax	Maximum number of iteration
X	Position of the particle
X_i	Position of ith particle
V	Velocity of the particle
V_i	Velocity of ith particle
P	Best position of the particle
P_i	Best position previously visited by ith particle
P_g	Best position visited by a particle
W	Inertia weight
C_1	Cognitive coefficient
C_2	Social coefficient
R_1 and R_2	Random number between 0 and 1
p	Dimension of search space
N	Number of bacteria in the population
N_c	Number of chemotaxi tactic steps
N_s	Number of swimming steps
N_{re}	Number of reproduction steps
N_{ed}	Number of elimination–dispersal steps
P_{ed}	Elimination-dispersal with probability
m	Counter swim length
$c(i)$	Direction of the tumble for bacterium i.

The fitness function for overshoot/undershoot minimization and settling time minimization is given by Equation 4.80:

$$F = (O_{sh} * 10,000) + t_{sh}^2 + \frac{0.001}{(\max dv)^2} \qquad (4.80)$$

where
O_{sh} is overshoot
t_{sh}^2 is settling time
max dv is maximum deviation

Minimization of F corresponds to a minimum overshoot (O_{sh}), minimum settling time (t_{sh}), and maximum deviation (max dv). Hence, a set of good control parameters K_P, K_I, and K_D is generated and yields a good response that will result in performance criteria minimization in the time domain (Korani et al. 2008). The optimization problem search space could be modeled as a social foraging environment where groups of parameters communicate cooperatively for finding solutions to difficult engineering problems. BF-PSO algorithm generates the optimum values of K_P, K_I, and K_D for the PID controller by running the M-file. In BF-PSO-based PID controller, the time taken for the computation of PID gains varies from 15.3 to 17.24 s depending on the load and regulation parameters. The optimized PID gain values of BF-PSO algorithm are obtained and applied in the Simulink model. This model depicts a plant that encloses an AVR and LFC loop within it and the PID controller getting a step input, and the regulated output is seen from the scope. If the low-frequency oscillations after a load disturbance in a power system are not noticed, the system is driven to instability. Hence, the controller should act immediately on improving the performance of the LFC/AVR and restore the system to stable state. The computational time taken by the algorithm is tabulated for comparing the performance of hybrid algorithms. The simulation results for AVR with BF-PSO-based PID controller under various loads are tabulated in Table 4.41.

As evidenced from Table 4.41, the settling time of AVR for $\Delta P_L = 0.1$ is reduced by 87% as compared to the conventional controller. For a single-area LFC system, performance analysis is made for change in load of 20% with regulation value as 20. As shown in Table 4.42, when compared to the conventional controller, it is observed that the settling time, peak overshoot, and oscillations of LFC are reduced by 77%, 39%, and 39%, respectively.

The two-area model is simulated for different regulations $R_1 = 20$ and $R_2 = 75$ for change in load of 0.1, 0.2, 0.3, and 0.8 p.u. The simulation results indicate that the BF-PSO-based PID controller improves searching capability and convergence characteristics.

TABLE 4.41

Performance Analysis of BF-PSO Based PID Controller for AVR

Parameter	$\Delta P_L = 0.1$	$\Delta P_L = 0.2$	$\Delta P_L = 0.8$
Computational time (s)	15.3	15.9	16.52
Settling time (s)	4.5	4.8	5.8
Overshoot (V)	0	0	0
Oscillation (V)	0–0.1	0–0.2	0–0.8

TABLE 4.42

Performance Analysis of BF-PSO Based PID Controller for LFC

Parameter	$R_1 = 20$		$R_2 = 75$	
	$\Delta P_L = 0.2$	$\Delta P_L = 0.8$	$\Delta P_L = 0.2$	$\Delta P_L = 0.8$
Computational time (s)	15.73	15.9	16.32	17.24
Settling time (s)	9.8	11.9	10.9	11.4
Overshoot (Hz)	−0.0092	−0.036	−0.0071	−0.035
Oscillation (Hz)	0–0.0092	0–0.036	0–0.0071	0–0.035

TABLE 4.43

Performance Analysis of BF-PSO Based PID Controller for Two Area LFC

Parameter	$R_1 = 20, R_2 = 75$			
	Computational Time = 15.56 s		Computational Time = 16.3 s	
	$\Delta P_{L1} = 0.1$	$\Delta P_{L2} = 0.2$	$\Delta P_{L1} = 0.3$	$\Delta P_{L2} = 0.8$
Settling time (s)	9.6	9.8	11.2	11.9
Overshoot (Hz)	−0.0045.	−0.015	−0.015	−0.023
Oscillation (Hz)	0–0.0051	0–0.0012	0–0.015	0–0.023

Table 4.43 shows the simulation results of BF-PSO-based PID control for a two-area LFC system. It is found that area 1 settles at 9.6 s and area 2 at 9.8 s for changes in load of 10% and 20% at their respective areas. The numerical analysis shows that the settling time, overshoot, and oscillations are reduced by 83%, 83.19%, and 82.8%, respectively, when compared to conventional PID controller. It is observed that the PID controller with proposed BF-PSO-based gain parameters has excellent optimization performance and improved convergence characteristics when compared to traditional PSO.

4.7.7 Hybrid Genetic Algorithm–Based PID Controller

Recent surveys of GA relating to improvements in the search process with respect to control system engineering problems can be found in Michalewicz (1999), Tsutsui and Goldberg (2002), and Yoshida et al. (2000). Intermixing the salient features of the algorithms may be found to be more effective in specific application areas such as power system operation and control. Recently, hybridization of the EA is getting popular due to their capabilities in handling several real-world problems involving complexity, noisy environment, imprecision, uncertainty, and vagueness (Grosan and Abraham 2007). Even though EA belongs to the same basic skeleton called evolution strategies, it is differentiated by the way the hybridization of the algorithm is developed. These optimization models could provide a social foraging environment where groups of parameters communicate cooperatively for finding solutions to engineering problems (Kim and Cho 2006). HGA is most suitable for parameter optimization problem, particularly when the control structure is provided. A typical task of Hybrid-GA is to find the best values of a predefined set of control parameters associated with process model. The major advantages of HGA are capability enhancement, improved quality, and efficiency in reaching the global solution.

The steps involved in HGA are as follows:

Step 1: Initialize the parameters such as number of bacteria in the population S, maximum number of swim length N_s, chemotactic steps N_c, the number of reproduction steps N_{re}, elimination–dispersal events N_{ed}, and elimination dispersal with probability P_{ed}.

Step 2: Evaluate all the chromosomes according to the fitness function. The fitness function exactly fits the problem of solution space and chromosome. The best fitness value is kept as the best solution set.

Step 3: In the chemotactic process, the bacteria climb the nutrient concentration, avoid noxious substances, and search for a way out of neutral media. The bacterium usually takes a tumble followed by a run. For N_c the direction of movement after a tumble is given by Equation 4.81:

$$\theta(i, j+1, k) = \theta(i, j+1, k) + C(i)\frac{\Delta(i)}{\sqrt{\Delta^T(i)\Delta(i)}} \tag{4.81}$$

If the cost $\theta(i, j + 1, k)$ is better, the bacterium takes another step of size $C(i)$ in that direction. This process will continue until the number of steps taken is not greater than N_s.

Step 4: After evaluating chemotaxis, the bacteria climb the swarming. Bacteria, in times of stress, release attractants to signal bacteria to swarm together. They, however, also release repellants to signal others to be at a minimum distance from one another. Thus, all of them will have a cell-to-cell attraction via attractants and cell-to-cell repulsion via repellants, as in Equation 4.81.

Step 5: In the selection process, a part of the chromosomes will be eliminated and the remaining chromosomes will be selected.

Step 6: The crossover operator produces two offspring (new candidate solutions) by recombining the information from two parents, as given in Equations 4.82 and 4.83:

$$\tilde{x}_j^u = \lambda_{\tilde{x}_j^v} + (1-\lambda)\bar{x}_j^u \tag{4.82}$$

$$\tilde{x}_j^v = \lambda_{\tilde{x}_j^v} + (1-\lambda)\bar{x}_j^v \tag{4.83}$$

Step 7: In mutation rate, several chromosomes will be selected from the population and then randomly change the value of parts of the dimensions. This will give the population a larger chance to generate new species. For optimization, it is the chance to get an abrupt evolution, as shown in Equation 4.84.

$$x_j = \begin{cases} \tilde{x}_j + \Delta(k, x_j^{(U)} - \tilde{x}_j), \tau = 0 \\ \tilde{x}_j - \Delta(k, \tilde{x}_j - x_j^{(L)}), \tau = 0 \end{cases} \tag{4.84}$$

Step 8: After the mutation steps have been covered, a reproduction step takes place. The fitness (accumulated cost) of the bacteria is sorted in ascending order. *Sr* (*Sr* = *S*/2) bacteria having higher fitness die, and the remaining *Sr* are allowed to split into two, thus keeping the population size constant, which is given by $ITSE^i_{health}$ in Equation 4.85:

$$ISTE^i_{health} = \sum_{j=1}^{N_i+1} ISTE(i,j,k,l) \tag{4.85}$$

Let i = 1, 2, ..., N be the health of the bacterium i. The *Sr* bacteria with the highest $ITSE_{health}$ values die, and the remaining *Sr* bacteria with the best values will split.

Step 9: Repeat steps 2 through 6 until the maximum iteration is reached; else go to step 3 and start the next generation of the chemotactic loop.

Step 10 Elimination–dispersal: For i = 1, 2, ..., N, with probability P_{ed}, eliminate and disperse each bacterium, which results in keeping the number of bacteria in the population constant. When an individual is eliminated, the course does not generate another one via the initialization process, but it generates a new individual via mutating all the dimensions from the eliminated one. If $1 < N_{ed}$, then go to step 3; otherwise, end. The variables used in the HGA and their definitions are given in Table 4.44.

The fitness function for HGA is defined in Equation 4.86. The objective function measures the performance of the system. The fitness function for HGA is defined as the integral of time multiplied by the absolute value of error (ITAE) of the corresponding system. Therefore, it becomes an unconstrained optimization problem to find a set of decision

TABLE 4.44

Variables and their Definitions Used in HGA Algorithm

Variable	Definition
p	Dimension of search space
N	Number of bacteria in the population
N_c	Number of chemotaxi tactic steps
N_s	Number of swimming steps
N_{re}	Number of reproduction steps
N_{ed}	Number of elimination -dispersal steps
P_{ed}	Elimination-dispersal with probability
$d_{attractant}$	Depth of the attractant
$w_{attractant}$	Width of the attractant
$L_{attractant}$	Depth of the attractant
$\widetilde{X}^u_j, \widetilde{X}^v_j$	Offspring's generation
$\overline{X}^u_j, \overline{X}^v_j$	Parent's generations
$\Delta(i)$	Random vector generated
λ	Multiplier
j	jth step of chromosome
$P(\phi)$	Position of the bacteria

variables by minimizing the objective function. The fitness of each particle is evaluated using the ITAE fitness function as in Equation 4.86:

$$\min F(K_p, K_i, K_d) = \frac{e - \beta\big((ts/\max(t)) + \alpha M_0\big)}{\alpha} + ess \tag{4.86}$$

where

$$\alpha = 1 - e - \beta^* \,|\, 1 - tr/\max(t)\,| \tag{4.87}$$

Also K_P, K_I, and K_D are the optimal gains of the PID controller, β is the weighting factor, M_0 is the overshoot, ts is settling time, and ess is the steady-state error.

If the weighting factor α increases, the rising time of the response curve is small, and when β decreases, the rising time also decreases (Kim and Park 2005). The advanced HGA for PID gain tuning has been better searching speed than the original GA.

The performance of the proposed HGA algorithm is evaluated by simulating with LFC and AVR models. The time taken by HGA algorithm in M-file to generate the optimum values of PID gains is 19.4 s. The models are simulated for different regulations and loads. The frequency deviation and terminal voltage response for a change in load of 0.2 p.u. and regulation of 20 are shown in Figures 4.65 and 4.66, respectively. The searching speed of the HGA is superior to conventional GA as evidenced from the analysis of the time taken by the algorithm for convergence. The computational times for different operating conditions of LFC and AVR are measured and tabulated for performance analysis. From Figure 4.65, it is found that the settling time of AVR with HGA-based integral controller is 4.1 s, and there is a no transient overshoot. It is observed from Figure 4.66 that the frequency deviation and the peak overshoot are minimum. The settling time for frequency deviation is 8.4 s, and the oscillation varies between −0.0094 and 0, which is very less as compared to conventional PID controller.

It is observed that the settling time, peak overshoot, and oscillations of LFC are reduced by 81.12%, 40.7%, and 40.7%, respectively. The settling time of AVR is reduced by 88.81% when compared to the conventional controller. The simulation results for AVR and LFC

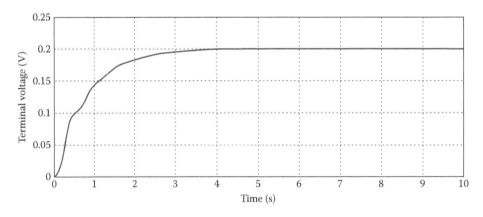

FIGURE 4.65
AVR with HGA-based PID controller for $\Delta P_L = 0.2$ p.u.

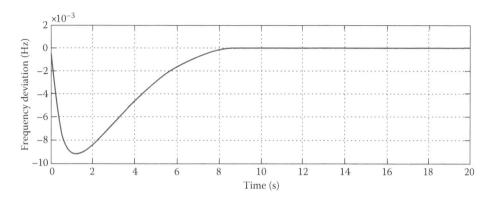

FIGURE 4.66
LFC with HGA-based PID controller for $R = 20$ and $\Delta P_L = 0.2$ p.u.

with HGA-based PID controller under various load changes and regulations are tabulated in Tables 4.45 and 4.46, respectively.

Tables 4.45 and 4.46 indicate the performance of the proposed algorithm for the loads varying from 0.1 to 0.8 p.u. It is observed that the proposed hybrid algorithm has more excellent optimization performance and searching speed even under different operating conditions. The results prove that the hybrid intelligent paradigms exhibit an ability to adapt itself for changing loads and regulations. The simulation response for HGA-based PID controller for the LFC of a two-area system is shown in Figure 4.67.

It is inferred that the settling times are 9.4 and 9.6 s for changes in load of 0.2 p.u. in area 1 and 0.2 p.u. in area 2, respectively. To test the durability of HGA-based PID controller, the model is tested for various load changes, and the results are depicted in Table 4.47.

When compared to conventional PID, the proposed algorithm achieves better results in terms of 84% reduction in settling time, 84.26% reduction in oscillation, and 84.1%

TABLE 4.45

Performance Analysis of HGA Based PID Controller for AVR

Parameters	$\Delta P_L = 0.1$	$\Delta P_L = 0.2$	$\Delta P_L = 0.5$	$\Delta P_L = 0.8$
Computational time (s)	19.4	19.53	19.92	20.24
Settling time (s)	4	4.1	5.4	6.2
Overshoot (V)	0	0	0	0
Oscillation (V)	0–0.1	0–0.2	0–0.5	0–0.8

TABLE 4.46

Performance Analysis of HGA Based PID Controller for LFC

| Parameters | $R_1 = 20$ | | $R_2 = 75$ | |
	$\Delta P_L = 0.2$	$\Delta P_L = 0.8$	$\Delta P_L = 0.2$	$\Delta P_L = 0.8$
Computational time (s)	19.23	19.78	20.12	20.86
Settling time (s)	8.4	8.9	10.8	11.5
Overshoot (Hz)	−0.0089	−0.035	−0.0071	−0.027
Oscillation (Hz)	0–0.0089	0–0.035	0–0.0071	0–0.027

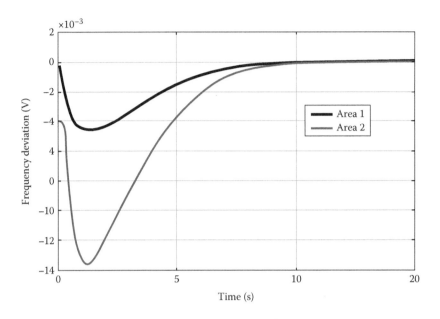

FIGURE 4.67
Two-area LFC with HGA-based PID controller.

TABLE 4.47

Performance Analysis of HGA Based PID Controller for Two Area LFC

	$R_1 = 20, R_2 = 75$			
	Computational Time = 19.4 s		Computational Time = 20.9 s	
Parameters	$\Delta P_{L1} = 0.1$	$\Delta P_{L2} = 0.2$	$\Delta P_{L1} = 0.3$	$\Delta P_{L2} = 0.8$
Settling time (s)	9.4	96	10.3	11.1
Overshoot (Hz)	−0.0047	−0.0138	−0.013	−0.024
Oscillation (Hz)	0–0.0047	0–0.0138	0–0.013	0–0.024

reduction in overshooting for changes in load of 0.1 and 0.2 p.u. in area 1 and area 2, respectively. These algorithms also establish the superiority of optimization and improved convergence rate as compared to the conventional controllers.

4.7.8 Performance Comparison of Single-Area System

The performance of the proposed EA-based controllers developed for AVR of the power generating system for various load changes is given in Tables 4.48 and 4.49 to indicate the efficiency of PSO algorithm for real-time applications and its suitability under varying load conditions. The computational time required for the algorithm to compute the values of PID gains is tabulated for performance comparison. The EA provides better convergence and reveals their superiority with respect to settling time, oscillations, and overshoot when compared to the standard PSO method.

TABLE 4.48

Performance Analysis of EA Based AVR

Change in Load (ΔP_L)	Fixed Parameters: $K_A = 10$, $\tau_A = 0.1$, $K_E = 1$, $\tau_E = 0.4$, $K_G = 1$, $\tau_G = 1$, $K_R = 1$, $\tau_R = 0.05$ Settling Time in Seconds						
	PSO	EPSO	MO-PSO	SPSO	FPSO	BF-PSO	HGA
0.1	9.03	4.5	4.8	4.7	4.5	4.5	4
0.2	11.2	4.9	5.0	5.1	4.7	4.8	4.1
0.6	12.9	5.2	5.3	5.4	5.0	5.5	5.6
0.8	14.6	5.5	5.7	5.7	5.3	5.8	6.2
Computational time (s)	26.8	10.2	14.52	17.9	10.73	15.9	19.53

With respect to the simulation results in Table 4.48, clearly the proposed hybrid EA-based PID controllers for AVR achieve improved dynamic performance with reduced computational time. When compared to PSO, the computational times of EPSO, MO-PSO, and SPSO algorithms are reduced by 61.94%, 46.26%, and 30.97%, respectively. For a change in load of 0.1 p.u, the settling times of AVR are reduced by 50.16%, 46.84%, and 47.95% when compared to conventional PSO. Similarly, the computational burdens of FPSO, BF-PSO, and HGA algorithms are reduced by 61.19%, 44%, and 28.35% with respect to basic PSO algorithm. The settling times of voltage are reduced by 57.14%, 58%, and 63.39% on applying hybrid algorithms in place of traditional PSO for ΔP_L value of 0.1 p.u. Tables 4.49 and 4.50 demonstrate the effectiveness of hybrid algorithms for the application of the single-area

TABLE 4.49

Performance Comparison of EA Based LFC for *R* Value of 20

Parameter	Fixed Parameters: $\tau_G = 1$, $\tau_T = 0.5$, $K_G = 1$, $H = 10$, $D = 0.8$, $R = 20$							
	$\Delta P_L = 0.2$				$\Delta P_L = 0.6$			
	PSO	EPSO	MO-PSO	SPSO	PSO	EPSO	MO-PSO	SPSO
Computational time (s)	28.4	10.2	14.52	17.9	29	10.7	14.9	18.3
Settling time (s)	10.4	9.1	9.7	99	13.4	11.6	11.8	11.9
Overshoot (Hz)	−0 0102	−0.0093	−0.0091	−0.0097	−0.042	−0.028	−0.031	−0.035
Oscillation (Hz)	0–0.0102	0–0.0093	0–0.0091	0–0.0097	0–0.042	0–0.028	0–0.031	0–0.035

TABLE 4.50

Performance Comparison of EA Based LFC for *R* Value of 75

Parameter	Fixed Parameters: $\tau_G = 1$, $\tau_T = 0.5$, $K_G = 1$, $H = 10$, $D = 0.8$, $R = 75$							
	$\Delta P_L = 0.2$				$\Delta P_L = 0.6$			
	PSO	EPSO	MO-PSO	SPSO	PSO	EPSO	MO-PSO	SPSO
Computational time (s)	28.9	10.4	15.52	18.52	29.8	10.9	15.9	18.94
Settling time (s)	12.2	11.5	11.8	11.9	13.4	12.5	12.8	12.8
Overshoot (Hz)	−0.0102	−0.0066	−0.0068	−0.0071	−0.042	−0.020	−0.068	−0.027
Oscillation (Hz)	0–0.0102	0–0.0066	0–0.0068	0–0.0071	0–0.042	0–0.020	0–0.068	0–0.027

TABLE 4.51

Performance Analysis of Hybrid EA for LFC

| | Fixed Parameters: $\tau_G = 1$, $\tau_T = 0.5$, $K_G = 1$, $H = 10$, $D = 0.8$, $R = 20$ | | | | | | | |
| | $\Delta P_L = 0.2$ | | | | $\Delta P_L = 0.8$ | | | |
Parameter	PSO	FPSO	BF-PSO	HGA	PSO	FPSO	BF-PSO	HGA
Computational time (s)	28.4	10.73	15.73	19.23	29.8	10.9	15.9	19.78
Settling time (s)	10.4	9.4	9.8	8.4	13.4	10.1	11.9	8.9
Overshoot (Hz)	−0.0102	−0.0089	−0.0092	−0.0089	−0.042	−0.037	−0.036	−0.035
Oscillation (Hz)	0–0.0102	0–0.0089	0–0.0092	0–0.0089	0–0.042	0–0.037	0–0.036	0–0.035

LFC system. Simulation is repeated for changes in load (ΔP_L) of 0.2 and 0.6 p.u. with regulation (R) values as 20 and 75.

From the simulation results in Tables 4.49 and 4.50, it can be found that these controllers can produce relatively better results with faster convergence rate and higher precision for single-area power systems. Numerical analysis is made on the performance of proposed controllers for a change in load of 0.2 p.u. and regulation value as 20. When compared to PSO algorithm, the settling times are reduced by 5.74%, 3.27%, and 2.5%, whereas overshoots are reduced by 35.3%, 32.68%, and 30.4% with EPSO, MO-PSO, and SPSO algorithms. The hybrid algorithms are selected in an effort to reduce the computational burden and in order to identify the changing dynamics of the power system. Table 4.51 indicates the performance comparison of FPSO, BF-PSO, and HGA for a change in load of 0.2 p.u. and R value of 20. Simulation is repeated for an increased load of 0.8 p.u. to analyze the efficiency of the proposed controller for higher loads.

The computational times taken by FPSO, BF-PSO, and HGA algorithms are reduced by 62.21%, 44.61%, and 32.2% when compared to basic PSO algorithm. Performance of the proposed hybrid EA is analyzed for 20% change in load perturbations and for regulation $R = 20$. When compared to conventional PSO algorithm, the settling times are reduced by 11.5%, 5.8%, and 19.23%, with the application of proposed FPSO, BF-PSO, and HGA algorithms, respectively. The overshoots are reduced by 9.8%, 9.8%, and 12.7% when compared to conventional PSO algorithm.

4.7.9 Performance Comparison of Two-Area System

A two-area power system is used as a test system to demonstrate the effectiveness of the proposed methods under various operating conditions and area load demand. By applying small and large load disturbances, the algorithm is tested for its validity for reducing the transient oscillations. The low-frequency oscillations, if not damped immediately after a sudden load in a power system, will drive the system to instability. Hence, the secondary controller employed in LFC has to manage efficiently for the increase in load and act dynamically to reduce the frequency of oscillations. Table 4.52 shows the simulation results of a two-area system for changes in load of $\Delta P_{L1} = 0.1$ and $\Delta P_{L2} = 0.2$ for R values of 20 and 75. Here the two control areas are interconnected; hence, the power flow between the areas and the tie-line frequency are together responsible for the LFC. The results in Table 4.53 validate the performance of the proposed algorithm for increased loading conditions.

The simulation results reveal that the hybrid EA method is a potential alternative to be implemented as a useful tool in solving power quality problems. As shown in Table 4.52,

TABLE 4.52

Performance Analysis of Two Area LFC for a Change in Load of $\Delta P_{L1} = 0.1$ and $\Delta P_{L2} = 0.2$

$R_1 = 20, R_2 = 75$

$\Delta P_{L1} = 0.1, \Delta P_{L2} = 0.2$

Parameters	PSO Computational Time = 29.6 s		EPSO Computational Time = 10.9 s		SPSO Computational Time = 17.6 s		MOPSO Computational Time = 14.6 s		FPSO Computational Time = 10.9 s		BF-PSO Computational Time = 15.56 s		HGA Computational Time = 19.4 s	
	Area 1	Area 2	Area 1	Area 2	Area 1	Area 2	Area 1	Area 2	Area 1	Area 2	Area 1	Area 2	Area 1	Area 2
Settling time (s)	12.2	12.5	9.5	9.8	9.8	10.1	9.9	10.7	10.2	10.5	9.6	9.8	9.4	9.6
Overshoot (Hz)	−0.0071	−0.0089	−0.005	−0.013	−0.0053	−0.013	−0.0045	−0.015	−0.013	−0.027	−0.0045	−0.015	−0.0047	−0.0138
Oscillation (Hz)	0–0.0071	0–0.0089	0–0.005	0–0.013	0–0.0053	0–0.013	0–0.0045	0–0.015	0–0.005	0–0.013	0–0.0045	0–0.015	0–0.0047	0–0.0138

TABLE 4.53

Performance Analysis of Two Area LFC for a Change in Load of $\Delta P_{L1} = 0.3$ and $\Delta P_{L2} = 0.8$

	$R_1 = 20, R_2 = 75$											
	$\Delta P_{L1} = 0.3, \Delta P_{L2} = 0.8$								$\Delta P_{L1} = 0.3, \Delta P_{L2} = 0.8$			
	EPSO		SPSO		MOPSO		FPSO		BF-PSO		HGA	
Parameters	Area 1	Area 2	Area 1	Area 2	Area 1	Area 2	Area 1	Area 2	Area 1	Area 2	Area 1	Area 2
Settling time (s)	10.4	11	11.3	11.7	11.4	11.8	10.9	11.7	11.2	11.9	10.3	11.1
Overshoot (Hz)	−0.016	−0.034	−0.017	−0.032	−0.015	−0.035	−0.026	−0.037	−0.015	−0.023	−0.013	−0.024
Oscillation (Hz)	0–0.016	0–0.034	0–C.017	0–0.032	0–0.015	0–0.035	0–0.026	0–0.037	0–0.015	0–0.023	0–0.013	0–0.024

the settling times are reduced by 21%, 13.28%, 13.3%, 14.84%, 12.5%, and 19.50% with the application of EPSO, SPSO, MO-PSO, FPSO, BF-PSO, and HGA algorithms, respectively. In terms of overshoot and oscillations, the proposed hybrid algorithms attain reduction of 15.15%, 18.2%, 12.13%, 21.2%, 30%, and 27.3% in comparison to PSO algorithm. The investigation is carried out for increased loads of 0.3 p.u. in area 1 and 0.8 p.u. in area 2 and the results in Table 4.53 explore the possibilities of proposed algorithms for different operating conditions. When a change in load is small, the overshoot and the settling time in the response are reduced. Therefore, we can choose appropriate gains based on the estimation to overshoot, oscillations, and settling time.

4.7.10 Computational Efficiency of EAs

The trend in present research and development of PID controller is to focus on the fast and reliable methods in order to obtain the best performance of existing PID control. Hence, intelligent optimization techniques are developed to decrease the computational time with increased reliability and efficiency of the controller. The inclusion of HEAs in PID gain tuning helps to automate the entire design process to a useful degree. The comparison of average computation time or time complexity of different hybrid EAs for combinatorial optimization of PID gains for AVR and LFC is shown in Figure 4.68.

The execution time is measured for different EAs with a number of iteration as 25 and swarm size as 50. As shown in Figure 4.68, the EPSO algorithm takes a computational time of 10.1 s, which is less when compared to other EAs proposed. The proposed hybrid algorithm has better searching speed than the original PSO algorithm. This technique puts the adaptively changing terms in original constant terms, so that the parameters of the original PSO algorithm change with the convergence rate, which is presented by objective function. As a result, the searching speed of advanced hybrid EA method is much faster than the original method. These algorithms find the best solution with fewer numbers of iterations, and there is a marginal difference in the time taken to converge to the best solution.

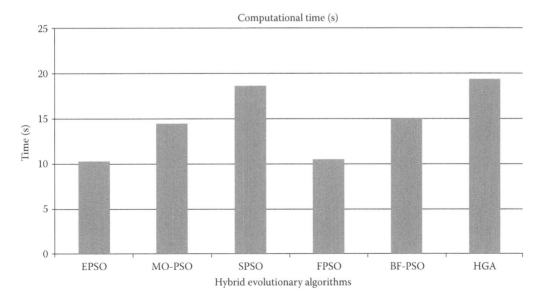

FIGURE 4.68
Comparative analysis of execution time for different evolutionary algorithms.

For higher values of iterations and swarm size, the computational efficiency and the program execution time are found to be increased. EA uses probabilistic transition rules to move in the search space (Zeng and Hu 2005). Furthermore, EA uses a parallel search through the search space; this increases the computational efficiency of the algorithms. The no-free-lunch (NFL) theorem states that there cannot exist any algorithm for solving all problems that are on average superior to any other algorithm. This theorem motivates research in new optimization algorithms, especially EA. Since the basic PSO method does not perform the selection and crossover operation in the evolutionary process, it can save computation time compared with the GA method, thus proving that the EA-based PID controller is superior. The EA search starts from a diverse set of initial points, which allows parallel search of a large area of the search space.

4.8 Summary

This chapter addresses the design and implementation of robust optimal controllers to the important power system problem of controlling single- and two-area interconnected bus system. The new proposed design utilizes iterative heuristic optimization algorithms such as FLC, GA, PSO, ACO, and hybrid EA to find the optimum gains of the PID controller. The optimized PID controller in the secondary control loop of LFC and AVR improves the performance by generating dynamic response characteristics. Traditionally, power systems have been operated and controlled in a centralized way with a considerable amount of human intervention. The modern power systems are much more complex, and hence the ability to maintain system stability in a deregulated power system environment is a major challenge. Stability phenomena can cause significant damage economically; thus, the limits of reliability and efficiency of the power system are very important issues. As the economic demand and environmental pressure continue to mount, large-scale power systems around the world require intelligent computing techniques to manage the challenging situations. In this chapter, different intelligent computing techniques such as FLC, GA, PSO, ACO, and hybrid EA for improving dynamic performance characteristics of LFC and AVR are investigated. These computational intelligence paradigms exhibit an ability to learn and adopt to new situations and facilitate intelligent behavior with the reduced computational burden. The simulation results proved that the hybrid EA-based PID controllers resulted in producing optimum response characteristics of frequency and voltage of the power generating system.

References

Abido, M.A. Optimal design of power-system stabilizers using particle swarm optimization, *IEEE Transactions on Energy Conversion* 17(3), 406–413, 2002.

Abul, H.R. and Sadrul, U.A.H.M. Design and implementation of a fuzzy controller based AVR for synchronous generator, *IEEE Transactions on Energy Conversion* 9(3), 550–557, 1994.

Adibi, M.M. and Fink, L.F. Overcoming restoration challenges associated with major power system disturbances, *IEEE Power and Energy Magazine* 4(5), 68–77, 2006.

Ahamed, T.P.I., Rao, P.S.N., and Sastry, P.S. A reinforcement learning approach to automatic generation control, *Transaction on Electric Power System Research* 63, 9–26, 2002.

Ali, M.H. *Wind Energy Systems: Solutions for Power Quality and Stabilization*, CRC Press, Boca Raton, FL, 2012.

ANSI/IEEE Std 421.1–1986, *IEEE Standard Definitions for Excitation Systems for Synchronous Machines.* Washington, DC: American National Standards Institute and The Institute of Electrical and Electronics Engineers, 1986.

ANSI/IEEE Std 122–1985, *IEEE Recommended Practice for Functional and Performance Characteristics of Control Systems for Steam Turbine-Generator Units.* Piscataway, NJ: The Institute of Electrical and Electronics Engineers, 1985.

Atlas, I.H. and Neyens, J. A fuzzy logic load-frequency controller for power systems, in *International Symposium on Mathematical Methods in Engineering MME06*, Turkey, pp. 27–29, 2006.

Awadallah, M.A. and Morcos, M.M. A fuzzy logic based AVR for a stand-alone alternator feeding heater load, *IEEE Power Engineering Review* 0272–1724/01, 53–56, 2001.

Back, T., Fogel, D.B., and Michalewicz, Z. *Handbook of Evolutionary Computation.* New York: Oxford University Press and Institute of Physics, 1997.

Back, T., Hammel, U., and Schwefel, H.-P. Evolutionary computation: Comments on the history and current state, *IEEE Transactions on Evolutionary Computation* 1(2), 3–17, 1997.

Banejad, M. and Housman, R. Optimal design of coefficients of PID controller in an AVR system using fuzzy particle swarm optimization algorithm, *Transactions on Evolutionary Computation* 5, 78–82, 2001.

Bansal, R.C. Bibliography on the fuzzy set theory applications in power systems (1994–2001), *IEEE Transactions on Power Systems* 18(4), 1291–1299, 2003.

Bianchi, L., Luca, M.G., and Dorigo, M. Ant Colony Optimization approach to the probabilistic traveling Salesman problem, in *Proceedings of PPSN-VII, Seventh International Conference on Parallel Problem Solving from Nature, Lecture Notes in Computer Science*, Springer Verlag, Germany, pp. 883–892, 2002.

Biswas, A., Dasgupta, S., Das, S., and Abraham, A. Synergy of PSO and bacterial foraging optimization—A comparative study on numerical benchmarks, *International Journal on Innovations in Hybrid Intelligent Systems, ASC* 44, 255–263, 2007.

Bollen, M.H.J. *Understanding Power Quality Problems: Voltage Sags and Interruptions.* Piscataway, NJ: IEEE Press, 2000.

Boubertakh, H., Tadjine, M., and Glorennec, P.Y. Tuning fuzzy PID controllers using ant colony optimization, in *Proceedings of 17th Mediterranean Conference on Control and Automation*, Greece, pp. 12–18, June 2009.

Buczak, A.L. and Uhrig, R.E. Hybrid fuzzy-genetic technique for multisensory fusion, *Information Sciences* 93(3–4), 265–281, 1996.

Cam, E. and Kocaarslan, I. A fuzzy gain scheduling PI controller application for an interconnected electrical power system, *Electrical Power System Research* 73, 267–274, 2005.

Chaturvedic, D.K., Satsangi, P.S., and Kalra, P.K. Load frequency control: A generalized neural network approach, *Journal of Electric Power and Energy Systems* 21, 405–415, 1999.

Chile, R.H., Waghmare, L.M., and Lingare, M.J. More efficient Genetic Algorithm for tuning optimal PID controller, in *Proceedings of Second National Conference on Computing for Nation Development*, New Delhi, India, pp. 521–522, 2008.

Chakrabarti, A. and Halder, S. *Power System Analysis—Operation and Control.* New Delhi, India: Prentice-Hall of India, 2006.

Chowdhury, S., Chowdhury, S.P., and Choudhuri, S. Advanced digital load frequency control with unknown deterministic power demand for interconnected power systems, *IE (I) Journal on Electrical Power Systems* 80, 87–95, 1999.

Chown, G.A. Design and experience with a fuzzy logic controller for automatic generation control (AGC), *IEEE Transaction on Power Systems* 13, 965–970, 1998.

Cirstea, M.N., Dinu, A., Khor, J.G. and McCormick, M. *Neural and Fuzzy Logic Control of Drives and Power Systems.* Burlington, VT: Newnes Publishers, 2002.

Demiroren, A. and Yesil, E. Automatic generation control with fuzzy logic controllers in the power system including SMES units, *Electrical Power and Energy Systems* 26, 291–305, 2004.

Devaraj, D. *Power System Operation and Control.* New Delhi, India: Vitasta Publishing Pvt. Ltd., 2008.

Djukanovic, M. Conceptual development of optimal load frequency control using artificial neural networks and fuzzy set theory, *International Journal Engineering Intelligent System Electrical Engineering and Communication* 3, 2–108, 1995.

Dorigo, M. and Blum, C. Ant colony optimization theory: A survey, *Theoretical Computer Science* 345, 243–278, 2005.

Dorigo, M., Caro, G.D., and Gambardella, L.M. Ant algorithms for discrete optimization, *Journal of Artificial Life* 5, 137–172, 1999.

Dorigo, M., Maniezzo, V., and Colorni, A. The ant system: An autocatalytic optimizing process, Technical Report TR91–016, Politecnico di Milano, 1991.

Driankov, D., Hellendoorn, H., and Reinfrank, M. *An Introduction to Fuzzy Control.* New Delhi, India: Narosa Publishing House, 2001.

Duan, H.-b., Wang, D.-B., and Yu, X.-f. Novel approach to nonlinear PID parameter optimization using Ant Colony Optimization algorithm, *Journal of Bionic Engineering* 3, 073–078, 2006.

Dugan, R.C., Santoso, S., and Granaghan, M.F. *Electrical Power Systems Quality.* New Delhi, India: Tata McGraw-Hill, 2nd edn, 2008.

Eberhart, R.C. and Shi, Y. Comparing inertia weights and constriction factors in particle swarm optimization, in *Proceedings of the Congress on Evolutionary Computation*, San Diego, CA, 2000, pp. 84–88.

Elgerd, O.E. *Electric Energy System Theory.* New York: McGraw Hill, 2nd edn, 1982.

El-Hawary, M.E. *Electric Power Applications of Fuzzy Systems.* Piscataway, NJ: IEEE Press, 2nd edn, 1998.

Ertugrul, C. and Kocaarslan, I. A fuzzy gain scheduling PI controller application for an interconnected electrical power system, *Journal of Electrical Power System Research* 73, 267–274, 2005.

Fang, H. and Chen, L. Application of enhanced PSO algorithm to optimal tuning of PID gains, IEEE Explore Reference 978-1-4244-2723-9/09, 35–39, 2009.

Fogel, D.B. *Evolutionary Computation: Toward a New Philosophy of Machine Intelligence.* New York: IEEE Press, 2nd edn, 2000.

Gaing, Z.L. A Particle swarm optimization approach for optimum design of PID controller in AVR system, *IEEE Transactions on Energy Conversion* 19(2), 384–391, 2004.

Glover, J.D., Sarma, M.S., and Overbye, T. *Power System Analysis and Design*, Thomson Learning, India Edition, 4th edn, 2007.

Goldberg, D.E. *Genetic Algorithms in Search, Optimization and Machine Learning.* Pearson Education (Singapore) Pvt. Ltd., 2nd edn, Indian Branch, 2003.

Gomez-Skarmeta, A.F., Valdes, M., Jimenez, F., and Marin-Blazquez, J.G. Approximate fuzzy rules approaches for classification with hybrid-GA techniques, *Information Sciences* 136(1–4), 193–214, 2001.

Griffin, I. On-line PID controller tuning using genetic algorithms, MSc Thesis, School of Electronic Engineering, Dublin City University, 2003.

Grosan, C. and Abraham, A. Hybrid evolutionary algorithms: Methodologies, architectures, and reviews, *Studies in Computational Intelligence* 75, 1–17, 2007.

Gupta, B.R. *Generation of Electrical Energy.* New Delhi, India: Eurasia Publishing House (Pvt) Ltd., 2003.

Ha, Q.P. and Negnevitsky, M. A robust modal controller with fuzzy tuning for multi-mass electromechanical systems, in *Proceedings of the Third Australian and New Zealand Conference on Intelligent Information System*, Perth, Western Australia, pp. 214–219, 1995.

Hadi, S. *Power System Analysis.* New Delhi, India: McGraw-Hill, 2nd edn, p. 62, 2002.

Hamid, A.T., Sadeh, J., and Ghazi, R. Design of augmented fuzzy logic power system stabilizers to enhance power systems stability, *IEEE Transactions on Energy Conversion* 11(1), 97–103, 1996.

Herrero, J.M., Balsco, X., Martinez, J., and Salcedo, V. Optimal PID tuning with genetic algorithms for non-linear process models, in *Proceedings of 15th Triennial World Congress*, Barcelona, Spain, 2002.

Ibraheem, I., Kumar, P., and Kothari, D.P. Recent philosophies of automatic generation control strategies in power systems, *IEEE Transactions on Power Systems* 20(1), 346–357, 2005.

IEEE Committee Report. Computer representation of excitation systems, *IEEE Transactions on Power Apparatus and Systems* PAS-87, 1460–1464, 1968.

IEEE Committee Report. Dynamic Models for steam and hydro turbines in power system studies, *IEEE Transactions on Power Apparatus and Systems* PAS-92(6), 1704–1915, 1973.

Ismail, A. Improving UAE power systems control performance by using combined LFC and AVR, in *The Seventh Annual U.A.E University Research Conference*, Osmania University, Hyderabad, India, pp. 50–60, 2006.

Juang, C.F. A hybrid of genetic algorithm and particle swarm optimization for recurrent network design, *IEEE Transactions on Systems, Man and Cybernetics, Part B* 34, 997–1006, 2004.

Juang, C.F. and Lu, C.F. Power system load frequency control by evolutionary fuzzy PI controller, in *IEEE International Conference on Fuzzy Systems*, Budapest, Hungary, pp. 1749–1755, 2004.

Kaimal, M.R, Dasgupta, S., and Harishankar, M. *Neuro-Fuzzy Control Systems*. Tamil Nadu, India: Narosa Publishing House, 1997.

Karnavas, K.L. and Papadopoulos, D.P. AGC for autonomous power system using combined intelligent techniques, *Journal of Electric Power System Research* 62, 225–239, 2002.

Kennedy, J. and Eberhart, R.C. Particle swarm optimization, *Proceedings of IEEE Conference on Neural Networks* 4, 1942–1948, 1995.

Kim, D.H., Abraham, A., and Cho, J.H. A hybrid genetic algorithm and bacterial foraging approach for global optimization, *Elsevier International Journal of Information Sciences* 177, 3918–3937, 2007.

Kim, D.H. and Cho, J.H. Adaptive tuning of PID controller for multivariable system using bacterial foraging based optimization, in *Third International Atlantic Web Intelligence Conference, Lodz, Poland, Advances in Web Intelligenc*e, pp. 231–238, 2005.

Kim, D.H. and Cho, J.H. A biologically inspired intelligent PID controller tuning for avr systems, *International Journal of Control, Automation, and Systems* 4(5), 624–636, 2006.

Kim, D.H. and Park, J.I. Intelligent tuning of PID controller for AVR system using a hybrid GA-PSO approach, *Lecture Notes in Computer Science*, Springer, 366–373, 2005.

Korani, W.M., Dorrah, H.T., and Emara, H.M. Bacterial foraging oriented by particle swarm optimization strategy for PID Tuning, *IEEE Transactions on Control System Magazine* 21(3), 68–73, 2008.

Kothari, D.P. and Nagrath, I.J. *Power System Engineering*, New Delhi, India: McGraw Hill, 2007.

Kumar, P. and Ibraheem, N. Dynamic performance evaluation of 2-area interconnected power systems, *Journal of Institution of Engineers (India)* 78, 199–209, 1998.

Kundur, P. *Power System Stability and Control*. New Delhi, India: McGraw Hill, 2nd edn, 1994.

LaMeres, B.J. and Nehrir, M.H. Fuzzy logic based voltage controller for a synchronous generator, *IEEE Computer Applications in Power* 12(2), 46–49, 1999.

Mahfouf, M., Chen, M., and Linkens, D.A. Adaptive weighted particle swarm optimisation for multi-objective optimal design of alloy steels, *Lecture Notes Computer Science* 3242, 762–771, 2004.

Masiala, M., Ghnbi, M., and Kaddouri, A. An adaptive fuzzy controller gain scheduling for power system load-frequency control, in *IEEE International Conference on Industrial Technology (ICIT)*, Hammamet, Tunisia, pp. 1515–1520, 2004.

Mathur, H.D. and Manjunath, H.V. Frequency stabilization using fuzzy logic based controller for multi-area power system, *The South Pacific Journal of Natural Science* 4, 22–30, 2007.

McArdle, M.G., Morrow, D.J., Calvert, P.A.J., and Cadel, O. A hybrid PI and PD type fuzzy logic controller for automatic voltage regulation of the small alternator, *Proceedings of Power Engineering Society Summer Meet* 3, 1340–1345, 2001.

Michalewicz, Z. *Genetic Algorithms + Data Structures = Evolutionary Programs*. New York: Springer-Verlag, 1999.

Mohammad, H.K., Mehdi, K., and Mohammad, B.M. Decentralized robust adaptive-output feedback controller for power system load frequency control, Springer-Verlag Electrical Engineering, 84, 75–83, 2002.

Momoh, J.A., Ma, X.W., and Tomsovic, K. Overview and literature survey of fuzzy set theory in power systems, Paper # 95 WM 208–9 PWRS, *Transaction of IEEE PES Winter Power Meeting*, New York, January/February, 1995.

Moorthy, R.G., Arumugam, S.M., and Loo, C.K. Hybrid particle swarm optimization with fine tuning parameters, *International Journal of Bio-Inspired Computations* 1(1/2), 14–31, 2009.

Mukherjee, V. and Ghosal, S.P. Velocity relaxed swarm intelligent tuning of Fuzzy based power system stabilizer, in *IEEE Power Engineering Society Conference*, Pittsburgh, PA, p. 85, 2008.

Murty, P.S.R. Operation and control in power systems, B.S. Publications, Hyderabad, IIIrd edition, 2005.

Nanda, J. and Mangla, A. Automatic generation control of an interconnected hydro-thermal system using conventional integral and fuzzy logic controller, in *IEEE International Conference on Electric Utility Deregulation, Restructuring and Power Technologies*, Hong Kong, People's Republic of China, April, 2004.

Oliveira, P.M., Cunha, J.B., and Coelho, J.O.P. Design of PID controllers using the particle swarm algorithm, in *Twenty-First IASTED International Conference: Modelling, Identification, and Control (MIC 2002)*, Innsbruck, Austria, 2002.

Oysal, Y. Fuzzy PID controller design for LFC using gain scaling technique, in *Power Tech Conference Proceedings*, Budapest, Hungary, 1999.

Padiyar, K.R. *Power System Dynamics- Stability and Control*. Hyderabad, India: BS Publications, 2nd edn, 2002.

Passino, K.M. Biomimicry of Bacterial foraging for distributed optimization and control, IEEE control systems magazine, 52–67, 2002.

Patel, R.N. Application of artificial Intelligence for tuning the parameters of an AGC, *International Journal of Mathematical, Physics and Engineering Sciences* 1(1), 34–40, 2007.

Popov, A., Farag, A., and Werner, H. Tuning of a PID controller using multi-objective optimization technique, in *Proceedings of the 44th IEEE International Conference on Decision and Control*, Seville, Spain, 2005.

Pratima Singh and Jha, A.N. Fuzzy logic and fuzzy-neuro based controllers for synchronous generator, in *Proceedings of the 25th National System Conference*, PSG College of Technology, India, pp. 263–269, 2001.

PSS/E User Handbook, *Speed Governor System Modeling*. PSS/E: Innsbruck, Austria.

Rabandi, I. and Nishimori, K. Optimal feedback control design using genetic algorithm in multimachine power system, *Electrical Power and Energy Systems* 23, 263–271, 2001.

Reddy, M.J. and Kumar, D.N. An efficient multi-objective optimization algorithm based on swarm intelligence for engineering design, *Engineering Optimization* 39(1), 49–68, 2007.

Sabahi, K., Sharifi, A., Aliyari Sh, M., Teshnehlab, M., and Aliasghary, M., Load frequency control in interconnected power system using multi-objective PID controller, *Journal of Applied Sciences* 8(20), 3676–3686, 2008.

Shanmugham, T., Sivanandam, S.N., and Mary, D. ANN based system identification of multi-area power system using MATLAB, in *Proceedings of National Conference on Advanced Computing*, Chennai, Tamil Nadu, India, 2001.

Shayegi, H., Shayanfar, H.A., and Malik, O.P. Robust decentralized neural networks based LFC in deregulated power system, *Electric Power Systems Research* 77, 541–251, 2008.

Shigenori Naka, Takamu Genji, Kenji Miyazato, and Yoshikazu Fukuyama, Hybrid particle swarm optimization using constriction factor approach, *IEEE Transactions on Energy Conversion* 19(2), 2004.

Shyh-Jier Huang, Enhancement of hydroelectric generation scheduling using ant colony system based optimization approaches, *IEEE Transactions on Energy Conversion* 16(3), 296–301, 2001.

Soundarrajan, A., Sumathi, S., and Sivamurugan, G. Voltage and Frequency control in power generating systems using hybrid evolutionary algorithms, *International Journal of Vibration and Control* 18(2), 214–227, 2012.

Stankovic, A.M., Tadmor, G., and Sakharuk, T.A. On robust control analysis and design for load frequency regulation, *IEEE Transactions on Power Systems* 13(2), 449–455, 1998.

Sumathi, S., Sivanandam, S.N., and Deepa, S.N. *Introduction to Fuzzy Logic Using MATLAB*. Berlin, Germany: Springer Verlag, 2007.

Takashi H. and Tomsovic, K. Current Status of Fuzzy System Applications in Power Systems, in *Proceedings of the IEEE, SMC99*, Tokyo, Japan, pp. 527–532, 1999.

Takashi, H., Yoshiteru, U., and Hiroaki, A. Integrated fuzzy logic controller generator for stability enhancement, *IEEE Transactions on Energy Conversion* 12(4), 400–406, 1997.

Talaq, J. and Al-Basri, F. Adaptive fuzzy gain scheduling for load frequency control, *IEEE Transaction on Power System* 14(1), 145–150, 1999.

Tsutsui, S. and Goldberg, D.E. Simplex crossover and linkage identification: single stage evolution Vs. multi-stage evolution, in *Proceedings in IEEE International Conference on Evolutionary Computation*, New York, pp. 974–979, 2002.

Uma Rao, K. *Computer Techniques and Models in Power Systems*. New York: I.K.International Publishing House pvt. ltd., 2007.

Wong, C.C., Li, S.A., and Wang, H.-Y. Hybrid evolutionary algorithm for PID controller design of AVR system, *Journal of the Chinese Institute of Engineers* 32(2), 251–264, 2009.

Wood, A.J. and Wollenberg, B.F. *Power Generation, Operation and Control*. New York: John Wiley & Sons, 2006.

Yang, T.C. and Ding Z.T. Decentralized power system load frequency control beyond the limit of diagonal dominance, *Electrical Power and Energy Systems* 24, 173–184, 2002.

Yesil, E. and Eksin, M.G. Self tuning fuzzy PID type load and frequency controller, *Journal of Energy Conversion Manage* 45, 377–390, 2004.

Yildiz, C., Yilmaz, A.S., and Bayrak, M. Genetic algorithm based PI controller for load frequency control in power systems, in *Proceedings of 5th International Symposium on Intelligent Manufacturing Systems*, pp. 1202–1210, 2006.

Yoshida, H. and Kenichi, K. A Particle swarm optimization for reactive power and Voltage control considering Voltage stability, in *IEEE International Conference on Intelligent System Applications to Power Systems* April 4–8, 1999.

Yoshida, H., Kawata, K., Fukuyama, Y., Takayama, S., and Nakanishi, Y. A particle swarm optimization for reactive power and voltage control considering Voltage security assessment, *IEEE Trans. Power Systems* 15(4), 1232–1239, 2000.

Yukita, K. Study of load frequency control using fuzzy theory by combined cycle power plant, *IEEE Power Engineering Society Winter Meeting*, 1, 422–428, 2000.

Zareiegovar, G. A new approach for tuning PID controller parameters of LFC considering system uncertainties, in *Proceedings of International Conference on Environment and Electrical Engineering*, Czech Republic, pp. 333–336, 2009.

Zeng, J.J. and Hu, J. Adaptive particle swarm optimisation guided by acceleration information, *International Conference on Computational Intelligence and Security* 1, 351–355, 2005.

Zhao, B., Guo, X., and Cao, Y.J. A multiagent based particle swarm optimization approach for optimal reactive power dispatch, *IEEE Transactions on Power Systems* 20(2), 1070–1078, 2005.

5

Job Shop Scheduling Problem

Learning Objectives

On completion of this chapter, the reader will have knowledge on

- Solving the job shop scheduling problem (JSSP) in terms of fuzzy processing time
- Step-by-step procedures of implementing GA, SPSO, ACO, and GSO for solving JSSPs
- Development of MATLAB® programs for solving JSSP using intelligent algorithms
- Computational parameters of JSSP instances with different job and machine sizes.

5.1 Introduction

Scheduling has been the subject of a vast amount of literature in the operations research (OR) field since the early 1950s. The main objective of scheduling is to determine an efficient allocation of shared resources over time to competing activities. Emphasis has been laid on investigating machine scheduling problems, where jobs represent activities and machines represent resources. Such problems are referred to as job shop scheduling problems (JSSPs). JSSP is not only NP-hard but also has a well-earned reputation of being one of the most computationally difficult combinatorial optimization problems considered to date (Sonmez and Baykasoglu 1998). The research on JSSP promotes not only the development of relative algorithms in the field of artificial intelligence but also the means of solutions and applications for complex JSSPs. JSSPs can be thought of as the allocation of resources over a specified time to perform a predetermined collection of tasks. JSSP can be considered as comprising two problems: first, assigning a proper machine from a set of machines to each operation, and second, sequencing each operation on every given machine. The former problem can be seen as a parallel machine problem, which is also an NP-hard problem, and the latter is similar to a classical job shop problem (Li et al. 2009).

The job shop problem is initially solved by "exact methods," such as the branch-and-bound method (BAB), and mathematical formulations that are based on exhaustive enumeration of a restricted region of solutions containing exact optimal solutions. Exact methods are theoretically important and have been successfully applied to benchmark instances, but sometimes they are quite time consuming even for moderate-scale problems. With the rapid progress in computer technology, it has become even more important to find practically acceptable solutions by "approximation methods," especially for large-scale problems, within a limited amount of time. Stochastic local search methods are such approximation methods for combinatorial optimization. They provide robust approaches to obtain high-quality solutions to problems of realistic sizes in reasonable amounts of time. Heuristics, meta-heuristics, and construction heuristics belonging to the class of stochastic local search methods have been proposed in analogy with processes in nature, such as biological evolution, and in the artificial intelligence context. They often work as an iterative process that guides and modifies the operations of subordinate heuristics; thus they are also called meta-heuristics. Meta-heuristics have been applied to a wide variety of combinatorial optimization problems with great successes. The demerits of exact procedures led to the application of various heuristic and meta-heuristic algorithms such as simulated annealing (SA), tabu search (TS), genetic algorithm (GA), particle swarm optimization (PSO), ant colony optimization (ACO), and several other bio-inspired algorithms (Ivers 2007).

Most of the proposed approximate methods solved the JSSP with the assumption that all the parameters were known exactly. The difficulty is in the estimation of the exact processing time for all jobs on the machines. The uncertainty and vagueness of the processing time was investigated using fuzzy logic by Fortemps (1997) and Ishii et al. (1992). The first initiative to the fuzzy-based approach for solving JSSPs was presented by Tsujimura et al. (1993), who used fuzzy set theory to model the processing times of a flow shop scheduling facility. Triangular fuzzy numbers (TFNs) were used to represent these processing times, with two TFNs for each job defining an upper and a lower bound. The objective of the JSSP to minimize the makespan was obtained using the BAB technique (Jones and Rabelo 1998). Decision-making approaches specifically consider fuzzy processing time and fuzzy due dates (McCahon and Lee 1990, Slowinski and Hapke 2000). In these methods, fuzzy numbers are used to compute the fuzzy makespan and fuzzy mean flow times.

Nakano and Yamada (1991) were among the first to apply a conventional GA that used binary representation of solutions to JSSP. Further, Yamada and Nakano (1992) presented a GA that used problem-specific representation of solutions with crossover and mutation, which were based on the Giffler and Thompson (GT) algorithm. Precisely, the researchers proved that a standard GA may not be flexible enough for practical applications, and this becomes increasingly apparent when the problem is complicated and involves conflict and multitasking (Morshed 2006). Several hybridized techniques have been developed to solve JSSPs, as GA suffered from genetic drift and premature convergence.

Cheung and Zhou (2001) developed a hybrid algorithm based on GA and a well-known dispatching rule for sequence-dependent setup time (SDST) job shops, where the setup times were separable. The first operations for each of the m machines were achieved by GA, while the subsequent operations on each machine were planned by the shortest processing time (SPT) rule. The primary objective was to minimize the makespan using the SPT rule. Balas and Vazacopoulos (1998) proposed a variable depth search procedure, called guided local search (GLS), based on an interchange scheme with the concept of neighborhood trees. The GLS was then embedded into a shifting bottleneck framework, and their results proved to be particularly efficient. A combination of tabu search and the shifting bottle neck procedure was proposed by Pezzella and Merelli (2000).

Wang et al. (2009) combined SA and GA in an attempt to regulate the convergence of GA. Zhou et al. (2006) proposed an immune algorithm that certified the diversity of the antibody to solve JSSPs. An effective PSO-based memetic algorithm (MA) for the permutation flow shop scheduling problem with the objective to minimize the completion time was proposed by Liu et al. 2007. A PSO algorithm to solve JSSP with new valid algorithm operators was presented by Lian et al. (2006a,b). PSO combined with SA was used to solve the problem of finding the minimum makespan in the job shop scheduling environment (Wei and Wu 2005). Using the shifting bottleneck procedure, Huang and Liao (2008) combined the ACO algorithm with tabu search. ACO provided the initial schedule, and was enhanced by the tabu search algorithm. Niu et al. (2008) addressed the problem of scheduling a set of jobs in a job shop environment with fuzzy processing time by combining GA and PSO. The algorithm reported by Niu et al. (2008) was tested on only 10 benchmark problems.

The primary focus of this chapter is the application of intelligent meta-heuristics, especially GA, stochastic PSO (SPSO), ACO, and hybrid GA SPSO, termed as genetic swarm optimization (GSO), to JSSPs with fuzzy processing time considering fuzzy numbers in the interval $(\lambda,1)$. In fuzzy logic, the concept of ranking the processing time is based on signed distance, which uses both negative and positive values to define ordering. The processing times are fuzzified, and schedules of the job shop are optimized using the bio-inspired intelligent meta-heuristics such as GA, SPSO, ACO, and GSO. GAs are reviewed, with a major emphasis on conventional binary models for combinatorial optimization of JSSPs. The solution is encoded into a binary string of fixed length with the job and machine sequences and genetic operators such as selection, crossover, and mutation. SPSO applied in this application is a modified version of the original PSO based on dynamic tuning of parameters such as the acceleration constants and the inertia weight during the run of the algorithm. ACO is modeled based on the communication principles and the cooperative work of real ants, thus inspiring the solution of diverse combinatorial optimization problems. The operations that correspond to the same job will be processed according to their technological sequence, and none of them will be able to begin its processing before the previous operation has finished, similar to the foraging behavior of ants.

A new hybrid mechanism is proposed by combining GA and SPSO, in order to determine the global optimum effectively and efficiently for JSSPs. The SPSO algorithm has some advantages such as convergence and robustness in obtaining excellent dynamic solutions, and it is suitable for research on population behavior. Combining the advantages of GA and SPSO algorithms, the hybrid of GA and SPSO, known as GSO, is used to solve the JSSP. GSO maintains the integrity of the two techniques during the entire run of the simulation. In each iteration, a few individuals are substituted by newly generated ones using GA, and the resulting set of solutions is moved to the solution space by SPSO. One of the major drawbacks in hybridization of global and local strategies—the problem of premature convergence—is eliminated by adopting the GSO algorithm. The main focus throughout this application is the minimum-makespan problem, in which makespan, which is the maximum completion time of all the operations, is used as an objective function to be minimized. The proposed algorithms are tested and validated on a set of standard benchmark instances with different sizes in terms of jobs and machines, obtained from the online OR library of Brunel University, London, UK (Beasley 1990). The makespan, average makespan, computational time, relative error, and mean relative error are computed and compared with state-of-the-art approaches available in existing literature.

5.2 Formulation of JSSP

Job shop scheduling is an activity that comprises a set of jobs to be processed on a set of machines. JSSP can be defined as the allocation of machines over time to perform a collection of minimizing or maximizing a specific performance measure while satisfying the operation precedence constraints, machine capacity constraint, processing time, and ready time requirements. The resource used in an industrial environment is known as machines, and the basic task module that performs the operation is called jobs (Muth and Thompson 1963). Each job may be comprised of several elementary tasks called operations, which are interrelated by precedence constraints. The processing of an operation requires the use of a particular machine for an uninterrupted duration, called the processing time of the operation. Each machine can process only one operation at a time. The routing, processing times, and precedence constraints are specified by a process plan. The main distinction between the classic flow shop and a job shop is that, in the former case each job passes the machines in the same order, whereas in the latter case the machine order may vary per job. Since workflow in a job shop is not unidirectional, scheduling becomes quite tedious. For a particular job shop process plan, several feasible schedules can be generated and measured. The processing order on each machine that minimizes the corresponding cost is desired by the objectives such as minimization of process cost, makespan, and flow time, or maximization of throughput, systems/resource utilization, and production rate. There are some assumptions to be made in solving job shop scheduling problem. They are as follows:

- Given a finite set of n jobs.
- Each job consists of a chain of operations.
- Given a finite set of m machines.
- Each machine can handle at most one operation at a time.
- Each operation needs to be processed during an uninterrupted period of a given length on a given machine.
- To find a reasonable schedule, the operations are allotted to machines based on minimal time interval length.

5.2.1 Problem Description

JSSP is NP-hard by nature. This complexity is further increased when additional constraints are added to solve real-world problems. The exact methods could solve only small size problems within acceptable time periods. Although they produce exact solutions, they often simplify the instances. Meta-heuristics are semistochastic approaches that can produce near-optimal solutions with less computational time. These approaches adapt to the problem situation in a dynamic manner based on the system constraints. The properties of these meta-heuristic algorithms, such as the use of a population of solutions and problem independence, enable them to be effectively used for job shop scheduling.

The stages of implementation of the intelligent heuristics for solving the JSSP in this chapter are shown in Figure 5.1. A total of 162 benchmark instances with different job and machine sizes are chosen to analyze the performance of meta-heuristic algorithms. The processing time of the JSSP instance is fuzzified using a λ-interval fuzzy approach to convert the uncertain values to fuzzy values using a triangular membership function. The objective of the study is to obtain a feasible schedule such that the makespan is minimized. The schedules with fuzzy processing time are further optimized using four meta-heuristics GA, SPSO, ACO, and GSO. Each of these algorithms evaluates the optimal makespan, average makespan, relative error, and the average computational time for all the bench mark instances. The performances of these intelligent algorithms are analyzed by comparing the evaluated parameters with state-of-the-art intelligent approaches existing in the literature.

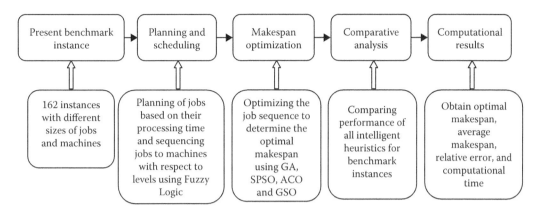

FIGURE 5.1
Block diagram for solving JSSP.

5.2.2 Mathematical Model of JSSP

A classical JSSP (Vaessens et al. 1996) consists of m machines and a set of n jobs. Each job is comprised of a sequence of operations, and each operation has to be processed on a given machine during an uninterrupted time period of a given length. Each machine can process at most one job at a time, which cannot be interrupted until its completion. It is assumed that successive operations of the same job can be processed on different machines. The assignment of operations to machines in different time intervals is referred to as scheduling. The goal of the JSSP is to determine a feasible schedule, thus optimizing the given objective. Consider three finite sets J, M, and O, where J represents a set of n jobs, M denotes a set of m machines, and O represents a set of N operations. The following notations are used in this section:

J_i: job to which operation i belongs

M_k: machine k on which operation i is to be processed

t_i: start time of operation i

p_i: processing time for operation i

C_{max}: makespan

$\rightarrow$: denotes a binary relation representing the precedence constraints of operations on the same job

For instance, $i \rightarrow j$ implies that operation i is the predecessor of operation j.

For optimal job shop scheduling, it is necessary to determine the starting time t_i for each operation $i \in O$, such that

$$\max_{i \in O}\left(t_i + p_i\right) \tag{5.1}$$

is minimized subject to

$$\forall i \in O : t_i \geq 0 \tag{5.2}$$

$$\forall i, j \in O, i \rightarrow j : t_j \geq t_i + p_i \tag{5.3}$$

$$\forall i, j \in O, i \neq j, M_i = M_j : (t_j \geq t_i + p_i) \vee (t_j \geq t_i + p_j) \tag{5.4}$$

Equation 5.3 represents the precedence constraints that help ensuring that the processing sequence of operations of each job conforms to the predetermined order, and Equation 5.4 represents the machine capacity constraints, which ensure that a machine processes only one job at a time and its processing cannot be interrupted. In order to determine a schedule, the described equations are to be modified into an integer programming form. This can be done by eliminating the binary relation $\rightarrow$ and the disjunctive operator $\vee$. While eliminating the binary relation, O is decomposed into subsets of operations corresponding to individual tasks. Then a sequence of consecutive integers based on the order of operation can be used to represent the operations. Let n_j represent the number of jobs in job j, and N_j represent the total number of jobs in the first j jobs. Obviously, the initial set of jobs is given by

$$N_0 = 0, \quad N_j = \sum_{k=1}^{j} n_k, \quad N = \sum_{k=1}^{n} n_k \tag{5.5}$$

Now, by assigning to n_j operations of task j numbers $N_{j-1} + 1 \ldots, N_{j-1} + n_j$, where $N_{j-1} + n_j = N_j$, for the operations of the first $j - 1$ jobs, Equation 5.3 can now be rewritten as

$$(\forall j \in J)(N_{j-1} + 1 \leq i \leq N_j - 1) : t_{i+1} \geq t_i + p_i \tag{5.6}$$

The makespan is now determined as the maximum completion times of the last operations in the jobs, expressed as

$$\forall j \in J : C_{\max} \geq t_{Nj} + p_{Nj} \tag{5.7}$$

The capacity constraints using binary variables $x_{ij} \in [0,1]$ are defined with respect to Equation 5.4 as

$$\forall i, j \in O, i \neq j, M_i = M_j : x_{ij} = \begin{cases} 1, t_j \geq t_i + p_i, \text{operation } i \text{ preceeds operation } j \\ 0, t_i \geq t_j + p_j, \text{operation } j \text{ preceeds operation } i \end{cases} \tag{5.8}$$

Let M_{UB} denote the upper bound of the makespan; then, Equation 5.4 can be modified using the inequalities in Equation 5.8 as

$$\forall i, j \in O, i \neq j, M_i = M_j : \begin{cases} t_j \geq t_i + p_i x_{ij} - M_{UB}(1 - x_{ij}) \\ t_i \geq t_j + p_j(1 - x_{ij}) - M_{UB} x_{ij} \end{cases} \tag{5.9}$$

Thus the objective of JSSP can be formulated as
    Minimize C_{max}
    subject to

$$\forall i \in O : t_i \geq 0 \tag{5.10}$$

$$(\forall j \in J)(N_{j-1} + 1 \leq i \leq N_j - 1) : t_{i+1} \geq t_i + p_i \tag{5.11}$$

$$\forall j \in J : C_{max} \geq t_{Nj} + p_{Nj} \tag{5.12}$$

$$\forall i, j \in O, i \neq j, M_i = M_j : \begin{cases} t_j \geq t_i + p_i x_{ij} - M_{UB}(1 - x_{ij}) \\ t_i \geq t_j + p_j(1 - x_{ij}) - M_{UB} x_{ij} \end{cases} \tag{5.13}$$

Thus the objective of the JSSP is to minimize the makespan subject to constraints given by Equations 5.10–5.13. Equation 5.10 ensures that the start time is always greater than zero, while Equation 5.11 guarantees the assignment of operations. Based on the start time and completion time of the last jobs, the makespan is defined according to Equation 5.12. Equation 5.13 guarantees the capacity constraints based on the upper bound of makespan.

5.2.3 Operation of the Job Shop Scheduling System

The general operation of the job shop scheduling system is explained in this section. The operation is divided into scheduling and makespan optimization. A feasible schedule has to be created in order to minimize the makespan. Consider a production plant that operates for 8 hours every day, 5 days a week. All jobs enter the system on a daily basis. For each day, a best schedule is to be established. If multiple copies of a job are available, they are scheduled as a single batch. On the very first day of scheduling, it is assumed that all jobs enter the system fresh and are free to be scheduled on any machine. However, for any subsequent day, there will be jobs carried forward from the previous day. Once a schedule is performed for the first day, the incomplete job operations for that day are stored and retrieved to be scheduled on the following day. For any following day, the partially completed operations for jobs from the previous day are unchanged from their machines and given first preference. This implies that, irrespective of any situation, all the jobs with partially completed operations are placed first in the scheduled sequence. The remaining operations are scheduled along with the new jobs entering on that day. This explains that all the incomplete operations on a particular job are carried over and remain on the same machine the next day, that is, no pre-emptions of operations for a job. Restricting jobs with partially completed operations from being reassigned reduces the setup changeovers on machines.

5.2.3.1 Scheduling of Job Sequences

The process of scheduling is defined as allocation of resources over time to perform a collection of tasks. It can be described as a decision-making process with the goal of optimizing one or more objectives such as to minimize the makespan and to reduce the job flow time. In manufacturing, the scheduling function interacts with other decision-making functions such as requirement planning within the plant. The four primary stages of scheduling decision are formulation, analysis, synthesis, and evaluation. Scheduling is concerned primarily with mathematical models that relate to the scheduling function.

TABLE 5.1

FT10 Problem with Job, Machine Sequence, and Processing Times

Machine Sequence (Processing Time in minutes)																			
Job 1	0	(29)	1	(78)	2	(9)	3	(36)	4	(49)	5	(11)	6	(62)	7	(56)	8	(44)	9	(21)
Job 2	0	(43)	2	(90)	4	(75)	9	(11)	3	(69)	1	(28)	6	(46)	5	(46)	7	(72)	8	(30)
Job 3	1	(91)	0	(85)	3	(39)	2	(74)	8	(90)	5	(10)	7	(12)	6	(89)	9	(45)	4	(33)
Job 4	1	(81)	2	(95)	0	(71)	4	(99)	6	(9)	8	(52)	7	(85)	3	(98)	9	(22)	5	(43)
Job 5	2	(14)	0	(6)	1	(22)	5	(61)	3	(26)	4	(69)	8	(21)	7	(49)	9	(72)	6	(53)
Job 6	2	(84)	1	(2)	5	(52)	3	(95)	8	(48)	9	(72)	0	(47)	6	(65)	4	(6)	7	(25)
Job 7	1	(46)	0	(37)	3	(61)	2	(13)	6	(32)	5	(21)	9	(32)	8	(89)	7	(30)	4	(55)
Job 8	2	(31)	0	(86)	1	(46)	5	(74)	4	(32)	6	(88)	8	(19)	9	(48)	7	(36)	3	(79)
Job 9	0	(76)	1	(69)	3	(76)	5	(51)	2	(85)	9	(11)	6	(40)	7	(89)	4	(26)	8	(74)
Job 10	1	(85)	0	(13)	2	(61)	6	(7)	8	(64)	9	(76)	5	(47)	3	(52)	4	(90)	7	(45)

A solution to a scheduling problem is any feasible resolution of the two types of constraints: limits on the capacity of available resources, and technological restrictions on machines in which jobs are to be performed. The complexity of scheduling increases with increasing number of jobs, resources, and constraints. Essentially, each scheduling problem is an optimization problem defined on the finite set of active schedules. In active schedules, no operation can be started earlier without violating a precedent constraint, or increasing the total processing time of any machine. Optimal schedules lie in the space of active schedules. It is possible to alter the operating sequence of the machines to produce a schedule with a smaller makespan and preserve the precedent constraints of the problem.

5.2.3.2 Makespan Optimization

The specific order of operations, or the order of machines that a job must visit, constitutes the precedent constraints or the technological constraints for that job. An example of the famous FT10 problem with 10 jobs and 10 machines is shown in Table 5.1. In the table, each row represents a job sequence J_i with processing time p_{ij} (job i operating on machine j). The processing times are given in parentheses. Each row is a permutation of numbers representing the sequence of machines the job must visit. For instance, in Table 5.1, job 5 should be processed on machine 2 first for 14 time units and then on machine 0 with 6 time units followed by machine 1 with 22 time units, and so on. The time taken by a job to wait in a queue is known as the waiting time. The sum of the processing time and waiting time is the completion time. Makespan is defined as the earliest time in which all the jobs are completed. The objective is to schedule all the jobs such that the makespan is minimized.

5.3 Computational Intelligence Paradigms for JSSP

The most popular model among scheduling applications is job shop scheduling since it is the best representation of the production engineering domain and has earned the reputation for being notoriously difficult to solve. Using exact procedures, the time requirement increases exponentially with a high degree polynomial for a linear increase in the problem size. Although approximation methods do not guarantee exact solutions, they are able to attain near-optimal solutions within moderate computing times and are therefore more suitable for larger problems. These approximate heuristics inspired by the natural phenomena have proved to be intelligent problem-solving approaches, especially for JSSP. In this application to solve JSSP, fuzzy logic is used to fuzzify the processing time of the jobs such that optimal makespan can be obtained by bio-inspired techniques such as GA, SPSO, ACO, and GSO. The step-by-step procedures to implement these algorithms are explained in this section.

5.3.1 Lambda Interval–Based Fuzzy Processing Time

Fuzzy logic, although a mathematical technique, defines its behavioral framework through a compact linguistic rule base. It has the ability to simultaneously consider multiple criteria and to model human experience in the form of simple rules. Furthermore, the advantage of the fuzzy logic system approach is that it incorporates both numerical and linguistic variables (Ross 2004).

The processing time of each operation is estimated depending on the nature of the machines used to perform the operations. The processing times are mostly uncertain, which are modeled by triangular membership functions represented usually in the form of triplets $\left(p_{ij}^{S}, p_{ij}^{M}, p_{ij}^{L} \right)$, where $p_{ij}^{S}, p_{ij}^{M}, p_{ij}^{L}$ denote the small, medium, and long processing times, respectively. During scheduling, the job is aligned in the order for machines with the data available in the benchmark instances. The time taken by a job to wait in a queue is calculated as the waiting time. The sum of the processing time and waiting time is the completion time. The job sequence after scheduling is optimized to rearrange the job sequence so that less waiting time and minimum makespan are obtained. Imprecision in job shop scheduling is based on fuzzification of (1) processing time, (2) due date, and (3) objective function or makespan minimization. In this application, the fuzzy processing time is considered and represented using fuzzy numbers. Some of the preliminaries used in formulating the fuzzy processing time in the interval $(\lambda, 1)$ are shown later:

5.3.1.1 Preliminaries

The following basic definitions of fuzzy sets are used to formulate the fuzzy JSSP based on fuzzy numbers (Gorzalezang 1981, Kaufmann and Gupta 1991, Yao and Wu 2000, Zimmermann, 1991). Consider a fuzzy set $\tilde{A}$ defined on $R \in [-\infty, \infty]$ with membership $\mu_{\tilde{A}(x)}$. Let A_λ denote the lambda cut-set, with $\lambda \in [0, 1]$ defined as $A_\lambda = \left\{ x \mid \mu_{\tilde{A}(x)} \geq \lambda \right\}$. Figure 5.2 shows the triangular membership with the λ interval. Here, $A_\lambda = [\tilde{A}_L(\lambda), \tilde{A}_R(\lambda)]$, where $\tilde{A}_L(\lambda)$ and $\tilde{A}_R(\lambda)$ denote the left and right points of the interval A_λ.

Definition 5.1

Let $\tilde{A} = \{a, b, c; \lambda\}$ be a triangular fuzzy number belonging to the family of lambda cut-set (membership grade = λ) fuzzy numbers $F_N(\lambda)$ defined as

$$\mu_{\tilde{A}}(x) = \begin{cases} \lambda(x-a)/(b-a), a \leq x \leq b \\ \lambda(c-x)/(c-b), b \leq x \leq c \\ 0, \text{otherwise} \end{cases} \tag{5.14}$$

where $a < b < c$ and $a, b, c \in R$.

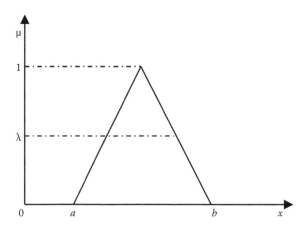

FIGURE 5.2
Triangular membership with (λ, 1) interval.

Definition 5.2

The interval-valued fuzzy set (Figure 5.3) is defined as $A_\lambda = [\tilde{A}_L, \tilde{A}_R]$. If $\tilde{A}_L = \{a, q, b; \lambda\}$ and $\tilde{A}_R = \{p, q, r; 1\}$, the membership function of A_λ is defined as

$$\mu_{\tilde{A}_L}(x) = \begin{cases} \lambda(x-a)/(q-a), a \leq x \leq q \\ \lambda(b-x)/(b-q), q \leq x \leq b \\ 0, \text{otherwise} \end{cases} \tag{5.15}$$

$$\mu_{\tilde{A}_R}(x) = \begin{cases} (x-p)/(q-p), p \leq x \leq q \\ (r-x)/(r-q), q \leq x \leq r \\ 0, \text{otherwise} \end{cases} \tag{5.16}$$

where $p < a < q < b < r$ and $a, b, p, q, r \in R$.

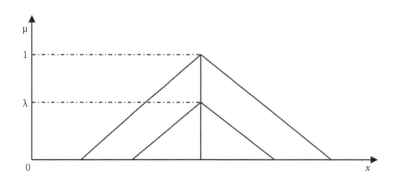

FIGURE 5.3
(λ, 1) interval-valued fuzzy number.

Definition 5.3

Let $\tilde{A} = \{a, b, c; \lambda\}$, and $\tilde{B} = \{p, q, r; \lambda\} \in F_N(\lambda)$. The binary operations performed on these two fuzzy numbers are

$$\tilde{A}(+)\tilde{B} = \{a+p, b+q, c+r; \lambda\} \tag{5.17}$$

$$\tilde{A}(-)\tilde{B} = \{a-p, b-q, c-r; \lambda\} \tag{5.18}$$

Definition 5.4

Let $\tilde{A} = \{a, b, c; \lambda\} \in F_N(\lambda)$, and $d(\tilde{A})$ be the defuzzified value of $\tilde{A}$ based on signed distance, defined by

$$d(\tilde{A}) = \frac{1}{2}\int_0^1 \left[\tilde{A}_L(\lambda) + \tilde{A}_R(\lambda) \right] d\lambda = \frac{1}{4}(a + 2b + c) \tag{5.19}$$

where
$\tilde{A}_L(\lambda) = a + (b-a)\lambda$, $\tilde{A}_R(\lambda) = c - (c-b)\lambda$, $\forall \, \lambda \in [0, 1]$

$d(\tilde{A})$ is the signed distance measured from 0 to $\tilde{A}$, also denoted as $d(\tilde{A}, \tilde{0})$

Definition 5.5

Let $\tilde{A} = \{a, b, c; \lambda\}$ and $\tilde{B} = \{p, q, r; \lambda\} \in F_N(\lambda)$. Ranking of lambda cut fuzzy numbers are defined in terms of the signed distance as

$$\tilde{A} \succ \tilde{B} \quad \text{if } d(\tilde{A}, \tilde{0}) > d(\tilde{B}, \tilde{0}) \tag{5.20}$$

$$\tilde{A} = \tilde{B} \quad \text{if } d(\tilde{A}, \tilde{0}) = d(\tilde{B}, \tilde{0}) \tag{5.21}$$

$$\tilde{B} \succ \tilde{A} \quad \text{if } d(\tilde{A}, \tilde{0}) < d(\tilde{B}, \tilde{0}) \tag{5.22}$$

5.3.1.2 JSSP with Fuzzy Processing Time

The processing time p_{ik} for job i on machine k denotes the mean of the fuzzy triangular membership, as shown in Figure 5.4. Given an interval, $[p_{ik} - \delta p_{ik1}, p_{ik} + \delta p_{ik2}]$ is more appropriate than using the single processing time p_{ik}. The parameters δp_{ik1} and δp_{ik2} are chosen such that they satisfy $0 < \delta p_{ik1} < p_{ik}$ and $0 < \delta p_{ik2}$. Similarly, considering the membership grade of p_{ik} within the interval $[\lambda, 1]$, we have $[p_{ik} - \delta p_{ik1}, p_{ik} + \delta p_{ik2}]$ for level with memberships grade 1, and $[p_{ik} - \delta p_{ik3}, p_{ik} + \delta p_{ik4}]$ for λ-level, as shown in Figure 5.5. The triangular fuzzy number corresponding to λ-level is given by

$$\tilde{p}_{ik} = \left[(p_{ik} - \delta p_{ik3}, p_{ik}, p_{ik} + \delta p_{ik4}; \lambda), (p_{ik} - \delta p_{ik1}, p_{ik}, p_{ik} + \delta p_{ik2}; 1) \right] \in F_N(\lambda, 1) \tag{5.23}$$

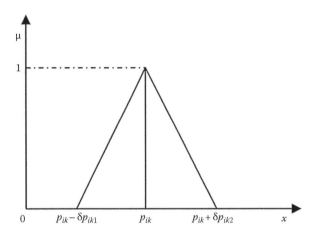

FIGURE 5.4
Representation of processing time as triangular fuzzy number.

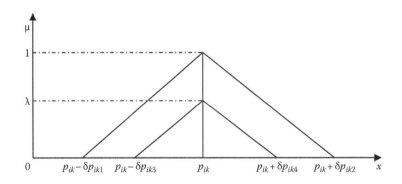

FIGURE 5.5
Representation of processing time as λ-level fuzzy number.

for $0 < \lambda < 1$, $0 < \delta p_{ik3} < \delta p_{ik1} < p_{ik}$ and $0 < \delta p_{ik4} < \delta p_{ik2}$, with $i = 1, 2, \ldots, n$; $k = 1, 2, \ldots, m$

Let c_{ik} denote the completion time of job i on machine k, defined for λ-level as

$$\tilde{c}_{ik} = \big[(c_{ik}, c_{ik}, c_{ik}; \lambda), (c_{ik}, c_{ik}, c_{ik}; 1)\big] \in F_N(\lambda, 1) \qquad (5.24)$$

where $i = 1, 2, \ldots, n$; $k = 1, 2, \ldots, m$.

From Definition 4, the signed distance from $\tilde{0}_1$ to $\tilde{c}_{ik}$ is $d^*(\tilde{c}_{ik}, \tilde{0}_1) = 2c_{ik}$, and the processing time based on the signed distance is computed as

$$p_{ik}^* = \tfrac{1}{2} d^*(\tilde{p}_{ik}, \tilde{0}_1)$$

$$= p_{ik} + \frac{1}{16}\big[\delta p_{ik4} - \delta p_{ik3} + (4 - 3\lambda)(\delta p_{ik2} - \delta p_{ik1})\big]$$

$$= \frac{1}{16}\big[\delta p_{ik4} + (p_{ik} - \delta p_{ik3}) + (4 - 3\lambda)\delta p_{ik2} + (4 - 3\lambda)(p_{ik} - \delta p_{ik1}) + (11 + 3\lambda)p_{ik}\big] \qquad (5.25)$$

Since $(4 - 3\lambda) = 1 + 3(1 - \lambda) > 0$, $0 < \delta p_{ik3} < \delta p_{ik1} < p_{ik}$, and $0 < \delta p_{ik4} < \delta p_{ik2}$ the signed distance $d^*(\tilde{p}_{ik}, \tilde{0}_1) > 0$, which is a positive distance from $\tilde{0}_1$ to $\tilde{p}_{ik}$. p_{ik}^* is the defuzzified value of $\tilde{p}_{ik}$.

The key factor in scheduling applications is ranking. Since each operand is a fuzzy set, fuzzy ranking methods are used. Let $\tilde{c}_{ik1}$ and $\tilde{c}_{ik2}$ denote two fuzzy completion times obtained from different job sequences; the rank is obtained using the difference of signed distance as

$$\Delta = d^*(\tilde{c}_{ik1}, \tilde{0}_1) - d^*(\tilde{c}_{ik2}, \tilde{0}_1) \tag{5.26}$$

If $\tilde{c}_{ik1} < \tilde{c}_{ik2}$, then the makespan obtained through sequence 1 is better than that obtained through sequence 2.

If $\tilde{c}_{ik1} = \tilde{c}_{ik2}$, then the makespan obtained through sequence 1 and 2 are equal.

If $\tilde{c}_{ik1} > \tilde{c}_{ik2}$, then the makespan obtained through sequence 1 is worse than that obtained through sequence 2.

More makespan can be ranked based on the signed distance measure.

The objective of a JSSP is to minimize the makespan with machine and job preceding constraints as follows:

$$\min\left(\max_{1 \leq k \leq m} \left(\max_{1 \leq i \leq n} (c_{ik}) \right) \right) \tag{5.27}$$

subject to the following machine and job preceding constraints:

Machine preceding constraints

$$c_{ik} - p_{ik}^* + P(1 - y_{ijk}) \geq c_{ij}, \quad i = 1, 2, \ldots, n; j, k = 1, 2, \ldots, m \tag{5.28}$$

If the job i processing on machine j precedes the job on machine k, then

$$c_{ik} - p_{ik}^* \geq c_{ij} \tag{5.29}$$

If the job on machine k precedes the job on machine j, then

$$c_{ij} - p_{ij}^* \geq c_{ik} \tag{5.30}$$

In Equation 5.28

$$y_{ijk} = \begin{cases} 1, & \text{job } i \text{ processing on machine } j \text{ precedes that on machine } k \\ 0, & \text{otherwise} \end{cases} \tag{5.31}$$

Equation 5.28 is derived from Equations 5.29–5.31.

Job preceding constraints

$$c_{lk} - c_{ik} + P(1 - z_{ilk}) \geq p_{lk}^*, \quad i, l = 1, 2, \ldots, n; k = 1, 2, \ldots, m \tag{5.32}$$

If job i comes before job l, then

$$c_{lk} - c_{ik} \geq p_{lk}^*$$ (5.33)

and if job l comes before job i, then

$$c_{ik} - c_{lk} \geq p_{ik}^*$$ (5.34)

In Equation 5.32

$$z_{ilk} = \begin{cases} 1, & \text{job } i \text{ preceeds job } l \text{ on machine } k \\ 0, & \text{otherwise} \end{cases}$$ (5.35)

Equation 5.32 is derived from Equations 5.33–5.35.

In all the previous cases $c_{ik} \geq 0$.

The problem definition of JSSP can be stated as "find a schedule to determine the operation sequences on the machines in order to minimize the total completion time" (Vassil et al. 2003), thus satisfying the constraints (1) each job must pass through a machine exactly once; (2) each job should be processed through the machine in a particular order; (3) each operation must be executed uninterrupted on a given machine; and (4) each machine can handle at most only one operation at a time. These job and machine preceding constraints are guaranteed by Equations 5.28 and 5.32, respectively.

5.3.2 Genetic Algorithm for JSSP

GAs are adaptive methods that may be used to solve search and optimization problems (Goldberg 1989). Over many generations, natural populations have evolved according to the principles of natural selection, thus following the survival of the fittest. Before a GA can run, a suitable encoding (or representation) for the problem must be devised. A fitness function is also required, which assigns a figure of merit to each encoded solution. During the run, parents must be selected for reproduction and recombined to generate offspring. It is assumed that a potential solution to a problem may be represented as a set of parameters. These parameters (known as genes) are joined together to form a string of values (chromosome or individual). The fitness of an individual depends on its chromosome and is evaluated by the fitness function. The individuals, during the reproductive phase, are selected from the population and recombined, producing offspring, which comprise the next generation. Parents are randomly selected from the population using a scheme that favors fitter individuals. The algorithm of the GA process for solving FJSSP is shown here:

- **Initial population:** The initial population is generated by localization following the approach of Kacem et al. (2002). This approach takes into account both the processing times (in this case, fuzzy processing time) and the workload of the machines, that is, the sum of the processing times of the operations assigned to each machine. The procedure consists in finding, for each operation, the machine with the minimum processing time, fixing that assignment, and then to add this time to every subsequent entry. The initial assignments are made on the basis of the random selection of job, most work remaining, and number of operations remaining.

- **Coding:** The schedule is represented in the form of a string. The string is formed by a triplet (i, j, k), one for each operation, where i represents the job that the operation belongs to, j is the progressive number of that operation within job i, and k is the machine assigned to perform the corresponding operation. The length of the string is equal to the total number of operations. Consider a sequence with three jobs and four machines, whose operational sequence is as follows:

  $S = (O_{11}, M_4), (O_{12}, M_2), (O_{21}, M_1), (O_{31}, M_1), (O_{32}, M_3)$. The solution for the GA in string form is represented as $[(1, 1, 4), (1, 2, 2), (2, 1, 1), (3, 1, 1), (3, 2, 3)]$.

- **Fitness function:** Evaluate the fitness $f(x)$ for each chromosome x in the population according to Equation 5.27. The fitness function of the chromosomes corresponds to the makespan of the solution they represent. Since the objective is to determine the least makespan values, GA evolves around chromosomes with lower fitness.

- **Selection:** The parent chromosomes are chosen for reproduction in the selection phase. In this approach, a rank-based selection mechanism is adopted, in which the individual is sorted according to its fitness, and a rank r_i is assigned to each of the N individuals in the population. The best individual gets rank N, while the worst individual is ranked 1. Based on the probability p_i given in Equation 5.36, the ith individual is chosen in the rank ordering.

$$p_i = \frac{2r_i}{N(N+1)}, \quad i = 1, 2, \ldots, N \tag{5.36}$$

- **Offspring generation:** The offspring for the next generation is generated by applying crossover and mutation to the selected parents. The initial population is modified by applying two-point crossover to the randomly selected $\rho c \times N$ members and a new population is obtained, where ρc is the crossover probability. If no crossover is performed, the offspring is an exact copy of the parents. The existing population is modified by applying bit-flip mutation to the randomly selected $\rho m \times N$ members of individuals, where ρm is the mutation probability.

- **Stopping criteria:** If the end condition is satisfied, the GA stops and returns the best solution in current population. The stopping conditions are usually the number of generations or achieving stable solutions.

5.3.3 Solving JSSP Using Stochastic Particle Swarm Optimization

In PSO algorithm, each particle in the swarm represents a solution to the problem and is defined with its position and velocity. Each particle has a position represented by a position vector x_i (i is the index of the particle) and a velocity represented by a velocity vector v_i. Each particle remembers its own best position so far in a vector $x_i^{\#}$, and its jth dimensional value is $x_{ij}^{\#}$. The best position vector among the swarm so far is then stored in a vector $x*$, and its jth dimensional value is x_j^{*}. During the iteration time t, the update of the velocity from the previous velocity to the new velocity is determined by Equation 5.37. The new position is then determined by the sum of the previous position and the new velocity by Equation 5.38:

$$v_{ij}(t+1) = wv_{ij}(t) + c_1 r_1 (x_{ij}^{*}(t) - x_{ij}(t)) + c_2 r_2 (x_j^{*}(t) - x_{ij}(t)) \tag{5.37}$$

$$x_{ij}(t+1) = x_{ij}(t) + v_{ij}(t+1) \tag{5.38}$$

where w is called as the inertia weight factor, r_1 and r_2 are random numbers that are used to maintain the diversity of the population and are uniformly distributed in the interval $[0, 1]$ for the jth dimension of the ith particle, c_1 is a positive constant, called the coefficient of the self-recognition component (cognitive component), and c_2 is a positive constant called the coefficient of the social component. From Equation 5.37, a particle decides where to move next by considering its own experience, which is the memory of its best past position, and the experience of its most successful particle in the swarm.

SPSO is applied to solve the fuzzy JSSP. Here, the acceleration constants c_1, c_2 and the inertia weight w vary during the algorithmic run, thus they are termed dynamic coefficients. The cognition component c_1 reflects the emphasis for the particles to move toward the current local optimal solution, and the social learning factor c_2 reflects the emphasis for the particles to move toward the global population best solution. Proper fine-tuning of these parameters results in faster convergence of the algorithm and alleviation of the local minima. Recent work reports that it might be even better to choose a larger cognitive parameter c_1 than the social parameter c_2, such that $c_1 + c_2 \leq 4$ (Arumugam and Rao 2008). In this chapter, the values of c_1 and c_2 are set such that $c_1 + c_2 \leq 3$ in order to obtain a better convergence rate toward the optimal makespan. The dynamic acceleration coefficients are initialized as in Equations 5.39 and 5.40:

$$c_1 = (c_{i1} - c_{f1}) * \left(\frac{max\_iter - iter}{max\_iter} \right) + c_{f1} \tag{5.39}$$

$$c_2 = (c_{i2} - c_{f2}) * \left(\frac{max\_iter - iter}{max\_iter} \right) + c_{f2} \tag{5.40}$$

where
  c_{i1} is the initial cognitive factor
  c_{i2} is the initial social factor
  c_{f1} is the final cognitive factor
  c_{f2} is the final social factor

Similarly, the inertia weight w of the particle is also updated during the iterations in a nonlinear manner according to Equation 5.41:

$$w = (w_{max} - w_{min}) * \left(\frac{max\_iter - iter}{max\_iter} \right) + w_{min} \tag{5.41}$$

where
  w_{max} is the maximum inertia weight
  w_{min} is the minimum inertia weight
  max_iter is the maximum number of iterations run by the SPSO
  iter is the value of the current iteration

For the fuzzy JSSP, the algorithm steps involved in the SPSO are as follows:

Step 1: Initialize the SPSO algorithm-dependent parameters such as the number of generations, number of dimensions, minimum inertia weight (w_{min}), maximum inertia weight (w_{max}), initial and final values for cognitive and social coefficients, and the maximum number of iterations.

Step 2: Initialize the operations of jobs and sequence of operation on machines as the initial set of particles based on the fuzzy processing time.

Step 3: Compute the dynamic acceleration constants according to Equations 5.39 and 5.40. Update the dynamic inertia weight nonlinearly based on Equation 5.41.

Step 4: Evaluate the fitness function based on Equation 5.27. The objective of the fuzzy JSSP is to compute the completion time of the jobs; hence for each particle in the swarm, the makespan is computed according to Equation 5.27.

Step 5: Determine the local best and the global best of the particles, and update the velocity based on Equations 5.37 and 5.38. During each iteration, the local and global best are updated using the fitness function, thus enhancing the solution toward the optimal value during each iteration.

Step 6: Repeat steps 3–5 until the maximum number of iterations have been reached.

5.3.4 Ant Colony Optimization for JSSP

Ant colonies exhibit interesting behaviors; though one specific ant has limited capabilities, the behavior of a whole ant colony is highly structured. They are capable of finding the shortest path from their nest to a food source without using visual cues but by exploiting pheromone information (Zhang et al. 2006). While walking, ants can deposit some pheromone on the path. The probability that the ants coming later choose the path is proportional to the amount of pheromone on the path, previously deposited by other ants. This theory is the basis for forming the ACO algorithm using artificial ants. The artificial ants are designed based on the behavior of real ants. They lay pheromone trails on the graph edges, and choose their path with respect to probabilities that depend on pheromone trails, which decrease progressively by evaporation.

At the end of each generation, each ant present in the population spawns a complete tour traversing all the nodes based on a probabilistic state transition rule. The nodes are chosen by the ants based on the order in which they appear in the permutation process. The node selection process involves a heuristic factor as well as a pheromone factor used by the ants. The heuristic factor, denoted by η_{ij}, and the pheromone factor, denoted by τ_{ij}, are indicators of how good it seems to have node j at node i of the permutation. The heuristic value is generated by some problem-dependent heuristics, whereas the pheromone factor stems from former ants that have found a good solution. The next node is chosen by an ant according to the following rule, which is called the pseudo-random proportional action choice rule (Merkle, 2003). With probability q_0, where $0 \leq q_0 < I$ is a parameter of the algorithm, the ant chooses a node from the set of nodes that have not been selected so far that maximizes $(\tau_{ij})^{\alpha}(\eta_{ij})^{\beta}$, where $\alpha \geq 0$ and $\beta \geq 0$ are constants that determine the relative influence of the pheromone values and the heuristic values on the decision of the ant. The probability of choosing the next node is chosen from the set S according to the probability distribution given by

$$P_{ij} = \frac{(\tau_{ij})^{\alpha}(\eta_{ij})^{\beta}}{\sum_{h \in S}(\tau_{ij})^{\alpha}(\eta_{ij})^{\beta}} \tag{5.42}$$

This probability, also known as the transition probability, is a tradeoff between the pheromone factor and the heuristic factor. The heuristic factor is computed as $\eta_{ij} = 1/F(X_j)$, $j \in S$, where $F(X)$ represents the cost function of X. While constructing its tour, an ant will modify the amount of pheromone on the passed edges by applying the local updating rule $\tau_{ij}(t) \leftarrow (1 - \rho)\tau_{ij}(t) + \rho\tau_0$, where $\tau_{ij}(t)$ is the amount of pheromone on the edge (i, j) at time t, ρ is a parameter governing pheromone decay such that $0 < \rho < 1$, and τ_0 is the initial value of pheromone on all edges. Once all ants have arrived at their destination, the amount of pheromone on the edge is modified again by applying the global updating rule: $\tau_{ij}(t) \leftarrow (1 - \rho)\tau_{ij}(t) + \rho\Delta\tau_{ij}(t)$, where $\Delta\tau_{ij}(t) = L^{-1}$ if this is the best tour, and otherwise $\Delta\tau_{ij}(t) = 0$. L indicates the length of the global best tour. The pheromone updating rule is meant to simulate the change in the amount of pheromone due both to the addition of new pheromone deposited by ants on the visited edges and to pheromone evaporation (Dorigo and Gambardella 1997). The algorithm stops iterating either when an ant has found a solution or when a maximum number of generations have been performed. The ACO procedure for the fuzzy JSSP is shown here:

Step 1: Initialization

    a. Let M_{ij} denote the machine index.

    b. Assign no. of ants = no. of jobs.

    c. Each job is a set of operations denoted by $\{O_{11}, O_{12}, ..., O_{JOBS,1}\}$.

    d. Initialize parameters α, β, and ρ.

    e. Record the current position of ants.

Step 2: Building ant tour

    a. If ant k is the first operation, then

    b. Assign the first operation O_{k1} of job 1 to ant k's first visited operation.

    c. Record the visited operation of ant k in its memory.

Else

    a. Choose the next operation O_{k2} based on constraints of the job.

    b. If next operation is not the last, then

    c. Assign M_{ij} after the next operation.

    d. Determine the transition probability according to Equation 5.42.

Step 3: Local pheromone updating rule

    a. Apply the local pheromone updating rule $\tau_{ij}(t) \leftarrow (1 - \rho)\tau_{ij}(t) + \rho\tau_0$, where $\tau_{ij}(t)$ is the amount of pheromone on the edge (i, j) at time t, ρ is a parameter governing pheromone decay such that $0 < \rho < 1$, and τ_0 is the initial value of pheromone on all edges.

Step 4: Global updating rule

    a. Apply the global updating rule $\tau_{ij}(t) \leftarrow (1 - \rho)\tau_{ij}(t) + \rho\Delta\tau_{ij}(t)$, where $\Delta\tau_{ij}(t) = L^{-1}$ if this is the best tour, otherwise $\Delta\tau_{ij}(t) = 0$, and L indicates the length of the global best tour. Determine the time taken by the ant k and compute the shortest time.

Step 5: If stopping condition is true, then stop; else continue from Step 2. The stopping condition is chosen as the number of iterations.

5.3.5 JSSP Based on Hybrid SPSO

In this section, a hybrid method based on GA and SPSO (termed GSO) to solve JSSP is proposed. SPSO has improved features in terms of convergence rate when compared with the basic PSO (Abido 2002, Fourie and Groenwold 2002). Because of the continuous nature of the particles, a standard encoding scheme cannot be defined for a JSSP. Thus SPSO is applied to the fuzzy JSSP to determine a suitable method between the job sequence and the position of particles. In GSO, the worst particles found by SPSO are replaced by the best particles obtained by GA.

By hybridizing GA and SPSO, the continuous problem-solving ability of SPSO is eliminated, thus finding the global optimum effectively and efficiently. These hybrid algorithms enlarge the application domain and improve the performance by combining the efficient features of both algorithms such as the mechanism involved in optimization, search pattern, mode of operation, and so on, thus increasing the optimization efficiency (Song 2010). Some of the basic ideas behind the design of this hybrid algorithm are as follows:

- Each feasible solution of the fuzzy JSSP is a point in the search space and is considered as an individual. The job sequence is not continuous; hence a suitable encoding scheme is adopted for SPSO to fit the problems.
- The infeasible solutions produced during the algorithmic run are adjusted by the hybrid combination of GA and SPSO.
- The chances of the algorithm getting trapped into the local optimization at a very early stage can be avoided by the GSO algorithm.

The procedure involved in the proposed hybrid algorithm is described as follows:

- **Coding:** The initial task in solving the fuzzy JSSP is to represent the solution. The chromosomes are represented in a string form, with the first part indicating the number of operations and the second part indicating the sequence of jobs. Consider an example with three jobs and four machines. Assume a string in the form [2, 3, 1, 4, 2, 1, 3, 4] representing the number of operations. From this string, it can be inferred that the operation $O1, 1$ of a job is operated on machine $M2$, while the operation $O1, 2$ is operated on machine $M3$, and so on. The length of the string is equal to the number of operations to be performed in a job. The second part of the chromosome denotes the sequence of jobs, for example, represented in a string like [3, 1, 2, 1, 2, 2, 1, 3]. This indicates that job $J3$ is processed first, followed by job $J1$, and then by job $J2$, and so on. Finally, the chromosome is represented in combined form as [2, 3, 1, 4, 2, 1, 3, 4, 3, 1, 2, 1, 2, 2, 1, 3] with machine assignment followed by the job sequence.
- **Initial population:** In optimization problems, the initial population is usually large enough to get near-optimum solutions. The chromosomes obtained from the coding part constitute the initial population or particles for the fuzzy GSO (FGSO) algorithm. The initial population is formed in a random manner.
- **Fitness evaluation:** The objective of the fuzzy JSSP is to compute the completion time of the jobs; hence for each chromosome the makespan is computed. For every generation, all the chromosomes are evaluated for fitness function based on Equation 5.27.

- **Offspring generation:** The new-generation chromosomes are produced by changing the assignment of the jobs on the machines and by changing the sequence of operation on the machines. In the evolutionary process, SPSO is used to generate new offspring based on new position of the particle, thus updating the velocity according to

$$x_{t+1}^i = x_t^i + \underbrace{w_t v_t^i + c_1 r_1 (p_t^i - x_t^i) + c_2 r_2 (p_t^g - x_t^i)}_{v_{t+1}^i} \tag{5.43}$$

The best position visited by the kth particle since the first time step is known as the local best position, denoted as pbest p_t^i, while gbest p_t^g represents the best position that the kth particle and its neighbors have visited from the beginning of the algorithm. During each iteration, pbest and gbest are updated based on the objective (fitness) value. If the current value of pbest is better, p_t^i and its objective value are updated with the current position and objective value. Similarly, if the objective value is better than the objective value of p_t^g, then p_t^g and its objective value are updated with the position and objective value of the current best particle. As a parallel process, the genetic operators of crossover and mutation are applied to update the particles according to Equation 5.44:

$$x_{t+1}^i = \hat{x}_t^i \oplus (p_t^i \otimes x_t^i) \oplus (p_t^g \otimes x_t^i) \tag{5.44}$$

Here, the symbol $\otimes$ denotes the crossover operation performed between the current best and the best individual obtained so far. The symbol $\oplus$ represents the optimal or best offspring resulting from $(p_t^i \otimes x_t^i)$, or $(p_t^g \otimes x_t^i)$ or $\hat{x}_t^i$. $(p_t^i \otimes x_t^i)$ denotes the crossover operation between the two parents p_t^i and x_t^i, and $\hat{x}_t^i$ is the mutated offspring of x_t^i, and x_{t+1}^i is the new particle obtained from genetic operators. The particles obtained through GA and SPSO are evaluated for their fitness, and the best individual is retained in the population for further generations. Here, two-point crossover and flip-bit mutation are applied.

The driving parameter involved in this hybridization process is the hybridization factor (HF), which represents the percentage of individuals in the population evolved through GA. In simple terms, if HF = 0, then the algorithm is a pure SPSO algorithm, and if HF = 1, then the algorithms is a pure GA. HF has the range [0, 1] in order to maintain the balance among the individuals evolved through GA and SPSO.

- **Stopping condition:** The algorithm terminates when the maximum number of iterations is reached, simultaneously obtaining the best optimal solution.

5.4 m-File Snippets and Outcome of JSSP Based on CI Paradigms

The proposed algorithms were implemented using MATLAB R2008b software on an Intel i3 CPU, 2.53 GHz, 4 GB RAM PC. Extensive experimental analysis was conducted to evaluate the efficiency and effectiveness of the proposed intelligent approaches GA, SPSO, ACO, and GSO on various sizes of benchmark instances with fuzzy processing time. This section describes the grouping of benchmark instances, the algorithm control parameters, and the experimental results of the intelligent algorithms for solving fuzzy-based JSSP.

5.4.1 Description of Benchmark Instances

The bio-inspired algorithms were experimented on 162 benchmark instances of various job and machine sizes obtained from the OR library Website (Beasley 1990) and Taillard Website (Taillard n.d.). The profiles of the JSSP instances are shown in Table 5.2. The test instances were classified in this work into the following groups.

- **Group A:** Five benchmark instances ABZ5–ABZ9 contributed by Adams et al. (1988), with two instances of size 10 × 10 and three instances of size 20 × 15; three instances symbolized as FT06 (6 × 6), FT10 (10 × 10), and FT20 (20 × 5) due to Fisher and Thompson (1963); 10 instances denoted as ORB01–ORB10 (10 × 10) due to Applegate and Cook (1991); 20 instances of three different sizes denoted as SWV01–SWV05 (20 × 10), SWV06–SWV10 (20 × 15), and SWV11–SWV20 (50 × 10); and 4 instances YN1–YN4 (20 × 20) due to Storer et al. (1992). Group A consists of a total of 42 benchmark instances.

- **Group B:** Forty benchmark instances denoted as LA01–LA05 (10 × 5), LA06–LA10 (15 × 5), LA11–LA15 (20 × 5), LA16–LA20 (10 × 10), LA21–LA25 (15 × 10), LA26–LA30 (20 × 10), LA31–LA35 (30 × 10), and LA36–LA40 (15 × 15), contributed by Lawrence (1984).

- **Group C:** Forty benchmark instances due to Taillard (1993) denoted by TA01–TA10 (15 × 15), TA11–TA20 (20 × 15), TA21–TA30 (20 × 20), and TA31–TA40 (30 × 15)

- **Group D:** Forty benchmark instances contributed by Taillard (1993): TA41–TA50 (30 × 20), TA51–TA60 (50 × 15), TA61–TA70 (50 × 20), and TA71–TA80 (100 × 20).

TABLE 5.2

Profile of JSSP Instances

Instance	$|J| \times |M|$	Instance	$|J| \times |M|$
ABZ5–ABZ6	10 × 10	LA16–LA20	10 × 10
ABZ7–ABZ9	20 × 15	LA21–LA25	15 × 10
FT06	6 × 6	LA26–LA30	20 × 10
FT10	10 × 10	LA31–LA35	30 × 10
FT20	20 × 10	LA36–LA40	15 × 15
ORB01–ORB10	10 × 10	TA01–TA10	15 × 15
SWV01–SWV05	20 × 10	TA11–TA20	20 × 15
SWV06–SWV10	20 × 15	TA21–TA30	20 × 20
SWV11–SWV20	50 × 10	TA31–TA40	30 × 15
YN1–YN4	20 × 20	TA41–TA50	30 × 20
LA01–LA05	10 × 5	TA51–TA60	50 × 15
LA06–LA10	15 × 5	TA61–TA70	50 × 20
LA11–LA15	20 × 5	TA71–TA80	100 × 20

5.4.2 MATLAB® m-File Snippets

The parameters are initialized based on the algorithm applied to compute makespan of the JSSP. The main program consists of the following stages:

- **Parameter Initialization:** The notations to parameters are assigned, and they are intialized to the optimum values.

- **Configuring the JSSP instances:** The JSSP instance data obtained from the web-sites are written in MATLAB in the following form:

  The example shown later is for the instance ABZ06, which is an instance of size 10 × 10 (10 jobs and 10 machines). Here "js" denotes the job sequence, and "T" denotes the processing time for each step of the job.

```
js=[8    9    6    4    5    7    3    1    10   2
    6    3    9    10   2    5    8    4    1    7
    2    8    7    3    10   1    5    4    9    6
    5    9    6    10   4    7    3    2    1    8
    9    10   8    4    5    1    2    6    7    3
    4    10   6    3    5    2    7    9    1    8
    6    3    8    7    10   1    5    9    4    2
    5    7    6    9    10   3    1    2    8    4
    10   7    6    9    8    1    4    2    5    3
    1    4    8    3    2    6    9    5    10   7];

T=[62   24   25   84   47   38   82   93   24   66
    47   97   92   22   93   29   56   80   78   67
    45   46   22   26   38   69   40   33   75   96
    85   76   68   88   36   75   56   35   77   85
    60   20   25   63   81   52   30   98   54   86
    87   73   51   95   65   86   22   58   80   65
    81   53   57   71   81   43   26   54   58   69
    20   86   21   79   62   34   27   81   30   46
    68   66   98   86   66   56   82   95   47   78
    30   50   34   58   77   34   84   40   46   44];
```

- **Fuzzifying the processing time:**

  Application of the CI algorithm

  The corresponding CI algorithm applied to solve the JSSP is executed with the fuzzy processing time. An example of applying the PSO to solve the JSSP instance ABZ06 is presented here:

  New generations are progressed based on the problem size. The following m-file snippet shows how a population for a single generation is formed.

```
newgen1=[subpop1,min2,min3,min4,min5,min6,min7,min8,min9,min10,optt1];
for i=1:psize
    [Ya,pop1(i,1:n)]=sort(newgen1(i,1:n));
    lz=ceil(pop1(i,1:n)/mnum);
    newgen1(i,n+1)=jsmakespan10c10(lz);
%%%%%%%%%%%%%%%%%%%%%%%%%%%%%%%%%%%%%%%%%% makespan10c10 %%%%%%%%%%%%%%%%
makespan20c5
    optt1(i)=newgen1(i,n+1);
end
```

```
[Ya,Ia]=sort(optt1);
optimy1(current_gen)=newgen1(Ia(1),n+1);
newgenp1=newgen1(1:psize,:);
newgeng1=newgen1(Ia(1),:);
xmin1=newgen1(Ia(1),1:n/10);%È«¾Ö×îÓÅÁ£×ÓÖÖÈ°
for i=1:psize
    min1(i,:)=xmin1;
end
```

From the new generations, the swarm is evaluated for fitness, and the makespan is computed. The best positon for each particle in the swarm is updated.

```
for i=1:psize
        [Ya,pop1(i,1:n)]=sort(newgen1(i,1:n));
        lz=ceil(pop1(i,1:n)/mnum);
         newgen1(i,n+1)=jsmakespan(lz);
         optt1(i)=newgen1(i,n+1);
         if newgen1(i,n+1)<=optimy1(current_gen)
             optimy1(current_gen)=newgen1(i,n+1);
             newgeng1=newgen1(i,:);
             xmin1=newgen1(i,1:n/jnum);
        end
    end
    % Updating the best position for each particle
    changeColumns=newgen1(:,n+1)<newgenp1(:,n+1);

    newgenp1(find(changeColumns),:)=newgen1(find(changeColumns),:);
    r=fix(rand(1,1)*psize)+1;
     newgen1(r,1:n/10)=xmin1;
     subpop1=newgen1(:,1:n/mnum);
    for i=1:psize
     min1(i,:)=xmin1;
    end
```

- **Computation of the makespan, relative error, and computational time:** The m-file snippet used to compute the makespan is given here:

```
if jp(gen(i))==1
        if mj(js(gen(i),1))==1
            starts(js(gen(i),1),1)=0;
            ends(js(gen(i),1),1)=T(gen(i),1);
        else
starts(js(gen(i),1),mj(js(gen(i),1)))=ends(js(gen(i),1),mj(js(gen(i),1))-1);
ends(js(gen(i),1),mj(js(gen(i),1)))=starts(js(gen(i),1),mj(js(gen(i),1)))
+T(gen(i),1);
        end
        else
        if mj(js(gen(i),jp(gen(i))))==1
starts(js(gen(i),jp(gen(i))),1)=ends(js(gen(i),jp(ge
n(i))-1),mjn(js(gen(i),jp(gen(i))-1),gen(i)));
ends(js(gen(i),jp(gen(i))),1)=starts(js(gen(i),jp(gen(i))),1)+T(gen(i),jp
(gen(i)));
        else
```

```
starts(js(gen(i),jp(gen(i))),mj(js(gen(i),jp(gen(i)))))=max(ends(js(gen(i),
jp(gen(i))),mj(js(gen(i),jp(gen(i))))-1),ends(js(gen(i),jp(gen(i))-1),
mjn(js(gen(i),jp(gen(i))-1),gen(i))));

ends(js(gen(i),jp(gen(i))),mj(js(gen(i),jp(gen(i)))))=starts(js(gen(i),
jp(gen(i))),mj(js(gen(i),jp(gen(i)))))+T(gen(i),jp(gen(i)));
        end
    end
```

5.4.3 Fuzzy Processing Time Based on (λ, 1) Interval Fuzzy Numbers

The procedure of obtaining fuzzy processing time from crisp processing time is demonstrated in detail for the FT06 benchmark instance in this section. Similar procedure is followed for all the other instances. The parameters used to obtain the fuzzy processing time are shown in Table 5.3. The parameters of the fuzzy processing time $\tilde{p}_{ik}$ are computed according to Equation 5.23. For each $\tilde{p}_{ik}$, δp_{ik1} and δp_{ik2} are randomly generated in the open interval $(0, p_{ik}/2)$, where p_{ik} is the crisp processing time given in the problem. The parameters δp_{ik3} and δp_{ik4} are carefully chosen to satisfy $0 < \delta p_{ik3} < \delta p_{ik1}$ and $0 < \delta p_{ik4} < \delta p_{ik2}$, respectively.

Table 5.4 shows the fuzzy processing times computed according to Equation 5.23. The terms $(p_{ik} - \delta p_{ik3}, p_{ik}, p_{ik} + \delta p_{ik4}; \lambda)$ and $(p_{ik} - \delta p_{ik1}, p_{ik}, p_{ik} + \delta p_{ik2}; 1)$ are computed and represented one below the other for each corresponding job and machine. For instance, job 3

TABLE 5.3

Parameter Settings to Obtain Fuzzy Processing Time

Parameters	Values
Interval parameter δp_{ik1}	Random in $(0, p_{ik}/2)$ such that $0 < \delta p_{ik1} < p_{ik}$
Interval parameter δp_{ik2}	Random in $(0, p_{ik}/2)$ such that $0 < \delta p_{ik2}$
Interval parameter δp_{ik3}	$0 < \delta p_{ik3} < \delta p_{ik1}$
Interval parameter δp_{ik4}	$0 < \delta p_{ik4} < \delta p_{ik2}$
Lambda	0.9

TABLE 5.4

Fuzzy Processing Time (in Minutes) for Instance FT06

Job	M_1	M_2	M_3	M_4	M_5	M_6
1	(2.4, 3, 4.2; 0.9)	(4.5, 6, 7.2; 0.9)	(0.7, 1, 1.5; 0.9)	(6.2, 7, 8.4; 0.9)	(5.3, 6, 6.8; 0.9)	(2.4, 3, 4.2; 0.9)
	(2.1, 3, 4.5; 1)	(4.2, 6, 7.8; 1)	(0.5, 1, 1.7; 1)	(5.6, 7, 8.8; 1)	(5.1, 6, 7.2; 1)	(2.1, 3, 4.5; 1)
2	(8.6, 10, 12; 0.9)	(7.1, 8, 9.1; 0.9)	(4.2, 5, 7.1; 0.9)	(3.8, 4, 4.8; 0.9)	(8.9, 10, 11.4; 0.9)	(9, 10, 11.5; 0.9)
	(8.2, 10, 12.4; 1)	(6.8, 8, 9.4; 1)	(4, 5, 7.5; 1)	(3.3, 4, 5.2; 1)	(8.3, 10, 11.9; 1)	(8.7, 10, 12.5; 1)
3	(8, 9, 10.7; 0.9)	(0.8, 1, 1.2; 0.9)	(4.1, 5, 5.9; 0.9)	(3.5, 4, 4.8; 0.9)	(6.2, 7, 7.7; 0.9)	(6.8, 8, 9.5; 0.9)
	(7.6, 9, 11.2; 1)	(0.6, 1, 1.6; 1)	(3.7, 5, 6.5; 1)	(3, 4, 5.2; 1)	(5.9, 7, 8.3; 1)	(6.1, 8, 9.9; 1)
4	(4.6, 5, 5.9; 0.9)	(3.8, 5, 5.9; 0.9)	(4.2, 5, 5.7; 0.9)	(2.7, 3, 3.6; 0.9)	(7.1, 8, 8.8; 0.9)	(8.2, 9, 10.2; 0.9)
	(4.2, 5, 6.3; 1)	(3.5, 5, 6.3; 1)	(3.9, 5, 6.3; 1)	(2.3, 3, 3.9; 1)	(6.6, 8, 9.3; 1)	(8, 9, 10.6; 1)
5	(2.6, 3, 3.6; 0.9)	(2.5, 3, 3.8; 0.9)	(7.8, 9, 10.8; 0.9)	(0.6, 1, 1.5; 0.9)	(4.2, 5, 5.6; 0.9)	(3.5, 4, 4.6; 0.9)
	(2.1, 3, 3.9; 1)	(2.2, 3, 4.2; 1)	(7.2, 9, 11.0; 1)	(0.3, 1, 1.8; 1)	(4.0, 5, 6.0; 1)	(3.2, 4, 5.0; 1)
6	(8.6, 10, 11.4; 0.9)	(2.3, 3, 3.7; 0.9)	(0.7, 1, 1.3; 0.9)	(2.5, 3, 3.6; 0.9)	(3.4, 4, 4.6; 0.9)	(8.4, 9, 10.2; 0.9)
	(8.2, 10, 12.2; 1)	(2.1, 3, 4.0; 1)	(0.5, 1, 1.5; 1)	(2.1, 3, 4.0; 1)	(3.0, 4, 5.0; 1)	(7.8, 9, 10.6; 1)

TABLE 5.5

Defuzzified Values of Fuzzy Processing Time for
Instance FT06 (in Minutes)

Job	M_1	M_2	M_3	M_4	M_5	M_6
1	3.279	6.469	1.148	7.424	6.273	3.242
2	10.469	8.329	5.459	4.279	10.486	10.475
3	9.468	1.106	5.318	4.260	7.281	8.479
4	5.238	5.351	5.287	3.172	8.318	9.346
5	3.315	3.239	9.503	1.178	5.251	4.203
6	10.524	3.242	1.119	3.213	4.258	9.340

processed on machine 4 produces a fuzzy processing time as (3.5, 4, 4.8; 0.9) and (3, 4, 5.2; 1). Here, in the first term the values 3.5, 4, and 4.8 represent the processing times $p_{ik} - \delta p_{ik3}$, p_{ik}, and $p_{ik} + \delta p_{ik4}$ at $\lambda = 0.9$, while the values in the second term, 3, 4, and 5.2 correspond to the processing times $p_{ik} - \delta p_{ik1}$, p_{ik}, and $p_{ik} + \delta p_{ik2}$ at $\lambda = 1$, respectively, in minutes. In a similar manner, all the values corresponding to the job and machine number are computed as presented in Table 5.4. Since the membership grade of p_{ik} is not always equal to 1, an interval $(\lambda, 1)$ is used to represent the membership grade. This kind of fuzzy computation of the processing time makes the job shop scheduling more suitable for practical problems.

The defuzzified values of the fuzzy processing times for the FT06 problem are shown in Table 5.5. These values are calculated from Equation 5.25 based on the signed distance obtained from Definition 4 in Section 5.3.1. The intelligent optimization heuristics is further applied on these defuzzified values of processing times to compute the optimal makespan.

5.4.4 Performance of GA-Based JSSP

It is very important to set the GA parameters properly in order to guarantee a good flow of information between the GA generations during the evolutionary process so as to obtain an effective cooperation between the individuals. The major parameters that affect the quality of the flow of information are considered and tuned properly to determine a good setting. The parameters and their settings are shown in Table 5.6. The population size is chosen according to the size of the problem as the product of the number of jobs and machines. The population size is also directly related to the number

TABLE 5.6

Parameter Setting for JSSP Based on GA

Parameter	Value
Population size	Instance dependent
Selection type	Rank-based mechanism
Selective pressure	1.5
Type of crossover	Two-point
Type of mutation	Flip-bit
Crossover rate	0.35
Mutation rate	0.002
Maximum no. of generations	Instance dependent

of generations. The next-generation offspring are generated using selection mechanism based on ranking.

The selection pressure is varied between 0 (less dependent) and 2 (strongly dependent). A weak selection pressure produces a bad individual with almost the same chance to reproduce as a good individual, whereas a strong selection pressure favors the reproduction of only good individuals. In this work, the selection pressure is set to 1.5 in order to maintain a proper balance between strong and weak dependency. Once the parents are selected, they are further treated with genetic operators such as crossover and mutation. A two-point crossover with a probability of 0.35 is chosen as the appropriate value based on several tests conducted on benchmark instances. Similarly, flip-bit type mutation with a rate of 0.002 is found to be the optimal setting.

Table 5.7 shows the experimental results for the instances belonging to Group A. The table lists the instance type along with the corresponding best known optimal makespan (OPT) value known so far. The computed makespan value (C_{max}), average makespan (AC_{max}, computed from several trials), relative error (RE, the percentage by which the obtained solution is more than OPT), and average computational time (T, from 10 trials for each benchmark instance) are computed and presented. The relative error is calculated as RE = ((C_{max}−OPT)/OPT)* 100. It is seen from Table 5.7 that the relative error is "nil" for five instances: ORB01, ORB05, SWV13, SWV16, and YN1. This implies that the makespan obtained through GA is equal to the optimal makespan value of the respective benchmark instances.

Parameters such as the population size and the number of generations increase linearly as the problem size increases. The instances of Group B—LA01–LA40—are tested with the fuzzy processing time using GA and the makespan, average makespan, relative error, and average computational time are computed. The results presented in Table 5.8 show the optimal makespan solved by GA for 10 instances: LA03, LA05, LA06, LA09, LA11, LA12, LA14, LA15, LA38, and LA40. The performance evaluation is based on the percentage deviation of the obtained makespan from the optimal makespan.

The GA is also tested on the Taillard instances TA01–TA40 (Group C), and the results are presented in Table 5.9. The performance of GA is not satisfactory on these instances because, after 10 trial runs for each benchmark instance, the proposed algorithm is unable to obtain the optimal makespan even for a single instance.

The computational results of GA for instances TA41–TA80—the computed makespan, average makespan, RE, and average computational time—are shown in Table 5.10. The instances in Group D are relatively easy to solve compared to the instances in Group C. The three instances TA51, TA56, and TA65 produced the optimal makespan (C_{max}) with 0% deviation from the known makespan (OPT). Since the sizes of the problem instances are too large, the computational times are observed to be very high for GA to converge toward the optimal makespan.

5.4.5 Analysis of JSSP Using SPSO

The fuzzy JSSP is analyzed using the SPSO based on stochastic process theory. Tuning the parameters to produce a set of optimal results is an intelligent task. The parameters and their optimal settings for the SPSO are shown in Table 5.11. The parameters such as acceleration constants and inertia weight are varied dynamically during the algorithmic run. The initial and final values of the acceleration constants (social and cognitive factors) are set to 1.2 and 2, respectively. During every iteration, it is ensured that $c_1 + c_2 \leq 3$. The inertia

TABLE 5.7

Experimental Results for Instances of Group A Using GA

Instance	OPT (min)	GA Optimal Results				Instance	OPT (min)	GA Optimal Results			
		C_{max} (min)	AC_{max} (min)	RE (%)	T (s)			C_{max} (min)	AC_{max} (min)	RE (%)	T (s)
ABZ5	1234	1277	1279	3.484603	36.17	SWV04	(1450, 1470)	1477	1477	0.47619	117.065
ABZ6	943	949	950	0.636267	36.98	SWV05	1424	1428	1428.75	0.280899	118.149
ABZ7	656	670	671	2.134146	749.885	SWV06	(1591, 1675)	1678	1678	0.179104	748.536
ABZ8	(645, 665)	672	673.5	1.052632	748.717	SWV07	(1446, 1594)	1606	1607	0.752823	749.524
ABZ9	(661, 679)	682	682.5	0.441826	748.376	SWV08	(1640, 1755)	1758	1758	0.17094	748.127
FT06	55	57	57.25	3.636364	5.987	SWV09	(1640, 1661)	1666	1666	0.301023	748.962
FT10	930	976	977.75	4.946237	36.178	SWV10	(1631, 1743)	1755	1756	0.688468	749.335
FT20	1165	1236	1237	6.094421	19.5245	SWV11	2983	2999	2999.5	0.536373	400.439
ORB01	1059	1059	1059	0	36.185	SWV12	(2972, 2979)	2989	2989	0.335683	401.552
ORB02	888	890	890	0.225225	36.0236	SWV13	3104	3104	3104	0	399.055
ORB03	1005	1012	1012	0.696517	36.0954	SWV14	2968	2977	2978	0.303235	399.854
ORB04	1005	1006	1007	0.099502	36.521	SWV15	(2885, 2886)	2908	2910	0.762301	400.457
ORB05	887	887	887	0	36.078	SWV16	2924	2924	2924	0	400.635
ORB06	1010	1023	1023.75	1.287129	36.127	SWV17	2794	2799	2799.75	0.178955	401.885
ORB07	397	400	400	0.755668	36.179	SWV18	2852	2880	2881	0.981767	402.324
ORB08	899	912	912.5	1.446051	36.254	SWV19	2843	2934	2935	3.200844	401.758
ORB09	934	944	945	1.070664	36.745	SWV20	2823	2888	2889	2.302515	401.665
ORB10	944	965	967	2.224576	36.154	YN1	(836, 884)	884	884	0	1174.85
SWV01	1407	1423	1423.75	1.137171	118.045	YN2	(861, 904)	909	909	0.553097	1195.45
SWV02	1475	1477	1478	0.135593	117.1245	YN3	(827, 892)	899	899.5	0.784753	1165.74
SWV03	(1369, 1398)	1409	1409.5	0.786838	117.854	YN4	(918, 968)	976	976	0.826446	1145.77

TABLE 5.8

Experimental Results for Instances of Group B Using GA

Instance	OPT (min)	GA Optimal Results					Instance	OPT (min)	GA Optimal Results			
		C_{max} (min)	AC_{max} (min)	RE (%)	T (s)				C_{max} (min)	AC_{max} (min)	RE (%)	T (s)
LA01	666	666	667	0.075075	9.4851		LA21	1046	1098	1098	4.971319	73.1247
LA02	655	656	656	0.152672	9.9586		LA22	927	951	951	2.588997	72.0254
LA03	597	597	597	0	9.8576		LA23	1032	1044	1044	1.162791	72.0456
LA04	590	592	592	0.338983	9.8453		LA24	935	981	982	4.919786	72.4792
LA05	593	593	593	0	9.7862		LA25	977	980	980	0.307062	72.5418
LA06	926	926	926	0	12.7676		LA26	1218	1222	1222	0.328407	117.0781
LA07	890	891	893	0.11236	11.9856		LA27	1235	1242	1242	0.566802	117.5268
LA08	863	866	866	0.347625	12.4213		LA28	1216	1218	1219	0.164474	118.1242
LA09	951	951	952	0	12.5512		LA29	1157	1164	1164	0.605013	118.0245
LA10	958	959	959.75	0.104384	12.4365		LA30	1355	1359	1376	0.295203	117.045
LA11	1222	1222	1222	0	18.9654		LA31	1784	1791	1791	0.392377	155.2345
LA12	1039	1039	1039	0	19.4521		LA32	1850	1913	1913	3.405405	155.8547
LA13	1150	1154	1155	0.347826	19.5263		LA33	1719	1753	1754	1.977894	155.9623
LA14	1292	1292	1292	0	19.452		LA34	1721	1734	1735	0.755375	156.4578
LA15	1207	1207	1207	0	19.2456		LA35	1888	1897	1899	0.476695	155.427
LA16	945	982	983	3.915344	36.4571		LA36	1268	1289	1299	1.656151	440.895
LA17	784	787	788	0.382653	36.4523		LA37	1397	1406	1407	0.644238	440.6352
LA18	848	861	861	1.533019	36.9685		LA38	1196	1196	1196	0	440.0245
LA19	842	855	855	1.543943	36.7582		LA39	1233	1236	1237	0.243309	440.0986
LA20	902	915	915	1.441242	36.1249		LA40	1222	1222	1222	0	440.0874

TABLE 5.9

Experimental Results for Instances of Group C Using GA

Instance	OPT (min)	GA Optimal Results			
		C_{max} (min)	AC_{max} (min)	RE (%)	T (s)
TA01	1231	1234	1234.2	0.243704	441.254
TA02	1244	1256	1267	0.96463	442.096
TA03	1218	1245	1247	2.216749	440.965
TA04	1175	1189	1189.5	1.191489	440.754
TA05	1224	1237	1237.75	1.062092	441.586
TA06	1238	1267	1268	2.342488	440.856
TA07	1227	1244	1246	1.385493	440.874
TA08	1217	1232	1236	1.232539	440.654
TA09	1274	1281	1281.5	0.549451	440.985
TA10	1241	1248	1248.45	0.564061	440.598
TA11	(1323,1361)	1377	1378.5	1.175606	749.2255
TA12	(1351,1367)	1372	1373	0.365764	748.448
TA13	(1282,1342)	1366	1389	1.788376	748.0968
TA14	1345	1355	1356	0.743494	749.124
TA15	1340	1367	1366	2.835821	749.118
TA16	(1302,1360)	1378	1379	8.676471	749.058
TA17	1462	1478	1479	0.341997	748.3255
TA18	1396	1467	1467	−1.2894	748.055
TA19	(1297,1335)	1378	1388	3.220974	748.663
TA20	(1318,1351)	1366	1368	1.110289	748.992

Instance	OPT (min)	GA Optimal Results			
		C_{max} (min)	AC_{max} (min)	RE (%)	T (s)
TA21	(1539, 1644)	1678	1688	2.068127	1152.52
TA22	(1511, 1600)	1655	1659	3.4375	1149.44
TA23	(1472, 1557)	1578	1598	1.348748	1148.73
TA24	(1602, 1647)	1656	1659	0.546448	1165.23
TA25	(1504, 1595)	1610	1617	0.940439	1181.46
TA26	(1539, 1645)	1677	1678.25	1.945289	1176.55
TA27	(1616, 1680)	1699	1703	1.130952	1174.22
TA28	(1591, 1614)	1666	1669	3.221809	1174.98
TA29	(1514, 1625)	1689	1697	3.938462	1175.86
TA30	(1473, 1584)	1607	1634	1.45202	1180.24
TA31	1764	1789	1789	1.417234	1461.6
TA32	(1774, 1796)	1834	1838	2.115813	1468.56
TA33	(1778, 1793)	1889	1895	5.354155	1462.225
TA34	(1828, 1829)	1867	1869	2.077638	1462.859
TA35	2007	2009	2010	0.099651	1461.785
TA36	1819	1823	1828	0.219901	1461.089
TA37	(1771, 1778)	1798	1799	1.124859	1462.857
TA38	1673	1689	1690.25	0.956366	1463.097
TA39	1795	1806	1809	0.612813	1461.125
TA40	(1631, 1674)	1799	1799.75	7.467145	1461.043

TABLE 5.10

Experimental Results for Instances of Group D Using GA

| Type | OPT (min) | GA Optimal Results | | | | Type | OPT (min) | GA Optimal Results | | | |
		C_{max} (min)	AC_{max} (min)	RE (%)	T (s)			C_{max} (min)	AC_{max} (min)	RE (%)	T (s)
TA41	(1859,2006)	2013	2014.5	0.348953	2574.52	TA61	2868	2877	2882.5	0.313808	15514
TA42	(1867,1945)	1955	1956	0.514139	2589.45	TA62	2869	2880	2889	0.383409	15598.52
TA43	(1809,1848)	1855	1857.5	0.378788	2588.75	TA63	2755	2766	2768	0.399274	15574.26
TA44	(1927,1983)	1992	1997	0.453858	2576.55	TA64	2702	2705	2707.75	0.111029	15544.25
TA45	(1997,2000)	2008	2012.75	0.4	2598.45	TA65	2725	2725	2729	0	15598.33
TA46	(1940,2008)	2018	2019.25	0.498008	2532.65	TA66	2845	2849	2855	0.140598	15552.51
TA47	(1789,1897)	1910	1911.5	0.685293	2498.44	TA67	2825	2930	2941	3.716814	15489.52
TA48	(1912,1945)	1951	1954	0.308483	2466.12	TA68	2784	2794	2799	0.359195	15575.52
TA49	(1915,1966)	1970	1971	0.203459	2478.47	TA69	3071	3077	3082.5	0.195376	15493.21
TA50	(1807,1924)	1930	1937	0.31185	2477.41	TA70	2995	3008	3012.25	0.434057	15515.54
TA51	2760	2760	2762	0	3903.5	TA71	5464	5467	5472.25	0.054905	22563.74
TA52	2756	2766	2768	0.362845	3905.68	TA72	5181	5189	5192.5	0.15441	22857.45
TA53	2717	2725	2730	0.294442	3905.23	TA73	5568	5575	5578.5	0.125718	22748.56
TA54	2839	2847	2849	0.281789	3908.221	TA74	5339	5349	5355	0.187301	22859.48
TA55	2679	2688	2689.5	0.335946	3908.787	TA75	5392	5401	5408.5	0.166914	22789.63
TA56	2781	2781	2783	0	3908.224	TA76	5342	5344	5356.75	0.037439	22859.56
TA57	2943	2953	2958	0.339789	3902.34	TA77	5436	5442	5449.25	0.110375	22758.79
TA58	2885	2899	2910.75	0.485269	3904.056	TA78	5394	5400	5402.25	0.111235	22749.44
TA59	2655	2667	2668	0.451977	3904.336	TA79	5358	5363	5366.5	0.093318	23564.77
TA60	2723	2730	2732.25	0.257069	3904.788	TA80	5183	5189	5195	0.115763	23564.14

TABLE 5.11

Parameter Settings for JSSP Based on SPSO

Parameter	Value
Population size	Instance dependent
Maximum inertia weight	0.9
Minimum inertia weight	0.4
Maximum velocity value	No. of jobs/5
Maximum position value	No. of jobs
Initial cognitive factor (c_{i1})	1.2
Initial social factor (c_{i2})	1.2
Final cognitive factor (c_{f1})	2
Final social factor (c_{f2})	2
Maximum no. of iterations	Instance dependent

weight w is decreased linearly from 0.9 to 0.4 based on the maximum iterations and current iteration. The position and velocity values are limited to number of jobs and number of jobs divided by 5, respectively. The population size and the number of iterations for the SPSO are based on the benchmark instance under consideration.

Based on the parameter setting for SPSO in Table 5.11, the algorithm was tested on the instances of Group A. The makespan, average makespan, relative error, and average computational time are computed. All the instances are run for 10 trials, and the computed makespan is presented in Table 5.12. The average computational time is the time taken by the algorithm to reach the optimal solution during 10 trials. While using the SPSO, it is observed that more instances in Group A have reached their optimal value compared to the performance of GA. From Table 5.12, it is found that eight instances—ORB01, ORB02, ORB04, ORB05, SWV02, SWV13, SWV16, and YN1—have 0% relative error. The average computational time varies based on the problem instance, and it is found that the time consumed by SPSO is less by 10.43% than the time consumed by GA.

The SPSO was tested on the Group B instances, and the results are shown in Table 5.13. Among the 40 instances, 18 instances produced optimal makespan values. This implies that 45% of the instances in Group B are able to converge toward the optimal solution with the application of SPSO.

The SPSO was run 10 times consecutively on instances available in Group C, and the results are shown in Table 5.14. The algorithm is able to compute only near-optimal makespan solutions rather than the best known values for the instances in this group. The instances TA01–TA10 are of size 15 × 15 (15 jobs and 15 machines), TA11–TA20 are of size 20 × 15, TA21–TA30 are of size 20 × 20, and TA31–TA40 are of size 30 × 15. It is observed from Table 5.14 that the average computational time (T) increases with increase in the problem size.

Table 5.15 shows the results obtained through the SPSO algorithm for instances TA41–TA80. Though the problem size of these instances is large, it is easier to solve these instances since the number of jobs $|J|$ is much higher than the number of machines $|M|$. Seven instances—TA51, TA56, TA65, TA67, TA71, TA76, and TA78—obtained the optimal makespan with null deviation from the original makespan.

TABLE 5.12

Experimental Results for Instances of Group A Using SPSO

Instance	OPT (min)	SPSO Optimal Results				Instance	OPT (min)	SPSO Optimal Results			
		C_{max} (min)	AC_{max} (min)	RE (%)	T (s)			C_{max} (min)	AC_{max} (min)	RE (%)	T (s)
ABZ5	1234	1256	1260	1.78282	34.1094	SWV04	(1450, 1470)	1472	1473	0.136054	113.674
ABZ6	943	946	954.75	0.318134	34.8438	SWV05	1424	1426	1426	0.140449	114.725
ABZ7	656	661	661	0.762195	676.994	SWV06	(1591, 1675)	1676	1679	0.059701	675.36
ABZ8	(645, 665)	670	670	0.75188	677.078	SWV07	(1446, 1594)	1600	1600	0.376412	675.099
ABZ9	(661, 679)	682	683	0.441826	678.121	SWV08	(1640, 1755)	1757	1757	0.11396	678.224
FT06	55	56	56.75	1.818182	4.768	SWV09	(1640, 1661)	1663	1663.25	0.120409	676.335
FT10	930	943	944	1.397849	34.0965	SWV10	(1631, 1743)	1750	1750	0.401606	676.859
FT20	1165	1187	1187.5	1.888412	18.114	SWV11	2983	2991	2993	0.268186	370.091
ORB01	1059	1059	1059	0	35.078	SWV12	(2972, 2979)	2980	2981	0.033568	370.115
ORB02	888	888	888	0	34.2541	SWV13	3104	3104	3104	0	370.142
ORB03	1005	1008	1008.25	0.298507	34.096	SWV14	2968	2970	2970	0.067385	370.668
ORB04	1005	1005	1005	0	34.078	SWV15	(2885, 2886)	2900	2902	0.4851	371.046
ORB05	887	887	887	0	34.758	SWV16	2924	2924	2924	0	370.095
ORB06	1010	1015	1016	0.49505	34.852	SWV17	2794	2798	2799	0.143164	371.995
ORB07	397	399	400	0.503778	34.178	SWV18	2852	2861	2863	0.315568	371.452
ORB08	899	902	903	0.333704	35.124	SWV19	2843	2847	2847	0.140696	370.584
ORB09	934	940	940	0.642398	35.098	SWV20	2823	2835	2835.75	0.42508	370.445
ORB10	944	953	954	0.95339	35.078	YN1	(836, 884)	884	884	0	985.41
SWV01	1407	1416	1416	0.639659	114.098	YN2	(861, 904)	909	909	0.553097	997.52
SWV02	1475	1475	1475	0	114.632	YN3	(827, 892)	896	896	0.44843	994.65
SWV03	(1369,1398)	1404	1405	0.429185	112.789	YN4	(918, 968)	977	977	0.929752	982.12

TABLE 5.13

Experimental Results for Instances of Group B Using SPSO

Instance	OPT (min)	SPSO Optimal Results				Instance	OPT (min)	SPSO Optimal Results			
		C_{max} (min)	AC_{max} (min)	RE (%)	T (s)			C_{max} (min)	AC_{max} (min)	RE (%)	T (s)
LA01	666	666	666	0	7.6563	LA21	1046	1067	1068	2.007648	71.8542
LA02	655	655	655	0	7.9688	LA22	927	946	946	2.049622	71.3654
LA03	597	597	597	0	7.4063	LA23	1032	1041	1042	0.872093	71.2546
LA04	590	590	590	0	7.5469	LA24	935	955	956.75	2.139037	69.7513
LA05	593	593	593	0	7.5	LA25	977	979	982	0.204708	69.5241
LA06	926	926	926	0	12.1719	LA26	1218	1220	1221	0.164204	113.764
LA07	890	890	890	0	11.9844	LA27	1235	1240	1240	0.404858	113.703
LA08	863	863	863	0	12.0469	LA28	1216	1216	1218	0	113.172
LA09	951	951	951	0	12.0156	LA29	1157	1158	1159.5	0.08643	114.813
LA10	958	958	958	0	11.9688	LA30	1355	1356	1356	0.073801	114.469
LA11	1222	1222	1222	0	18.2541	LA31	1784	1789	1790	0.280269	153.046
LA12	1039	1039	1040	0	18.1245	LA32	1850	1897	1897	2.540541	152.954
LA13	1150	1151	1153	0.086957	18.9654	LA33	1719	1727	1727	0.465387	153.742
LA14	1292	1292	1292	0	18.8569	LA34	1721	1767	1789	2.672865	153.012
LA15	1207	1207	1207	0	18.2345	LA35	1888	1889	1889.75	0.052966	153.422
LA16	945	956	956.5	1.164021	35.4536	LA36	1268	1270	1270	0.157729	424.235
LA17	784	785	785.75	0.127551	36.1241	LA37	1397	1406	1406	0.644238	425.124
LA18	848	855	856	0.825472	35.9687	LA38	1196	1196	1196	0	424.037
LA19	842	852	852	1.187648	35.8546	LA39	1233	1233	1233	0	425.743
LA20	902	913	913	1.219512	35.1452	LA40	1222	1222	1222	0	424.089

TABLE 5.14

Experimental Results for Instances of Group C Using SPSO

Instance	OPT (min)	SPSO Optimal Results			
		C_{max} (min)	AC_{max} (min)	RE (%)	T (s)
TA01	1231	1235	1236	0.324939	425.098
TA02	1244	1246	1247	0.160772	424.865
TA03	1218	1228	1229.6	0.821018	425.045
TA04	1175	1180	1180.25	0.425532	424.652
TA05	1224	1229	1229	0.408497	424.127
TA06	1238	1244	1245.2	0.484653	424.852
TA07	1227	1234	1236	0.570497	425.069
TA08	1217	1225	1227	0.657354	424.854
TA09	1274	1281	1282	0.549451	425.092
TA10	1241	1246	1247	0.402901	424.856
TA11	(1323,1361)	1372	1373	0.808229	675.118
TA12	(1351,1367)	1372	1379	0.365764	676.234
TA13	(1282,1342)	1357	1368	1.117735	676.985
TA14	1345	1352	1353	0.520446	676.15
TA15	1340	1359	1361	1.41791	675.984
TA16	(1302,1360)	1376	1374	1.176471	675.033
TA17	1462	1469	1477	0.478796	675.224
TA18	1396	1444	1446	3.438395	675.874
TA19	(1297,1335)	1354	1365	1.423221	675.889
TA20	(1318,1351)	1359	1367	0.592154	676.66

Instance	OPT (min)	SPSO Optimal Results			
		C_{max} (min)	AC_{max} (min)	RE (%)	T (s)
TA21	(1539,1644)	1657	1658.5	0.790754	987.52
TA22	(1511,1600)	1623	1630	1.4375	945.11
TA23	(1472,1557)	1566	1569	0.578035	968.25
TA24	(1602,1647)	1655	1657.75	0.485732	974.25
TA25	(1504,1595)	1599	1562	0.250784	933.02
TA26	(1539,1645)	1654	1659.5	0.547112	935.41
TA27	(1616,1680)	1687	1689.75	0.416667	938.48
TA28	(1591,1614)	1656	1659.25	2.60223	985.42
TA29	(1514,1625)	1646	1649	1.292308	988.05
TA30	(1473,1584)	1599	1605	0.94697	978.55
TA31	1764	1767	1768	0.170068	1289.44
TA32	(1774,1796)	1809	1812	0.723831	1290.35
TA33	(1778,1793)	1836	1866	2.398215	1290.77
TA34	(1828,1829)	1855	1859	1.421542	1288.05
TA35	2007	2008	2008.5	0.049826	1288.45
TA36	1819	1820	1826	0.054975	1289.7
TA37	(1771,1778)	1786	1788	0.449944	1289.88
TA38	1673	1688	1689	0.896593	1290.23
TA39	1795	1798	1802	0.167131	1291.41
TA40	(1631,1674)	1699	1705	1.493429	1288.05

TABLE 5.15

Experimental Results for Instances of Group D Using SPSO

| Instance | OPT (min) | SPSO Optimal Results | | | | Instance | OPT (min) | SPSO Optimal Results | | | |
		C_{max} (min)	AC_{max} (min)	RE (%)	T (s)			C_{max} (min)	AC_{max} (min)	RE (%)	T (s)
TA41	(1859,2006)	2012	2019.5	0.299103	2396.71	TA61	2868	2872	2875	0.13947	13884.5
TA42	(1867,1945)	1951	1952.5	0.308483	2385.74	TA62	2869	2875	2878.25	0.209132	13779.4
TA43	(1809,1848)	1852	1853.5	0.21645	2385.96	TA63	2755	2762	2763.5	0.254083	13782.1
TA44	(1927,1983)	1986	1990.75	0.151286	2377.41	TA64	2702	2704	2707.25	0.074019	13689.5
TA45	(1997,2000)	2004	2010	0.2	2356.77	TA65	2725	2725	2728	0	13854.4
TA46	(1940,2008)	2013	2017	0.249004	2398.45	TA66	2845	2848	2853	0.281195	13987.1
TA47	(1789,1897)	1902	1905	0.263574	2384.74	TA67	2825	2825	2832	0	13889.4
TA48	(1912,1945)	1950	1952.5	0.257069	2398.47	TA68	2784	2789	2792.25	0.179598	13854.6
TA49	(1915,1966)	1969	1972	0.152594	2374.88	TA69	3071	3076	3078.25	0.162813	13882.6
TA50	(1807,1924)	1928	1931.5	0.2079	2356.85	TA70	2995	3000	3002.5	0.166945	13859.2
TA51	2760	2760	2761	0	3847.1	TA71	5464	5464	5469.5	0	21660.5
TA52	2756	2760	2765	0.145138	3897.22	TA72	5181	5182	5185.75	0.019301	21743.5
TA53	2717	2719	2720.75	0.073611	3905.07	TA73	5568	5573	5577.5	0.089799	21653.3
TA54	2839	2843	2844.5	0.140895	3885.12	TA74	5339	5345	5349.75	0.112381	21856.8
TA55	2679	2683	2684.75	0.149309	3889.16	TA75	5392	5396	5400.5	0.074184	21779.7
TA56	2781	2781	2786	0	3884.79	TA76	5342	5342	5350.25	0	21859.6
TA57	2943	2951	2955	0.271831	3856.75	TA77	5436	5439	5446.5	0.055188	21453.8
TA58	2885	2891	2892.5	0.207972	3877.86	TA78	5394	5394	5399	0	21798.6
TA59	2655	2665	2667	0.376648	3843.26	TA79	5358	5360	5364.25	0.037327	21789.5
TA60	2723	2725	2728.75	0.073448	3886.1	TA80	5183	5184	5189.5	0.019294	21524.9

5.4.6 Outcome of JSSP Based on ACO

In JSSP, the ants construct a sequence by first choosing a job for the first position, then a job for the second position, and so on, until all jobs are scheduled. The parameters of the ant system for solving the benchmark problems are divided into two groups: those that influence the state transition (*a* and *b*), and those that determine the pheromone update (the evaporation constant *r* and the number of ants *m*). From the simulation, it appears that the parameters *Q* (pheromone allocation per unit distance) and τ0 (initial pheromone level) are of very little importance to the algorithm's performance. The ACO parameters and their values are shown in Table 5.16.

The ACO algorithm for JSSP was tested on all benchmark instances in Group A, and the makespan, relative error, average makespan, and the computational time are as shown in Table 5.17. It is observed that 0% relative error is obtained for five instances—ORB01, ORB05, SWV13, SWV16, and YN1—indicating that the computed makespan is equal to the original value.

The ACO algorithm was applied with fuzzy processing time to the instances in Group B: LA01–LA20. With 10 trials performed on each instance, the average makespan and the average computational time were computed along with the makespan and relative error, as shown in Table 5.18. The results are more similar to GA, yielding optimal makespan for 10 instances: LA03, LA05, LA06, LA09, LA11, LA12, LA14, LA15, LA38, and LA40.

The performance of ACO on instances belonging to Group C are shown in terms of C_{max}, AC_{max}, RE, and *T* in Table 5.19. The ACO based on fuzzy processing time is not able to attain the optimal solution for the instances TA01–TA40. Though near-optimal solutions in terms of makespan are obtained, it is observed that ACO failed to obtain 0% deviation from the original optimal value even for a single instance in this group.

The feasibility of ACO in solving the instances of Group D can be observed from the results shown in Table 5.20. Optimal solutions in terms of the makespan and average makespan are obtained by ACO for three instances in this group: TA51, TA56, and TA65. For the other instances, ACO is capable of producing only near-optimal makespan. Though the trials were repeated 10 times for each instance, it is found that ACO is not able to obtain optimal solutions.

5.4.7 Investigation of GSO on JSSP

The choice of appropriate parameters is a tedious and time-consuming process, which, under several circumstances, depend on the problem instance. Table 5.21 lists the GSO parameters and their values used in this experiment. The initial values of acceleration

TABLE 5.16

Parameter Settings for JSSP Based on ACO

Parameter	Value
No. of ants	Instance dependent
Weight of pheromone trail, α	1
Weight of heuristic information, β	1
Pheromone evaporation parameter, ρ	0.7
Constant for pheromone updating, Q	10
Initial pheromone level, τ0	10

TABLE 5.17

Experimental Results for Instances of Group A Using ACO

Instance	OPT (min)	ACO Optimal Results				Instance	OPT (min)	ACO Optimal Results			
		C_{max} (min)	AC_{max} (min)	RE (%)	T (s)			C_{max} (min)	AC_{max} (min)	RE (%)	T (s)
ABZ5	1234	1269	1284	2.836305	34.55912	SWV04	(1450, 1470)	1475	1479.5	0.340136	113.8917
ABZ6	943	948	954	0.530223	36.57346	SWV05	1424	1428	1433.5	0.280899	117.0189
ABZ7	656	669	674	1.981707	717.0902	SWV06	(1591, 1675)	1677	1689	0.119403	736.5012
ABZ8	(645, 665)	672	678.25	1.052632	723.0435	SWV07	(1446, 1594)	1604	1615	0.627353	736.0464
ABZ9	(661, 679)	682	689	0.441826	739.0018	SWV08	(1640, 1755)	1758	1764.25	0.17094	713.5749
FT06	55	57	62	3.636364	5.649938	SWV09	(1640, 1661)	1664	1668.75	0.180614	699.3808
FT10	930	975	978.5	4.83871	35.1355	SWV10	(1631, 1743)	1752	1759.25	0.516351	719.5028
FT20	1165	1201	1211.25	3.090129	19.07167	SWV11	2983	2996	3004	0.435803	388.0378
ORB01	1059	1059	1072	0	35.58017	SWV12	(2972, 2979)	2986	2998	0.234978	395.3899
ORB02	888	890	892	0.225225	35.59688	SWV13	3104	3104	3118.25	0	373.4525
ORB03	1005	1011	1018.75	0.597015	35.70363	SWV14	2968	2972	2992	0.134771	382.6984
ORB04	1005	1006	1014.25	0.099502	34.85184	SWV15	(2885, 2886)	2907	2918	0.727651	393.3898
ORB05	887	887	893.5	0	35.18303	SWV16	2924	2924	2934	0	398.1459
ORB06	1010	1023	1029.5	1.287129	35.41885	SWV17	2794	2799	2812.5	0.178955	391.907
ORB07	397	400	408.25	0.755668	36.01246	SWV18	2852	2869	2875.25	0.596073	396.9022
ORB08	899	909	917	1.112347	36.23908	SWV19	2843	2864	2869.75	0.738656	372.0067
ORB09	934	944	954	1.070664	36.49049	SWV20	2823	2837	2846	0.495926	392.9829
ORB10	944	965	969	2.224576	35.29392	YN1	(836, 884)	884	895.75	0	1024.053
SWV01	1407	1417	1420	0.710732	116.0916	YN2	(861, 904)	909	917.25	0.553097	1168.135
SWV02	1475	1476	1489	0.067797	115.5477	YN3	(827, 892)	899	910	0.784753	1135.23
SWV03	(1369, 1398)	1409	1415.25	0.786838	117.452	YN4	(918, 968)	977	986	0.929752	1000.683

TABLE 5.18

Experimental Results for Instances of Group B Using ACO

| Instance | OPT (min) | ACO Optimal Results | | | Instance | OPT (min) | ACO Optimal Results | | | |
		C_{max} (min)	AC_{max} (min)	RE (%)	T (s)			C_{max} (min)	AC_{max} (min)	RE (%)	T (s)
LA01	666	667	674	0.15015	8.232699	LA21	1046	1092	1110	4.397706	72.51419
LA02	655	656	668	0.152672	8.872195	LA22	927	951	963	2.58997	71.92786
LA03	597	597	601	0	8.873694	LA23	1032	1044	1058	1.162791	71.50053
LA04	590	592	598	0.338983	7.679441	LA24	935	959	967	2.566845	70.12914
LA05	593	593	597	0	9.667327	LA25	977	980	996	0.307062	69.82891
LA06	926	926	935	0	12.65126	LA26	1218	1221	1238	0.246305	114.1581
LA07	890	891	902	0.11236	11.98532	LA27	1235	1242	1249	0.566802	113.9673
LA08	863	865	871	0.23175	12.09133	LA28	1216	1217	1222	0.082237	114.7782
LA09	951	951	956	0	12.05802	LA29	1157	1163	1175	0.518583	115.3627
LA10	958	959	963	0.104384	12.05313	LA30	1355	1358	1364	0.221402	114.6479
LA11	1222	1222	1230	0	18.26275	LA31	1784	1791	1799	0.392377	153.5278
LA12	1039	1039	1048	0	18.42847	LA32	1850	1911	1924	3.297297	153.2821
LA13	1150	1154	1168	0.347826	19.31163	LA33	1719	1728	1733	0.52356	153.8259
LA14	1292	1292	1300	0	19.28114	LA34	1721	1742	1759	1.220221	153.1296
LA15	1207	1207	1218	0	19.17541	LA35	1888	1895	1905	0.370763	154.3908
LA16	945	964	975	2.010582	36.18973	LA36	1268	1280	1298	0.946372	432.5226
LA17	784	786	792	0.255102	36.41426	LA37	1397	1406	1419	0.644238	430.4684
LA18	848	857	868	1.061321	36.551	LA38	1196	1196	1209	0	428.2863
LA19	842	854	861	1.425178	36.09239	LA39	1233	1234	1257	0.081103	439.8128
LA20	902	913	924	1.219512	35.38637	LA40	1222	1222	1229	0	434.1312

TABLE 5.19

Experimental Results for Instances of Group C Using ACO

| Instance | OPT (min) | ACO Optimal Results | | | | Instance | OPT (min) | ACO Optimal Results | | | |
		C_{max} (min)	AC_{max} (min)	RE (%)	T (s)			C_{max} (min)	AC_{max} (min)	RE (%)	T (s)
TA01	1231	1235	1247	0.324939	427.0674	TA21	(1539, 1644)	1674	1699	1.824818	1116.757
TA02	1244	1253	1267	0.723473	439.742	TA22	(1511, 1600)	1638	1657	2.375	989.0095
TA03	1218	1238	1258	1.642036	431.8789	TA23	(1472, 1557)	1576	1586	1.220295	1043.99
TA04	1175	1181	1196	0.510638	429.2668	TA24	(1602, 1647)	1656	1692.25	0.546448	1048.409
TA05	1224	1232	1247	0.653595	431.717	TA25	(1504, 1595)	1614	1625	1.19122	1169.774
TA06	1238	1263	1281	2.019386	427.7895	TA26	(1539, 1645)	1656	1675	0.668693	1111.437
TA07	1227	1236	1258	0.733496	434.767	TA27	(1616, 1680)	1693	1708	0.77381	994.7243
TA08	1217	1227	1234	0.821693	436.1074	TA28	(1591, 1614)	1661	1669	2.91202	1095.754
TA09	1274	1281	1289	0.549451	434.1537	TA29	(1514, 1625)	1654	1668	1.784615	1102.177
TA10	1241	1247	1259	0.483481	431.0656	TA30	(1473, 1584)	1607	1618	1.45202	1053.908
TA11	(1323, 1361)	1377	1381	1.175606	718.4828	TA31	1764	1768	1773	0.226757	1359.724
TA12	(1351, 1367)	1372	1383	0.365764	710.2076	TA32	(1774, 1796)	1815	1824	1.057906	1465.339
TA13	(1282, 1342)	1361	1379	1.415797	698.3278	TA33	(1778, 1793)	1839	1856	2.565533	1378.858
TA14	1345	1353	1385	0.594796	700.8805	TA34	(1828, 1829)	1865	1876	1.968289	1448.803
TA15	1340	1367	1384	2.014925	741.8037	TA35	2007	2009	2018	0.099651	1430.073
TA16	(1302, 1360)	1377	1386	1.25	677.6692	TA36	1819	1822	1830	0.164926	1430.328
TA17	1462	1476	1483	0.957592	748.2457	TA37	(1771, 1778)	1792	1801.75	0.787402	1324.156
TA18	1396	1461	1476	4.65616	703.8938	TA38	1673	1689	1699.5	0.956366	1313.244
TA19	(1297, 1335)	1361	1375	1.947566	691.0449	TA39	1795	1801	1817	0.334262	1446.84
TA20	(1318, 1351)	1362	1385	0.814212	736.0825	TA40	(1631, 1674)	1757	1769	4.958184	1353.168

TABLE 5.20

Experimental Results for Instances of Group D Using ACO

Instance	OPT (min)	ACO Optimal Results				Instance	OPT (min)	ACO Optimal Results			
		C_{max} (min)	AC_{max} (min)	RE (%)	T (s)			C_{max} (min)	AC_{max} (min)	RE (%)	T (s)
TA41	(1859, 2006)	2013	2025.25	0.348953	2562.662	TA61	2868	2874	2883.25	0.209205	14751.83
TA42	(1867, 1945)	1952	1966.25	0.359897	2487.527	TA62	2869	2878	2885	0.313698	13985.55
TA43	(1809, 1848)	1854	1867	0.324675	2536.339	TA63	2755	2765	2778.5	0.362976	15414.87
TA44	(1927, 1983)	1991	1205	0.403429	2559.088	TA64	2702	2705	2719.25	0.111029	14028.09
TA45	(1997, 2000)	2005	2021	0.25	2558.293	TA65	2725	2725	2738	0	14305.8
TA46	(1940, 2008)	2014	2029	0.298805	2493.826	TA66	2845	2849	2867	0.140598	15457.65
TA47	(1789, 1897)	1904	1915	0.369004	2390.777	TA67	2825	2901	2919	2.690265	14697.95
TA48	(1912, 1945)	1951	1962	0.308483	2455.538	TA68	2784	2790	2799.25	0.215517	15523.45
TA49	(1915, 1966)	1970	1979	0.203459	2433.276	TA69	3071	3077	3084	0.195376	14922.15
TA50	(1807, 1924)	1929	1938.75	0.259875	2471.986	TA70	2995	3004	3019.5	0.300501	14105.68
TA51	2760	2760	2766.5	0	3864.569	TA71	5464	5466	5487.25	0.036603	21918.38
TA52	2756	2764	2778	0.290276	3905.335	TA72	5181	5185	5194	0.077205	22573.41
TA53	2717	2720	2738	0.110416	3905.137	TA73	5568	5574	5586	0.107759	22541.36
TA54	2839	2845	2859	0.211342	3904.866	TA74	5339	5348	5359	0.168571	22186.52
TA55	2679	2688	2697	0.335946	3904.621	TA75	5392	5398	5410	0.111276	22226.86
TA56	2781	2781	2796	0	3894.905	TA76	5342	5343	5384	0.01872	22041.32
TA57	2943	2953	2967	0.339789	3897.005	TA77	5436	5441	5459	0.091979	21821.68
TA58	2885	2895	2918	0.34662	3879.107	TA78	5394	5400	5418	0.111235	22281.29
TA59	2655	2667	2679	0.451977	3900.436	TA79	5358	5363	5374	0.093318	22391.43
TA60	2723	2726	2738	0.110173	3888.126	TA80	5183	5189	5196.25	0.115763	22857.37

TABLE 5.21

Parameter Setting for GSO-Based JSSP

Parameter	Value
Population size	Instance dependent
Maximum inertia weight	0.9
Minimum inertia weight	0.4
Initial velocity	Number of jobs/5
Initial position	Number of jobs
Initial cognitive factor (c_1)	1.2
Final cognitive factor (c_1)	2
Initial social factor (c_2)	1.2
Final social factor (c_2)	2
Type of crossover	Two-point
Type of mutation	Flip-bit
Crossover rate	0.35
Mutation rate	0.002
Maximum no. of iterations	Instance dependent
Hybridization factor	0.5

constants c_1 and c_2 are set to 1.2, and initial population of swarm is varied with respect to the jobs and machines on each level. The inertia weight w determines the search behavior of the algorithm.

Large values for w facilitate searching new locations, whereas small values provide a finer search in the current area. A balance can be established between global and local exploration by decreasing the inertia weight during the execution of the algorithm. In the experiments, the initial value of inertia weight is set to 0.9 for all benchmark instances. In the GA module, two-point crossover with a crossover rate of 0.35 and flip-bit mutation with a rate of 0.002 are set. The effectiveness of the proposed technique is validated based on HF, which, if set to 1 implies a wholesome GA approach, and if set to 0 implies a wholesome SPSO approach.

The outcome of the GSO in Table 5.22 has been found satisfactory for the benchmark instances of Group A. Among the solved 42 instances, 30 instances resulted in the optimal makespan. In addition, the average computational time is also found to be less than that of GA, SPSO, and ACO algorithms. This is due to the fact that GA and SPSO work in parallel, thus resulting in a faster convergence rate.

The impact of GSO on instances LA01–LA40 is shown in Table 5.23. It is seen that 90% (36/40 instances) of the instances in Group B attained the optimal makespan. In addition, though the instances LA21, LA24, LA25, and LA27 failed to reach the optimal results, the relative error shows that there is very little deviation from the original OPT value.

The instances in Group C are solved by the intelligent heuristics GA, SPSO, and ACO in Sections 5.4.3–5.4.5 respectively. But none of these techniques is able to obtain optimal solutions in terms of makespan even for a single problem instance. From Table 5.24, it is observed that GSO is capable of solving 13 instances: TA01, TA02, TA04, TA05, TA08, TA10, TA14, TA22, TA24, TA31, TA35, TA36, and TA39. The results are satisfactory since the proposed hybrid heuristic obtained optimal solution for 32.5% of the instances in Group C.

Instances TA41–TA80 in Group D were also solved using the FGSO, and the results are presented in Table 5.25. Among the 40 instances, nearly 55% (22 instances) in this group resulted in optimal makespan value. The relative error of the remaining instances is also found to be very small, thus proving the capability of GSO in providing optimal solutions.

TABLE 5.22

Experimental Results for Instances of Group A Using GSO

Instance	OPT (min)	GSO Optimal Results				Instance	OPT (min)	GSO Optimal Results			
		C_{max} (min)	AC_{max} (min)	RE (%)	T (s)			C_{max} (min)	AC_{max} (min)	RE (%)	T (s)
ABZ5	1234	1242	1242	0.648298	33.451	SWV04	(1450, 1470)	1470	1470	0	109.564
ABZ6	943	943	943	0	33.024	SWV05	1424	1428	1428	0.280899	109.098
ABZ7	656	659	659	0.457317	655.3242	SWV06	(1591, 1675)	1675	1675	0	655.8125
ABZ8	(645, 665)	668	668	0.451128	655.8859	SWV07	(1446, 1594)	1594	1594	0	655.984
ABZ9	(661, 679)	680	680	0.147275	655.4783	SWV08	(1640, 1755)	1755	1755	0	655.635
FT06	55	55	55	0	4.012	SWV09	(1640, 1661)	1666	1666	0.301023	657.088
FT10	930	930	930	0	33.057	SWV10	(1631, 1743)	1743	1743	0	655.458
FT20	1165	1165	1167	0	17.854	SWV11	2983	2983	2983.5	0	347.568
ORB01	1059	1059	1059	0	33.416	SWV12	(2972, 2979)	2979	2979	0	346.9852
ORB02	888	888	888	0	33.0931	SWV13	3104	3104	3104	0	347.1255
ORB03	1005	1005	1005	0	33.127	SWV14	2968	2968	2968	0	347.0154
ORB04	1005	1005	1005	0	33.078	SWV15	(2885, 2886)	2890	2891	0.1386	346.228
ORB05	887	887	887	0	33.546	SWV16	2924	2924	2924	0	346.9812
ORB06	1010	1011	1011	0.09901	32.941	SWV17	2794	2794	2794	0	347.055
ORB07	397	397	397	0	32.854	SWV18	2852	2853	2853	0.035063	347.886
ORB08	899	899	900	0	32.564	SWV19	2843	2843	2843	0	345.985
ORB09	934	934	934	0	33.014	SWV20	2823	2823	2823	0	346.883
ORB10	944	944	944	0	32.078	YN1	(836,884)	884	884	0	847.22
SWV01	1407	1407	1407	0	109.056	YN2	(861, 904)	905	905	0.110619	896.45
SWV02	1475	1475	1475	0	109.43	YN3	(827, 892)	892	892	0	856.35
SWV03	(1369, 1398)	1399	1399	0.071531	108.273	YN4	(918, 968)	969	969	0.103306	821.45

TABLE 5.23

Experimental Results for Instances of Group B Using GSO

| Instance | OPT (min) | GSO Optimal Results | | | Instance | OPT (min) | GSO Optimal Results | | | |
		C_{max} (min)	AC_{max} (min)	RE (%)	T (s)			C_{max} (min)	AC_{max} (min)	RE (%)	T (s)
LA01	666	666	666	0	6.88	LA21	1046	1050	1050	0.382409	68.9063
LA02	655	655	655	0	6.99	LA22	927	927	928	0	67.8281
LA03	597	597	597	0	6.63	LA23	1032	1032	1032	0	68.6406
LA04	590	590	590	0	6.72	LA24	935	936	936	0.106952	66.8543
LA05	593	593	593	0	6.92	LA25	977	978	979	0.102354	68.7586
LA06	926	926	926	0	11.5246	LA26	1218	1218	1219	0	109.9219
LA07	890	890	890	0	11.4623	LA27	1235	1236	1236	0.080972	108.9688
LA08	863	863	863	0	11.4587	LA28	1216	1216	1216	0	108.1245
LA09	951	951	951	0	11.7542	LA29	1157	1157	1157	0	109.0254
LA10	958	958	958	0	11.6545	LA30	1355	1355	1355	0	108.2351
LA11	1222	1222	1222	0	17.524	LA31	1784	1784	1785	0	151.5862
LA12	1039	1039	1039	0	17.5689	LA32	1850	1850	1857	0	151.0558
LA13	1150	1150	1150	0	18.4536	LA33	1719	1719	1724	0	151.0452
LA14	1292	1292	1292	0	18.1456	LA34	1721	1721	1724	0	151.0369
LA15	1207	1207	1207	0	17.9852	LA35	1888	1888	1888	0	151.453
LA16	945	945	945	0	33.9574	LA36	1268	1268	1268	0	420.2478
LA17	784	784	784	0	34.1452	LA37	1397	1397	1399	0	421.0253
LA18	848	848	848	0	33.8569	LA38	1196	1196	1196	0	420.1573
LA19	842	842	842	0	34.2134	LA39	1233	1233	1233	0	420.0365
LA20	902	902	905	0	34.1578	LA40	1222	1222	1222	0	421.0891

TABLE 5.24

Experimental Results for Instances of Group C Using GSO

| Instance | OPT (min) | GSO Optimal Results | | | | Instance | OPT (min) | GSO Optimal Results | | | |
		C_{max} (min)	AC_{max} (min)	RE (%)	T (s)			C_{max} (min)	AC_{max} (min)	RE (%)	T (s)
TA01	1231	1231	1231.75	0	420.987	TA21	(1539, 1644)	1649	1655	0.304136	890.09
TA02	1244	1244	1244	0	421.056	TA22	(1511, 1600)	1600	1605	0	863.24
TA03	1218	1219	1226	0.082102	421.083	TA23	(1472, 1557)	1558	1560	0.064226	869.44
TA04	1175	1175	1175	0	420.853	TA24	(1602, 1647)	1647	1649	0	845.99
TA05	1224	1224	1225	0	420.975	TA25	(1504, 1595)	1596	1599	0.062696	859.65
TA06	1238	1239	1240	0.080775	421.085	TA26	(1539, 1645)	1649	1650	0.243161	853.22
TA07	1227	1228	1229	0.0815	421.126	TA27	(1616, 1680)	1686	1686	0.357143	874.58
TA08	1217	1217	1217	0	420.856	TA28	(1591, 1614)	1616	1619	0.123916	847.09
TA09	1274	1275	1274	0.078493	420.887	TA29	(1514, 1625)	1626	1629	0.061538	855.25
TA10	1241	1241	1241	0	420.889	TA30	(1473, 1584)	1589	1592.5	0.315657	865.12
TA11	(1323, 1361)	1363	1365	0.146951	655.022	TA31	1764	1764	1764	0	1099.451
TA12	(1351, 1367)	1370	1374	0.219459	655.741	TA32	(1774, 1796)	1804	1812	0.445434	1100.542
TA13	(1282, 1342)	1346	1347	0.298063	655.009	TA33	(1778, 1793)	1810	1814	0.948132	1100.741
TA14	1345	1345	1345	0	655.089	TA34	(1828, 1829)	1830	1834	0.054675	1098.462
TA15	1340	1342	1349	0.149254	656.335	TA35	2007	2007	2008	0	1098.77
TA16	(1302, 1360)	1361	1365	0.073529	657.809	TA36	1819	1819	1819	0	1098.745
TA17	1462	1465	1469	0.205198	657.304	TA37	(1771, 1778)	1780	1786	0.112486	1098.245
TA18	1396	1400	1404	0.286533	655.387	TA38	1673	1675	1679	0.119546	1099.556
TA19	(1297, 1335)	1340	1347	0.374532	655.454	TA39	1795	1795	1796	0	1100.4
TA20	(1318, 1351)	1352	1355	0.074019	655.882	TA40	(1631, 1674)	1684	1687	0.597372	1100.883

TABLE 5.25

Experimental Results for Instances of Group D Using GSO

Instance	OPT (min)	GSO Optimal Results				Instance	OPT (min)	GSO Optimal Results			
		C_{max} (min)	AC_{max} (min)	RE (%)	T (s)			C_{max} (min)	AC_{max} (min)	RE (%)	T (s)
TA41	(1859, 2006)	2007	2008.75	0.04985	2154.85	TA61	2868	2868	2870.5	0	11859.25
TA42	(1867, 1945)	1946	1947.5	0.051414	2156.85	TA62	2869	2870	2871.25	0.034855	9985.25
TA43	(1809, 1848)	1848	1849	0	2188.54	TA63	2755	2756	2758	0.036298	11235.52
TA44	(1927, 1983)	1984	1987	0.050429	2196.52	TA64	2702	2702	2703.5	0	11954.85
TA45	(1997, 2000)	2000	2002	0	2174.75	TA65	2725	2725	2726.5	0	11982.55
TA46	(1940, 2008)	2009	2010	0.049801	2184.96	TA66	2845	2846	2848.25	0.035149	11874.52
TA47	(1789, 1897)	1897	1899	0	2166.23	TA67	2825	2825	2827.5	0	11994.52
TA48	(1912, 1945)	1947	1949.5	0.102828	2155.96	TA68	2784	2785	2786.25	0.03592	11987.41
TA49	(1915, 1966)	1967	1968.75	0.050865	2085.74	TA69	3071	3071	3073.5	0	11452.87
TA50	(1807, 1924)	1925	1926.5	0.051975	2063.52	TA70	2995	2996	2998	0.033389	9995.44
TA51	2760	2760	2761.5	0	3654.142	TA71	5464	5464	5466	0	19954.25
TA52	2756	2756	2757	0	3655.852	TA72	5181	5181	5183.75	0	19742.96
TA53	2717	2718	2719.25	0.036805	3659.244	TA73	5568	5569	5572.5	0.01796	19745.36
TA54	2839	2839	2840	0	3657.559	TA74	5339	5339	5342	0	19854.25
TA55	2679	2680	2681.5	0.037327	3659.124	TA75	5392	5393	5399.75	0.018546	19988.32
TA56	2781	2781	2781.75	0	3654.055	TA76	5342	5342	5344.5	0	19909.17
TA57	2943	2943	2944	0	3654.099	TA77	5436	5438	5439.5	0.036792	19946.87
TA58	2885	2885	2885.25	0	3659.885	TA78	5394	5394	5398.25	0	19912.77
TA59	2655	2658	2659	0.112994	3657.011	TA79	5358	5358	5362.25	0	19219.49
TA60	2723	2723	2724.5	0	3655.155	TA80	5183	5183	5185.25	0	19615.84

5.5 Discussion

A statistical analysis is performed to prove that the proposed intelligent meta-heuristics allow the search process to be efficient in finding feasible schedule for the JSSP instances, thus minimizing the makespan. This section compares the results obtained by GA, SPSO, ACO, and GSO in terms of makespan, mean relative error, and computational efficiency with approaches reported the in literature, as discussed later.

5.5.1 Optimal Makespan

The makespans obtained by the proposed GA, SPSO, ACO, and GSO algorithms based on fuzzy processing time are compared with those using approaches in the literature such as parallel greedy randomized adaptive search procedure with path-relinking (GPPR) (Aiex et al. 2003), tabu search with shifted bottleneck (TSSB) (Pezzella and Merelli 2000), ACO with most work remaining heuristic (ACO-MWR) (Huang and Liao 2008), ACO with time remaining heuristic (ACO-TR) (Huang and Liao 2008), greedy randomized adaptive search procedure (GRASP) (Aiex et al. 2003), Shifting Bottleneck (SB) (Adams et al. 1988) procedure, fast taboo search (FTS) (Nowicki and Smutnicki 1996), TSSB, hybrid GA (HGA) (Goncalves 2005), and hybrid PSO (HPSO) (Sha and Hsu 2006).

Twelve instances were considered for comparison in Group A and D while 40 instances are considered in Group B and C, and the results are shown in Table 5.26. For Group A, the makespan of 12 instances—ABZ5–ABZ9, FT10, FT20, and ORB01–ORB05—computed by the proposed approaches are compared with GPPR, TSSB, ACO-MWR, ACO-TR, and GRASP. It is observed that the proposed GSO solved 8/12 instances, which is better than TSSB (5/12), GRASP (0/12), GA (2/12), SPSO (3/12), and ACO (2/12) algorithms.

TABLE 5.26

Comparison of Optimal Makespan

Instance	Heuristic Technique	A/B[a]	Instance	Heuristic Technique	A/B[a]
Group A	GPPR	9/12	Group B	SB	18/40
	TSSB	5/12		FTS	33/40
	ACO-MWR	9/12		TSSB	32/40
	ACO-TR	9/12		HGA	27/40
	GRASP	0/12		HPSO	37/40
	GA	2/12		GA	10/40
	SPSO	3/12		SPSO	18/40
	ACO	2/12		ACO	10/40
	GSO	8/12		GSO	36/40
Group C	ACO-MWR	9/40	Group D	ACO-MWR	1/12
	ACO-TR	11/40		ACO-TR	1/12
	TSSB	4/40		TSSB	0/12
	HPSO	11/40		HPSO	0/12
	GA	0/40		GA	0/12
	SPSO	0/40		SPSO	1/12
	ACO	0/40		ACO	0/12
	GSO	13/40		GSO	4/12

[a] A/B = No. of instances resulted in optimal solutions/no. of instances validated.

The makespan obtained by the proposed algorithms for all the 40 instances of Group B were compared with approaches in the literature such as SB, FTS, TSSB, HGA, and HPSO. From Table 5.26, it is found that the proposed GSO is able to find the optimal makespan for 36 instances in this group, which is higher than with the SB (18/40), FTS (33/40), TSSB (32/40), HGA (27/40), GA (10/40), SPSO (18/40), and ACO (10/40) approaches. The number of instances in Group C that resulted in optimal makespan is 13/40 while applying the GSO algorithm, whereas none of the instances was solved in this group using GA, SPSO, and ACO algorithms. The approaches ACO-MWR, ACO-TR, TSSB, and HPSO solved 9/40, 11/40, 4/40, and 11/40 instances, respectively, in Group C, which are less than that of the proposed GSO.

Similarly, 12 instances of Group D—TA41–TA50, TA62 and TA67—are compared with ACO-MWR, ACO-TR, TSSB, and HPSO. ACO-MWR and ACO-TR are able to find optimal makespan for only one instance, while TSSB, HPSO, GA and ACO were not able to obtain the optimal makespan even for a single instance in this group. The proposed GSO approach found optimal makespan for four instances in Group D: TA43, TA45, TA47, and TA67—in Group D. Thus it can be concluded that the ability of random search in GA and large-scale search of the swarm optimization approach has increased the ability of GSO to find optimal solutions significantly better than GA, SPSO, and ACO.

5.5.2 Computational Efficiency

The average computational time required for the meta-heuristics to converge toward the optimal solution for instances group-wise is shown in Table 5.27. JSSP instances were evaluated by carrying out two experiments: the first was based on the optimal makespan and computational time to test the proposed algorithms on a set of small and medium-size instances (Groups A, B, and C), thus trying to determine a good tradeoff between the optimal solutions and computational efficiency. In the second experiment, the analysis was performed by testing the proposed algorithms on a set of bigger instances (Group D), trying to approximate the best known optimum as much as possible, irrespective of the computational time. This is apparent from the increased values of average computational time obtained for Group D, as shown in Table 5.27.

The computational time of GSO (312.7233 s) for Group A instances is less than that of the proposed GA (375.4365 s), SPSO (336.2583 s) and ACO (358.679 s) algorithms. Likewise, in Group B, C, and D, it is seen that GSO computed the optimal makespan with less time than GA, SPSO, and ACO algorithms. The computational time of GSO is apparently improved, which makes it a suitable approach for JSSPs. Thus it can be concluded that, among the proposed algorithms, GSO is found to be capable of generating optimal schedules in a computationally efficient manner.

TABLE 5.27

Comparison of Average Computational Time

Instance Group	Average Computational Time (in Seconds) of Proposed Bio-Inspired Algorithms			
	GA	SPSO	ACO	GSO
Group A	375.4365	336.2583	358.679	312.7233
Group B	108.0297	104.5581	106.0362	102.4
Group C	955.0799	838.4513	903.1417	759.7074
Group D	11230.18	10454.29	10848.15	9257.638

5.5.3 Mean Relative Error

The robustness was assessed based on the mean relative error (MRE), which is the average of percentage deviation of the computed makespan from the best known optimal makespan. Table 5.28 shows the comparative analysis based on MRE for the proposed algorithms with existing approaches in the literature: GPPR, TSSB, GRASP, ACO-MWR, ACO-TR, SB, FTS, HGA, and HPSO. The MRE of the proposed algorithms for instances in Group A is compared with those of heuristic algorithms in the literature such as GPPR, TSSB, ACOFT-MWR, ACOFT-TR, and GRASP. Twelve instances in Group A—ABZ5–ABZ9, FT10, FT20, and ORB01–ORB05—are chosen for comparative analysis of MRE. The MRE of GSO (0.142002%) is less than that of GPPR (2.228333%), TSSB (1.1475%), ACO-MWR (0.625667%), ACO-TR (0.688583%), GRASP (3.34%), GA (1.650948%), SPSO (0.636802%), and ACO (1.307773%) algorithms, which implies that GSO is able to achieve optimal makespan for the instances considered in this group.

The MRE of the proposed techniques for instances in Group B is compared with those of SB, FTS, TSSB, HGA, and HPSO algorithms. The MRE of the GSO is 0.0168%, which is lesser than the MRE of the other algorithms. The MRE obtained for instances in Group C are compared with that of ACO-MWR, ACO-TR, TSSB, and HPSO. The MRE of the proposed GSO, on average, for the 40 instances under Group C is 0.149%, which is less than that of ACO-MWR (2.3763%), ACO-TR (2.372475%), TSSB (0.989511%), and HPSO (0.466688%). The MRE of 12 instances in Group D—TA41–TA50, TA62 and TA67—are compared with ACO-MWR, ACO-TR, TSSB, and HPSO algorithms. The MRE of GSO (0.036835%) for the chosen benchmark instances in Group D is less than that of ACO-MWR (4.279%), ACO-TR (4.0776667%), TSSB (2.15399%), HPSO (1.766718%), GA (0.6835878%), SPSO (0.50453%), and ACO (0.5108786%) algorithms.

TABLE 5.28

Comparison of Mean Relative Error

Instance	Heuristic Technique	MRE (%)	Instance	Heuristic Technique	MRE (%)
Group A	GPPR	2.228333	Group B	SB	1.45972868
	TSSB	1.1475		FTS	0.0538081
	ACO-MWR	0.625667		TSSB	0.1091098
	ACO-TR	0.688583		HGA	0.4209445
	GRASP	3.34		HPSO	0.0170562
	GA	1.650948		GA	0.8939106
	SPSO	0.636802		SPSO	0.4718308
	ACO	1.307773		ACO	0.688612
	GSO	0.142002		GSO	0.0168172
Group C	ACO-MWR	2.3763	Group D	ACO-MWR	4.279
	ACO-TR	2.372475		ACO-TR	4.0776667
	TSSB	0.9895113		TSSB	2.1539975
	HPSO	0.4666883		HPSO	1.766718
	GA	1.8859505		GA	0.6835878
	SPSO	0.8329595		SPSO	0.5045348
	ACO	1.2880706		ACO	0.5108786
	GSO	0.1490132		GSO	0.0368348

TABLE 5.29

Estimation of Algorithms for JSSP

Advantages	Performance of Algorithms
Optimal makespan	GSO obtained optimal makespan for 101 instances in Group A, B, C, and D.
Robustness	In all the four groups A, B, C, and D, the mean relative error of GSO is comparatively low, thus signifying robustness.
Computational time	The computational time of GSO is less than that of GA, SPSO, and ACO for instances in Group A, B, C, and D.

In this work, though the performance of GA and ACO was less satisfactory than that of SPSO and GSO, they were still able to reach the makespan much closer to the best known optimum when compared against existing approaches in the literature. It can be concluded that the MRE of GSO is reduced to an acceptable level compared to other listed heuristics, thus confirming its robustness to solving JSSPs.

5.6 Advantages of CI Paradigms

The optimal makespan, robustness, and computational time are presented for all the JSSP benchmark instances with the exact algorithms in Table 5.29.

For all the 162 instances considered in the JSSP problem, the computational time of GSO is reduced by 21.44% over GA, by 12.47% over SPSO, and by 17.1% over ACO. The merits and demerits of the optimization techniques in terms of the computational time are as follows:

- **GA:** Since GA is more likely to converge toward global minima, it was applied in this chapter to solve JSSPs. The computational time was high as a result of premature convergence in the local minima.

- **SPSO:** In SPSO, the acceleration constants and the inertia weight varied during the algorithmic run, thus improving the convergence rate and alleviating the algorithm from the local minima. Such dynamic update of parameters increases convergence, thus improving the computational efficiency.

- **ACO:** ACO adapts to new changes dynamically, such as new schedules in the JSSP, and hence it is applied to solve JSSPs. The convergence is usually uncertain in ACO, thus increasing the computational overhead in solving JSSPs.

- **GSO:** The hybrid of GA and SPSO tries to increase the ability of the algorithm to control convergence in the local and global optima, thus reducing the computational time.

5.7 Summary

In this chapter, JSSPs were solved in a fuzzy sense using the optimization algorithms GA, SPSO, ACO, and GSO. The processing time for each job is often imprecise in many real-world applications, which in turn is critical for the scheduling procedures. Therefore, JSSP with fuzzy processing time was addressed using signed distance for ranking of fuzzy numbers in the $(\lambda, 1)$ interval using triangular fuzzy numbers.

GA was applied to solve JSSP, trying to improve the searching efficiency and thus improving results in terms of optimal makespan. The information sharing ability between swarms was enhanced by applying SPSO based on stochastic process theory. The algorithm-dependent parameters of SPSO were tuned dynamically, thus improving the searching ability and in turn reducing the computational time. ACO was used to solve JSSP, following the communication principles and cooperative work of real ants. It was found that the algorithm was capable of obtaining better solutions very close to the best known makespan. GSO applied the evolutionary searching mechanism of SPSO characterized by population cooperation and competition to effectively perform exploration. On the other hand, GSO utilized adaptive local search of GA to perform exploitation. Thus in the proposed GSO, the balance between the global exploration and the local exploitation was maintained.

The performance of the meta-heuristic algorithms was analyzed by conducting a series of experiments on a standard set of 162 benchmark instances of various job and machine sizes. For each instance, the optimal makespan, average makespan from 10 trials, relative error between the obtained and the best known makespan, and average computational time from 10 trials were computed. An extensive comparative analysis was also performed with state-of-the-art heuristic approaches in literature: GRASP, GPPR, TSSB, SB, FTS, ACO-MWR, ACO-TR, HGA, and HPSO. The results presented in Tables 5.26–5.28 in terms of optimal makespan, computational efficiency, and MRE demonstrates the effectiveness of the proposed GSO in solving JSSP. From a practical perspective of the conducted experiments, it was found that all the proposed algorithms—GA, SPSO, ACO, and GSO—were capable of solving JSSP instances. In addition, the performance of all the algorithms on instances in the OR library and Taillard's class showed that optimal makespan could be achieved even for problems of large job and machine dimensions. In future, the research on JSSP can be extended by applying bio-inspired techniques such as bee colony optimization, immune systems, biogeography-based optimization, and multiobjective evolutionary algorithms using fuzzy processing times.

References

Abido, M., Optimal design of power system stabilizers using particle swarm optimization, *IEEE Transactions on Energy Conversion* 17(3), 406–413, 2002.

Adams, J., Balas, E., and Zawack, D., The shifting bottleneck procedure for job shop scheduling, *Management Science* 34, 391–401, 1988.

Aiex, R.M., Binato, S., and Resende, M.G.C., Parallel GRASP with path-relinking for job shop scheduling, *Parallel Computing* 29, 393–430, 2003.

Ali, A. and Fawaz, S.A., A PSO and Tabu search heuristics for the assembly scheduling problem of the two-stage distributed database application, *Computers and Operations Research* 33(4), 1056–1080, 2006.

Applegate, D. and Cook, W., A computational study of the job-shop scheduling problem, *ORSA Journal on Computing* 3, 149–156, 1991.

Arumugam, M.S. and Rao, M.V.C., On the improved performances of the particle swarm optimization algorithms with adaptive parameters, cross-over operators and root mean square (RMS) variants for computing optimal control of a class of hybrid systems, *Applied Soft Computing* 8(1), 324–336, 2008.

Balas, E. and Vazacopoulos, A., Guided local search with shifting bottleneck for job shop scheduling, *Management Science* 44, 262–275, February 1998.

Balujia, S. and Davies, S., Fast probabilistic modeling for combinatorial optimization, in *Proceedings of the American Association of Artificial Intelligence:AAAI'98*, Menlo Park, CA, pp. 469–476, 1998.

Beasley, J.E., OR-Library: Distributing test problems by electronic mail, *Journal of the Operational Research Society* 41(11), 1069–1072, 1990. http://people.brunel.ac.uk/~mastjjb/jeb/orlib/jobshop info.html.

Bing, W. and Zhen, Y., A particle swarm optimization algorithm for robust flow-shop scheduling with fuzzy processing times, in *Proceedings of the IEEE International Conference on Automation and Logistics, 2007*, Jinan, China, August 18–21, 2007, pp. 824–828.

Blum, C. Beam ACO – hybridizing ant colony optimization with beam search: an application to open shop scheduling, *Computers and Operations Research* 32, 1565–1591, 2005.

Blum, C., and Sampels, M., An ant colony optimization algorithm for shop scheduling problems, *Journal of Mathematical Modelling and Algorithms* 3, 285–308, 2004.

Cheung, W., and Zhou, H., Using genetic algorithms and heuristics for job shop scheduling with sequence dependant set-up times, *Annals of Operational Research* 107, 65–81, 2001.

Colorni, A., Dorigo, M., Maniezzo, V., and Trubian, M., Ant system for job shop scheduling, *Belgian Journal of Operations Research, Statistics and Computer Science* 34(1), 39–53, 1994.

Davis, L., Job shop scheduling with genetic algorithms, in *Proceedings of the First International Conference on Genetic Algorithms*, Pittsburgh, July 24–26, 1985, pp. 136–140.

Dorigo, M. and Gambardella, L.M., Ant colony system: a cooperative learning approach to the traveling salesman problem, *IEEE Transactions on Evolutionary Computation* 1(1), 53–66, 1997.

Fisher, H. and Thompson, G.L., *Probabilistic Learning Combinations of Local Job-Shop Scheduling Rules*, Englewood Cliffs, NJ: Prentice-Hill, 1963.

Fortemps, P., Jobshop scheduling using imprecise durations: A fuzzy approach, *IEEE Transactions on Fuzzy Systems* 5(4), 557–569, 1997.

Goldberg, D.E., *Genetic Algorithms in Search Optimization and Machine Learning*. Boston, MA: Addison Wesley Longman Publishing Co., 1989.

Goncalves, J.F., A hybrid genetic algorithm for the job shop scheduling problem, *European Journal of Operational Research* 167(1), 77–95, 2005.

Gorzalezang, M.B., A method of inference in approximate reasoning based on interval-valued fuzzy set, *Fuzzy Sets and Systems* 21, 1–17, 1981.

Grefenstette, J.J., Optimization of control parameters for genetic algorithms, *IEEE Transactions on Systems, Man and Cybernetics* 16(1), 122–128, 1986.

Huang, K.L. and Liao, C.J., Ant colony optimization combined with taboo search for the job shop scheduling problem, *Computers and Operations Research* 35, 1030–1046, 2008.

Husbands, P., Genetic algorithms for scheduling, *AISB Quarterly* 89, 38–45, 1996.

Ishii, H., Tada, M., and Masuda, T., Two scheduling problems with fuzzy due dates, *Fuzzy Sets and Systems* 46, 339–347, 1992.

Ivers, B., Job shop optimization through multiple independant swarms, in *Proceedings of the IEEE Congress on Evolutionary Computation: CEC 2007*, Singapore, September 25–28, 2007, pp. 3361–3368.

Jones, A. and Rabelo, L.C., Survey of job shop scheduling techniques, Technical report—NISTIR, National Institute of Standards and Technology, Gaithersburg, MD, 1998.

Kacem, I., Hammadi, S., and Borne, P., Approach by localization and multiobjective evolutionary optimization for flexible job-shop scheduling problems, *IEEE Transactions on Systems, Man, and Cybernetics, Part C* 32(1), 1–13, 2002.

Kaufmann, A. and Gupta, M.M., *Introduction to Fuzzy Arithmetic Theory and Applications*. New York: van Nostrand Reinhold, 1991.

Lawrence, S., Resource constrained project scheduling: An experimental investigation of heuristic scheduling techniques, Supplement PA1984, Graduate School of Industrial Administration, Carnegie Mellon University, Pittsburgh, 1984.

Li, J.Q., Pan, Q.K., Xie, S.X., Li, H., Jia, B.X., and Zhao, C.S., An effective hybrid particle swarm optimization algorithm for flexible job-shop scheduling problem, *MASAUM Journal of Computing* 1, 69–74, 2009.

Lian, Z., Gu, X., and Jiao, B., A similar particle swarm optimization algorithm for job-shop scheduling to minimize makespan, *Applied Mathematics and Computation* 183, 1008–1017, 2006a.

Lian, Z.G., Gu, X.S., and Jiao, B., A novel particle swarm optimization algorithm for permutation flow-shop scheduling to minimize makespan, *Chaos, Solitons and Fractals* 35(5), 851–861, July 2006b.

Liao, C.J., Tseng, C.T., and Luarn, P., A discrete version of particle swarm optimization for flow-shop scheduling problems, *Computers and Operations Research* 34(10), 3099–3111, 2007.

Liu, B., Wang, L., and Jin, Y.H., An effective PSO-based memetic algorithm for flow shop scheduling, *IEEE Transactions on Systems, Man and Cybernetics—Part B: Cybernetics* 37(1), 18–27, 2007.

McCahon, C.S. and Lee, E.S., Job sequencing with fuzzy processing times, *Computers and Mathematics with Applications* 19(7), 31–41, 1990.

Merkle, D., An Ant Algorithm with a new pheromone evaluation rule for total tardiness problem, *Proceedings of the Real World Applications of Evolutionary Computing: EvoWorkshops 2000*, pp. 290–299. Springer Berlin Heidelberg, 2000.

Morshed, M.S., A hybrid model for job shop scheduling, PhD Dissertation, University of Birmingham, Birmingham, U.K., 2006.

Muth, J. and Thompson, G., *Industrial Scheduling*, Prentice Hall, Englewood Cliffs, NJ, 1963.

Nakano, R. and Yamada, T., Conventional genetic algorithm for job shop problems, in *Proceedings of the Third International Conference on Genetic Algorithms*, R. Belew, and L. Booker, Eds. Morgan Kaufman, San Mateo, CA, 1991.

Niu, Q., Jiao, B., and Gu, X., Particle swarm optimization combined with genetic operators for job shop scheduling problem with fuzzy processing time, *Applied Mathematics and Computation* 205(1), 148–158, 2008.

Nowicki, E. and Smutnicki, C., A fast taboo search algorithm for the job shop problem, *Management Science* 42(6), 797–813, 1996.

Pezzella, F. and Merelli, E., A tabu search method guided by shifting bottleneck for the job shop scheduling problem, *European Journal of Operational Research* 120(2), 297–310, 2000.

Ross, T., *Fuzzy Logic with Engineering Applications*. Hoboken, NJ: John Wiley and Sons, 2004.

Sha, D.Y. and Hsu, C.Y., A hybrid particle swarm optimization for job shop scheduling problem, *Computers and Industrial Engineering* 51(4), 791–808, 2006.

Slowinski, R. and Hapke, M., *Scheduling Under Fuzziness*, Physica-Verlag, Heidelberg, 2000.

Song, X., Hybrid particle swarm algorithm for job shop scheduling problems, in *Future Manufacturing Systems*, T. Aized, Ed. InTech Publishers, Europe, 2010.

Sonmez, A.I. and Baykasoglu, A. A new dynamic programming formulation of (n*m) flowshop sequencing problems with due dates, *International Journal of Production Research* 36(8), 2269–2283, 1998.

Storer, R.H., Wu, S.D., and Vaccari, R., New search spaces for sequencing instances with application to job shop scheduling, *Management Science* 38(10), 1495–1509, 1992.

Surekha, P. and Sumathi, S., Solving fuzzy based job shop scheduling problems using GA and ACO, *International Journal of Emerging Trends in Computing and Information Sciences, E-ISSN 2218–6301* 1(2), 95–102, October 2010.

Surekha, P. and Sumathi, S., PSO and ACO based approach for solving combinatorial fuzzy job shop scheduling problem, *International Journal of Computer Technology and Applications*, 2(1), 112–120, January–February 2011.

Surekha, P. and Sumathi, S., Solution to job shop scheduling problem using hybrid genetic swarm optimization based on $(\lambda, 1)$ fuzzy processing time, *European Journal of Scientific Research* 64(2), 168–188, November 2011.

Taillard, E.D., Benchmarks for basic scheduling problems, *European Journal of Operational Research* 64(2), 108–117, 1993.

Taillard, É.D., University of Applied Sciences of Western Switzerland, http://mistic.heig-vd.ch/taillard/problemes.dir/ordonnancement.dir/ordonnancement.html, n.d.

Tsujimura, Y., Park, S.H., Chang, I.S., and Gen, M., An effective method for solving flow shop scheduling problems with fuzzy processing times, *Computers and Industrial Engineering* 25(1–4), 239–242, 1993.

Vaessens, R., Aarts, E., and Lenstra, J., Job shop scheduling by local search, *INFORMS Journal on Computing* 8(3), 302–317, 1996.

Vassil, N.A. et al., eds., *Computational Science—Proceedings of ICCS 2001: International Conference*, San Francisco, CA, May 28–30, vol. 2073, Springer, 2003.

Wang, B., Yang, X.F., and Li, Q.Y., Genetic simulated-annealing algorithm for robust job shop scheduling, *Advances in Intelligent and Soft Computing* 62, 817–827, 2009.

Wei, X.J. and Wu, Z.M., An effective hybrid optimization approach for multi-objective flexible job-shop scheduling problems, *Computers and Industrial Engineering* 48(2), 409–425, 2005.

Yamada, T. and Nakano, R., A genetic algorithm applicable to large-scale job-shop problems, in *Proceedings of the Second International Conference on Parallel Problem Solving from Nature: PPSN'92*, San Diego, CA, pp. 281–290, 1992.

Yao, J.S. and Wu, K.M., Ranking fuzzy numbers based on decomposition principle and signed distance, *Fuzzy Sets and Systems* 116(2), 275–288, 2000.

Zhang, J., Hu, X., Tan, X., Zhong, J.H., and Huang, Q., Implementation of an ant colony optimization for job shop scheduling problem, *Transactions of the Institute of Measurement and Control* 28(1), 93–108, 2006.

Zhang, Z. and Lu, C., Application of the improved particle swarm optimizer to vehicle routing and scheduling problems, in *Proceeding of the IEEE International Conference on Grey Systems and Intelligent Services: GSIS 2007*, Nanjing, China, November 18–20, 2007, pp. 1150–1152.

Zhou, Y., Beizhi, L., and Yang, J., Study on job shop scheduling with sequence-dependent setup times using biological immune algorithm, *International Journal of Advanced Manufacturing Technology* 30(1–2), 105–111, 2006.

Zimmermann, H.J., *Fuzzy Set Theory and Its Applications*. Boston, MA: Kluwer, 1991.

Zwaan, S. and Marques, C., Ant colony optimization for job shop scheduling, in *Proceedings of the Third workshop on Genetic Algorithms and Artificial Life: GAAL'99*, 1999.

6

Multidepot Vehicle Routing Problem

Learning Objectives

On completion of this chapter, the reader will have knowledge on

- Modeling a multidepot vehicle routing problem (MDVRP) for given set of customers and depots.
- Grouping the customers based on the distance between them and the depots.
- Assigning customers of the same depot to several routes using the Clarke and Wright saving method and to sequence each route in the scheduling phase.
- Determining optimal route using the optimization algorithms GA, MPSO, ABC, GSO, and IGSO.
- Choosing the suitable optimization technique for solving MDVRP by performing a comparative analysis in terms of optimal solution, robustness, computational time, and algorithmic efficiency.

6.1 Introduction

The vehicle routing problem (VRP) is one of the most challenging combinatorial optimization tasks in real-time logistics applications. VRP can be defined as a problem of finding the optimal routes of delivery or collection from one or several depots to a number of cities or customers while satisfying capacity and time constraints. In a real-world environment, drivers choose the shortest path to reach a destination, and because of this the distance traveled and the cost are minimized. Collection of household garbage, trucks delivering gasoline, goods distribution, snow ploughing, street cleaning, school bus routing, dial-a-ride systems, transportation for handicapped persons, routing of salespeople, and mail delivery are the most common applications of VRP. VRP plays a vital role in the distribution, logistics, and supply chain management. A large amount of research effort has been devoted to studying VRP since 1959, when Dantzig and Ramser described the problem as a generalized traveling salesman problem (TSP).

VRP consists in designing an optimal set of routes for a fleet of vehicles in order to serve a given set of customers. The interest in VRP is motivated by its practical relevance but also by its considerable difficulty. During the past two decades, an increasing number of optimization techniques based on operations research have been proposed for the effective management of the provision of goods and services in distribution systems. VRPs have several variants based on their operational mechanism and mathematical modeling. Based on the constraints and characteristics of VRPs, they are classified into capacitated VRP, multidepot VRP (MDVRP), periodic VRP, split delivery VRP, stochastic VRP, VRP with backhauls, VRP with pick-up and delivery, and VRP with time window.

MDVRP, which is an extension of the classical VRP, is an NP-hard problem for simultaneously determining the routes for several vehicles from multiple depots to a set of customers and then returning to the same depot. The objective of the problem is to find routes for the vehicles such that all the customers are served at minimal cost in terms of total travel distance without violating the capacity and travel time constraints of the vehicles. During the past three decades, classical VRPs have been paid much attention from a research perspective, but the number of research projects on MDVRP is relatively few. Salhi and Sari (1997) addressed a multilevel composite heuristic with two reduction tests. The initial feasible solutions were constructed in the first level, while the intradepot and the interdepot routes were improved in the second and third levels. Wu et al. (2002) reported a simulated annealing (SA) heuristic for solving the multidepot location routing problem (MDLRP). To solve the problem on a larger scale, the original problem was divided into two sub-problems: the location-allocation problem and the general vehicle routing problem. Each sub-problem was then solved in a sequential and iterative manner by the simulated annealing algorithm embedded in the general framework for the problem-solving procedure. Giosa et al. (2002) developed a "cluster first, route second" strategy for the MDVRP with time windows (MDVRPTW), which is an extension of the MDVRP. Considering the operational nature of the MDVRPTW, "the cluster first, route second" technique focuses more on minimizing the computational time. Haghani and Jung (2005) presented a formulation for solving the dynamic vehicle routing problem with time-dependent travel times using a genetic algorithm. The performance of the algorithm was evaluated by comparing its results with the exact solutions and lower bounds for randomly generated test problems. For small-sized problems with up to 10 demands, GA provided same results as the exact solutions. For problems with 30 demand nodes, GA results were found to have less than 8% gap with lower bounds.

Lee et al. (2006) handled the MDVRP by formulating the problem as deterministic dynamic programming (DP) with finite-state and action spaces, and then using a shortest path heuristic search algorithm. Creviera et al. (2007) proposed a heuristic combining the tabu search method and integer programming for MDVRP, in which vehicles may be replenished at intermediate depots along their route. Jeon et al. (2007) suggested a hybrid genetic algorithm (HGA) for MDVRP, which considered the improvement of generation for an initial solution, three different heuristic processes, and a float mutation rate for escaping from the local solution in order to find the best solution. In order to solve the MDVRP efficiently, two hybrid GAs (HGA1 and HGA2) are developed by Ho et al. (2008). In their approach, the initial solutions were generated randomly in HGA1. The Clarke and Wright saving method and the nearest neighbor heuristic were incorporated into HGA2 for the initialization procedure. Results proved that HGA2 was superior to HGA1 in terms of the total delivery time.

Chen and Xu (2008) developed a hybrid GA with simulated annealing for solving the MDVRP. Since the MDVRP integrates hard optimization problems, three improvement heuristic techniques were introduced by Mirabi et al. (2010). Each hybrid heuristic combined elements from both constructive heuristic search and improvement techniques.

The improvement techniques used in this method were deterministic, stochastic, and simulated annealing (SA). These techniques outperformed Giosa's (2002) "cluster first, route second" method in terms of minimum delivery time. Lau et al. (2010) considered the cost due to the total traveling distance and the cost due to the total traveling time for solving MDVRP. They employed a stochastic search technique called the fuzzy-logic-guided genetic algorithm (FLGA) to solve the problem.

MDVRP solutions were provided by Renaud et al. (1996) in three phases: fast improvement, intensification, and diversification, for 23 benchmark instances using the tabu search algorithm. They considered MDVRPs with capacity and route length restrictions. GA for MDVRP proposed by Ombuki-Berman and Hanshar (2009) employed an indirect encoding and an adaptive interdepot mutation exchange strategy for the MDVRP with capacity and route-length restrictions. Thangiah and Salhi (2001) presented a genetic clustering technique for MDVRP where a generalized clustering mechanism was applied to GAs as a post-optimizer. Su (1999) presented a real-time dynamic vehicle control and scheduling system for multidepot physical distribution. To perform the system objectives effectively, their system had five major modules: the global information collection system, depot controller, route planner, vehicle scheduler, vehicle route, and the time table feedback system. This method was more suitable for scheduling based on real-time status of the system.

Wenjing and Ye (2010) introduced a modified PSO algorithm with the mutation operator and improved inertia weight for solving MDVRP. The simulation results showed that this modified method could not only avoid premature automatically according to the convergence level but also get a better optimal solution than the standard basic PSO. Wang et al. (2008) used a particle position matrix based on goods for PSO to solve a multidepot single VRP model, and every matrix column corresponded to one good. Matrix elements were random number between 0 and 1, and matrix element sorting rules were established to get the single-vehicle route, thus satisfying the objective function.

The artificial bee colony (ABC) algorithm is a new population-based meta-heuristic approach proposed by Karaboga (2005). This approach is inspired by the intelligent foraging behavior of a swarm of honeybees. Karaboga and Basturk (2007) demonstrated that ABC could outperform GAs and PSO in multivariable function optimization. Brajevic (2011) presented the ABC algorithm for the capacitated VRP. In general, ABC algorithms are applied for continuous optimization problems. Since VRPs are combinatorial optimization problems, certain modifications were implemented by Ivona. Though ABC is a fairly new approach introduced only a few years ago, it has not yet been applied to solve MDVRP. Hence, it is worthwhile evaluating the performance of ABC algorithm for solving MDVRP.

In this chapter, the performance of heuristics such as GAs, modified PSO (MPSO), ABC optimization, hybrid MPSO with GA, the genetic swarm optimization (GSO), and improved GSO (IGSO) on MDVRP benchmark instances is addressed. The solution to MDVRP is obtained in four stages: grouping, routing, scheduling, and optimization. Customers are clustered based on the distance between them and depots in the grouping phase. In routing, customers of the same depot are assigned to several routes by the Clarke and Wright saving method, and each route is sequenced in the scheduling phase. Better routing and scheduling can result in shorter delivery distance, shorter time spent in serving all customers, a higher level of efficiency, and lower delivery cost. The scheduled routes are optimized using the optimization algorithms in order to obtain a global optimum solution with optimal distance, improved robustness, reduced computational time, and better algorithmic efficiency.

A set of five different Cordeau's benchmark instances (p01, p02, p03, p04, p06) from the online resource of University of Malaga, Spain, were experimented using MATLAB® R2008b. Several investigations are conducted on the benchmark instances using the proposed heuristics. The results are evaluated in terms of depot's route length, optimal route, optimal distance, computational time, and the number of vehicles.

The basic concepts of MDVRP along with the mathematical modeling and architecture of the proposed approach are discussed in this chapter. The implementation strategies of the proposed bio-inspired intelligent heuristics such as GA, MPSO, ABC, GSO, and IGSO for solving the MDVRP are focused. The MATLAB m-file snippets, simulation results, and experimental analysis of the MDVRP instances using heuristic algorithms are also discussed in this chapter.

6.2 Fundamental Concepts of MDVRP

In MDVRP, the number and locations of the depots are predetermined. Each depot is large enough to store all the products ordered by the customers. Each vehicle starts and finishes at the same depot. The location and demand of each customer is also known in advance, and each customer is visited by a vehicle exactly once. Figure 6.1 shows an example of the MDVRP with two depots and 10 customers. Since there are additional depots for storing the products, the decision makers have to determine depots through which the customers are served (Ho et al. 2008). The decision-making stages are classified into grouping, routing, scheduling, and optimization, as shown in Figure 6.2.

In grouping, customers are clustered based on the distance between customers and depots. In the example, customers 1, 5, 9, 4, and 8 are assigned to depot A, while customers 7, 10, 3, 6, and 2 are assigned to depot B. In depot A, customers 1, 5, and 9 are in the first route, while customers 4 and 8 are in the second route. The customers of the same depot are assigned to several routes in the routing phase by the Clarke and Wright saving method, and each route is sequenced in the scheduling phase. The aim of routing is to minimize the number of routes without violating the capacity constraints. Since there are

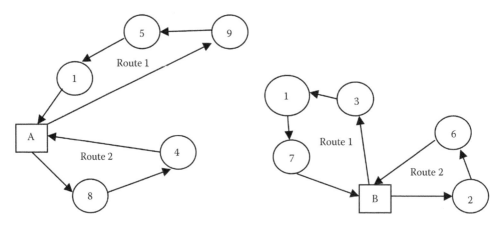

FIGURE 6.1
Example of an MDVRP with two depots and 10 customers.

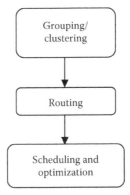

FIGURE 6.2
Decision making in MDVRP.

two depots, the minimum number of routes can be limited to two. More number of routes increase the number of vehicles required, thus reducing the distance. In general, the objective of the MDVRP is to minimize the total delivery distance or time spent in serving all customers, thus utilizing efficiently the number of vehicles.

6.2.1 Mathematical Formulation of MDVRP

MDVRP is formulated with the objective of forming a sequence of customers on each vehicle route. The time required to travel between customers along with the depot and demands is known in advance. It is assumed that all vehicles have the same capacity, and each vehicle starts its travel from a depot, and upon completion of service to customers it has to return to the depot. The notations used and the mathematical model are as follows:

- *Sets*
  - I: Set of all depots
  - J: Set of all customers
  - K: Set of all vehicles
- *Indices*
  - i: depot index
  - j: customer index
  - k: route index
- *Parameters*
  - N: Number of vehicles
  - C_{ij}: Distance between point i and j, $i, j \in I \cup J$
  - V_i: Maximum throughput at depot i
  - d_j: Demand of customer j
  - Q_k: Capacity of vehicle (route) k

- *Decision variables*
    - $x_{ijk} = \begin{cases} 1, & \text{if } i \text{ immediately proceeds } j \text{ on route } k \\ 0, & \text{otherwise} \end{cases}$
    - $z_{ij} = \begin{cases} 1, & \text{if customer } j \text{ is allotted to depot } i \\ 0, & \text{otherwise} \end{cases}$
    - U_{lk}: auxiliary variable for sub-tour elimination constraints in route k
- *Mathematical model*
    - The objective *function* is to minimize the total distance of all vehicles given by

$$\min \sum_{i \in I \cup J} \sum_{j \in I \cup J} \sum_{k \in K} C_{ij} x_{ijk} \tag{6.1}$$

    - Each customer has to be assigned a single route according

$$\sum_{k \in K} \sum_{i \in I \cup J} x_{ijk} = 1, \quad j \in J \tag{6.2}$$

    - The capacity constraint for a set of vehicles is given by

$$\sum_{j \in J} d_j \sum_{i \in I \cup J} x_{ijk} \leq Q_k, \quad k \in K \tag{6.3}$$

    - Equation 6.4 gives the new sub-tour elimination constraint set as

$$U_{lk} - U_{jk} + N x_{ijk} \leq N - 1, \quad l, j \in J, k \in K \tag{6.4}$$

    - The flow conservation constraints are expressed as

$$\sum_{j \in I \cup J} x_{ijk} - \sum_{j \in I \cup J} x_{ijk} = 0, \quad k \in K, i \in I \cup J \tag{6.5}$$

    - Each route can be served at most once according to

$$\sum_{i \in I} \sum_{j \in J} x_{ijk} \leq 1, \quad k \in K \tag{6.6}$$

    - The capacity constraints for the depots are given by

$$\sum_{j \in J} d_i z_{ij} \leq V_i, \quad i \in I \tag{6.7}$$

- Constraints in Equation 6.8 specify that a customer can be assigned to a depot only if there is a route from that depot going through that customer:

$$-z_{ij} + \sum_{u \in I \cup J} (x_{iuk} + x_{ujk}) \leq 1, \quad i \in I, j \in J, k \in K \tag{6.8}$$

- The binary requirements on the decision variables are given by Equations 6.9 and 6.10:

$$x_{ijk} \in \{0, 1\}, \quad i \in I, j \in J, k \in K \tag{6.9}$$

$$z_{ij} \in \{0, 1\}, \quad i \in I, j \in J \tag{6.10}$$

- The positive values of the auxiliary variable is defined as

$$U_{lk} \geq 0, \quad l \in J, k \in K \tag{6.11}$$

The objective function according to Equation 6.1 minimizes the total delivery distance. The constraint in Equation 6.2 ensures that each customer is allotted only one route, while the constraint in Equation 6.3 guarantees the capacity limit of the vehicles. Similarly, the sub-tour avoidance is imposed by Equation 6.4, and the flow is limited by Equation 6.5. The route to be served and the limit on the depots are given by Equations 6.6 and 6.7, respectively. Thus the MDVRP aims at minimizing the total delivery distance by satisfying the constraints mentioned before.

6.2.2 Grouping Assignment

Group assignment, also referred to as cyclic assignment, requires each depot to attract one customer every time until all customers are assigned. The cyclic assignment assigns one customer at a time in a cycle based on the location of the depot heads. First, the head of each depot is set as the depot itself. Then, for each depot, the closest customer to the depot head is assigned to it, and the depot head is updated by the closest customer if the vehicle of the depot has enough serving capacity. The earlier procedure is repeated until all customers are assigned.

The first step is to assign customers to each of the n links, known as the grouping problem. Because the objective here is to minimize the total delivery time spent in distribution, customers are assigned to the nearest depot. Grouping assignment procedure is given as follows:

Step 1: Calculate the distance between the customers.

Step 2: Calculate the distance between each customer and each depot.

Step 3: Assign the customers to the nearest depot.

Step 4: The customers are divided into M groups, where M is equal to the number of depots.

For example, in a distribution network, a supplier owns two depots (d_A and d_B) to deliver the products to a set of customers. Each customer ci should be assigned to a single depot exactly.

Equation 6.12 shows that, if the customer ci is located near the depot A, then the customer ci is assigned to depot A.

$$\text{If } D(ci, dA) < D(ci, dB), \quad \text{then assign } ci \text{ to } dA \tag{6.12}$$

Equation 6.13 indicates that, if the customer ci is located near the depot B, then the customer ci is assigned to depot B.

$$\text{If } D(ci, dA) > D(ci, dB), \quad \text{then assign } ci \text{ to } dB \tag{6.13}$$

Equation 6.14 ensures that, if the distance between customer ci and depot A is equal to the distance between customer ci and depot B, then customer ci is assigned to either depot A or depot B.

$$\text{In case } D(ci, dA) = D(ci, dB), \text{ select a depot arbitrarily.} \tag{6.14}$$

Equation 6.15 represents the distance between customer i and depot k.

$$D(ci, dk) = \sqrt{(x_{ci} - x_{dk})^2 + (y_{ci} - y_{dk})^2} \tag{6.15}$$

In Equation 6.15, x_{ci} and y_{ci} denote the x and y coordinates of the customer ci, x_{dk}, and y_{dk} indicate the x and y coordinates of depot k. The customers are grouped and served by either depot A or depot B.

6.2.3 Routing Algorithm

In the past decade, since the evolution of supply chain management, the system design problem considering the location of distribution facilities has become very significant. In the routing phase, customers in each group are divided into different routes. The aim of routing is to minimize the number of routes, or vehicles, used while not violating the vehicle capacity constraint. A key element of many distribution systems is the routing and scheduling of vehicles through a set of customers requiring service. Better routing decision can result in a higher level of customer satisfaction. A wrong grouping assignment solution will result in routes of higher total cost (distance) than with a better grouping assignment. So the routing phase is strongly dependent on the grouping assignment.

In 1964, Clarke and Wright published an algorithm for solving the classical vehicle routing problem. The savings algorithm is a heuristic algorithm, and therefore it does not provide an optimal solution to the problem with certainty. The method does, however, often yield a relatively good solution, which deviates only slightly from the best known solution. In the first step of the savings algorithm, the savings for all pairs of customers are calculated, and all pairs of customer points are sorted in descending order of the savings. Subsequently, from the top of the sorted list of point pairs, one pair of points is considered at a time. When a pair of points i–j is considered, the two routes that visit i and j are combined (such that j is visited immediately after i on the resulting route) if this can be done without deleting a previously established direct connection between two customer points and if the total demand on the resulting route does not exceed the vehicle capacity. There are two versions of the savings algorithm: sequential and parallel. In the sequential version, exactly one route is built at a time, while in the parallel version more than one route may be built at a time.

Initially, in Figure 6.3a, customers i and j are visited on separate routes. An alternative to this is to visit the two customers on the same route: for example, in the sequence i–j as

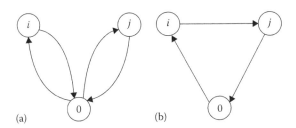

FIGURE 6.3
Illustration of the savings concept. (a) Separate route. (b) Same route.

illustrated in Figure 6.3b. Since the transportation costs are given, the savings that would result from taking the route in Figure 6.4b instead of the two routes in Figure 6.4a can be calculated. Denoting the transportation cost between two given points i and j by c_{ij}, the total transportation cost D_a in Figure 6.4a is

$$D_a = c_{0i} + c_{i0} + c_{0j} + c_{j0} \tag{6.16}$$

Equivalently, the transportation cost D_b in Figure 6.4b is

$$D_b = c_{0i} + c_{ij} + c_{j0} \tag{6.17}$$

By combining the two routes, the savings S_{ij} can be obtained as

$$S_{ij} = D_a - D_b = c_{i0} + c_{0j} - c_{ij} \tag{6.18}$$

The customers with a larger saving value are grouped in the same route without violating the vehicle capacity constraint.

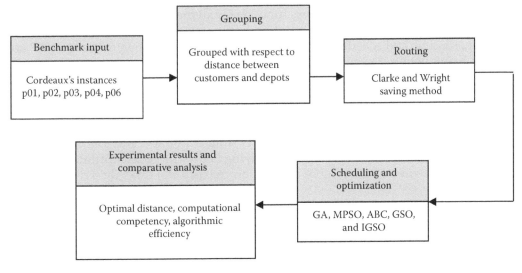

FIGURE 6.4
Proposed block diagram for MDVRP.

6.3 Computational Intelligence Algorithms for MDVRP

MDVRP is more challenging and sophisticated than the single-depot VRPs. In addition, MDVRP is NP-hard, which means that an efficient algorithm for solving the problem to optimality is unavailable. Therefore, MDVRP is difficult to solve with exact procedures such as branch-and-bound and branch-and-cut algorithms. To deal with the problem efficiently and effectively, heuristic algorithms such as GA, MPSO, ABC, GSO, and IGSO are applied in this work. Figure 6.4 illustrates the block diagram of the proposed MDVRP.

The well-known Cordeaux's instances (p01, p02, p03, p04, p06) are chosen as benchmark problems. The customers are clustered based on the distance between them and the depots. The customers of the same depot are assigned to several routes using the Clark and Wright saving method (Clarke and Wright 1964). The routes obtained from the Clarke and Wright saving method is scheduled and optimized using heuristic algorithms such as GA, MPSO, ABC, GSO, and the IGSO. The results of all the previous optimization techniques are evaluated with respect to the number of customers serviced, the number of vehicles required, the optimal route, the optimal distance, and the computational time. A comparative analysis of the proposed heuristic techniques is performed based on the optimal distance, robustness, computational competency, and algorithmic efficiency to identify the most suitable optimization algorithm for solving MDVRP.

Because of the complexity of the problem, solving the MDVRP to optimality is extremely time consuming. In order to tackle the problem efficiently, researchers have preferred heuristic methods over exact methods such as branch-and-bound and branch-and-cut algorithms (Ho et al. 2008). In this chapter, the application of GA, MPSO, ABC, GSO, and IGSO is proposed to solve MDVRP. The step-by-step procedure required for implementing these intelligent algorithms is delineated in this section.

6.3.1 Solution Representation and Fitness Function

The solution representation and fitness function for all the proposed bio-inspired techniques are similar, and the feasible solution is generated based on three basic steps: grouping, routing, and scheduling. The individuals for the solution of the MDVRP are encoded using path representation, in which the customers are listed in the order in which they are visited. Consider an MDVRP instance with six customers designated 1–6. If the path representation for this instance is (0 2 4 1 0 3 6 5 0), then two routes are required by the vehicles to serve all the customers. The first route starts from the depot at 0 and travels to customers 2, 4, and 1, and upon serving, the vehicle returns to the depot. Similarly, the second route starts from depot at 0, services customers 3, 6, and 5, and returns to the depot. While applying heuristics-based optimization techniques, each individual in the initial population consists of n links for n depots in the MDVRP.

- **Grouping:** In this stage, the customers are assigned to each of the n links. The objective of the MDVRP is to minimize the total delivery time, and hence customers are assigned to the nearest depots. In the example, there are two depots A and B, each customer c_i has to be assigned to a single depot exactly.

This process of grouping is done based on the distance computation according to the following rule:

- If $D(c_i, A) < D(c_i, B)$, then customer c_i is assigned to depot A.
- If $D(c_i, A) > D(c_i, B)$, then customer c_i is assigned to depot B.
- If $D(c_i, A) = D(c_i, B)$, then customer c_i is assigned to a depot chosen arbitrarily between A and B.

In the previous cases, $D(c_i, k) = \sqrt{(x_{c_i} - x_k)^2 + (y_{c_i} - y_k)^2}$ represents the distance between the customer c_i and the depot k.

- **Routing:** The customers in the same link are assigned to several routes using the Clarke and Wright saving method. The routing is based on the distance traveled by the vehicles for serving the customers. A saving matrix $S(c_i, c_j)$ is constructed for every two customers i and j in the same link. Further, the customers with a large saving value are grouped in the same route without violating the capacity constraints. The saving matrix is constructed according to $S(c_i, c_j) = D(k, c_i) + D(k, c_j) - D(c_i, c_j)$.

- **Scheduling:** Starting from the first customer, the delivery sequence is chosen such that the next customer is as close as the previous customer. This process is repeated until all the unselected customers are sequenced. The scheduling is performed by the respective optimization technique based on the fitness function. The fitness function for the MDVRP is described in this section. At the end of the scheduling phase, a feasible solution of the MDVRP example problem (Figure 6.1) is constructed, as shown in Figure 6.5.

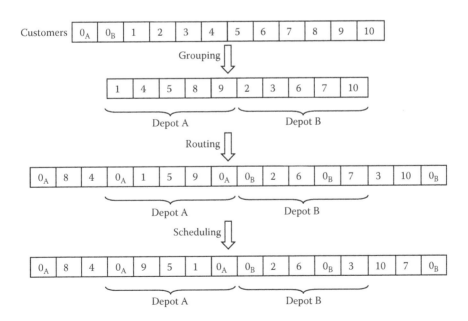

FIGURE 6.5
Solution representation and initial population.

- **Fitness evaluation:** For the MDVRP, the objective function is to minimize the maximum delivery time spent among the n depots. The delivery operations start at the same time in every depot, while it takes different times to complete serving the assigned customers. Some vehicles belonging to a depot may complete the delivery faster, while some others may complete their task in a longer duration. Let D_t be the total delivery time required by a depot k, and let $\min(D_t)$ represent the minimum delivery time spent by all n depots. Then

$$D_t = \sum_{k=1}^{m_k}\left[d\big[c(m_c),c(0)\big] + \sum_{i=1}^{m_c} d[c(i-1),c(i)] \right] \tag{6.19}$$

where

$d(a,b) = \dfrac{\sqrt{(x_b - x_a)^2 + (y_b - y_a)^2}}{V}$ is the travel time of a vehicle from customer a

     to customer b

V is the speed of the vehicle

$c(i)$ is the location of the ith customer

$c(0)$ is the initial position of the depot

m_c is the number of customers in route r

m_k is the number of routes in depot k

Thus the fitness function is defined as $F = \sum \min(D_t)$. This function acts as the objective function to solve MDVRPs using the proposed intelligent techniques.

6.3.2 Implementation of MDVRP Using GA

GA is based on a parallel search mechanism (Goldberg 1989), which makes it more efficient than other classical optimization techniques such as branch-and-bound, tabu search, and simulated annealing. The basic idea of GA is to maintain a population of candidate solutions that evolves under selective pressure. GA imitates the mechanism of natural selection and the survival of the fittest as in natural evolution. GA can avoid getting trapped in a local optimum by tuning the genetic operators crossover and mutation. Because of its high potential for global optimization, GA has received great attention in solving MDVRPs. The general scheme of the genetic algorithm for MDVRP is explained in this section.

6.3.2.1 Initial Population

The initial population is formed based on the routes, depots, and customers according to the description available in Section 6.3.1.

6.3.2.2 Selection

During each generation, the parents are selected for mating and reproduction. In this MDVRP application, tournament selection is applied (Renaud et al. 1996) to generate new individuals in the population. This selection strategy is based on the fitness evaluation given in Section 6.3.1. The selection procedure is as follows:

Step 1: Select a set of g individuals from the population in a random manner to form the tournament set.

Step 2: Choose a random number r_n in the range [0,1].

Step 3: If r_n < *threshold*, select the fittest individual from the tournament set for reproduction.

   Else, choose any two chromosomes at random from the tournament set for reproduction.

Step 4: Apply elitism to guarantee that the best individuals are selected.

6.3.2.3 Crossover

A problem-specific crossover technique, the best cost route crossover (BCRC) developed by Ombuki-Berman and Hanshar (2009) for vehicle routing problem with time windows (VRPTW) was applied in this work for MDVRP with slight improvements. The steps involved in BCRC are shown as follows:

Step 1: Choose the parents from tournament selection.

Step 2: Select a route from each parent in a random manner.

Step 3: Remove all customers belonging to route 1 from parent 1.

Step 4: For every customer belonging to route 1,

- Compute the cost of insertion of route 1 into each location of parent 2 and store the costs in an ordered list.
- For each insertion location, check whether the insertion is feasible or not.
- Generate a random number $r_n \in [0,1]$.
- Choose the first feasible insertion location if r_n < *threshold*.
- Else, if r_n > *threshold*, choose the first entry in the ordered list, despite the feasibility.

Step 5: Repeat Step 4 for customer belonging to route 2.

6.3.2.4 Mutation

The flip-bit mutation operator is used in finding a solution to MDVRP using GA. A substring is selected from the parent in a random manner and is flipped to form an offspring. The route reversal mutation works on only one chromosome at a time.

   The design steps used for the implementation of MDVRP based on GA are as follows:

Step 1: Generate randomly the initial population of GA based on routes, depots, and customers.

Step 2: Evaluate the fitness function for each chromosome in the population according to Equation 6.19.

Step 3: Calculate the total fitness value $\sum fi$, which is represented as the summation of individual fitness of each chromosome.

Step 4: Repeat the following steps until new chromosomes have been created:

- Select a pair chromosome from current population, based on tournament selection.
- With the crossover probability of 0.6, exchange the pair using BCRC to form two new chromosomes.
- Mutate the two new chromosomes at each row with the mutation probability 0.02 and place the resulting chromosomes in the new population.

Step 5: Replace the current population with the new population.

Step 6: If the fixed number of generations is reached, stop and return the best solution in current population; else, go to step 2.

The chromosomes in the population evolve through successive iterations, called generations, thus evaluating the measure of fitness based on Equation 6.19. The fitter the chromosomes, the higher the probability of being selected to perform the genetic operations such as crossover and mutation. Selected individuals are chosen for reproduction (or crossover) at each generation, with an appropriate mutation factor to randomly modify the genes of an individual, in order to develop the new population. The current population is replaced with the new population until a fixed number of generations is reached.

6.3.3 Solving MDVRP Using MPSO

PSO is a population-based search technique consisting of potential solutions known as particles, duplicating a flock of birds. The particles are initialized randomly based on the optimization problem, and they freely fly across the multidimensional search space. Each particle updates its own velocity and position, thus driving the swarm towards the best fit regions. Eventually, all the particles gather around the point with the highest objective function. In contrast to GA, PSO does not have any genetic operators such as crossover and mutation. In GA, the chromosomes share the information with each other, and the whole population moves towards the optimal solution, while in PSO only the best particle shares the information with others. Because of this one-way information sharing mechanism, the evolution of PSO is always towards the best solution. Moreover, all particles tend to converge to the best solution quickly without getting trapped in the local minima.

6.3.3.1 Solution Representation and Fitness Evaluation

The particles for the solution of the MDVRP are encoded using the path representation, and the fitness function is evaluated based on distance according to the explanation given in Section 6.3.1.

6.3.3.2 Updating Particles

In the evolutionary process, MPSO is used to generate new offspring based on new position of the particle, thus updating the velocity according to

$$x_{t+1}^i = x_t^i + \underbrace{w_t v_t^i + c_1 r_1(p_t^i - x_t^i) + c_2 r_2(p_t^g - x_t^i)}_{v_{t+1}^i} \tag{6.20}$$

The best position visited by the kth particle since the first time step is known as the local best position, denoted as pbest p_t^i, while gbest p_t^g represents the best position that the kth particle and its neighbors have visited from the beginning of the algorithm. In this work, since the distances are real coded, the velocities and position update rules are slightly modified as

$$v_{t+1}^i = \text{int}\left\{w_t v_t^i + c_1 r_1(p_t^i - x_t^i) + c_2 r_2(p_t^g - x_t^i)\right\} \tag{6.21}$$

$$x_{t+1}^{i} = abs\left\{x_{t}^{i} + v_{t+1}^{i}\right\}$$ (6.22)

where

abs {} denotes the absolute value function

int {} is the integral function

r_1 and r_2 are two independent random numbers uniformly distributed in the range {0,1}

c_1 and c_2 are the weighting factors, also known as cognitive and social parameters, respectively

During each iteration, the p_{best} and g_{best} are updated based on the objective (fitness) value. If the current value of p_{best} is better, p_{best} and its objective value are updated with the current position and objective value. Similarly, if the objective value is better than that of g_{best}, then g_{best} and its objective value are updated with the position and objective value of the current best particle.

6.3.3.3 Algorithm

The step-by-step procedure for implementing the MDVRP using MPSO is shown as follows:

Step 1: Initialize the parameters of MPSO to suitable values. Set the value of the maximum number of iterations.

Step 2: Initialize the population of particles according to the delivery sequence of customers.

Step 3: Evaluate the fitness function for all particles, and choose the best local and global positions p_{best} and g_{best} of the particles.

Step 4: Update the particles according to Equations 6.21 and 6.22.

Step 5: Compute the fitness function of all the updated particles and determine the best fit particle. Compare the fitness and update the particles accordingly.

Step 6: Test for the stopping condition. If maximum number of iterations is reached, then stop; else, continue with Step 3.

6.3.4 Artificial Bee Colony-Based MDVRP

The ABC optimization algorithm is becoming more popular recently and is based on the foraging behavior of honeybees. ABC is a population-based search technique, in which the individuals, known as the food positions, are modified by the artificial bees during the course of time. The objective of the bees, in turn, is to discover food sources with high nectar concentration. The colony of artificial bees is grouped into employed bees, onlooker bees, and scout bees. During the initialization phase, the objective of the problem is defined along with the ABC algorithmic control parameters. An employed bee is assigned for every food source available in the problem. In employed bee phase, the employed bee stays on a food source and stores the neighborhood of the source in its memory. During the onlooker phase, onlooker bees watch the waggle dance of the employed bees within the hive to choose a food source. The employed bee whose food source has been abandoned becomes a scout bee. Scout bees search for food sources randomly during the scout phase. Thus the local search is carried out by the employed bees and the onlooker bees, while the global search is performed by the onlooker and the scout bees, thus maintaining a balance between the exploration and exploitation process. In this section, the step-by-step procedure to implement ABC technique for MDVRP is discussed.

6.3.4.1 Generation of Initial Solutions

The initial solutions to the ABC problem are formed using the grouping and routing stages. In this mode, ABC algorithm generates randomly distributed initial solutions, due to which the constraints satisfaction may not be guaranteed. Therefore during initialization, every solution has to be checked according to each route as follows:

- If the value of more than one position in the corresponding positions of all routes in the solution is 1, one position is selected in random and its value is set to 1, while the value of others is set to 0.
- If the values of the corresponding positions of all routes in the solution are all 0, one position is selected in random and its value is set to 1, and the others remain unchanged.

This kind of an initialization ensures for every initial solution that each customer is served exactly once by exactly one vehicle. If the vehicle starts and ends at different depots, then the solution is infeasible.

6.3.4.2 Constraints Handling

In order to handle the constraints, the ABC algorithm employs Deb's rules (Deb 2000), which are used in constrained optimization problems. The method uses a tournament selection operator, where two solutions are compared at a time by applying the following criteria:

- Any feasible solution satisfying all constraints is preferred to any infeasible solution violating any of the constraints.
- Among two feasible solutions, the one having better fitness value is preferred.
- Among two infeasible solutions, the one having the smaller constraint violation is preferred.

Because initialization with feasible solutions is a very time-consuming process and in some cases it is impossible to produce a feasible solution randomly, the ABC algorithm does not consider the initial population to be feasible. Structure of the algorithm already directs the solutions to a feasible region in the running process due to Deb's rules.

6.3.4.3 Neighborhood Operators

In order to produce new solutions for the employed and onlooker bees, two neighborhood operators are applied (Eiben and Smith 2003). The first neighborhood operator is called SwapMutation. It was introduced as a neighborhood operator for solving the TSP and is called "2-change." The idea of the mutation operation is to randomly mutate the tour and thereby produce a new solution *g* that is not very far from the original one *f*. In this example, the mutation operator is designed to conduct customer exchanges in a random manner. The steps for the SwapMutation operator are as follows:

- Randomly select two routes from the solution *f*, and randomly select two customers from each selected route.
- Exchange the customers in the different routes and generate the new solution *g*.

The second neighborhood operator, also based on random changes, is called Insert Mutation. The steps for the InsertMutation operator are as follows:

- Randomly select the routes from the solution *f*, and randomly select one customer from one selected route.
- Remove the selected customer from one route to the other and generate the new solution *g*.

After the new solution is produced, the selection process is applied based on Deb's method. If the new solution *g* is accepted instead of the solution *f*, SwapMutation operator is applied in the mutated routes in order to improve the new solution. SwapMutation is applied in the following way:

- Compare all possible pairwise exchanges of customer locations in the route to find the exchange that produces the shortest distance.
- When the pair of customers whose exchange produces the shortest distance is found, the route is rearranged. If such a pair of customers is not determined, the route stays unchanged.

In the employed bee phase, the SwapMutation operator is applied two times. The onlooker bee phase has one difference from the employed bee phase: it applies one SwapMutation and one InsertMutation operator instead of two SwapMutation operators.

The pseudocode of the proposed ABC algorithm is given here:

Construct initial employed bee colony solutions;
Evaluate fitness value for each solution;
iteration = 1;

While iteration ≠ maximum iteration count.

6.3.4.3.1 *Employed Bee Phase*
For each employed bee, apply two times:

- SwapMutation operator
- Selection process based on Deb's method
- If the new solution is accepted, improve the new solution and evaluate them.

6.3.4.3.2 Onlooker Bee Phase

Calculate the probability values for the solutions:
  For the onlookers selected based on probabilities, the following is applied two times:

- First time Swap Mutation operator, second time Insert Mutation operator
- Selection process based on Deb's method
- If the new solution is accepted, improve the new solution and evaluate them.

6.3.4.3.3 Scout Bee Phase

Determine every infeasible solution for the scout and replace them with new produced solutions

  iteration = iteration + 1
  Memorize the best solution achieved so far
  End *While*

6.3.5 GSO for MDVRP

The performance of GA and PSO is much faster in terms of converging toward the optimum compared to evolutionary algorithms (Abido 2002, Fourie and Groenwold 2002). Because of the continuous nature of the particles, a standard encoding scheme cannot be defined for MDVRP. Thus MPSO is applied to the MDVRP to determine a suitable method between the scheduled sequence and the position of particles. In GSO, the worst particles found by MPSO are replaced by the best particles obtained by GA. Application of a single algorithm for obtaining the solution to MDVRP problem is not feasible, and hence hybrid algorithms are used as effective methods. These hybrid algorithms enlarge the application domain and improve the performance by combining the efficient features of algorithms, such as the mechanism involved in optimization, search pattern, and mode of operation, thus increasing the optimization efficiency. Some of the basic ideas behind the design of GSO are as follows:

- Each feasible solution of the MDVRP is a point in the search space and is considered as an individual. The route sequence is not continuous; hence a suitable encoding scheme is adopted for MPSO to fit the problems.
- The infeasible solutions produced during the algorithmic run are adjusted by the hybrid combination of GA and MPSO.
- The algorithm getting trapped in a local optimum at a very early stage can be avoided by using the GSO algorithm.

The overall structure of the proposed hybrid algorithm is described as follows:

6.3.5.1 Solution Coding and Fitness Evaluation

The description of solution representation and fitness evaluation is provided in Section 6.1.3.

6.3.5.2 Offspring Generation

The new generation of chromosomes is produced by changing the route of the vehicles to the customers. In the evolutionary process, MPSO is used to generate new offspring based on new position of the particle, thus updating the velocity according to

$$v_{t+1}^i = \text{int}\left\{w_t v_t^i + c_1 r_1 (p_t^i - x_t^i) + c_2 r_2 (p_t^g - x_t^i)\right\} \tag{6.16}$$

$$x_{t+1}^i = abs\left\{x_t^i + v_{t+1}^i\right\} \tag{6.17}$$

where abs {} denotes the absolute value function, int {} is the integral function, r_1 and r_2 are two independent random numbers uniformly distributed in the range {0,1}, and c_1 and c_2 are the weighting factors, also known as cognitive and social parameters, respectively. As a parallel process, the genetic operators of crossover and mutation are applied to update the particles according to Equation 6.18:

$$x_{t+1}^i = \hat{x}_t^i \oplus (p_t^i \otimes x_t^i) \oplus (p_t^g \otimes x_t^i) \tag{6.18}$$

Here, the symbol $\otimes$ denotes the crossover operation performed between the current best and the best individual obtained so far. The symbol $\oplus$ represents the optimal or best offspring resulting from $(p_t^i \otimes x_t^i)$, or $(p_t^g \otimes x_t^i)$ or $\hat{x}_t^i$. $(p_t^i \otimes x_t^i)$ denotes the crossover operation between the two parents p_t^i and x_t^i, and $\hat{x}_t^i$ is the mutated offspring of x_t^i. x_{t+1}^i is the new particle obtained from genetic operators. The particles obtained through GA and through MPSO are evaluated for their fitness, and the best individual is retained in the population for further generations. Here, the best cost route crossover and flip-bit mutation are applied. The key parameter that controls the percentage of individuals in the population evolved through GA and MPSO in the combined approach is the hybridization factor (HF). In simple terms, if HF = 0, then the algorithm is a pure MPSO algorithm, and if HF = 1 then the algorithms is a pure GA. The range of HF is [0,1] in order to maintain the balance between the individuals evolved through GA and MPSO.

6.3.5.3 Stopping Condition

The algorithm terminates with the number of iterations and computation of the best solution.

The flowchart of the proposed GSO approach is shown in Figure 6.6. The initial swarm consists of the scheduled sequence based on the fitness function of the MDVRP. The MPSO solutions are checked for feasibility, and the infeasible solutions are reproduced by GA to generate feasible solutions through best cost route crossover and flip-bit mutation. The feasible solutions generated by MPSO are combined with the feasible solutions generated from GA, and this new swarm is evaluated for fitness. The process is repeated for a fixed number of iterations, and the optimal results for the MDVRP instances are determined.

6.3.6 IGSO for MDVRP

Hybrid algorithms are used as effective methods for obtaining optimal solution to large-scale, constraint-based optimization problems. In this case, solution to MDVRP is achieved using a hybrid combination of GA and MPSO, termed as GSO. In GSO,

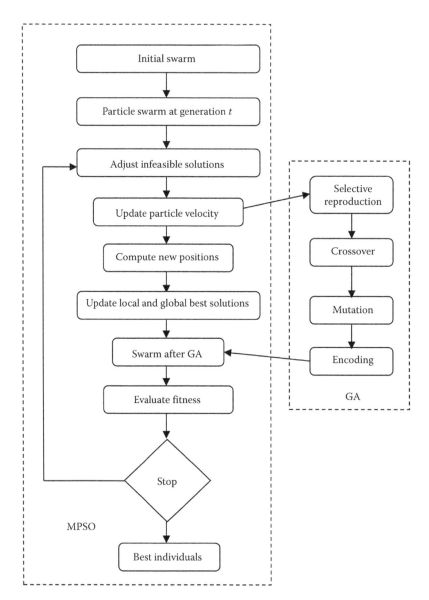

FIGURE 6.6
Flowchart of the proposed GSO algorithm.

GA and MPSO are hybridized to replace the worst particles found by MPSO by the best particles obtained by GA, and vice versa. The GSO is modified with an enhancement step using MPSO, denoted as improved GSO (IGSO) in order to improve the optimal distance and computational time of MDVRP instances obtained by GSO. The basic ideas behind the implementation of IGSO are as follows:

- The infeasible solutions produced during the algorithmic run are adjusted by enhancing the elites in the hybrid combination of GA and MPSO.
- The chances of the algorithm getting trapped into the local optimization at a very early stage can be avoided by the IGSO algorithm.

The step-by-step implementation of the proposed hybrid algorithm for solving MDVRP is described as follows:

6.3.6.1 Solution Coding and Fitness Evaluation

The structure of the initial solutions for the MDVRP based on IGSO is framed on the basis of the grouping and routing techniques. The routes obtained using the Clarke and Wright saving method is further scheduled based on fitness function using IGSO. The description of solution representation and fitness evaluation are provided in Section 6.3.1.

6.3.6.2 Enhancement

In each generation, half of the best individuals, termed as elites, are selected based on the fitness evaluation. The elites are enhanced instead of reproducing the next generation of offspring directly. The enhancement of the elites is performed by updating the velocity and position of particles in MPSO according to Equations 6.23 and 6.24, respectively:

$$v_{t+1}^i = \text{int}\left\{w_t v_t^i + c_1 r_1 (p_t^i - x_t^i) + c_2 r_2 (p_t^g - x_t^i)\right\} \tag{6.23}$$

$$x_{t+1}^i = abs\left\{x_t^i + v_{t+1}^i\right\} \tag{6.24}$$

6.3.6.3 Selection

While the elites are enhanced using MPSO, the GA operations are performed in parallel. In order to select parents from the selected best half elites, tournament selection is applied. Two enhanced elites are selected randomly, and their fitness values are compared to select the one with better fitness as a parent and placed in the mating pool.

6.3.6.4 Offspring Generation

Parents are selected randomly from the mating pool in groups of two, and two offspring are created by performing crossover on the parent solutions. Single individual is selected based on the mutation probability, and a new genetic offspring is generated to maintain diversity in the population. In this work, best cost route crossover (BCRC) and route reversal mutation are applied to increase the diversity in the population such that better elites are produced in the next generation. The offspring generated from GA and the enhanced elites from MPSO form a new population, and their fitness is evaluated and compared in order to select the elites for the next generation.

6.3.6.5 Stopping Condition

The algorithm terminates with the number of iterations and computation of the best solution.

The flowchart of the proposed IGSO approach is shown in Figure 6.7. The initial swarm consists of the scheduled sequence based on the fitness function of the MDVRP. The enhanced elites using MPSO and offspring generated using GA form a new

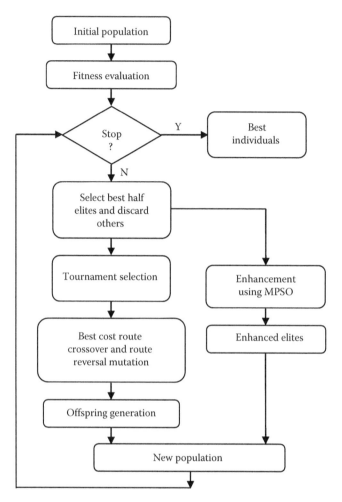

FIGURE 6.7
Flowchart of the proposed IGSO algorithm.

population for further generations and evaluated for fitness. The process is repeated for a fixed number of iterations, and the optimal results for the MDVRP instances are determined.

6.4 MATLAB® m-File Snippets for MDVRP Based on CI Paradigms

In this section, the simulation results of stages involved in solving MDVRP, such as grouping, routing, and optimization, are analyzed. The main objective of MDVRP is to minimize the total delivery distance, the number of vehicles required, and the computation time. The performance of the MDVRP using GA, MPSO, ABC, GSO, and IGSO is evaluated using a set of five Cordeau's instances: p01, p02, p03, p04, and p06 taken from NEO (n.d.). The simulation is done using MATLAB R2008b on an Intel i3 CPU, 2.53 GHz, 4 GB RAM PC.

6.4.1 Experimental Benchmark Instances

Table 6.1 shows the specifications of five Cordeau's instances: p01 (four depots and 50 customers), p02 (four depots and 50 customers), p03 (five depots and 75 customers), p04 (two depots and 100 customers), and p06 (three depots and 100 customers). In this work, the depots and customers are represented in the form of coordinates, and each depot has a limited number of vehicles. Each vehicle has a constant loading capacity, and each vehicle starts and finishes at the same depot. The demand of each customer is also known in advance, and each customer is visited by a vehicle exactly once.

6.4.2 Grouping and Routing

In grouping, the customers are assigned to adjacent depots so that the distance traveled by the vehicle is shorter. The customers are clustered based on the minimum distance between them and the depots. The MATLAB m-file snippet is shown here for grouping of instance p01 based on distance:

```
for(j=1:length(depm(:,1)))%no of suppliers
    for(i=1:length(cusm(:,1)))%no of customers
        p(j,i)=sqrt (((cusm(i,2)-depm(j,1))^2)+((cusm(i,3)-depm(j,1))^2));
    end
end
for i=1:length(p(1,:))
    [v ps]=min(p(:,i));
    cus_min(i,:)=[v ps];
end
disp('----------------Grouping output-----------------')
cus1_p=find(cus_min(:,2)==1)
cus2_p=find(cus_min(:,2)==2)
cus3_p=find(cus_min(:,2)==3)
cus4_p=find(cus_min(:,2)==4)
```

In the grouping phase, the Euclidean distance between the customer and the depot is computed and, based on the minimum distance, the Cordeau's instances p01, p02, p03, p04, and p05 are grouped; the results are shown in Table 6.2.

In p01, 11 customers (customer IDs: 4, 13, 17, 18, 19, 25, 40, 41, 42, 44, and 45) are assigned to depot A whose location is at (20,20) indicating the x and y coordinates; 19 customers (customer IDs: 5, 6, 7, 10, 12, 14, 15, 23, 24, 27, 33, 37, 38, 39, 43, 46, 47, 48, and 49) are served by depot B located at x coordinate 30, y coordinate 40; 16 customers (customer IDs: 1, 2, 8, 9, 11, 16, 21, 22, 26, 28, 29, 30, 31, 32, 34, and 50) are assigned to depot C located at x coordinate 50,

TABLE 6.1

Specifications of Benchmark Instances

Parameters/Instances	p01	p02	p03	p04	p06
Total no. of customers	50	50	75	100	100
Total no. of depots	4	4	5	2	3
Total no. of vehicles	32	20	35	24	30
No. of vehicles in each depot	8	5	7	12	10
Capacity of each vehicle	80	100	140	100	100
Best known distance (km)	576.87	473.53	641.19	1001.59	876.5

TABLE 6.2

Assignment of Customers to Depot for Benchmark Instances

Depot	Customers Allotted	No. of Customers
Problem instance: p01		
A(20,20)	4-13-17-18-19-25-40-41-42-44-45	11
B(30,40)	5-6-7-10-12-14-15-23-24-27-33-37-38-39-43-46-47-48-49	19
C(50,30)	1-2-8-9-11-16-21-22-26-28-29-30-31-32-34-50	16
D(60,50)	3-20-35-36	4
Problem instance: p02		
A(20,20)	4-13-17-18-19-25-40-41-42-44-45	11
B(30,40)	5-6-7-10-12-14-15-23-24-27-33-37-38-39-43-46-47-48-49	19
C(50,30)	1-2-8-9-11-16-21-22-26-28-29-30-31-32-34-50	16
D(60,50)	3-20-35-36	4
Problem instance: p03		
A(40,40)	3-4-5-6–9-12-15-17-18-20-25-26-27-29-30-32-34-37-39-40-44-45-47-48-50-51-55-60-67-68-70-75	32
B(50,22)	7- 8-13-35-46-52-57-58-72	9
C(55,55)	10-11-14-19-31-38-53-54-59-65-66	11
D(25,45)	2-16-21-24-28-33-36-49-62-63-69-71-73-74	14
E(20,20)	1-22-23-41-42-43-56-61-64	9
Problem instance: p04		
A(15,35)	2-5-6-7-8-13-14-15-16-17-18-21-22-23-36-37-38-40-41-42-43-44-45-46-47-48-52-53-56-57-58-59-60-61-72-73-74-75-82-83-84-85-86-87-89-91-92-93-94-95-96-97-98-99-100	55
B(35,55)	1-3-4-9-10-11-12-19-20-24-25-26-27-28-29-30-31-32-33-34- 35-39-49-50-51-54-55-62-63-64-65-66-67-68-69-70-71-76-77-78-79-80-81-88-90	45
Problem instance: p06		
A(15,20)	5-14-15-16-17-37-38-41-42-43-44-45-46-57-59-61-84-85-86-87-91-92-93-95-96-97-98-99-100	29
B(50,20)	1-3-9-10-11-12-20-24-25-29-30-32-33-34-35-50-51-54-55-63- 64-65-66-68-70-71-76-77-78-79-80-81-90	33
C(35,35)	2-4-6-7-8-13-18-19-21-22-23-26-27-28-31-36-39-40-47-48-49-52-53-56-58-60-62-67-69-72-73-74-75-82-83-88-89-94	38

y coordinate 30; and remaining customers (customer IDs: 3, 20, 35, and 36) are assigned to depot D located at x coordinate 60, y coordinate 50). Similarly, in the other benchmark problems p02, p03, p04, and p06, the customers are grouped to the associated depots, as shown in Table 6.2.

In the routing phase, the customers in each group are divided into different routes. The aim of routing is to minimize the number of routes, or vehicles used, without violating the vehicle capacity constraint. The Clark and Wright saving method is applied to solve the routing process. There are two versions of the savings algorithm: a sequential version and a parallel version. In the sequential version, exactly one route is built at a time (excluding routes with only one customer), while in the parallel version more than one route may be built at a time. In this work, the parallel version is used to compute the routes at a faster rate. It is worth noting that the number of routes may be reduced during the process of the parallel version of the savings algorithm. For example, the two routes 0-1-2-0 and 0-3-4-0 will be combined into one route if the connection from depot A to B is established; in that case, the resulting route becomes 0-1-2-3-4-0.

Fully loaded percentage (FLP) is defined as the number of products loaded in each vehicle in every depot, and is calculated according to Equation 6.25:

$$FL = \frac{Q - LQ}{Q} \times 100 \tag{6.25}$$

where
  FL is the fully loaded percentage of each vehicle
  Q is the capacity of each vehicle
  LQ is the loading quantity of each vehicle during the distribution process

The code snippet for routing customer 1 in instance p01 and the route assignment to depot A is shown here:

```
%----------routing customer 1------------%
for i=1:length(cus1_p_vl(:,1))
    limt1=Temp;
    sum_rt=sum_rt+cus1_p_vl(i,4);
    if sum_rt<limt1
        rts=[rts cus1_p_vl(i,1)];
    else
        cus1_roots(rt).route=rts;
        sum_rt=0;
        rt=rt+1;
        rts=[];
        rts=cus1_p_vl(i,1);
        sum_rt=cus1_p_vl(i,4);
    end
    if (i==length(cus1_p_vl(:,1))) && (sum_rt<limt1)
        cus1_roots(rt).route=rts;
    end
end
%----------route assignment for Depot A------------%
Depot_A_Route1=cus1_roots(1,1)
Depot_A_Route2=cus1_roots(1,2)
Depot_A_Route3=cus1_roots(1,3)
Depot_A_Route4=cus1_roots(1,4)
Depot_A_Route5=cus1_roots(1,5)
%----------Scheduled output------------%
for i=1:length(opt_roots)
    rots=sedul(i).data;
    for j=1:length(opt_roots(i).data)
        act_val=rots(1,j).data;
        ind=opt_roots(i).data(j);
        poss=ind.data-1;
        poss=poss(find(poss));
        cus_op_in{i,j}=[act_val(poss(1:end-1)) poss(end)];
    end
end
%----------Optimized Route without CI algorithms------------%
 Depot_A_Optimized_Route1= cus_op_in{1,1}
 Depot_A_Optimized_Route2= cus_op_in{1,2}
```

```
Depot_A_Optimized_Route3= cus_op_in{1,3}
Depot_A_Optimized_Route4= cus_op_in{1,4}
Length_1_2_3_4=cus_op_in{1,1}(1,4)+cus_op_in{1,2}(1,3)+cus_op_in{1,3}
(1,4)+cus_op_in{1,4}(1,4);
Depot_A_Route_Length=Length_1_2_3_4
```

In the routing phase, the route of vehicles, fully loaded percentage (FLP), distance (D) and the number of vehicles (NV) served by a depot are computed and the results are tabulated. The route allocation of customers for instance p01 using the Clark and Wright saving method is shown in Table 6.3. The depots along with their coordinates are specified, and the set of routes followed by the customers with the distance is tabulated. It can be observed that, for depot B located at (50, 22), there are six set of routes generated for six different vehicles. The total distance for the assigned vehicles to start from the depot B, serve all the customers, and return to the depot B is 261.65 km.

Table 6.4 shows the results obtained for the benchmark instance p02. From the experiments carried out, it is seen that four vehicles are used by depot A, five by depot B, four by depot C, and one by depot D. The distances are computed using Euclidean distance but they do not guarantee the optimal values.

The routing algorithm is run for the p03 benchmark instance consisting of five depots and 75 customers, and the results are shown in Table 6.5. The highest fully loaded percentage of depot A is 96.82%, which implies that the vehicle serves customers through the route A-20-25-26-27-29-A in an efficient manner. Each route is followed by a vehicle, seven vehicles serve the customer requirements in depot A, while two vehicles serve depot B, three vehicles in depot C and D, and two vehicles serve depot E, following the routes shown in Table 6.5.

TABLE 6.3

Route Allocation for Customers in Instance p01

Depot	Route	FLP (%)	D (km)	NV
A(40,40)	A-4-13-17-A	85.85	28.825	5
	A-18-19-A	78.69	21.21	
	A-25-40-A	86.82	20.65	
	A-41-42-44-A	92.5	34.28	
	A-45-A	89.65	16.25	
B(50,22)	B-5-6-7-B	75.64	43.2	6
	B-10-12-14-B	74.78	47.28	
	B-15-23-24-27-B	93.75	56.225	
	B-33-37-38-B	83.5	28.425	
	B-39-43-46-47-B	93.5	53.82	
	B-48-49-B	64.85	32.7	
C(55,55)	C-1-2-C	78.51	28.46	5
	C-8-9-11-C	86.84	42.3	
	C-16-21-22-26-28-29-C	92.4	73.3	
	C-30-31-32-C	86.5	27.405	
	C-34-50-C	72.65	19.325	
D(25,45)	D-3-20-D	76.56	16.42	2
	D-35-36-D	85.6	14.88	

TABLE 6.4

Route Allocation for Customers in Instance p02

Depot	Route	FLP (%)	D (km)	NV
A(40,40)	A-4-13-17-A	85.85	28.825	4
	A-18-19-A	78.69	21.21	
	A-25-40-41-A	86.82	20.65	
	A-42-44-45A	92.5	34.28	
B(50,22)	B-5-6-7-10-B	75.64	43.2	5
	B-12-14-15-B	74.78	47.28	
	B-23-24-27-33-B	93.75	56.225	
	B-37-38-39-43-46-B	83.5	28.425	
	B-47-48-49-B	93.5	53.82	
C(55,55)	C-1-2-8-C	78.51	28.46	4
	C-9-11-16-21-22-26-C	86.84	42.3	
	C-28-29-30-31-32C	92.4	73.3	
	C-34-50-C	86.5	27.405	
D(25,45)	D-3-20-35-36-D	76.56	16.42	1

TABLE 6.5

Route Allocation for Customers in Instance p03

Depot	Route	FLP (%)	D (km)	NV
A(40,40)	A-3-4-5-6-A	85.85	43.49	7
	A-9-12-15-17-18-A	78.69	52.21	
	A-20-25-26-27-29-A	96.82	74.2	
	A-30-32-34-37-39-A	93.5	68.4	
	A-40-44-45-47-A	89.65	53.065	
	A-48-50-51-55-60-A	93.65	79.27	
	A-67-68-70-75-A	68.51	46.77	
B(50,22)	B-7-8-13-35-46-52-B	95.64	54.29	2
	B-57-58-72-B	74.78	29.82	
C(55,55)	C-10-11-14-C	78.51	34.475	3
	C-19-31-38-53-C	86.84	48.32	
	C-54-59-65-66-C	90.4	41.79	
D(25,45)	D-2-16-21-D	76.56	48.34	3
	D-24-28-33-36-D	85.6	54.87	
	D-49-62-63-69-71-73-74-D	94.65	74.165	
E(20,20)	E-21-22-23-41-42-43-E	91.84	50.78	2
	E-56-61-64-E	81.35	23.445	

TABLE 6.6

Route Allocation for Customers in Instance p04

Depot	Route	FLP (%)	D (km)	NV
A (15,35)	A-2-5-6-7-8-13-A	90.62	67.835	11
	A-14-15-16-17-18-21-A	93.68	100.99	
	A-22-23-36-37-38-A	89.37	83.725	
	A-40-41-42-43-44-45-46-A	93.5	99.28	
	A-47-48-52-A	76.7	49.325	
	A-53-56-57-58-59-60-A	92.5	81.15	
	A-61-72-73-74-75-A	85.55	50.1	
	A-82-83-84-85-A	76.5	46.725	
	A-86-87-89-91-92-A	87.6	44.945	
	A-93-94-95-A	78.96	39.24	
	A-96-97-98-99-100-A	81.5	53.39	
B (35,55)	B-1-3-4-9-10-B	90.5	48.965	10
	B-11-12-19-20-24-25-B	94.5	94.885	
	B-26-27-28-29-30-B	93.5	104.395	
	B-31-32-33-34-B	88.6	72.38	
	B-35-39-49-B	87.6	43.325	
	B-50-51-54-55-62-63-B	92.5	117.46	
	B-64-65-66-67-B	76.5	52.435	
	B-68-69-70-71-76-B	96.61	78.085	
	B-77-78-79-80-81-B	95.5	86.585	
	B-88-90-B	58.67	34.37	

The benchmark instance p04 consisting of two depots and 100 customers, considered as one of the large-size problems, is run using the routing algorithm, and the results are shown in Table 6.6. It is observed that the total distance required by depot A with 11 vehicles to serve customers is 708.51 km and, similarly, depot B used 10 vehicles with a total distance of 733.94 km such that all customers are served.

The route allocation for customers in p06 instance with three depots and 100 customers is shown in Table 6.7. Depot A is allocated with a set of six routes with a total distance of 216.49 km with loading capacities of 93.75%, 95%, 68.75%, 95%, 88.75%, and 73.75%, respectively. Similarly, depot B is allocated with seven routes whose total distance is 397.22 km, and depot C is allocated with nine routes with a total distance of 476.36 km.

6.4.3 Impact of GA on MDVRP

In GA, each chromosome is represented by the route of each vehicle. The set of chromosomes form the initial population thus forms the search space. The population size decides the number of chromosomes in a single generation. A larger population size slows down the GA run, while a smaller value leads to exploration of a small search space. A reasonable range of the population size is [50,100]. Based on the encoding of chromosomes in this work, the population size is set to 50. GAs generate a new route sequence by selecting two individuals in the population to which the genetic operators crossover and mutation are applied. The best cost route crossover technique is applied to exploit a better solution for MDVRP. The choice of mutation is route-reversal type, which explores a wider search space for MDVRP. The parameters used in solving MDVRP using GA for five benchmark problems are shown in Table 6.8.

TABLE 6.7

Route Allocation for Customers in Instance p06

Depot	Route	FLP (%)	D (km)	NV
A (15,20)	A-5-14-15-16-17-A	93.75	44.905	6
	A-37-38-41-42-43-44-45-46-A	95	72.725	
	A-57-59-61-84-A	68.75	39.605	
	A-85-86-A	95	19.46	
	A-87-91-92-93-95-A	88.75	26.77	
	A-96-97-98-99-100-A	73.75	13.025	
B (50,20)	B-1-3-9-10-11-B	83.75	54.055	7
	B-12-20-24-25-29-30-B	83.75	88.05	
	B-32-33-34-35-50-51-B	98.75	54.06	
	B-54-55-63-64-65-B	73.75	83.555	
	B-66-68-70-B	82.5	37	
	B-71-76-77-78-79-80-B	92.5	58.5	
	B-81-90-B	56.25	22	
C (35,35)	C-2-4-6-7-8-13-18-C	97.5	108.86	9
	C-19-21-22-23-C	93.75	57.965	
	C-26-27-28-31-C	95	69.2	
	C-36-39-40-47-C	90	24.805	
	C-48-49-52-C	93.75	32.23	
	C-53-56-58-60-62-C	75	79.6	
	C-67-69-72-73-74-C	91.25	47.1	
	C-75-82-83-88-89-C	86.25	47.6	
	C-94-C	53.25	9	

TABLE 6.8

Parameter Settings for GA-Based MDVRP

Parameters	Settings
Population size	Based on the number of customers
Selection	Tournament selection
Crossover	Best Cost Route crossover
Mutation	Route reversal mutation
Crossover probability	0.6
Mutation probability	0.02
Elite	4

The m-file snippet for evaluation of the GA routine is presented here:

```
% Genetic Algorithm Operators
    rand_pair = randperm(pop_size);
    for p = 4:4:pop_size
        rtes = pop(rand_pair(p-3:p),:);
        dists = total_dist(rand_pair(p-3:p));
        [ignore,idx] = min(dists);
```

```
        best_of_4_rte = rtes(idx,:);
        ins_pts = sort(ceil(n*rand(1,2)));
        I = ins_pts(1);
        J = ins_pts(2);
        for k = 1:4% Mutate the Best to get Three New Routes
            tmp_pop(k,:) = best_of_4_rte;
            switch k
                case 2% Flip
                    tmp_pop(k,I:J) = fliplr(tmp_pop(k,I:J));
                case 3% Swap
                    tmp_pop(k,[I J]) = tmp_pop(k,[J I]);
                case 4% Slide
                    tmp_pop(k,I:J) = tmp_pop(k,[I+1:J I]);
                otherwise % Do Nothing
            end
        end
        new_pop(p-3:p,:) = tmp_pop;
    end
% Evaluate fitness function (Distance)
    for p = 1:pop_size
        d = dmat(pop(p,n),pop(p,1)); % Closed Path
        for k = 2:n
            d = d + dmat(pop(p,k-1),pop(p,k));
        end
        total_dist(p) = d;
    end
% To find the Best Route in the Population
    [min_dist,index] = min(total_dist);
    dist_history(iter) = min_dist;
    if min_dist < global_min
        global_min = min_dist;
        opt_rte = pop(index,:);
    end
```

The routes are scheduled and optimized using GA to determine the number of customers serviced (NC) and the number of vehicles required (NVR), and the optimal route. NVA indicates the total number of vehicles available in each depot for the corresponding benchmark instance. These parameters are computed and presented in Table 6.9 for benchmark instances p01, p02, p03, p04, and p06. A total of 11 customers are serviced by depot Λ in instance p01 with four optimal routes A-17-4-13-A, A-19-18-A, A-25-41-40-A, and A-44-42-45-A. Likewise, depot B serves 19 customers, depot C serves 16 customers, and depot D serves 4 customers with the optimal routes as shown in table. Instance p02 has four depots, each of which serves 11, 19, 16, and 4 customers with 3, 5, 3, and 1 optimal route, respectively. A set of 16 optimal routes is generated by instance p03 using GA to serve 75 customers from five depots. Though instance p04 has only two depots, the number of vehicles required is too large since there are 100 customers to be served from these depots. Depot A has 10 optimal routes to serve 55 customers, while depot B has 8 optimal routes to supply 45 customers. Instance p06 has three depots each generating 5, 5, and 7 optimal routes to serve 29, 33, and 38 customers, respectively.

The computational results obtained through GA for the five benchmark instances are shown in Table 6.10 in terms of computed optimal distance, computational time, and

TABLE 6.9

Optimal Route Using GA

Depot	NVA	NC	NVR	Optimal Route
Instance p01				
A(20,20)	8	11	4	A-17–4–13-A, A-19–18-A, A-25–41–40-A, A-44–42–45-A
B(30,40)	8	19	5	B-7–10–5–6-B, B-12–14–15-B, B-23–27–33–37–24, B-39–38–46–43–47-B, B-48–49-B
C(50,30)	8	16	4	C-8–2–9–1-C, C-28–26–11–21–16–22-C, C-29–31–32–30-C, C-50–34-C
D(60,50)	8	4	1	D-35–20–3–36-D
Instance p02				
A(20,20)	5	11	3	A-13–18–4–17-A, A-25–41–40–19-A, A-44–45–42-A
B(30,40)	5	19	5	B-6–7–10–5-B, B- 15–12–23–14-B, B-27–24–37–33–38-B, B- 46–39–47–43–48-B, B-49-B
C(50,30)	5	16	3	C-9–2–8–1-C, C- 11–16–21–29–28–26–22-C, C-34–50–31–32–30-C,
D(60,50)	5	4	1	D-3–20–35–36-D
Instance p03				
A(40,40)	7	32	6	A-9–3–6–5–4-A, A-18–17–20–15–26–12–25-A, A-34–27–37–29–30–32-A, A- 40–39–44–47–45-A, A-48–60–67–55–50–51–68-A, A-75–70-A
B(50,22)	7	9	2	B-52–46–7–35–8–13–57-B, B-72–58-B
C(55,55)	7	11	3	C-19–14–11–10-C, C-59–65–31–38–53–54-C, C-66-C
D(25,45)	7	14	3	D-2–24–16–21-D, D-49–36–71–69–28–62–73–33–63-D, D-74-D
E(20,20)	7	9	2	E-61–22–42–41–56–23–43–1-E, E-64-E
Instance p04				
A(15,35)	12	55	10	A-8–7–6–13–2–14–5-A, A-22–15–16–17–18–21-A, A-38–43–41–23–40–36–37–42-A, A-45–46–47–44-A, A-57–56–58–53–52–48-A, A-73–72–74–59–61–60-A, A-82–83–84–85–75-A, A-91–86–89–87–92-A, A-97–95–94–96–93-A, A-99–98–100-A
B(35,55)	12	45	8	B-1–4–3–9–11–10-B, B-27–19–20–24–25–12–26-B, B-29–30–31–28-B, B-33–32–35–34–39-B, B-49–51–55–54–50–62-B, B-67–65–66–64–63-B, B-69–70–71–78–77–68–76-B, B-88–90–81–79–80-B
Instance p06				
A(15,20)	10	29	5	A-17–5–37–15–14–38–16-A, A-41–59–45–46–44–43–42–57-A, A-61–84–86–85-A, A-96–95–87–97–92–91–93-A, A-100–98–99-A
B(50,20)	10	33	5	B-24–12–3–20–9–11–10–1-B, B-34–29–25–30–32–33–35-B, B-55–65–64–63–51–50–54-B, B-70–66–71–68–76-B, B-81–79–78–80–77–90-B
C(35,35)	10	38	7	C-7–19–8–18–6–13–4–2-C, C-26–23–22–21–27-C, C-31–36–40–39–28-C, C-49–47–48-C, C-52–53–56–67–58–60–62-C, C-69–82–83–72–75–74–73-C, C-88–94–89-C

TABLE 6.10

Computational Results for Benchmark Instances Using GA

Benchmark Instance Type	No. of Customers	No. of Depots	Best Known Distance (km)	Optimal Distance (km)	Computational Time (s)	Gap between Best Known and Optimal Distance (%)
p01	50	4	576.87	598.45	4.0692	3.740877
p02	50	4	473.53	478.65	3.4207	1.081241
p03	75	5	641.19	699.23	6.8128	9.051919
p04	100	2	1001.59	1011.36	10.1081	0.975449
p06	100	3	876.50	882.48	9.3177	0.682259

TABLE 6.11

Parameter Settings for MPSO-Based MDVRP

Parameter	Setting
Swarm size	Based on the number of customers
Cognitive factor, c_1	1.28
Social coefficient, c_2	1.28
First inertia weight, w_{int}	0.9
Last inertia weight, w_{end}	0.4
No. of iterations	300

gap between the best known and optimal distance. The gap between the best known distance and optimal distances (P_d) is calculated using Equation 6.26:

$$P_d = \frac{V_o - V_{bk}}{V_{bk}} \times 100 \qquad (6.26)$$

where

V_o is the optimal distance obtained by the optimization technique

V_{bk} is the best known distance of the instance

From Table 6.10, it is observed that the optimal distances computed by GA for all the instances are near the optimal and best known solution. These distances are computed based on a total of 14 routes for instance p01, 12 routes for instance p02, 16 routes for instance p03, 18 routes for instance p04, and 17 routes for instance p06. The differences between the best known and optimal distances interpret near-optimal solutions for instances p04 (0.975449%) and p06 (0.682259%). Relatively small differences are obtained for instances p01 and p02 with values 3.740877% and 1.081241%, respectively, while for instance p03 GA is not capable of finding the optimal distance since the difference 9.051919% is quite large.

6.4.4 Evaluation of MPSO for MDVRP

The parameters of MPSO are set as shown in Table 6.11 based on empirical studies and studies from the literature. The acceleration constants c_1 and c_2 are set to 1.28, and initial population of the swarm is varied with respect to the customers and depots in each instance. The inertia weight w determines the search behavior of the algorithm. Large values for w facilitate searching new locations, whereas small values provide a finer search in the current area. A balance can be established between global and local exploration by decreasing the inertia weight during the execution of the algorithm.

The routes for each vehicle are taken as the swarm particles, and the MPSO algorithm iterates to determine the p_{best} and g_{best} to optimize the delivery distance. The evaluation of the fitness function of the swarm is performed using MATLAB m-file as shown as follows:

```
% the best fitness of the swarm
[curFitBest,curFitBestPos] = min(curIndiFitBest);
swarmBestEver = indiBestEver(curFitBestPos,:);
minFitness = min(curIndiFitBest);
maxFitness = max(curIndiFitBest);
pMutation = 0.0;
curPath = curSwarm(curFitBestPos,:);
curOutput = [curPath curPath(1)];
```

```
global_min = Inf;
total_dist = zeros(1,ini_par);
dist_history = zeros(1,num_iter);
tmp_par = zeros(4,n);
new_par = zeros(ini_par,n);
for iter = 1:num_iter

    for p = 1:ini_par
        d = dmat(par(p,n),par(p,1));
        for k = 2:n
            d = d + dmat(par(p,k-1),par(p,k));
        end
        total_dist(p) = d;
    end
    [min_dist,index] = min(total_dist);
    dist_history(iter) = min_dist;
    if min_dist < global_min
        global_min = min_dist;
        opt_rte = par(index,:);
    end
```

The scheduled routes for the five benchmark instances p01, p02, p03, p04, and p06 are optimized using MPSO, and the route length, the optimal route, the number of customers serviced, and the number of vehicles required are evaluated; the results are shown in Table 6.12.

Instance p01 has generated a total of 13 routes from four depots, thus servicing all the 50 customers. In instance p02 with four depots, depot A produces three optimal routes, depot B generates four optimal routes, while depot C framed three optimal routes and depot D one optimal route. A total of 75 customers are served in instance p03 with five depots requiring 14 vehicles among the available 35 vehicles. Similarly, instance p04 has generated 15 optimal routes with two depots and instance p06 constructed 17 optimal routes with three depots.

The results obtained through MPSO are shown in Table 6.13 in terms of the optimal distance, computational time, and gap between the best known distance and optimal distance. The gap between the best known distance and optimal distances obtained by MPSO is computed according to Equation 6.27. It can be inferred from the results that the MPSO algorithm is capable of obtaining near-optimal solutions when compared with the best known distance. For instance, the best known distance of p02 is 473.53 km, while the optimal value is 475.47 km with a gap of 0.41%. It can also be observed that the computational time increases with the size of the problem.

6.4.5 Solution of MDVRP Based on ABC

In ABC, a colony of artificial forager bees acts as agents searching for artificially rich food sources, analogous to good solutions for a given problem. In this work, the position of a food source represents a set of possible routes to MDVRP. The parameters that govern the ABC algorithm are the colony size, number of food sources, limit, number of employed bees, number of onlooker bees, and maximum number of iterations. The colony size is based on the number of customers in each MDVRP instance. The smaller the colony size, the faster the convergence, and vice versa. The number of employed bees and onlooker bees are usually set to one-half of the colony size for best results. The number of food sources is always chosen to be equal to the number of employed bees. The ABC algorithm is run for a maximum of 800 iterations irrespective of the size of the problem. The parameters and their values used for running the ABC algorithm for MDVRP are shown in Table 6.14.

TABLE 6.12

Optimal Route Using MPSO

Depot	NVA	NC	NVR	Optimal Route
Instance p01				
A(20,20)	8	11	3	A-13–18–4–17-A, A-40–41–25–19-A, A-44–45–42–A
B(30,40)	8	19	5	B-7–6–5–10-B, B-15–12–23–14-B, B-33–37–24–27–38-B, B-39–46–48–43–47-B, B-49-B
C(50,30)	8	16	4	C-9–1–8–2-C, C-26–28–29–21-C, C-16–11–22-C, C-30–34–50–31–32-C
D(60,50)	8	4	1	D-20–3–36–35-D
Instance p02				
A(20,20)	5	11	3	A-13–19–17–4–18-A, A-25–41–40–42-A, A-45–44–A
B(30,40)	5	19	4	B-6–7–10–5-B, B-12–15–14–24–23-B, B-39–38–27–43–37–33-B, A-47–48–46–49-A
C(50,30)	5	16	3	C-9–1–8–2-C, C-26–28–29–21–16–11–22-C, C-30–34–50–31–32-C
D(60,50)	5	4	1	D-3–20–35–36-D
Instance p03				
A(40,40)	7	32	5	A-12–9–3–6–5–4, A-25–18–17–29–20–15–27–26-A, A-39–34–37–30–32-A, A-40–44–47–48–45-A, A-67–75–70–60–68–51–50–55-A,
B(50,22)	7	9	2	B-52–46–7–35–8–13–57-B, B-72–58-B
C(55,55)	7	11	3	C-11–14–19–10-C, C-59–54–53–38–31–65-C, C-66-C
D(25,45)	7	14	2	D-21–2–24–16–28-D, D-74–36–71–69–62–73–33–63–49-D
E(20,20)	7	9	2	E-56–41–42–22–61–1–43–23-E, E-64-E
Instance p04				
A(15,35)	12	55	8	A-8–5–14–15–2–13–6–7-A, A-37–16–17–36–18–21–23–22-A, A-38–44–45–46–47–40–41–42–43-A, A-52–48–57–56–58–53-A, A-60–73–72–75–74–59–61-A, A-84–85–82–83-A, A-93–92–87–89–86–91-A, A-94–95–97–100–98–99–96-A
B(35,55)	12	33	7	B-11–10–1–12–4–3–9-B, B-26–28–27–19–20–29–24–25-B, B-30–31–33–34–35–32-B, B-51–50–54–55–39–49-B, B-67–62–64–63–66–65-B, B-78–71–70–69–76–68–77-B, B-79–81–90–88–80-B
Instance p06				
A(15,20)	10	29	5	A-5–37–15–14–38–16–17-A, A- 45–61–44–43–41–57–42–59–46-A, A-86–84–85-A, A- 91–92–97–87–95–96–93-A, A-100–98–99-A
B(50,20)	10	33	5	B-20–11–10–1–12–24–3–9-B, B-29–25–33–30–32–35–34-B, B-55–54–50–51–65–64–63-B, B-68–71–66–70–76-B, B- 80–77–79–78–81–90-B
C(35,35)	10	38	7	C-6–13–2–4–19–7–8–18-C, C-26–23–22–21–27-C, C-31–36–40–39–28-C, C-49–47–48–52-C, C-67–56–58–60–62–69–53-C, C-88–82–83–73–74–75–72-C, C-94–89-C

TABLE 6.13

Computational Results for Benchmark Instances Using MPSO

Benchmark Instance Type	No. of Customers	No. of Depots	Best Known Distance (km)	Optimal Distance (km)	Computational Time (s)	Gap between Best Known and Optimal Distance (%)
p01	50	4	576.87	580.35	3.6319	0.603255
p02	50	4	473.53	475.47	2.9434	0.409689
p03	75	5	641.19	687.65	5.1899	7.245902
p04	100	2	1001.59	1008.55	7.7211	0.694895
p06	100	3	876.50	880.35	8.1418	0.439247

TABLE 6.14

ABC Parameters for MDVRP

S. No.	Parameters	Value
1	Colony size	Based on no. of customers
2	No. of food sources	No. of employed bees
3	Food source limit	No. of onlooker bees × dimension of the problem
4	No. of employed bees	No. of customers/2
5	No. of onlooker bees	No. of customers/2
6	Maximum No. of iterations	800

The optimal route was evaluated using ABC along with the number of vehicles required and the number of customers serviced for the problem instances using MATLAB. The phases of the ABC algorithm and its implementation are depicted in the form as m-file snippets as shown as follows:

```
%Calculate fitness
fFitness=zeros(size(fObjV));
ind=find(fObjV>=0);
fFitness(ind)=1./(fObjV(ind)+1);
ind=find(fObjV<0);
Fitness=ind;

%reset trial counters
trial=zeros(1,FoodNumber);

%/*The best food source is memorized*/
BestInd=find(ObjVal==min(ObjVal));
BestInd=BestInd(end);
GlobalMin=ObjVal(BestInd);
GlobalParams=Foods(BestInd,:);
pop_size=col_size;
pop_size = 4*ceil(pop_size/4);
num_iter = max(1,round(real(num_iter(1))));
pop = zeros(pop_size,n);
for k = 1:pop_size
    pop(k,:) = randperm(n);
end
global_min = Inf;
total_dist = zeros(1,pop_size);
dist_history = zeros(1,num_iter);
tmp_pop = zeros(4,n);
new_pop = zeros(pop_size,n);
for iter = 1:num_iter

    for p = 1:pop_size
        d = dmat(pop(p,n),pop(p,1));
        for k = 2:n
            d = d + dmat(pop(p,k-1),pop(p,k));
        end
        total_dist(p) = d;
    end
```

```
    [min_dist,index] = min(total_dist);
    dist_history(iter) = min_dist;
    if min_dist < global_min
        global_min = min_dist;
        opt_rte = pop(index,:);
    end

% EMPLOYED BEE PHASE
    for i=1:(FoodNumber)

        Param2Change=fix(rand*D)+1;
        neighbour=fix(rand*(FoodNumber))+1;
        while(neighbour==i)
                neighbour=fix(rand*(FoodNumber))+1;
        end;

        sol=Foods(i,:);
sol(Param2Change)=Foods(i,Param2Change)+(Foods(i,Param2Change)-Foods(neig
hbour,Param2Change))*(rand-0.5)*2;

        ind=find(sol<lb);
        sol(ind)=lb(ind);
        ind=find(sol>ub);
        sol(ind)=ub(ind);

        %evaluate new solution
        ObjValSol=Foods;
        FitnessSol=ObjValSol;
    end;

%ONLOOKER BEE PHASE
while(t<FoodNumber)
    if(rand<Foods)
        t=t+1;
        sol=Foods(i,:);

sol(Param2Change)=Foods(i,Param2Change)+(Foods(i,Param2Change)-Foods(neig
hbour,Param2Change))*(rand-0.5)*2;
        ind=find(sol<lb);
        sol(ind)=lb(ind);
        ind=find(sol>ub);
        sol(ind)=ub(ind);

        %evaluate new solution
        ObjValSol=sol;
        FitnessSol=ObjValSol;
    end;
end;
%Memorize the best food source*/
        ind=find(ObjVal==min(ObjVal));
        ind=ind(end);
        if (ObjVal(ind)<GlobalMin)
        GlobalMin=ObjVal(ind);
        GlobalParams=Foods(ind,:);
        end;
```

```
% SCOUT BEE PHASE
%A check on the "limit" value.
ind=find(trial==max(trial));
ind=ind(end);
if (trial(ind)>limit)
    Bas(ind)=0;
    sol=(ub-lb).*rand(1,D)+lb;
    ObjValSol=sol;
    FitnessSol=ObjValSol;
    Foods(ind,:)=sol;
    Fitness(ind)=FitnessSol;
    ObjVal(ind)=ObjValSol;
end;
```

The computational results are presented in Table 6.15. In instance p01, Depot A requires three vehicles for generating three optimal routes A-13-17-4-18-A, A-40-41-25-19-A, and A-42-45-44-A, respectively. In the same way, from Table 6.15 it can be seen that depot B requires five vehicles for generating the optimal routes B-6-7-10-5-B, B-15-12-23–14-B, B-33-37-24-27-38-B, B-39-46-48-43-47-B, and B-49-B. Likewise depots C and D produce optimal routes with three and one vehicles, respectively. Similarly, for instance p02, ABC locates the optimal routes for depots A, B, C, and D each serving 11, 19, 16, and 4 customers with constructive number of vehicles such as 3, 4, 3, and 1, respectively. ABC on instance p03 with five depots produces five optimal routes from depot A, two from depot B, three from depot C, two from depot D, and two from depot E.

Instance p04 has a total of 24 vehicles available from two depots, and the optimal routes framed require only 15 vehicles shared as 8 and 7 among depots A and B, correspondingly. It is also observed from Table 6.15 that problem instance p06 with three depots has 30 vehicles available, among which 50% of the vehicles are only used to serve all the customers.

The evaluated results in terms of optimal distance and computational time are shown in Table 6.16. The optimal distance obtained by instance p01 for serving all customers from four depots with a total of 12 vehicles is 584.72 km. Instance p02 has generated 11 optimal routes with the near-optimal distance of 477.91 km. For instance p03 with 75 customers and five depots, the best known distance reported is 641.19 km, while the proposed ABC is capable of achieving an optimal distance of 668.84 km. For instances p04 and p06, the gap between the best known distance and the obtained optimal is found to be 0.9065586% and 0.635482%, respectively. Comparatively, ABC takes a longer time to converge in spite of fewer tuning parameters. This fact is evident from the very high computational time tabulated for the instances considered in this work.

6.4.6 MDVRP based on GSO and IGSO

A hybrid of GA and MPSO delivered better optimal distance when compared to the individual performance of both GA and MPSO for the p01, p02, p03, p04, and p06 benchmark instances. The parameters of GSO and IGSO are set as shown in Table 6.17. The size of population is based on the customers in each MDVRP instance. The social and cognitive factors are set according to $0 < c_1 + c_2 < 4$. The inertia weight is varied in order to maintain a proper balance between the exploration and exploitation of the algorithm and is set such that $\frac{c_1 + c_2}{2} - 1 < \omega < 1$. In the GA constituent of IGSO, the best cost route crossover with a

TABLE 6.15

Optimal Route Using ABC

Depot	NVA	NC	NVR	Optimal Route
Instance p01				
A(20,20)	8	11	3	A-13-17-4-18-A, A-40-41-25-19-A, A-42-45-44-A
B(30,40)	8	19	5	B-6-7-10-5-B, B-15-12-23-14-B, B-33-37-24-27-38-B, B-39-46-48-43-47-B, B-49-B
C(50,30)	8	16	3	C-2-9-1-8-C, C-11-22-26-28-29-21-16-C,C-30-32-31-50-34-C,
D(60,50)	8	4	1	D-20-3-36-35-D
Instance p02				
A(20,20)	5	11	3	A-19-13-18-4-17-A, A-42-40-41-25-44-A, A-45-A
B(30,40)	5	19	4	B-7-5-10-12-6-B, B-27-33-15-14-24-23-B, B-47-43-48-46-38-39-37-B, B-49-B
C(50,30)	5	16	3	C-1-11-9-2-8-C, C-28-31-26-22-16-30-21-29-C, C-50-32-34-C
D(60,50)	5	4	1	D-36-35-20-3-D
Instance p03				
A(40,40)	7	32	5	A-6-3-9-12-4-5-A, A-25-18-17-29-20-15-27-26-A, A-32-39-34-37-30-40-A, A-45-60-47-48-51-44-50-55-A, A-67-75-68-70-A
B(50,22)	7	9	2	B-52-46-7-35-8-13-57-B, B-72-58-B
C(55,55)	7	11	3	C-10-19-14-11-C, C-65-31-38-53-54-59-C, C-66-C
D(25,45)	7	14	2	D-2-24-16-28-21-D, D-49-74-36-71-69-62-73-33-63-D
E(20,20)	7	9	2	E-22-42-41-56-23-43-1-61-E, E-64-E
Instance p04				
A(15,35)	12	55	8	A-8-7-6-13-2-15-14-5-A, A-22-23-21-18-36-17-16-37-A, A-46-45-44-38-43-42-41-40-47-A, A-52-48-57-56-58-53-A, A-73-72-75-74-59-61-60-A, A-85-84-83-82-A, A-89-86-91-93-92-87-A, A-96-99-98-100-97-95-94-A
B(35,55)	12	33	7	B-3-9-11-10-1-12-4-B, B-24-25-26-28-27-19-20-29-B, B-31-30-32-35-34-33-B, B-51-50-54-55-39-49-B,B- 67-65-66-63-64-62-B, B-78-77-68-76-69-70-71-B, B-80-88-90-81-79-B
Instance p06				
A(15,20)	10	29	4	A-17-16-38-14-15-41-42-37-5-A, A-59-46-45-84-61-44-43-57-A, A-91-86-85-92-87-A, A-95-96-99-93-100-98-97-A
B(50,20)	10	33	5	B-20-11-10-1-12-25-24-3-9-B, B-29-50-33-51-30-32-35-34-B, B-63-55-54-65-66-64-B, B-68-76-70-71-78-79-77-B, B-90-81-80-B
C(35,35)	10	38	6	C-19-7-8-18-6-13-2-21-4-C, C-27-22-23-26-28-C, C-47-40-39-31-36-C,C-52-48-49-56-53-C, C-67-69-62-60-58-73-72-C,C- 89-83-82-88-75-74-94-C

TABLE 6.16

Computational Results for Benchmark Instances Using ABC

Benchmark Instance Type	No. of Customers	No. of Depots	Best Known Distance (km)	Optimal Distance (km)	Computational Time (s)	Gap between Best Known and Optimal Distance (%)
p01	50	4	576.87	584.72	8.1456	1.3607919
p02	50	4	473.53	477.91	8.9547	0.9249678
p03	75	5	641.19	668.84	12.6472	1.1930941
p04	100	2	1001.59	1010.67	17.5541	0.9065586
p06	100	3	876.50	882.07	18.0546	0.635482

TABLE 6.17

Parameter Settings for GSO and IGSO

Parameters	GSO	IGSO
Population size	Instance dependent	Instance dependent
Maximum inertia weight	0.9	0.9
Minimum inertia weight	0.4	0.4
Initial velocity	0	0
Initial position	Random	Random
Cognitive factor (c_1)	1.28	1.24
Social factor (c_2)	1.28	1.24
Error gradient	1e-25	1e-25
Type of crossover	Best cost route	Best cost route
Type of mutation	Flip-bit	Route reversal
Crossover rate	0.6	0.7
Mutation rate	0.02	0.028
Maximum number of iterations	Instance dependent	Instance dependent
Hybridization factor	0.5	—

crossover rate of 0.7 and route reversal mutation with a rate of 0.028 are chosen as the optimal settings. The hybridization parameter applicable in GSO is set to 0.5 to ensure that individuals of GA and MPSO participate equally in the hybrid combination.

For the problem instances, the number of vehicles required and the optimal route are computed by GSO and IGSO, and the results are presented in Tables 6.18 and 6.19. From Table 6.18, for instance p03, Depot A requires five vehicles for generating five optimal routes A-5-6-3-9-12-4-15-A, A-20-29-30-17-18-25-26-27-A, A-40-44-32-39-34-37-A, A-47-60-45-48-55-50-51-A, and A-70-68-75-67-A, respectively. In the same way, from Table 6.18 it can be seen that depot B requires one vehicle for generating the optimal route B-13-8-35-7-58-72-46-52-57-B. Likewise, depot C, D, and E require two vehicles each for finding the optimal route.

Based on the results obtained in Table 6.19, to serve all the customers, instance p01 with 50 customers and four depots requires 11 vehicles, while instance p02 requires 9 vehicles. Instance p03 has five depots and the optimal number of routes from each depot A, B, C, D, and E is 5, 1, 2, 2, and 1, respectively. The number of optimal routes generated by instance p04 is 7 and 5, corresponding to depots A and B. Instance p06 serves 29 customers from depot A with four vehicles, 33 customers from depot B with four vehicles, and 38 customers from depot C with six vehicles.

The experimental results obtained through GSO and IGSO for the benchmark instances considered in this work are shown in Table 6.20 in terms of the optimal distance, computational time, and gap between best known distance and optimal distance. The gap between the best known distance (BKD) and optimal distance obtained by IGSO is calculated according to Equation 6.26. The optimal distances computed through IGSO for instances p01, p02, p03, p04, and p06 are 576.94, 474.03, 644.29, 1001.94, and 876.68 km, respectively. The gap between the BKD and the computed optimal distance for IGSO is 0.012134% in the case of instance p01, 0.10559% for p02, 0.483476% for p03, 0.034944% for p04, and 0.020536% for p06. The reduction in gap for instances in IGSO shows that the enhancement step introduced in the hybrid combination has improved the optimal distance when compared to GSO. The computational time obtained shows that the IGSO algorithm is capable of converging at a faster rate towards the optimal solution than the GSO algorithm.

TABLE 6.18

Optimal Route Using GSO

Depot	NVA	NC	NVR	Optimal Route
Instance p01				
A(20,20)	8	11	3	A-13-19-17-4-18-A, A-44-42-40-41-25-A, A-45-A
B(30,40)	8	19	4	B-7-6-12-10-5-B, B-15-14-24-23-27-B,B-46-38-39-33-37-43-B,B- 48-47-49-B
C(50,30)	8	16	3	C-9-11-1-8-2-C, C-28-31-26-22-16-30-21-29-C, C-32-34-50-C
D(60,50)	8	4	1	D-20-3-36-35-D
Instance p02				
A(20,20)	5	11	2	A-13-19-17-4-18-A, A-40-42-45-44-25-41-A
B(30,40)	5	19	3	B-7-5-10-12-6-B, B-33-15-37-14-24-23-27-B, B-46-48-43-47-39-49-38-B
C(50,30)	5	16	3	C-11-1-8-2-16-9-C, C-29-22-28-31-26-32-30-21-C, C-34-50-C
D(60,50)	5	4	1	D-3-20-35-36-D
Instance p03				
A(40,40)	7	32	5	A-5-6-3-9-12-4-15-A, A-20-29-30-17-18-25-26-27-A, A-40-44-32-39-34-37-A, A-47-60-45-48-55-50-51-A, A-70-68-75-67-A
B(50,22)	7	9	1	B-13-8-35-7-58-72-46-52-57-B
C(55,55)	7	11	2	C-11-14-19-10-31-C, C-38-65-66-59-54-53-C
D(25,45)	7	14	2	D-2-21-28-16-24-D, D-69-71-36-74-49-63-33-73-62-D
E(20,20)	7	9	2	E-41-42-22-61-1-43-23-56-E, E-64-E
Instance p04				
A(15,35)	12	55	8	A-13-6-7-8-5-14-15-2-A, A-18-21-23-22-37-16-17-36-A, A-43-42-41-40-47-46-45-44-38-A, A-57-56-58-53-52-48-59-A, A-72-75-74-61-84-83-82-60-73-A, A-87-85-91-86-89-A, A-98-93-99-96-94-95-97-92-A, A-100-A
B(35,55)	12	45	6	B-3-9-11-10-1-12-4-B, B-30-19-27-28-26-25-24-29-20-B, B-33-34-35-32-31-39-B,B-51-55-54-50-62-49-64-63-B, B-69-67-68-65-66-70-B, B-80-76-88-90-71-81-78-79-77-B
Instance p06				
A(15,20)	10	29	4	A-14-38-43-15-41-42-37-5-17-16-A, A-61-44-57-59-46-45-84-A, A-91-87-92-85-86-A
B(50,20)	10	33	4	B-11-10-1-12-25-24-29-3-9-20-B, B-32-30-51-33-50-54-34-35-B, B-65-68-55-70-63-64-66-B, B-77-79-78-81-71-90-76-80-B
C(35,35)	10	38	6	C-19-7-8-18-6-13-2-21-4-C, C-27-22-23-26-28-C,C-47-40-39-31-36-C,C-56-58-60-48-49-52-53-C, C-67-69-62-73-72-74-75-C, C-83-94-89-88-82-C

6.5 Discussions

The experimental results obtained on MDVRP instances using the heuristic optimization techniques GA, MPSO, ABC, GSO, and IGSO were compared and discussed in terms of optimal distance, robustness, computational efficiency, and algorithmic efficiency as follows:

6.5.1 Optimal Distance

The optimal distance obtained for the benchmark instances using GA, MPSO, ABC, GSO, and IGSO algorithms were compared with the existing approaches in the literature such as genetic clustering (GC) (Thangiah and Salhi 2001) and GA (Ombuki-Berman and Hanshar 2009).

TABLE 6.19

Optimal Route Using IGSO

Depot	NVA	NC	NVR	Optimal Route
Instance p01				
A(20,20)	8	11	3	A-19-13-18-4-17-A, A-40-41-25-44-42-A, A-45-A
B(30,40)	8	19	4	B-7-5-10-12-6-B, B-15-27-23-24-14-B, B-46-38-39-33-37-43-B, B-49-47-48-B
C(50,30)	8	16	3	C-2-8-1-11-9-C, C-29-28-31-26-22-16-30-21-C, C32-34-50-C
D(60,50)	8	4	1	D-3-36-35-20-D
Instance p02				
A(20,20)	5	11	2	A-19-13-18-4-17-A, A-40-41-25-44-45-42-A
B(30,40)	5	19	3	B-7-5-10-12-6-B, B-37-15-33-27-23-24-14-B, B-43-48-46-38-49-39-47-B
C(50,30)	5	16	3	C-9-16-2-8-1-11-C, C-31-26-32-30-21-29-22-28-C, C-50-34-C
D(60,50)	5	4	1	D-3-36-35-20-D
Instance p03				
A(40,40)	7	32	5	A-12-9-3-6-5-15-4-A, A-17-18-25-26-27-20-29-30-A, A-40-44-32-39-34-37-A, A-45-60-47-48-51-50-55-A, A-75-68-70-67-A
B(50,22)	7	9	1	B-72-46-52-57-13-8-35-7-58-B
C(55,55)	7	11	2	C-14-19-10-31-11-C, C-53-38-65-66-59-54-C
D(25,45)	7	14	2	D-21-2-24-16-28-D, D-69-71-36-74-49-63-33-73-62-D
E(20,20)	7	9	1	E-41-42-22-61-1-43-23-56-64-E
Instance p04				
A(15,35)	12	55	7	A-5-8-7-6-13-2-15-14-A, A-21-23-22-37-16-17-36-18-A, A-45-44-38-43-42-41-40-47-46-A, A-57-56-58-53-52-48-59-A, A-74-75-72-73-60-82-83-84-61-A, A-86-91-85-87-89-A, A-98-93-99-100-96-94-95-97-92-A
B(35,55)	12	45	5	B-11-9-3-4-12-1-10-25-26-28-27-19-30-20-29-24-B, B-33-34-35-32-31-39-B, B-51-55-54-50-62-49-64-63-B, B-67-69-70-66-65-68-B, B-79-78-81-71-90-88-76-80-77-B
Instance p06				
A(15,20)	10	29	4	A-41-15-43-14-38-16-17-5-37-42-A, A-44-57-59-46-45-84-61-A, A-91-87-92-85-86-A, A-95-96-99-93-100-98-97-A
B(50,20)	10	33	4	B-20-11-10-1-12-25-24-29-3-9-B, B-32-30-51-33-50-54-34-35-B, B-65-68-55-70-63-64-66-B, B-76-90-71-81-78-79-77-80-B
C(35,35)	10	38	6	C-6-18-8-7-19-4-21-2-13-C, C-27-22-23-26-28-C, C-36-31-39-40-47-C, C-58-56-53-52-49-48-60-C, C-75-67-69-62-73-72-74-C, C-88-82-83-94-89- C

TABLE 6.20

Computational Results for MDVRP Instances Using GSO and IGSO

				GSO			IGSO		
Instance	No. of Customers	No. of Depots	BKD (km)	Optimal Distance (km)	CT (s)	Gap (%)	Optimal Distance (km)	CT (s)	Gap (%)
p01	50	4	576.87	578.52	1.5316	0.286026	576.94	1.3725	0.012134
p02	50	4	473.53	474.85	1.4259	0.278757	474.03	0.9750	0.10559
p03	75	5	641.19	653.78	2.3659	1.963537	644.29	2.2773	0.483476
p04	100	2	1001.59	1005.73	3.0191	0.413343	1001.94	2.3918	0.034944
p06	100	3	876.50	877.19	3.0976	0.078722	876.68	2.4094	0.020536

The optimal distance obtained for the five MDVRP instances using the existing approaches and the proposed algorithms are presented in Table 6.21. For instance p03, the distance obtained by the proposed GA, MPSO, ABC, GSO, and IGSO is small compared to that obtained by GA (Ombuki-Berman and Hanshar 2009) in literature by 1.08%, 2.72%, 5.38%, 7.51%, and 8.85%, respectively. MPSO yields the minimum distance for instances p01, p02, p03, p04, and p06 when compared to GA, with a difference of 3.02%, 0.66%, 1.66%, 0.28%, and 0.24% respectively. Though ABC took more time to generate the optimal route, the distances are reduced when compared to the proposed GA. For instance p01, the distance obtained by ABC is 2.29% smaller than that obtained by GA. In the same way, the optimal distance obtained by ABC is 0.15% (instance p02), 4.34% (instance p03), 0.07% (instance p04), and 0.05% (instance p06) smaller than GA. The improvement in optimal distance obtained by IGSO over GSO is 0.273% in p01, 0.173% in p02, 1.45% in p03, 0.377% in p04, and 0.058% in p06. The enhancement step introduced in IGSO has shown a significant improvement in terms of optimal distance, specifically in instance p03. Thus from the comparative analysis in terms of the optimal distance, it can be seen that IGSO is a better approach to solve the chosen MDVRP instances.

6.5.2 Robustness

The robustness of the heuristic techniques applied in this work is evaluated based on the effective vehicle management without violating the capacity limit of the vehicles. Computational results in terms of the number of vehicles required (NVR) and the total number of vehicles available (NVA) are shown in Table 6.22. Among the proposed intelligent algorithms, it is found from Table 6.22 that GA utilized more number of vehicles for all the instances to determine the optimal distance, thus increasing the vehicle utilization cost. For test instances p02, p03, and p04, it is observed that MPSO and ABC used the required vehicles in a similar manner without any difference. ABC exploited fewer vehicles in a more efficient way in the case of instance p01 and p06 with 37.5% and 50%, respectively, while MPSO utilized 40.63% and 56.67% of the vehicles available. This implies that ABC is robust against MPSO in spite of achieving the optimal distance.

While comparing GSO and IGSO, no significant difference in vehicle utilization efficiency is found in instances p01, p02, and p06. IGSO has shown significant improvement over GSO in utilizing vehicles specifically for instances p03 (31.43%) and p04 (50%). Overall, for all the test instances, GSO and IGSO used fewer vehicles when compared with GA, MPSO, and ABC algorithms. Considering the vehicle utilization efficiency for instances p03 and p04, it can be seen that IGSO is more robust compared to GA, MPSO, ABC, and GSO algorithms.

6.5.3 Computational Time

The computational efficiency of the bio-inspired heuristics is evaluated in terms of the time taken for the algorithm to converge. The computational time of the proposed techniques for the benchmark instances are presented in Table 6.23. It can be seen from the table that computational time of IGSO (2.3918 s) for instance p04 is smaller than that of GA (10.1081 s), MPSO (7.7211 s), ABC (17.5541 s), and GSO (3.0191 s). Overall, for all instances, it is found that ABC consumes more time to determine the optimal route and distance. On average, the computational time of ABC is 48.39% higher than GA, 57.73% higher than MPSO, 85.58% higher than GSO, and 82.5% higher than IGSO. The longer time for convergence required by ABC leads to a setback in applying the ABC technique to solve practical

TABLE 6.21

Comparative Analysis in terms of Optimal Distance

Instance Type	No. of Customers	No. of Depots	Best known Distance (km)	Distance Reported in Literature (km)		Distance Obtained by the Proposed Approaches (km)					
				GC (2001)	GA (2009)	GA	MPSO	ABC	GSO	IGSO
p01	50	4	576.87	591.73	622.18	598.45	580.35	584.72	578.52	576.94
p02	50	4	473.53	463.15	480.04	478.65	475.47	477.91	474.85	474.03
p03	75	5	641.19	694.49	706.88	699.23	687.65	668.84	653.78	644.29
p04	100	2	1001.59	1062.38	1024.78	1011.36	1008.55	1010.67	1005.73	1001.94
p06	100	3	876.50	976.02	908.88	882.48	880.35	882.07	877.19	876.68

TABLE 6.22

Comparative Analysis of Vehicle Utilization Efficiency

Instance Type	No. of Customers	No. of Depots	NVA	NVR (%)				
				GA	MPSO	ABC	GSO	IGSO
p01	50	4	32	43.75	40.63	37.50	34.38	34.38
p02	50	4	20	60.00	55.00	55.00	45.00	45.00
p03	75	5	35	45.71	40.00	40.00	34.29	31.43
p04	100	2	24	75.00	62.50	62.50	58.33	50.00
p06	100	3	30	56.67	56.67	50.00	46.67	46.67

TABLE 6.23

Comparative Analysis of Computational Time

Instance Type	No. of Customers	No. of Depots	Computational Time (s)				
			GA	MPSO	ABC	GSO	IGSO
p01	50	4	4.0692	3.6319	8.1456	1.5316	1.3725
p02	50	4	3.4207	2.9434	8.9547	1.4259	0.9750
p03	75	5	6.8128	5.1899	12.6472	2.3659	2.2773
p04	100	2	10.1081	7.7211	17.5541	3.0191	2.3918
p06	100	3	9.3177	8.1418	18.0546	3.0976	2.4094

MDVRP instances. Computational results indicate that the time taken by IGSO is smaller since the hybrid combination of GA and MPSO minimizes the chances of the algorithm getting trapped in the local minima.

6.5.4 Algorithmic Efficiency

The algorithmic efficiency of the heuristic algorithms used to solve MDVRP in this work was calculated using BigO notation in terms of the estimated time and calculated time of the algorithm. The algorithmic efficiency of the proposed intelligent heuristics was evaluated for the problem instances, and the results are presented in Table 6.24. Based on the size of the code, the efficiency varies significantly. The large number of programming lines in the code leads to a decrease in the efficiency of the algorithm. The algorithmic efficiency of instance p02 for GA is 88.21%, for MPSO 90.27%, for ABC 95.88%, for GSO 93.11%, and for IGSO 92.81%. On average, the algorithmic efficiency for all the MDVRP instances in

TABLE 6.24

Comparison of Algorithmic Efficiency

Instance Type	No. of Customers	No. of Depots	Algorithmic Efficiency (%)				
			GA	MPSO	ABC	GSO	IGSO
p01	50	4	89.77	90.48	94.52	92.56	92.27
p02	50	4	88.21	90.27	95.88	93.11	92.81
p03	75	5	89.09	91.12	94.37	93.71	92.64
p04	100	2	89.56	90.55	95.06	93.29	92.98
p06	100	3	89.14	91.79	96.11	93.43	92.86

this work is 89.15% for GA, 90.84% for MPSO, 95.19% for ABC, 93.22% for GSO, and 92.71% for IGSO. It is observed that due to the fewer control parameters in ABC, the algorithmic efficiency is improved over GA, MPSO, GSO, and IGSO algorithms.

6.6 Advantages of CI Paradigms

The analysis of MDVRP instances p01, p02, p03, p04, and p06 in terms of optimal distance, robustness, computational time, and algorithmic efficiency with the proposed algorithms is presented in Table 6.25.

In MDVRP, the computational time for all the five Cordeau's instances computed by IGSO is less by 72.05% over GA, 65.88% over MPSO, 85.58% over ABC, and 17.61% over GSO. The merits and demerits of the optimization techniques in terms of the computational time are as follows:

- GA: Due to the efficiency in searching capability, GA is applied in this chapter to solve MDVRP. Sometimes, GA converges faster without producing the optimal solution, thus leading to premature convergence. This fact accounts for the increase in computational time.

- MPSO: In MPSO, absolute and integral value functions are used to update the velocity and particles. Occasionally, when the velocity reaches a maximum value and saturates, the particle will continue to conduct searches within a hypercube and will probably remain in the optima but will not converge in the local area, thus increasing the computational time in the MDVRP problem.

- ABC: Due to the ability in handling constraints, ABC algorithm is applied to solve MDVRP. The possibility of missing relevant information on the behavior of the objective function is high, and hence the performance of the algorithm is slow, thus increasing the computational time.

- GSO: The hybrid combination of GA and MPSO resulted in better ability to reach the global optimum with stable convergence characteristics. The parallelization of GA and MPSO has increased the convergence rate, thus minimizing the computational time.

- IGSO: In IGSO, the offspring from GA and enhanced elites from MPSO produce the next-generation individuals. Such parallel operation in IGSO reduces the probability of producing infeasible solutions, thus converging faster than the GSO algorithm.

TABLE 6.25

Investigation of Algorithms on MDVRP

Advantages	Performance of Algorithms
Optimal distance	IGSO resulted in optimal distance for instances p01, p02, p03, p04, and p06.
Robustness	IGSO utilized vehicles in an efficient manner in all instances p01, p02, p03, p04, and p06 when compared with GA, MPSO, ABC, and GSO algorithms.
Computational time	The computational time consumed by IGSO is less than with GA, MPSO, ABC, and GSO for all instances p01, p02, p03, p04, and p06.
Algorithmic efficiency	The algorithmic efficiency of ABC is improved over GA, MPSO, GSO, and IGSO in cases p01, p02, p03, p04, and p06.

6.7 Summary

Intelligent heuristics such as GA, MPSO, ABC, GSO, and IGSO were proposed in this chapter for solving MDVRP. Initially, the customers were grouped based on their distance from the depot. Further, the customers of the same depot were assigned to routes using the Clarke and Wright saving method. This procedure also ensured that two routes could be easily merged into a single route. The routes were later scheduled and optimized by the proposed intelligent heuristics. The objective of MDVRP was to find routes for vehicles to service all the customers at a minimal cost in terms of number of routes and total travel distance, without violating the capacity and depot limits.

The effectiveness of the proposed techniques was tested on a set of five different Cordeau's benchmark instances—p01, p02, p03, p04, and p06—in the MATLAB R2008b environment. The intelligent heuristics were applied to evaluate the optimal routes, number of vehicles required, and optimal distance of the MDVRP instances. In addition to the problem-dependant evaluations such as optimal distance and robustness, additional assessments in terms of computational time and algorithmic efficiency of the proposed algorithms were also estimated.

Considering the optimal distance, it was observed (Table 6.21) that IGSO showed improvements compared to the other proposed algorithms GA, ABC, MPSO, and GSO. From Table 6.22, it was seen that IGSO utilized fewer available vehicles compared to GA, MPSO, ABC, and GSO, thus implying robustness. Likewise, the computational efficiency of IGSO was better, thus demonstrating earlier convergence rate, which was evident from the results presented in Table 6.23. The algorithmic efficiency of ABC (Table 6.24) was reasonably high due to the fewer control parameters in the algorithm. Thus based on the investigations carried out on the proposed intelligent algorithms, it was observed that IGSO is an appropriate approach for solving practical MDVRP instances based on the optimal solution, robustness, and computational efficiency.

In future, it would be of considerable interest to incorporate several practical constraints such as urgency in delivery, random service time, and nonuniform vehicle capacity to the MDVRP. This work may be extended with new optimization techniques such as bacterial foraging optimization (BFO), biogeography-based optimization (BBO), and artificial immune systems (AIS). The experimental results may be used to analyze, compare, and determine better optimization technique in the future.

References

Abido, M., Optimal design of power system stabilizers using particle swarm optimization, *IEEE Transactions on Energy Conversion* 17(3), 406–413, 2002.

Allaoua, B., Optimal power flow solution using ant manners for electrical network, *Advances in Electrical and Computer Engineering* 9, 34–40, 2009.

Blum, C., Beam ACO—hybridizing ant colony optimization with beam search: an application to open shop scheduling, *Computers and Operations Research* 32, 1565–1591, 2005.

Blum, C., and Sampels, M., An ant colony optimization algorithm for shop scheduling problems, *Journal of Mathematical Modelling and Algorithms* 3, 285–308, 2004.

Boato, G., Conotter, V., and De Natale, F.G.B., GA-based robustness evaluation method for digital image watermarking, in *Digital Watermarking*, Y.Q. Shi, H.-J. Kim, and K. Stefan, Eds. Springer-Verlag, New York, 2008.

Boato, G., Conotter, V., and De Natale, F.G.B., Watermarking robustness evaluation based on perceptual quality via genetic algorithms, *IEEE Transactions on Information Forensics and Security* 4(2), 207–216, 2009.

Brajevic, I., Artificial bee colony algorithm for the capacitated vehicle routing problem, in *Proceedings of the fifth European Conference on European Computing Conference: ECC'11*, Paris, France, April 28–30, 2011, pp. 239–244.

Brown, C., Liebovitch, L.S., and Glendon, R., Levy flights in Dobe Ju/'hoansi foraging patterns, *Human Ecology* 35, 129–138, 2007.

Burt, P.J., and Adelson, E., The Laplacian Pyramid as a compact image code, *IEEE Transactions on Communications* 31(4), 532–540, 1983.

Chen, C-C., and Lin, C-S., A genetic algorithm based nearly optimal image authentication approach, *International Journal of Innovative Computing, Information and Control* 3(3), 631–640, 2007.

Chen, P., and Xu, X., A hybrid algorithm for multi-depot vehicle routing problem, in *Proceedings of the IEEE International Conference on Service Operations Logistics, and Informatics: IEEE/SOLI 2008*, Beijing, China, October 12–15, 2008, pp. 2031–2034.

Chen, Q., Kotani, K., Lee, F., and Ohmi, T., Scale-invariant feature extraction by VQ-based local image descriptor, in *Proceedings of the International Conference on Computational Intelligence for Modeling, Control and Automation*, Vienna, Austria, December 10–12, 2008, pp. 1217–1222.

Chen, W.C., and Wang, M.S., A fuzzy c-means clustering-based fragile watermarking scheme for image authentication, *Expert Systems with Applications* 36(2), 1300–1307, 2009.

Cheung, W., and Zhou, H., Using genetic algorithms and heuristics for job shop scheduling with sequence dependant set-up times, *Annals of Operational Research* 107, 65–81, 2001.

Ching-Tzong Su and Chien-Tung Lin, New approach with a hopfield modeling framework to economic dispatch, *IEEE Transactions on Power Systems* 15(2), 541–545, 2000.

Chiou, J.P., Variable scaling hybrid differential evolution for large scale economic dispatch problems, *Electrical Power Systems Research* 77(1), 212–218, 2007.

Christian, R., and Jean-Luc, D., A survey of watermarking algorithms for image authentication, *EURASIP Journal on Applied Signal Processing* 2002(6), 613–621, 2002.

Clarke, G., and Wright, J.W., Scheduling of vehicles from a central depot to a number of delivery points, *Operations Research*, 12, 568–581, 1964.

Colorni, A., Dorigo, M., Maniezzo, V., and Trubian, M., Ant system for job shop scheduling, *Belgian Journal of Operations Research, Statistics and Computer Science* 34(1), 39–53, 1994.

Comaniciu, D., and Meer, P., Mean shift: a robust approach toward feature space analysis, *IEEE Transactions on Pattern Analysis and Machine Intelligence* 24(5), 603–619, 2002.

Cong, J., Affine invariant watermarking algorithm using feature matching, *Digital Signal Processing* 16(3), 247–254, 2006.

Cox, I., and Kilan, J., Secure spread spectrum watermarking for images, audio and video, in *Proceedings of the IEEE International Conference on Image Processing*, Lausanne, Switzerland, September 16–19, 1996, pp. 243–246.

Cox, I., Miller, M., and Bloom, J., *Digital Watermarking*. San Mateo, CA: Morgan Kaufmann, 2002.

Creviera, B., Cordeaua, J.F., and Laporte, G., The multi-depot vehicle routing problem with inter-depot routes, *European Journal of Operational Research* 176, 756–773, 2007.

Dantzig, G.B., and Ramser, J.H., The truck dispatching problem, *Management Science* 6(1), 80–91, 1959.

Davis, L., Job shop scheduling with genetic algorithms, in *Proceedings of the First International Conference on Genetic Algorithms*, Pittsburgh, PA, July 24–26, 1985, pp. 136–140.

Deb, K., An efficient constraint-handling method for genetic algorithms, *Computer Methods in Applied Mechanics and Engineering* 186(2–4), 311–338, 2000.

Delaigle, J., De Vleeschouwer, C., and Macq, B., Psychovisual approach to digital picture watermarking, *Journal of Electronic Imaging* 7(3), 628–640, July 1998.

Dempster, A., Laird, N., and Rubin, D., Maximum likelihood from incomplete data via the EM algorithm, *Journal of the Royal Statistical Society: Series B* 39(1), 1–38, 1977.

Dongeun, L., Tackyung, K., Seongwon, L., and Joonki, P., Genetic algorithm based watermarking in discrete wavelet transform domain, in *Intelligent Computing*, G. Goos, J. Hartmanis, and J. van Leeuwen, Eds. Springer-Verlag, Berlin, Germany, 2006.

Dorigo, M., and Gambardella, L.M., Ant colony system: a cooperative learning approach to the traveling salesman problem, *IEEE Transactions on Evolutionary Computation* 1(1), 53–66, 1997.

Dos, L., Coelho, S., and Mariani, V.C., Combining of chaotic differential evolution and quadratic programming for economic dispatch optimization with valve-point effect, *IEEE Transactions on Power Systems* 21(2), 989–996, 2006.

Eberhart, R., and Kennedy, J., A new optimizer using particles swarm theory, in *Proceedings of the Sixth IEEE International Symposium on Micro Machine and Human Science: MHS '95*, Nagoya, Japan, 1995, pp. 39–43.

Eiben, E., and Smith, J.E., *Introduction to Evolutionary Computing*, Natural computing series. New York: Springer-Verlag, 2003.

Emek, S., and Pazarci, M., A cascade DWT-DCT based digital watermarking scheme, in *Proceedings of the 13th European Signal Processing Conference: EUSIPCO'05*, Antalya, Turkey, September 4–8, 2005, pp. 492–495.

Felzenszwalb, P., and Huttenlocher, D., Efficient graph-based image segmentation, *International Journal of Computer Vision* 59(2), 167–181, 2004.

Fisher, H., and Thompson, G.L., *Probabilistic Learning Combinations of Local Job-Shop Scheduling Rules*. Englewood Cliffs, NJ: Prentice-Hill, 1963.

Fogel, D.B., Phenotypes, genotypes, and operators in evolutionary computation, in *Proceedings of the IEEE International Conference on Evolutionary Computation 1995*, Perth, Australia, November 29–December 1, 1995, pp. 193–199.

Fortemps, P., Jobshop scheduling using imprecise durations: a fuzzy approach, *IEEE Transactions on Fuzzy Systems* 5(4), 557–569, 1997.

Fourie, P., and Groenwold, A., The particle swarm optimization algorithm in size and shape optimization, *Structural and Multidisciplinary Optimization* 23, 259–267, 2002.

Fu, Y.G., and Shen, R.M., Color image watermarking scheme based on linear discriminant analysis, *Computer Standards and Interfaces* 30(3), 115–120, 2008.

Gaing, Z-L., Particle swarm optimization to solving the economic dispatch considering the generator constraints, *IEEE Transactions on Power Systems* 18(3), 1187–1195, August 2003.

Gibbons, D.W., Brood parasitism and cooperative nesting in the moorhen, *Gallinula chloropus*, *Journal on Behavioral Ecology and Sociobiology* 19, 221–232, 1986.

Giosa, D., Tansini, I.L., and Viera, I.O., New assignment algorithms for the multi-depot vehicle routing problem, *Journal of the Operational Research Society* 53, 977–984, 2002.

Goldberg, D.E., *Genetic Algorithms in Search Optimization and Machine Learning*. Boston, MA: Addison Wesley Longman Publishing Co., 1989.

Goncalves, J.F., A hybrid genetic algorithm for the job shop scheduling problem, *European Journal of Operational Research* 167(1), 77–95, 2005.

Gorzalezang, M.B., A method of inference in approximate reasoning based on interval-valued fuzzy set, *Fuzzy Sets and Systems* 21, 1–17, 1981.

Grefenstette, J.J., Optimization of control parameters for genetic algorithms, *IEEE Transactions on Systems, Man and Cybernetics* 16(1), 122–128, 1986.

Haghani, A., and Jung, S., A dynamic vehicle routing problem with time-dependent travel times, *Computers and Operations Research* 32, 2959–2986, 2005.

Hartley, H., Maximum likelihood estimation from incomplete data, *Biometrics* 14(2), 174–194, June 1958.

Hathaway, R.J., and Bezdek, J.C., Fuzzy c-means clustering of incomplete data, *IEEE Transactions on Systems, Man, and Cybernetics-Part B: Cybernetics* 31(5), 735–744, 2001.

Hemamalini, S., and Sishaj, P.S., Economic/emission load dispatch using artificial bee colony algorithm, in *Proceedings of the International Conference on Control, Communication and Power Engineering: CCPE-2010*, Chennai, India, July 28–29, 2010, pp. 338–343.

Hernandez, J.R., Amado, M., and Perez-Gonzalez, F., DCT-Domain watermarking techniques for still images: Detector performance analysis and a new structure, *IEEE Transactions on Image Processing* 9(1), 55–68, 2000.

Ho, W., Ho, G.T.S., Ji, P., and Lau, H.C.W., A hybrid genetic algorithm for the multi-depot vehicle routing problem, *Journal of Engineering Applications of Artificial Intelligence* 21(4), 548–557, 2008.

Huang, K.L., and Liao, C.J., Ant colony optimization combined with taboo search for the job shop scheduling problem, *Computers and Operations Research* 35, 1030–1046, 2008.

Huang, C.-M., and Wang, F.-L., An RBF network with OLS and EPSO algorithms for real-time power dispatch, *IEEE Transactions on Power Systems* 22(1), 96–104, February 2007.

Husbands, P., Genetic algorithms for scheduling, *AISB Quarterly* 89, 38–45, 1996.

Kim, H., Lee, H.-Y., and Lee, H.-K., Robust image watermarking using local invariant features, *Optical Engineering* 45(3), 1–11, 2006.

Iba, H., and Nomana, N., Differential evolution for economic load dispatch problems, *Electrical Power Systems Research*, 78(3), 1322–1331, 2008.

Ishii, H., Tada, M., and Masuda, T., Two scheduling problems with fuzzy due dates, *Fuzzy Sets and Systems* 46, 339–347, 1992.

Ishtiaq, M., Sikandar, B., Jaffar, M.A., and Khan, A., Adaptive watermark strength selection using particle swarm optimization, *ICIC Express Letters* 4(5), 1–6, 2010.

Ivers, B., Job shop optimization through multiple independent swarms, in *Proceedings of the IEEE congress on Evolutionary Computation: CEC 2007*, Singapore, September 25–28, 2007, pp. 3361–3368.

Jagadeesh, B., Kumar, S.S., and Rajeswari, K.R., A genetic algorithm based oblivious image watermarking scheme using singular value decomposition, in *Proceedings of the IEEE International Conference on Networks and Communications: NETCOM '09*, Chennai, India, December 27–29, 2009, pp. 224–229.

Jeon, G., Leep, H.R., and Shim, J.Y., A vehicle routing problem solved by using a hybrid genetic algorithm, *Computers and Industrial Engineering* 53, 680–692, 2007.

Jiriwibhakorn, S., and Khamsawang, S., Solving the economic dispatch problem by using differential evolution, *International Journal of Electrical and Electronics Engineering* 3(10), 641–645, 2009.

Jones, A., and Rabelo, L.C., Survey of job shop scheduling techniques, Technical report—NISTIR, National Institute of Standards and Technology, Gaithersburg, MD, 1998.

Kacem, I., Hammadi, S., and Borne, P., Approach by localization and multiobjective evolutionary optimization for flexible job-shop scheduling problems, *IEEE Transactions on Systems, Man, and Cybernetics, Part C* 32(1), 1–13, 2002.

Karaboga, D., An idea based on honey bee swarm for numerical optimization, Technical Report TR06, Erciyes University, Kayseri, 2005.

Karaboga, D., and Basturk, B., A powerful and efficient algorithm for numerical function optimization: artificial bee colony (ABC) algorithm, *Journal of Global Optimization* 39(3), 359–371, 2007.

Kaufmann, A., and Gupta, M.M., *Introduction to Fuzzy Arithmetic Theory and Applications*. New York: van Nostrand Reinhold, 1991.

Ketcham, M., and Vongpradhip, S., Intelligent audio watermarking using genetic algorithm in DWT domain, *International Journal of Intelligent Technology* 2(2), 135–140, 2007.

Khan, A., Mirza, A.M., and Majid, A., Optimizing perceptual shaping of a digital watermark using genetic programming, *Iranian Journal of Electrical and Computer Engineering* 3(2), 1251–1260, 2004.

Khan, A., Mirza, A.M., and Majid, A., Intelligent perceptual shaping of a digital watermark: exploiting characteristics of human visual system, *International Journal of Knowledge-based Intelligent Engineering Systems* 9, 1–11, 2005.

Kundur, D., and Hatzinakos, D., A robust digital image watermarking method using wavelet—based fusion, in *Proceedings of the IEEE International Conference on Image Processing 1997*, Santa Barbara, CA, October 26–29, 1997, Vol. 1, pp. 544–547.

Labbi, Y., and Attous, B.D., A hybrid GA–PS method to solve the economic load dispatch problem, *Journal of Theoretical and Applied Information Technology* 15(1), 61–68, 2010.

Lau, H.C.W., Chan, T.M., Tsui, W.T., and Pang, W.K., Application of genetic algorithms to solve the multidepot vehicle routing problem, *IEEE Transactions on Automation Science and Engineering* 7(2), 383–392, 2010.

Lawrence, S., Resource constrained project scheduling: an experimental investigation of heuristic scheduling techniques, Supplement PA1984, Graduate School of Industrial Administration, Carnegie Mellon University, Pittsburgh, 1984.

Lee, C.G., Epelman, M.A., White, C.C., and Bozer, Y.A., A shortest path approach to the multiple-vehicle routing problem with split pick-ups, *Transportation Research Part B: Methodological* 40, 265–284, 2006.

Lee, H.-Y., and Lee, H.-K., Copyright protection through feature-based watermarking using scale-invariant keypoints, in *Proceedings of the International Conference on Consumer Electronics: ICCE '06*, January 7–11, 2006, pp. 225–226.

Lee, Z.J., Lin, S.W., Su, S.F., and Lin, C.Y., A hybrid watermarking technique applied to digital images, *Applied Soft Computing* 8(1), 798–808, 2008.

Li, J.Q., Pan, Q.K., Xie, S.X., Li, H., Jia, B.X., and Zhao, C.S., An effective hybrid particle swarm optimization algorithm for flexible job-shop scheduling problem, *MASAUM Journal of Computing* 1, 69–74, 2009.

Li, W.T., Shi, X.W., and Xu, L., Improved GA and PSO culled hybrid algorithm for antenna array pattern synthesis, *Progress in Electromagnetics Research* 80, 461–476, 2008.

Lian, Z., Gu, X., and Jiao, B., A similar particle swarm optimization algorithm for job-shop scheduling to minimize makespan, *Applied Mathematics and Computation* 183, 1008–1017, 2006.

Lian, Z.G., Gu, X.S., and Jiao, B., A novel particle swarm optimization algorithm for permutation flow-shop scheduling to minimize makespan, *Chaos, Solitons and Fractals* 35(5), 851–861, July 2006.

Liao, C.J., Tseng C.T., and Luarn, P., A discrete version of particle swarm optimization for flow-shop scheduling problems, *Computers and Operations Research* 34(10), 3099–3111, 2007.

Lin, P.L., Hsieh, C.K., and Huang, P.W., A hierarchical digital watermarking method for image tamper detection and recovery, *Pattern Recognition* 38(12), 2519–2529, 2005.

Liu, B., Wang, L., and Jin, Y.H., An effective PSO-based memetic algorithm for flow shop scheduling, *IEEE Transactions on Systems, Man and Cybernetics—Part B: Cybernetics* 37(1), 18–27, 2007.

Lowe, D.G., Distinctive image features from scale-invariant keypoints, *International Journal on Computer Vision* 60(2), 91–110, 2004.

Mairgiotis, A., Chantas, G., Galatsanos, N., Blekas, K., and Yang, Y., New detectors for watermarks with unknown power based on student-t image priors, in *Proceedings of the IEEE 9th Workshop on Multimedia Signal Processing: MMSP 2007*, Crete, Greece, October 1–3, 2007, pp. 353–356.

Mariani, V.C., Coelho, L., and Dos, L., Correction to combining of chaotic differential evolution and quadratic programming for economic dispatch optimization with valve-point effect, *IEEE Transactions on Power Systems* 21(3), 1465, 2006.

McCahon, C.S., and Lee, E.S., Job sequencing with fuzzy processing times, *Computers and Mathematics with Applications* 19(7), 31–41, 1990.

Meng, K., Dong, Z.Y., Wang, D.H., and Wong, K.P., A self-adaptive RBF neural network classifier for transformer fault analysis, *IEEE Transactions on Power Systems* 25(3), 1350–1360, August 2010.

Merkle, D., An ant algorithm with a new pheromone evaluation rule for total tardiness problem, in *Proceedings of the Real World Applications of Evolutionary Computing: EvoWorkshops 2000*, S. Cagnoni, Ed. Springer-Verlag, Berlin, Germany, 2003.

Miller, M.L., Doerr, G.J., and Cox, I.J., Applying informed coding and embedding to design a robust, high capacity watermark, *IEEE Transaction on Image Processing* 13(6), 792–807, 2004.

Mirabi, M., Ghomi, S.M.T.F., and Jolai, F., Efficient stochastic hybrid heuristics for the multi-depot vehicle routing problem, *Robotics and Computer Integrated Manufacturing* 26, 564–569, 2010.

Mohaghegi, S., del Valle, Y., Venayagamoorthy, G.K., and Harley, R.G., A comparison of PSO and backpropagation for training RBF neural networks for identification of a power system with STATCOM, in *Proceedings of the IEEE Swarm Intelligence Symposium: SIS 2005*, Pasadena, CA, June 8–10, 2005, pp. 381–384.

Mohamed, F.K., and Abbes, R., RST robust watermarking schema based on image normalization and DCT decomposition, *Malaysian Journal of Computer Science* 20(1), 77–90, 2007.

Morshed, M.S., A hybrid model for job shop scheduling, PhD dissertation, University of Birmingham, Birmingham, U.K., 2006.

Muth, J., and Thompson, G., *Industrial Scheduling*. Englewood Cliffs, NJ: Prentice Hall, 1963.

Nakano, R., and Yamada, T., Conventional genetic algorithm for job shop problems, in *Proceedings of the Third International Conference on Genetic Algorithms*, R. Belew and L. Booker, Eds. Morgan Kaufman, San Mateo, CA, 1991.

Nayak, S.K., Krishnanand, K.R., Panigrahi, B.K., and Rout, P.K., Application of artificial bee colony to economic load dispatch problem with ramp rate limits and prohibited operating zones, in *Proceedings of the World Congress on Nature and Biologically Inspired Computing: NaBIC 2009*, Coimbatore, India, December 9–11, 2009, pp. 1237–1242.

NEO, Description for files of Cordeau's Instances, http://neo.lcc.uma.es/vrp/vrp-instances/description-for-files-of-cordeaus-instances/, online resource of University of Malaga, Spain, n.d.

Nguyen, P.B., Luong, M., and Beghdadi, A., Statistical analysis of image quality metrics for watermark transparency assessment, in *Proceedings of the Conference on Advances in Multimedia Information Processing: PCM 2010*, Shanghai, China, September 21–24, 2010, pp. 685–696.

Niu, Q., Jiao, B., and Gu, X., Particle swarm optimization combined with genetic operators for job shop scheduling problem with fuzzy processing time, *Applied Mathematics and Computation* 205(1), 148–158, 2008.

Nowicki, E., and Smutnicki, C., A fast taboo search algorithm for the job shop problem, *Management Science* 42(6), 797–813, 1996.

Ombuki-Berman, B., and Hanshar, F., Using genetic algorithms for multi-depot vehicle routing, in *Bio-Inspired Algorithms for the Vehicle Routing Problem*, F.B. Pereira and J. Tavares, Eds. Springer–Verlag, Berlin, Germany, 2009.

Orero, S.O., and Irving, M.R., Scheduling of generators with a hybrid genetic algorithm, presented at the *First International Conference on Genetic Algorithms in Engineering Systems: Innovations and Applications GALESIA*, Sheffield, U.K., September 12–14, 1995, pp. 200–206.

Pan, J.S., Huang, H.C., and Jain, L.C., *Intelligent Watermarking Techniques (Innovative Intelligence)*. River Edge, NJ: World Scientific Press, 2004.

Park, J.-B., Jeong, Y.-W., Shin, J.-R., and Lee, K.Y., An improved particle swarm optimization for nonconvex economic dispatch problems, *IEEE Transactions on Power Systems* 25(1), 156–166, February 2010.

Pavlyukevich, I., Levy flights, non-local search and simulated annealing, *Journal of Computational Physics*, 226(2), 1830–1844, 2007.

Payne, R.B., Sorenson, M.D., and Klitz, K., *The Cuckoos*. Oxford: Oxford University Press, 2005.

Perez-Guerrero, R.E., and Cedenio-Maldonado, R.J., Economic power dispatch with non-smooth cost functions using differential evolution, in *Proceedings of the 37th Annual North American Power Symposium*, Ames, IA, October 23–25, 2005, pp. 183–190.

Pezzella, F., and Merelli, E., A tabu search method guided by shifting bottleneck for the job shop scheduling problem, *European Journal of Operational Research* 120(2), 297–310, 2000.

Pitas, I., A method for signature casting on digital images, in *Proceedings of the IEEE International Conference on Image Processing*, Lausanne, Switzerland, September 16–19, 1996, pp. 215–218.

Piva, A., Barni, M., Bartolini, F., and Cappellini, V., Mask building for perceptually hiding frequency embedded watermarks, in *Proceedings of the Fifth IEEE International Conference on Image Processing ICIP'98*, Chicago, IL, October 4–7, 1998, pp. 450–454.

Potdar, V., Han, S., and Chang, E., A survey of digital image watermarking techniques, in *Proceeding of the Third IEEE-International Conference on Industrial Informatics*, Perth, Australia, August 10–12, 2005, pp. 709–716.

Prabhakar, K.S., Palanisamy, K., Jacob, R.I., and Kothari, D.P., Security constrained UCP with operational and power flow constraints, *International Journal of Recent Trends in Engineering* 1(3), 106–114, 2009.

Price, K.V., Storn, R.M., and Lampinen, J.A., *Differential Evolution: A Practical Approach to Global Optimization.* Berlin, Germany: Springer, 2005.

Rahnamayan, S., Tizhoosh, H.R., and Salama, M.M.A., Opposition based differential evolution, *IEEE Transactions on Evolutionary Computation* 12(1), 64–79, 2008.

Rajan, C.C.A., Genetic algorithm based simulated annealing method for solving unit commitment problem in utility system, in *Proceedings of the IEEE Transmission and Distribution Conference and Exposition: IEEE PES 2010*, New Orleans, LA, April 19–22, 2010, pp. 1–6.

Renaud, J., Laporte, G., and Boctor, F.F., A tabu search heuristic for the multi-depot vehicle routing problem, *Computers and Operations Research* 23(3), 229–235, 1996.

Revathy, K., Applying EM algorithm for segmentation of textured images, in *Proceedings of the World Congress on Engineering (WCE 2007)* Vol. 1, London, U.K., July 2–4, 2007.

Reynolds, A.M., and Frye, M.A., Free-flight odor tracking in Drosophila is consistent with an optimal intermittent scale-free search, *PLoS One* 2(4), e354, 2007.

Ross, T., *Fuzzy Logic with Engineering Applications.* Hoboken, NJ: John Wiley and Sons, 2004.

Rudolf, A., and Bayrleithner, R., A genetic algorithm for solving the unit commitment problem of a hydro-thermal power systems, *IEEE Transactions on Power Systems* 14(4), 1460–1468, 1999.

Ruxton, G.D., and Broom, M., Intraspecific brood parasitism can increase the number of eggs that an individual lays in its own nest, *Proceedings of Biological Science* 269(1504), 1989–1992, 2002.

Saber, A.Y., and Venayagamoorthy, G.K., Economic load dispatch using bacterial foraging technique with particle swarm optimization biased evolution, in *Proceedings of the IEEE Swarm Intelligence Symposium SIS 2008*, St. Louis, MO, September 21–23, 2008, pp. 1–8.

Salhi, S., and Sari, M., A multi-level composite heuristic for the multidepot vehicle fleet mix problem, *European Journal of Operational Research* 103(1), 95–112, 1997.

Saneifard, S., Prasad, N.R., and Smolleck, H.A., A fuzzy logic approach to unit commitment, *IEEE Transactions on Power Systems* 12(2), 988–995, 1997.

Sasaki, H., Watanabe, M., and Yokoyama, R., A solution method of unit commitment by artificial neural networks, *IEEE Transactions on Power Systems* 7(3), 974–981, 1992.

Senjyu, T., Yamashiro, H., Shimabukuro, K., Uezato, K., and Funabashi, T., A fast solution technique for large scale unit commitment problem using genetic algorithm, in *Proceedings of the IEEE/PES Transmission and Distribution Conference and Exhibition*, Asia Pacific, October 6–10, 2002, pp. 1611–1616.

Sha, D.Y., and Hsu, C.Y., A hybrid particle swarm optimization for job shop scheduling problem, *Computers and Industrial Engineering*, 51(4), 791–808, 2006.

Sheble, G.B., Maifeld, T.T., Brittig, K., Fahd, G., and Fukurozaki-Coppinger, S., Unit commitment by genetic algorithm with penalty methods and a comparison of Lagrangian search and genetic algorithm economic dispatch example, *International Journal of Electrical Power and Energy Systems* 18(6), 339–346, 1996.

Shia, X.H., Lianga, Y.C., Leeb, H.P., Lub, C., and Wanga, L.M., An improved GA and a novel PSO-GA based hybrid algorithm, *Information Processing* 93(5), 255–261, 2005.

Shieh, C.S., Huang, H.C., Pan, J.S., and Wang, F.H., Genetic watermarking based on transform domain technique, *Pattern Recognition* 37(3), 555–565, 2004.

Sikander, B., Ishtiaq, M., Jaffar, M.A., and Mirza, A.M., Adaptive digital image watermarking of images using genetic algorithm, in *Proceedings of the IEEE International Conference on Information Science and Applications: ICISA 2010*, Seoul, Korea, April 21–23, 2010, pp. 1–8.

Skok, M., Skrlec, D., and Krajcar, S., The genetic algorithm method for multiple depot capacitated vehicle routing problem solving, in *Proceedings of the Fourth International Conference on Knowledge Based Intelligent Engineering Systems and Allied Technologies*, Brighton, U.K., August 30–September 1, 2000, pp. 520–526.

Slowinski, R., and Hapke, M., *Scheduling Under Fuzziness.* Heidelberg, Germany: Physica-Verlag, 2000.

Sonmez, A.I., and Baykasoglu, A., A new dynamic programming formulation of (n*m) flowshop sequencing problems with due dates, *International Journal of Production Research* 36(8), 2269–2283, 1998.

Starkweather, T., Whitley, D., and Cookson, B., A genetic algorithm for scheduling with resource consumption, in *Proceedings of the Joint German/US Conference on Operations Research in Production Planning and Control*, Hagen, Germany, June 25–26, 1992, pp. 567–583.

Storer, R.H., Wu, S.D., and Vaccari, R., New search spaces for sequencing instances with application to job shop scheduling, *Management Science* 38(10), 1495–1509, 1992.

Storn, R., and Price, K.V., Differential evolution a simple and efficient heuristic for global optimization over continuous spaces, *Journal of Global Optimization* 11(4), 341–359, 1997.

Su, C.T., Dynamic vehicle control and scheduling of a multi-depot physical distribution system, *Integrated Manufacturing Systems* 10, 56–65, 1999.

Sumpavakup, C., Srikun, I., and Chusanapiputt, S., A solution to the optimal power flow using artificial bee colony algorithm, in *Proceedings of the International Conference on Power System Technology: POWERCON 2010*, Hangzhou, China, October 24–28, 2010, pp. 1–5.

Sun, Z., and Ma, J., DWT-domain watermark detection using Gaussian mixture model with automated model selection, in *International Symposium on Computer Network and Multimedia Technology: CNMT 2009*, Wuhan, China, January 18–20, 2009, pp. 1–4.

Surekha, P., and Sumathi, S., Solution to multi-depot vehicle routing problem using genetic algorithms, *World Applied Programming* 1(3), 118–131, August 2011.

Sviatoslav, V., Frederic, D., and Thierry, P., Content adaptive watermarking based on a stochastic multiresolution image modeling, *Proceedings of the Tenth European Signal Processing Conference: EUSIPCO 2000*, Tampere, Finland, September 5–8, 2000, pp. 5–8.

Swarup, K.S., and Yamashiro, S., Unit commitment solution methodology using genetic algorithm, *IEEE Transactions on Power Systems* 17(1), 87–91, 2002.

Szeto, W.Y., Wu, Y., and Ho, S.C., An artificial bee colony algorithm for the capacitated vehicle routing problem, *European Journal of Operational Research* 215(1), 126–135, 2011.

Taillard, E.D., Benchmarks for basic scheduling problems, *European Journal of Operational Research* 64(2), 108–117, 1993.

Tang, C.W., and Hang, H.M., A feature-based robust digital image watermarking scheme, *IEEE Transactions on Signal Processing* 51(4), 950–959, 2003.

Tao, H., Zain, J.M., Abd Alla, A.N., and Hongwu, Q., An implementation of digital image watermarking based on particle swarm optimization, *Communications in Computer and Information Science* 87(2), 314–320, 2010.

Thangiah, S.R., and Salhi, S., Genetic clustering: An adaptive heuristic for the multidepot vehicle routing problem, *Applied Artificial Intelligence* 15(4), 361–383, 2001.

Thitithamrongchai, C., Self-adaptive differential evolution based optimal power flow for units with non-smooth fuel cost functions, *Journal of Electrical Systems* 3, 88–99, 2007.

Tran, T., Nguyen, T.T., and Nguyen, H.L., Global optimization using levy flight, in *Proceedings of the International Conference on Information and Communication Technology: ICT.rda'04*, Hanoi, Vietnam, 2004, September 24–25, pp. 1–12.

Tsujimura, Y.S., Park, S.H., Chang, I.S., and Gen, M., An effective method for solving flow shop scheduling problems with fuzzy processing times, *Computers and Industrial Engineering* 25(1–4), 239–242, 1993.

Vaessens, R., Aarts, E., and Lenstra, J., Job shop scheduling by local search, *INFORMS Journal on Computing* 8(3), 302–317, 1996.

Vallabha, H., Multiresolution watermark based on wavelet transform for digital images, Technical Report, Cranes Software International Ltd, Bangalore, India, 2003.

Viswanathan, G.M., Buldyrev, S.V., Havlin, S., da Luz, M.G.E., Raposo, E.P., and Stanley, H.E., Optimizing the success of random searches, *Nature* 401, 911–914, 1999.

Vlachogiannis, J.G., and Lee, K.Y., Quantum-inspired evolutionary algorithm for real and reactive power dispatch, *IEEE Transactions on Power Systems* 23(4), 1627–1636, 2008.

Voloshynovskiy, S., Pereira, S., Iquise, V., and Pun, T., Attack modeling: Towards a second generation watermarking benchmark, *Journal of Signal Processing* 80(6), 1177–1214, 2001.

Walsh, M.P., and Malley, M.J.O., Augmented hopfield network for unit commitment and economic dispatch, *IEEE Transactions on Power Systems* 12(4), 1765–1774, 1997.

Wang, B., Yang, X.F., and Li, Q.Y., Genetic simulated-annealing algorithm for robust job shop scheduling, *Advances in Intelligent and Soft Computing* 62, 817–827, 2009.

Wang, M.S., and Chen, W.C., A hybrid DWT-SVD copyright protection scheme based on k-means clustering and visual cryptography, *Computer Standards and Interfaces* 31(4), 757–762, 2009.

Wang, S., Wang, L., and Wang, D., Particle swarm optimization for multi-depots single vehicle routing problem, in *Proceedings of the Chinese Control and Decision Conference: CCDC 2008*, Yantai, Shandong, July 2–4, 2008, pp. 4659–4661.

Wang, S.K., Chiou, J.P., and Liu, C.W., Non-smooth/non-convex economic dispatch by a novel hybrid differential evolution algorithm, *IET Generation Transmission and Distribution* 1(5), 793–803, 2007.

Wang, W., Men, A., and Yang, B., A feature-based semi-fragile watermarking scheme in DWT domain, in *Proceedings of the Second IEEE International Conference on Network Infrastructure and Digital Content*, Beijing, China, September 24–26, 2010, pp. 768–772.

Wang, Y., Doherty, J.F., and Van Dyck, R.E., A wavelet-based watermarking algorithm for ownership verification of digital images, *IEEE Transactions on Image Processing* 11(2), 77–88, 2002.

Wang, Y., Zhou, J., Xiao, W., and Zhang, Y., Economic load dispatch of hydroelectric plant using a hybrid particle swarm optimization combined simulation annealing algorithm, in *Proceedings of the Second WRI Global Congress on Intelligent Systems: GCIS*, Wuhan, China, December 16–17, 2010, pp. 231–234.

Wang, Y.-R., Lin, W.-H., and Yang, L., An intelligent PSO watermarking, in *Proceedings of the International Conference on Machine Learning and Cybernetics: ICMLC*, Qingdao, China, July 11–14, 2010, pp. 2555–2558.

Wang, Z., Bovik, A.C., and Lu, L., Why is image quality assessment so difficult? in *Proceedings of the IEEE International conference on Acoustics, Speech and Signal Processing: ICASSP*, Orlando, FL, May 13–17, 2002, vol. 4, pp. 3313–3316.

Watson, A.B., Visual optimization of DCT quantization matrices for individual images, in *Proceedings of AIAA Computing in Aerospace 9*, San Diego, CA, October 19–21, 1993, pp. 286–291.

Wei, X.J., and Wu, Z.M., An effective hybrid optimization approach for multi-objective flexible job-shop scheduling problems, *Computers and Industrial Engineering* 48(2), 409–425, 2005.

Wenjing, Z., and Ye, J., An improved particle swarm optimization for the multi-depot vehicle routing problem, in *Proceedings of the International Conference on E-Business and E-Government: ICEE*, Guangzhou, China, May 7–9, 2010, pp. 3188–3192.

Wu, L.H., Wang, Y.N., Yuan, X.F., and Zhou, S.W., Environmental/economic power dispatch problem using multi-objective differential evolution algorithm, *Electric Power Systems Research* 80(9), 1171–1181, 2010.

Wu, T.H., Low, C., and Bai, J.W., Heuristic solutions to multi-depot location-routing problem, *Computers and Operations Research*, 29(10), 1393–1415, 2002.

Yamada, T., and Nakano, R., A genetic algorithm applicable to large-scale job-shop problems, in *Proceedings of the Second International Conference on Parallel Problem Solving from Nature: PPSN'92*, Brussels, Belgium, 1992, pp. 281–290.

Yang, X.-S., and Deb, S., Cuckoo search via levy flights, in *Proceedings of the World Congress on Nature and Biologically Inspired Computing: NaBIC 2009*, Coimbatore, India, 2009, pp. 210–214.

Yao, J.S., and Wu, K.M., Ranking fuzzy numbers based on decomposition principle and signed distance, *Fuzzy Sets and Systems* 116(2), 275–288, 2000.

Yare, Y., Venayagamoorthy, G.K., and Saber, A.Y., Heuristic algorithms for solving convex and nonconvex economic dispatch, in *Proceedings of the IEEE International Conference on Intelligent System Applications to Power Systems: ISAP'09*, Curitiba, Brazil, November 8–12, 2009, pp. 1–8.

Yi-leh, W., Agrawal, D., and Abbadi, A.E., A comparison of DFT and DWT based similarity search in time-series databases, in *Proceedings of the Ninth International Conference on Information and Knowledge Management: CIKM'00*, McLean, VA, November 6–11, 2000, pp. 488–495.

Yuryevich, J., Evolutionary programming based optimal power flow algorithm, *IEEE Transaction on Power Systems* 14, 1245–1250, November 1999.

Zhang, F., and Zhang, H., Applications of a neural network to watermarking capacity of digital image, *Neurocomputing* 67, 345–349, 2005.

Zhang, J., Hu, X., Tan, X., Zhong, J.H., and Huang, Q., Implementation of an ant colony optimization for job shop scheduling problem, *Transactions of the Institute of Measurement and Control*, 28(1), 93–108, 2006.

Zhang, Z., and Lu, C., Application of the improved particle swarm optimizer to vehicle routing and scheduling problems, in *Proceeding of the IEEE International Conference on Grey Systems and Intelligent Services: GSIS 2007*, Nanjing, China, November 18–20, 2007, pp. 1150–1152.

7

Digital Image Watermarking

Learning Objectives

On completion of this chapter, the reader will have knowledge on

- Pre-watermarking stages such as image segmentation, feature extraction, orientation assignment, and image normalization.
- Watermark embedding and extraction based on discrete wavelet transforms (DWTs).
- Applying GA, PSO, and HPSO to digital image watermarking (DIWM) in the DWT domain to obtain the peak signal-to-noise ratio (PSNR) and normalized cross-correlation (NCC) among the watermarked and the extracted images.
- Testing and validating the bio-inspired algorithms on gray-scale images and color images.
- Evaluating perceptual transparency, robustness, and computational time, thus performing a comparative analysis to determine the suitable technique for DIWM.

7.1 Introduction

Effective digital image copyright protection methods have become a vital and instantaneous necessity in multimedia applications because of the increasing unauthorized manipulation and reproduction of original digital objects. With digital multimedia distribution over the World Wide Web, intellectual property rights (IPRs) are threatened more than ever due to the possibility of unlimited copying (Christian and Jean-Luc 2002, Potdar et al. 2005). One solution would be to restrict access to the data using standard encryption techniques. However, encryption does not provide overall protection. Once the encrypted data are decrypted, they can be freely distributed or manipulated. The earlier problem can be solved by hiding some ownership data into the multimedia data, which can be extracted later to prove the ownership. For example, in bank currency notes, a watermark is embedded that is used to check the genuineness of the note. The same "watermarking" concept may be used in multimedia digital contents for checking the authenticity of the original content. Among several protection methodologies such as steganography, IP security, secure wireless transmission, and disk encryption, digital watermarking is the leading approach that is used to protect against illegal manipulation of data (Tang and Hang 2003). Digital image watermarking (DIWM) is a technique to embed a secret message or valuable information (watermark) within an ordinary image (host image) and extract the image at the destination, thereby protecting the image from image processing attacks during the transmission process. The watermark can be a random signal, an organization's trademark symbol, or a copyright message for copy control and authentication. The chosen watermark to be embedded in the host image should be resilient to standard manipulations of unintentional as well as intentional nature, should be statistically unremovable, and must be capable of withstanding multiple watermarking to facilitate traitor tracing (Hernandez et al. 2000).

The type of manipulations and the signal processing attacks on the watermark depend upon the specific application of DIWM. The most important characteristics of digital watermarking, such as imperceptibility, robustness, inseparability, security, permanence, data capacity, and fidelity, allow the technique to be applicable in several security areas such as owner identification, copyright protection, image authentication, broadcast monitoring, transaction tracking, and usage control.

The frequently used watermarking techniques are spatial-domain watermarking and frequency-domain watermarking. In spatial-domain watermarking, the perceptual information about the image is obtained and used to embed the watermarking key in predefined intensity regions of the image. Embedding an invisible watermark is simpler and more effective in the spatial domain, but when subject to image alterations, the robustness is poor (Vallabha 2003). In the frequency domain, the watermark is transformed into the frequency domain by application of the Fourier transform, the discrete cosine transform, or the discrete wavelet transform (DWT). The watermarks are added to the transform coefficients of the image instead of modifying the pixels, thus making it difficult to remove the embedded watermark. Compared to the spatial-domain technique, frequency-domain techniques are more robust and have a high range of control in maintaining the perceptual quality of the watermark.

The most commonly used frequency-domain transform techniques include the discrete Fourier transform (DFT), the discrete cosine transform (DCT), the DWT, and the Hadamard transform. It is of great interest to determine the best transformation domain to devise robust watermark embedding methods, given that the watermarked data undergo lossy compression prior to watermark detection or extraction. DWT has several advantages over other transform domain techniques such as DCT, DFT, and the Hadamard transform (Emek and Pazarci 2005). DWT has exceptional properties such as excellent localization in time and frequency domains, symmetric spread distributions, and multi-resolution characteristics (Wu et al. 2000), which led to the development of various DWT-based algorithms.

Image segmentation and feature extraction are introduced into watermarking with the aim of obtaining high robustness against lossy compression. Segmentation methods can be broadly classified into mean-shift segmentation (Comaniciu and Meer 2002), graph-based segmentation (Felzenszwalb and Huttenlocher 2004), and hybrid segmentation techniques. The expectation maximization (EM) algorithm belongs to the class of mean-shift segmentation, which is a popular image segmentation technique based on probabilistic models. Though the EM algorithm is formulated by Hartley (1958), it was formalized and the proof of convergence was provided by Dempster et al. (1977). Mairgiotis et al. (2007) developed an approach by adding new detectors for additive watermarks when the power of the watermark was unknown. EM algorithm is used in their work to obtain ML estimates of the unknown parameters. Sun and Ma (2009) presented a DWT domain watermark detection approach using the Gaussian mixture model (GMM) with automated model selection. In order to fix the number of model components in advance, the component-wise EM algorithm is used to realize automatic mixture model selection.

Feature extraction schemes are incorporated into DIWM to provide a high degree of resistance against geometric attacks. A robust DIWM scheme that combines image feature extraction and image normalization was developed by Tang and Hang (2003). The goal was to resist both geometric distortion and signal processing attacks. A feature extraction method called the Mexican Hat wavelet scale interaction was adopted, and it was observed that the normalized image was nearly invariant with respect to rotations. Lee and Lee (2006) presented a feature-based watermarking method based on scale-invariant keypoints for copyright protection. The feature points were extracted by using a scale-invariant keypoint extractor, and then these points were decomposed into a set of disjoint triangles.

The "difference of Gaussians" (DoG) is a close approximation to the Laplacian transform; these Laplacian or DoG pyramids were first identified by Burt and Adelson (1983), and since then they have been adopted for a variety of applications in image processing. One of the most popular applications of DoG was to extract features based on the difference between blurred and original images. Lowe (2004) provided a method for extracting distinctive invariant features from images that could be used to perform reliable matching between different views of an object. The features were invariant to image scale and rotation, thus making it more suitable for watermarking applications. Zheng and Zhao (2007) presented a rotation-invariant feature and image-normalization-based image watermarking algorithm. The subregions centered at those feature points were used for watermark embedding and detection. Image normalization was applied to those subregions to achieve the scaling invariance.

In order to resist geometric distortions, Kim et al. (2006) used the local invariant feature of the image called the scale-invariant feature transform (SIFT), which is invariant to translation and scaling distortions. The watermark is inserted into the circular patches generated by the SIFT. In this method, rotation invariance is achieved using the translation property of the polar-mapped circular patches. Wang et al. (2010a) presented a scheme that extracted content-based image features from the approximation subband in the wavelet domain to generate the watermark. The watermark was then embedded into the middle frequency subband in the wavelet domain. Results showed that their scheme was robust to common content-preserving image processing. Mohamed and Abbes (2007) presented a new robust watermarking schema based on logo embedding in the DCT-transformed domain using image normalization techniques. Image normalization was used for calculating the affine transform parameters so that the watermark embedding and detection was performed in the original coordinate system. Their algorithm was found to be robust to attacks such as low-pass median Gaussian noise, aspect ratio change, rotation, scaling, and JPEG compression. A rotation- and scaling-invariant image watermarking scheme in the DCT domain was presented by Zheng et al. (2009) based on rotation-invariant feature extraction and image normalization. Each homogeneous region was approximated as a generalized Gaussian distribution, with the parameters estimated using EM.

Presently, intelligent digital watermarking techniques are adopted because most of the multimedia applications require imperceptible and robust watermarks. In digital watermarking, genetic algorithms (GAs) are used to design several optimized algorithms for better tradeoff between imperceptibility, robustness, and security. Ketcham and Vongpradhip (2007) developed an innovative DWT watermarking scheme based on GAs for audio signals. The optimal localization and intensity were obtained using GA, and the method was found to be robust against cropping, low-pass filtering, and additive noise.

Al-Haj (2007) described an imperceptible and robust DIWM scheme based on a combination of DWT and DCT. Similarly, Barni et al. (2001) provided a DWT-based technique for the evaluation of fidelity and robustness. Both these algorithms were capable of extracting the watermark, but suffered from the problems of unsatisfactory fidelity and robustness to attacks focused in their work.

Wei et al. (2006) presented an algorithm that yielded a watermark that was invisible to human eyes and robust to various image manipulation. The results showed that only some specific positions were best suited for embedding the watermark. The authors applied GA to train the frequency set for embedding the watermark and compared their approach with the Cox et al.'s method (2002) to prove robustness. The analysis of GA was restricted to JPEG compression attack in this method. Cong (2006) developed a scheme that did not require the original image for watermarking because the information from the shape specific points of the original image were memorized by the neural network. This scheme applied the shape-specific points technique and featured the point-matching method by GA for resisting geometric attacks. Boato et al. (2009) developed a new, flexible, and effective evaluation tool based on GAs to test the robustness of DIWM techniques. Given a set of possible attacks, the method finds the best possible un-watermarked image in terms of weighted peak signal-to-noise ratio (WPSNR). Shieh et al. (2004) addressed an innovative watermarking scheme based on GA in the transform domain considering the watermarked image quality.

Lin et al. (2005) presented a spatial-fragile watermarking scheme in which the original image was divided into several blocks and permuted based on a secret key. The method was capable of restoring the tampered images and was resistant to counterfeiting attacks. Wang et al. (2009) provided a copyright protection scheme that first extracted the image features from the host image by using the DWT and singular value decomposition (SVD). Using the k-means clustering algorithm, the extracted features were clustered, resulting in a master share. The master share and a secret image were used to build an ownership share according to a two-out-of-two visual cryptography (VC) method. A novel color image watermarking scheme was proposed by Fu and Shen (2008) based on linear discriminant analysis. The watermark combined with a reference was embedded into all the RGB channels of the color image. Using the reference watermark, a linear discriminant matrix was developed to retrieve the correct watermark after attacks. A hybrid watermarking scheme based on GA and particle swarm optimization (PSO) was presented by Lee et al. (2008). In this hybrid scheme, the parameters of the perceptual lossless ratio (PLR), which are defined by just noticeable difference (JND) for two complementary watermark modulations, were derived. Aslantas et al. (2009) embedded watermark bits in the least significant bits (LSBs) of some DCT coefficients, and adjusted the inverse IDCT coefficients governed by an embedding rule associated with a set of translation maps trained by a GA.

Khan et al. (2004) optimized the perceptual shaping of the watermark in the complete DCT domain using genetic programming (GP). A constant watermarking strength was obtained from spatial activity masking, and GP was used for all the selected coefficients during embedding. To improve the quality of the marked image, Pan et al. (2004) used GA to find the most suitable and favorable positions for embedding in the block-based watermarking scheme. To shape the watermark according to the original image before the embedding process, Cox et al. (2002) and Hernandez et al. (2000) used Watson's perceptual model (1993). But both of them used an 8 × 8 block-based DCT domain watermarking. Ishtiaq et al. (2010) investigated a new method based on PSO for adaptive watermark strength optimization in the DCT domain. The watermark strength was intelligently selected through PSO, which helped in perceptual shaping of the watermark such that it was less perceivable to the human visual system.

There has been a considerable amount of research on the applications of DWT in DIWM systems by virtue of its excellent and exceptional properties, but the scope of optimization in this area is very poor. In this chapter, intelligent image watermarking algorithms in the DWT domain based on GA, PSO, and hybrid PSO (HPSO) are presented. The input images are initially segmented into a number of homogeneous regions using the EM algorithm, and the feature points are extracted based on the DoG algorithm. Then the circular regions based on image normalization and orientation assignment are defined for the watermark embedding and extraction processes. An optimized DWT for DIWM is capable of producing perceptual transparency and robustness between the watermarked and extracted images. During the past few years, bio-inspired techniques such as GA and PSO have shown excellent performances in optimization problems (Lee et al. 2008, Shia et al. 2005). Moreover, watermark techniques based on these intelligent algorithms seemed to improve the security, robustness, and quality of the watermarked images (Pan et al. 2004). Thus an extensive approach by applying GA, PSO, and HPSO to DIWM in the DWT domain is proposed. The performance of these algorithms is tested and validated on six gray-scale images—Peppers, Mandrill, Lena, Barbara, Boat, and Cameraman—and one real-time camera-captured image, Flowers. The results are simulated and parameters—such as robustness, transparency, CPU time, and fitness value—are evaluated.

This chapter deals with the preprocessing schemes of the DIWM approach, including image segmentation, feature extraction, orientation assignment, and image normalization. The DWT-based watermark embedding and extraction procedures including the evaluation metrics are enlightened. The optimization techniques such as GA, PSO, and HPSO, as well as their operators and the procedure for DIWM, are elaborated in later sections of this chapter. The MATLAB® m-file snippets are also presented, along with a discussion based on the performance metrics.

7.2 Basic Concepts of Image Watermarking

A watermarking algorithm is capable of embedding watermark in various types of data, including text, image, audio, and video. The process of embedding is performed by employing a private key to identify the locations to embed the watermark. The watermarked object is then subjected to an unintentional attack, and the objective is to recover the image in a robust manner. To check the watermark, the private key that was used to embed the watermark is shared with the owner, using which he or she is able to detect the watermark. During the transmission process, the image may be subjected to attacks, and the detected image may not be the original one. Hence it is required to compare the extracted image with the watermarked image to detect the strength of the watermarking scheme. This detection is performed by a correlation operator.

7.2.1 Properties of Digital Watermarking Technique

Some of the major properties of watermarking include:

- **Transparency or fidelity:** The quality of the original image should be maintained after the embedding of the watermark and its extraction. In case of visible distortions to the watermark, the commercial value of the watermark is reduced.

- **Robustness:** During the extraction and detection process, watermarks should be robust against geometric and nongeometric attacks.
- **Capacity or data payload:** The amount of data that can be embedded so that the watermark can be detected successfully is known as the data payload.

7.2.2 Watermarking Applications

The main applications of digital watermarking are as follows:

- **Owner identification:** In copyright infringement, identifying the owner is a complex task. The copyright to the owner can be included as a watermark in the data or image itself.
- **Copyright protection:** Any kind of redistribution or reproduction of data over the Internet or peer-to-peer (P2P) networks can be avoided by digital watermarking.
- **Image authentication:** The ownership of an image can be identified in cryptography-based applications.
- **Broadcast monitoring:** This application is commonly used while broadcasting data in satellite communication. Here, watermarking can be used to monitor whether the data that is supposed to be broadcasted was really broadcasted or not.
- **Transaction tracking:** The watermark embedded in a digital work can be used to record one or more transactions taking place in the history of a copy of this work.
- **Usage control:** Watermarking can be used to limit the usage of copyright-protected data.
- **Content archiving:** A digital object identifier can be included in the form of a watermark to archive digital contents. Further, this identifier can be applied to classify and organize digital contents, thus reducing the possibility of tampering.
- **Meta-data insertion:** Meta-data usually refers to data that defines large data. In this application, images are labeled with its content, which in turn is mainly used in search engines.
- **Tamper detection:** Tamper detection is a very important aspect in satellite imagery or medical imagery, where highly sensitive data are transmitted.

7.3 Preprocessing Schemes

The proposed watermarking scheme is illustrated in Figure 7.1. In this scheme, first, the images are segmented into a number of homogeneous regions by EM algorithm in order to identify the embedding strength of the segmented region. Further the DoG filter is used to extract the feature points (Zheng et al. 2009), such that adequate accurate data are chosen for further analysis, thus avoiding the involvement of complex data.

The images are then subjected to orientation assignment such that the circular regions extracted are invariant to rotation. Scaling invariance is achieved by performing image normalization before embedding the watermark. The watermark image is embedded into the host image using the DWT domain. The optimization techniques applied in this

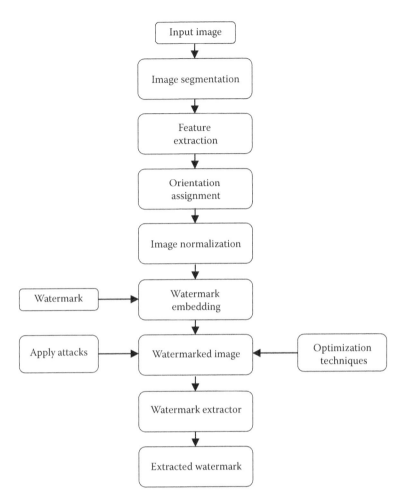

FIGURE 7.1
Watermark embedding and extraction scheme.

chapter—GA, PSO, and HPSO—improve the watermark amplification factor, thereby increasing the robustness of the algorithm against attacks. Both geometric and nongeometric attacks are applied before extracting the watermark from the embedded image. In this section, the phases of the watermarking scheme are discussed.

7.3.1 Image Segmentation

Image segmentation is a well-designed technique in which a signal is decomposed into segments with different time and frequency resolutions. The goal of segmentation is to simplify and change the representation of an image in terms of different homogeneous regions, which is more meaningful and easier to analyze. EM algorithm is used for image segmentation and hence to determine the embedding strength of the segmented homogenous region (Belongie et al. 1998, Do and Batzoglou 2008).

The EM algorithm is a general method of finding the maximum-likelihood estimate of the parameters of an underlying distribution from a given dataset when the data is incomplete or has missing values. There are two main applications of the EM algorithm.

The first is when the data indeed has missing values as a result of problems or limitations of the observation process. The second occurs when optimizing the likelihood function is analytically intractable but it can be simplified by assuming the existence of values for additional but *missing* (or *hidden*) parameters. The latter application is more common in the computational pattern recognition community (Revathy 2007).

EM is an iterative procedure alternating between an expectation (E) step and a maximization (M) step. The E step computes an expectation of the log likelihood with respect to the current estimate of the distribution for the latent variables. The M step computes the parameters that maximize the expected log likelihood found in the E step (Belongie et al. 1998). These parameters are then used to determine the distribution of the latent variables in the next iterative E step. The EM algorithm used to segment images is as follows:

Given

- Instances from the image X generated by a mixture of k Gaussian distributions, and
- Unknown means $(\mu_1, \ldots, \mu_k)$ of the k Gaussians

Determine

- The maximum likelihood estimates of the means $(\mu_1, \ldots, \mu_k)$.

Procedure

- Select a random pair of means, $h = (\mu_1, \mu_2)$.
- Perform E step: Compute the expected value $E[z_{ij}]$ of each hidden variable z_{ij}, assuming the current hypothesis $h = (\mu_1, \mu_2)$:

$$E[z_{ij}] = \frac{p(x = x_i \mid \mu = \mu_j)}{\sum_{n=1}^{2} p(x = x_i \mid \mu = \mu_n)} = \frac{e^{-\frac{1}{2\sigma^2}(x_i - \mu_j)^2}}{\sum_{n=1}^{2} e^{-\frac{1}{2\sigma^2}(x_i - \mu_n)^2}} \tag{7.1}$$

- Perform M step: Calculate the new maximum likelihood hypothesis $h' = (\mu_1', \mu_2')$, assuming that the value taken on by each hidden variable z_{ij} is the expected value $E[z_{ij}]$ calculated before. Replace $h = (\mu_1, \mu_2)$ by $h' = (\mu_1', \mu_2')$, according to Equation 7.2:

$$\mu_j \leftarrow \frac{\sum_{i=1}^{m} E[z_{ij}] x_i}{\sum_{i=1}^{m} E[z_{ij}]} \tag{7.2}$$

Though several image segmentation algorithms are available, the author applies EM since it provides a simple, easy-to-implement, and efficient tool for learning the parameters of a model, and it also presents a mechanism for building and training rich probabilistic models for image processing applications.

7.3.2 Feature Extraction

The huge set of data available in images is simplified for analysis by the feature extraction technique. The number of variables in the large dataset often causes several problems while analyzing complex data. The complex data variables in turn require a large amount of memory and computation power, which overfits the training sample and generalizes poor samples. Thus, feature extraction is one of the efficient methods of constructing combinations of the data variables and maintaining the data with sufficient accuracy.

DoG algorithm proposed by David Lowe (Lowe 2004) is a gray-scale image enhancement algorithm that involves the subtraction of one blurred version of an original gray-scale image from another less blurred version of the original image. The original gray-scale images are convolved with Gaussian kernels with different standard deviations to obtain the blurred images. This process of blurring suppresses only high-frequency spatial information but retains the other information. The spatial information within the range of frequencies is preserved in the blurred images. This kind of a technique is similar to a band-pass filter that discards all but a handful of spatial frequencies that are present in the original gray-scale image.

The major stages of computation to generate the set of image features are the following:

- **Scale-space extrema detection:** The overall scales and the image locations are determined in the initial state. DoG algorithm is applied to identify potential interest points that are invariant to scale and orientation.
- **Keypoint localization:** At each candidate location, a detailed model is fit to determine the location and scale from which the keypoints are selected based on measures of their stability.

To efficiently detect stable keypoint locations, scale-space extrema in the DoG function is convolved with the image $D(x, y, \sigma)$, which can be computed from the difference of two nearby scales separated by a constant multiplicative factor k by using Equation 7.3:

$$D(x,y,\sigma) = (G(x,y,k\sigma) - G(x,y,\sigma)) * I(x,y) = L(x,y,k\sigma) - L(x,y,\sigma) \tag{7.3}$$

The steps of the DoG algorithm are as follows:

Step 1: Input the original and blurred version of the image.

Step 2: Determine the Gaussian filter for the original and blurred image.

Step 3: Compute the DoG-filtered image based on Equation 7.3.

Step 4: Compute the DoG further from successive filtered images.

Step 5: For each upsampling and downsampling level of the DoG image, repeat Steps 2 to 4.

The local extrema in the DoG are determined from the algorithm, and those with strong edge responses are removed, thus resulting in the final feature selection points.

7.3.3 Orientation Assignment

Orientation assignment is the key step in achieving invariance to rotation, as the keypoint descriptor can be represented relative to the orientation of the image (Chen et al. 2008). Each circular region is made rotation invariant by defining a window centered at the chosen feature point. For all the pixels in the selected window, the gradients are computed and histogram of the gradient is determined. The peak of the histogram is selected as the orientation of the feature point $\Theta(x, y)$ (Zheng et al. 2009). The scale of the keypoint is used to select the Gaussian-smoothed image L with the closest scale, so that all computations are performed in a scale-invariant manner. For each image sample $L(x, y)$ at this scale, the gradient magnitude $m(x, y)$ and orientation $\Theta(x, y)$ are precomputed using pixel differences according to Equations 7.4 and 7.5, respectively:

$$m(x,y) = \sqrt{(L(x+1,y)-L(x-1,y))^2 + (L(x,y+1)-L(x,y-1))^2} \tag{7.4}$$

$$\theta(x,y) = \tan^{-1}\frac{(L(x,y+1)-L(x,y-1))}{(L(x+1,y)-L(x-1,y))} \tag{7.5}$$

The magnitude and direction calculations for the gradient are done for every pixel in a neighboring region around the keypoint in the Gaussian-blurred image L. In the case of multiple orientations being assigned, an additional keypoint is created that has the same location and scale as the original keypoint for each additional orientation.

7.3.4 Image Normalization

Image scaling is considered one of the fatal geometric attacks the image may undergo. Scaling can be either symmetric or nonsymmetric, and the scaling factor in the x direction is different from the that in the y direction (Alghoniemy and Tewfik 2000). The normalized image is assumed to have a predefined area and a unit aspect ratio. The aspect ratio γ of an image $f(x, y)$ is defined as

$$\gamma = \frac{I_y}{I_x} \tag{7.6}$$

where I_y and I_x are the height and the width of $f(x, y)$, respectively. Let $f((x/a),(y/b))$ be the rescaled image with $\gamma = 1$ and area $\alpha = (a/x)(b/y)$, where a and b are the required scaling factors; then

$$aI_x = bI_y \tag{7.7}$$

where a and b are determined as

$$a = \sqrt{\frac{\beta\gamma}{m_{0,0}}}, \; b = \sqrt{\frac{\beta}{\gamma m_{0,0}}} \tag{7.8}$$

where β and $m_{0,0}$ are the zero-order moment of $f((x/a), (y/b))$ and $f(x, y)$, respectively.

Transforming the image into its standard form requires translating the origin of the image into its centroids. The image in the new coordinates system has aspect ratio $\gamma = 1$ and area α. By using Equation 7.9, the coordinates (x, y) are changed into (x', y'):

$$x' = \frac{x - \bar{x}}{a}, y' = \frac{y - \bar{y}}{b} \tag{7.9}$$

7.4 Discrete Wavelet Transform for DIWM

The concept of DIWM is to add a watermark image to the host image such that the watermark image is unobtrusive and secure, and is capable of recovering partially or completely using appropriate cryptographic measures. A perceptibility criterion is applied to ensure the imperceptibility of the changes caused by the watermark embedding, which may be implicit or explicit, fixed or adaptive to the host data. As a result of this, the samples such as the pixels or the transform coefficients responsible for the watermarking can only be customized by a relatively small amplitude (Barni et al. 2001).

The novelty of the DWT algorithm resides in the manner the robustness and the invisibility are improved on the watermark image (Ketcham and Vongpradhip 2007). The major objective of the wavelet transform is to decompose the input image in a hierarchical manner into a series of successive low-frequency subbands and their associated detailed subbands. The low-frequency subband and the detailed subbands contain the information required to reconstruct the low-frequency approximation at the next higher resolution level (Barni et al. 2001). Such kind of an excellent space and frequency energy compaction is provided by wavelet techniques, and hence DWT has received incredible interest in several signal- and image-processing applications.

The watermark amplification factor is modulated on the basis of the local image characteristics, in a pixel-by-pixel manner. Most of the DWT-based watermarking concepts concentrate on the subbands or block-based techniques, whereas here the watermark amplification factor is adjusted pixel-wise. As a consequence, the gray-level sensibility, isofrequency masking, non-isofrequency masking, noise sensibility, and so on, are taken into account (Ketcham and Vongpradhip 2007). Because of the excellent spatial frequency localization property of DWT, it is easier to identify the image areas in which a watermark can be hidden more conveniently (Piva et al. 1998). In contrast to the DFT/DCT watermarking techniques, if a DWT coefficient is modified, only the region of the image corresponding to that coefficient will be modified. Thus DWT is applied in this application for watermark embedding and extraction.

7.4.1 Watermark Embedding

In DWT watermark embedding, a bit stream of length L is transformed into a sequence $W(1), \ldots, W(L)$ by replacing 0 by -1, and the sequence $W(K) \in \{-1,1\}$ $(k = 1, \ldots, L)$ is used as the watermark. The original image is first decomposed using a Haar filter into several bands using the DWT with the pyramidal structure. The watermark is added to the largest coefficients in all bands of details that represent the high and middle frequencies of the image (Sviatoslav et al. 2000).

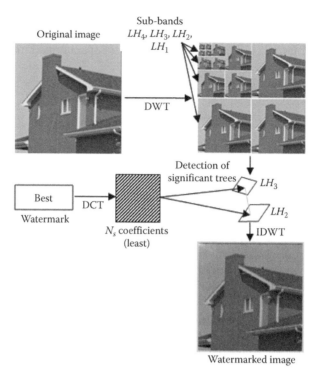

FIGURE 7.2
Block diagram of the DWT watermark embedding procedure.

The procedure for watermark embedding (Figure 7.2) using DWT is explained here:

Step 1: *Decomposition of DWT subbands*: The original image is decomposed into four levels using the 2D-DWT technique, and from the resulting subbands two are used for embedding the watermark. The levels used in this application are LH_3 and LH_2 or HL_3 and HL_2 since the coefficients in these subbands survive the most common attacks.

Step 2: *Detection of Significant Trees*: The number of significant trees N_s is detected on the basis of thresholds T_1 and T_2 satisfying the boundary condition

$$\left| x_{i,j}^{LH_3} \right| \geq T_1 \left| x_{k,l}^{LH_2} \right| \geq T_2, \quad \forall k = 2*i, 2*i-1, l = 2*j, 2*j-1 \tag{7.10}$$

where
　$x_{i,j}^{LH_3}$ is the (i, j)th coefficient in LH_3
　$x_{k,l}^{LH_2}$ is the (k, l)th coefficient in LH_2

The threshold levels are given by

$$T_1 = \frac{\text{Median}(|x_1|, |x_2|, |x_3|, \ldots, |x_N|)}{N}, \quad x_i \in LH_3$$

$$T_2 = \frac{\text{Median}(|x_1|, |x_2|, |x_3|, \ldots, |x_M|)}{M}, \quad x_i \in LH_2$$

where N and M denote the number of coefficients in the subbands LH_3 and LH_2, respectively. One coefficient in the LH_3 subband (parent) and four coefficients in the LH_2 subband (children) constitute one significant tree. Once all the significant trees are detected, the total number of trees is denoted as N_s.

Step 3: *Application of DCT to watermarking*: The watermark image is transformed into the frequency domain by arranging all the DCT coefficients in a zigzag manner starting from the lowest frequency component to the highest. The first DCT coefficients are used for embedding into the original image.

Step 4: *Sorting Significant Trees*: The sum of the absolute parent coefficient in LH_3 subband and the maximum absolute value of the children in the LH_2 subband is computed. Based on the sum, the significant trees are arranged in descending order. The sum is computed as

$$\left| x_{i,j}^{LH_3} \right| + \max_{k,l}\left(\left| x_{k,l}^{LH_2} \right| \right), \quad k = 2*i, 2*i-1, l = 2*j, 2*j-1 \tag{7.11}$$

Step 5: *Watermark Embedding*: The sorted significant trees are chosen, and the lower DCT coefficients are embedded into these sorted positions according to

$$wx_q^{LH_3} = x_q^{LH_3} + \alpha w_q, \quad q = 1, \ldots, N_s \tag{7.12}$$

$$wx_q^{LH_2} = \bar{x}_q^{LH_2} + \alpha w_q, \quad q = 1, \ldots, N_s \tag{7.13}$$

where

$x_q^{LH_3}$ is parent of the qth sorted significant tree

$\bar{x}_q^{LH_2}$ is the child of the qth significant tree

whose absolute value is maximum among the four children. Here, α is the embedding strength factor, which is inversely proportional to DC component of the watermark image and is given by

$$\alpha = \frac{f}{DC} \tag{7.14}$$

where f is the embedding energy of the watermark.

Step 6: *Application of inverse DWT*: The DWT coefficients are transformed by inverse DWT to get the watermarked image.

7.4.2 Watermark Extraction

The watermark extraction process requires the original image and the watermarked image. Both images are decomposed, and significant trees are detected from the LH_3 and

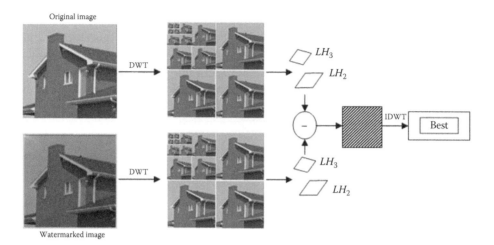

FIGURE 7.3
DWT watermark extraction process.

LH_2 subbands of the original image. Once the trees are sorted, then the DCT coefficients are extracted using Equation 7.15:

$$W_q^{LH_3} = \overline{Y}_q^{LH_3} - Y_q^{LH_3}$$
$$W_q^{LH_2} = \overline{Y}_q^{LH_2} - Y_q^{LH_2} \quad q = 1, \ldots, N_s \tag{7.15}$$

$$W_q = \frac{\left(W_q^{LH_3} + W_q^{LH_2}\right)}{2\alpha}, \quad q = 1, \ldots, N_s \tag{7.16}$$

where W_q is the qth coefficient of the watermark image and is found by determining the average of extracted coefficient from the parent element and the extracted coefficient from maximum child element of the qth sorted significant trees. All the extracted coefficients are then rearranged in a zigzag manner with a matrix size equal to N_s. By applying inverse DCT, the extracted watermark image is obtained. The extraction process is shown in Figure 7.3.

7.5 Performance Metrics

The performance of DIWM is assessed in terms of quality metrics that are classified into two groups: subjective metrics and objective metrics. The vital metric used to evaluate the quality of images received by the end-users are subjective metrics. In these metrics, the users evaluate the quality of the received digital image. Though these measurements are realistic, subjective metrics are difficult to evaluate, since they require multiple tests with several subjects in real-time applications. As an alternative, objective metrics are used to provide a set of quality metrics that can predict the perceived quality of the image from the user's perspective, without accessing the original image (Wang et al. 2002b). In this section, the metrics used in the intelligent DIWM for validating the perceptual quality and robustness are delineated.

7.5.1 Perceptual Image Quality Metrics

The two metrics for determining the quality of a watermarked image are the mean square error (MSE) and the PSNR (Voloshynovskiy et al. 2001). MSE and PSNR are widely used because they are simple to calculate, have clear physical meanings, and are mathematically easy to deal with for optimization purposes.

The signal-to-noise ratio (SNR) metric is used to measure the amount of noise that corrupts a signal. The SNR is computed by comparing the level of a desired signal to the level of background noise. After getting the level of the noise in each image (the original and the watermarked), the SNR compares the level of noise in both images and shows the differences between the two images in order to know the quality of the watermarked image after embedding the hidden data (Ali et al. 2010). The SNR is represented by Equation 7.17:

$$SNR = 10 * \log_{10} \frac{\sum_{i=1}^{n} \sum_{j=1}^{m} (I_{ij})^2}{\sum_{i=1}^{n} \sum_{j=1}^{m} (I_{ij} - I_{ij}')^2} \tag{7.17}$$

where I_{ij} and I_{ij}' represent a single pixel in the original image and the watermarked image, respectively. MSE is computed as the cumulative squared error between the original image and the watermarked image. A lower value of MSE indicates that the error between the two images is less. MSE is computed as

$$MSE = \frac{1}{mn} \sum_{i=1}^{n} \sum_{j=1}^{m} (I_{ij} - I_{ij}')^2 \tag{7.18}$$

where m and n denote the height and width of the image.

The PSNR (Ali et al. 2010) is a metric which computes the peak signal-to-noise ratio, in decibels, between two images. This ratio is often used as a measure of the quality of the original image compared to the watermarked image. The higher the value of the PSNR, the better the quality of the watermarked image. It is normal in watermarking to measure the PSNR in the original image and in the watermarked image after embedding the hidden data. The PSNR is computed as

$$PSNR = 10 \log_{10} \frac{I_{peak}^2}{\frac{1}{mn} \sum_{i=1}^{n} \sum_{j=1}^{m} (I_{ij} - I_{ij}')^2} \tag{7.19}$$

PSNR and MSE are related as follows:

$$PSNR = 10 \log_{10} \frac{I_{peak}^2}{MSE} \tag{7.20}$$

where I_{peak} represents the peak intensity level in the original image, which is most commonly taken as 255 for an 8-bit gray-scale image.

7.5.2 Robustness Evaluation

Robustness is a measure of the immunity of the watermark against attempts at modification and manipulation of the image, such as compression, filtering, rotation, scaling, collision attacks, resizing, cropping, and so on. Robustness depends on the information capacity of the watermark, the watermark strength/visibility, and the detection statistics (threshold). Robustness is also influenced by the choice of images (size, content, color depth). Most watermarking schemes are based on thresholding a correlation between an extracted vector and a pseudo-random sequence. With decreasing threshold, the probability of missed detections also decreases, and the robustness increases. But at the same time, the rate of false detections also increases.

Robustness is usually tested using typical image-processing operations and can be divided into two groups: gray-scale manipulations such as filtering, noise addition, lossy compression, gamma correction, color quantization, and color truncation to a finite palette, and geometric transformations such as scaling, cropping, rotation, affine transforms, and general rubber sheet deformations of the StirMark type. It is much easier to achieve robustness with respect to gray-scale transformations than geometrical transformations. Robustness of a watermarked image is generally measured in terms of the correlation factor. The correlation factor, also known as the NCC, measures the similarity and difference between the original watermark and the extracted watermark. The NCC value generally lies between 0 and 1. Ideally, it should be 1, but a value of 0.75 and above is acceptable (Gunjal 2014). NCC is given by

$$NCC = \frac{\sum_{k=1}^{N} w_k w_k'}{\sqrt{\sum_{k=1}^{N} w_k^2 \sum_{k=1}^{N} w_k'^2}} \tag{7.21}$$

where
  N denotes the number of pixels in the watermark
  w_k and w_k' are the original and the extracted watermark, respectively

7.6 Application of CI Techniques for DIWM

Determining the optimal tradeoff between transparency and robustness is a very challenging problem in DIWM because the payload varies for different types of images. In intelligent watermarking, bio-inspired techniques such as GA and PSO have been proposed to determine the optimal embedding parameters for each specific image. The basic principle is to evolve a population of potential embedding parameters through time using a combination of robustness and quality metrics as the objective function. Although PSO has a high rate of convergence, many authors have shown that PSO has difficulty in jumping out of the local optima. This drawback is overcome by incorporating GAs into PSO, by merging genetic operators such as selection, mutation, and crossover to PSO, to form the hybrid technique HPSO. In this section, the step-by-step procedures involved in the application of intelligent techniques of GA, PSO, and HPSO are explained. These heuristics are used for the optimization of the watermark amplification factor in DIWM to improve the robustness against geometric and non-geometric attacks.

7.6.1 Genetic Algorithm

GA (Goldberg 1989) is a heuristic search technique for determining the global maximum/minimum solutions for problems in the area of evolutionary computation (Boato et al. 2008). Any optimization problem is modeled in GA by defining the chromosomal representation and fitness function and applying the GA operators. The GA process begins with a few randomly selected genes in the first generation, called the population. Each individual in the population corresponding to a solution in the problem is called a chromosome, which consists of finite-length strings. The objective of the problem, called fitness function, is used to evaluate the quality of each chromosome in the population. Chromosomes that possess good quality are said to be fit, and they survive to form a new population of the next generation. The three GA operators—selection, crossover, and mutation—are applied to the chromosomes repeatedly to determine the best solution over successive generations (Shieh et al. 2004). In DIWM using the DWT domain, the value of the watermark amplification factor α balances the imperceptibility and the robustness. This balance is obtained through GA optimization technique.

In DIWM, the population is initialized by choosing a set of random positions in the cover image and inserting the watermark image into the selected positions. The optimal solutions for digital watermarking using DWT are obtained based on two key factors: the DWT subband and the value of the watermark amplification factor (Abu-Errub and Al-Haj 2008). The GA algorithm searches its population for the best solution with all possible combinations of the DWT subbands and watermark amplification factors. The GA procedure will attempt to find the specific subband that will provide simultaneous perceptual transparency and robustness. In order to improve the robustness of the algorithm against attacks, the watermark strength or the amplification factor α should be optimized, but this factor varies on each subband.

The input image is first encoded through a binary string encoding scheme. The 1's in the binary string indicate the position of the watermarks. Once all the chromosomes are encoded, the objective function is evaluated. The objective function, also known as the fitness function, is a combination of the PSNR and the correlation factor ($\rho = \alpha * NC$) and is given by

$$\text{Fitness function} = PSNR + 100 * \rho \tag{7.22}$$

where *PSNR* is computed according to Equation 7.19.

Here, the correlation factor is the product of the normal correlation (NC) and the watermark strength factor α. The fitness function increases proportionately with the PSNR value. Since NC is the key factor contributing to the robustness ultimately, the fitness value increases with the robustness measure. The correlation factor ρ has been multiplied by 100, since its normal values fall in the range 0–1, whereas PSNR values may reach the value of 100.

The fitness function is evaluated for all the individuals in the population, and the best fit individual along with the corresponding fitness value are obtained. Genetic operators such as crossover and mutation are performed on the selected parents to produce new offspring, which are included in the population to form the next generation. The entire process is repeated for several generations until the best solutions are obtained. The correlation factor ρ measures the similarity between the original watermark and the one extracted from the attacked watermarked image (robustness). The procedure for implementing DIWM using GA is shown here.

- Initialize the watermark amplification factor α between 0 and 1; initialize the population size, the number of iterations, crossover rate, and mutation rate.
- Generate the first generation of GA individuals based on the parameters specified by performing the watermark embedding procedure. A different watermarked image is generated for each individual.
- **While** max iterations have not reached
  - Evaluate the perceptual transparency of each watermarked image by computing the corresponding PSNR value.
  - Apply a common attack on the watermarked image.
  - Perform the watermark extraction procedure on each attacked watermark image.
  - Evaluate robustness by computing the correlation between the original and extracted watermarks.
  - Evaluate the fitness function for the PSNR and ρ values.
  - Select the individuals with the best fitness values.
  - Generate a new population by performing the crossover and mutation functions on the selected individuals.
- **End While**

7.6.2 Particle Swarm Optimization

PSO is a randomized search technique inspired by the social behavior of fish schooling or bird flocking, developed by Eberhart and Kennedy (1995). A swarm can be defined as a group of mobile agents that show a collective behavior in attaining a specified goal. Each candidate solution in the swarm is known as a particle. The initial population of particles is generated in a random manner, and the search for optimal solution is performed in an iterative manner. Each ith particle moves in the m-dimensional search space with a velocity $V^i = (v_1^i, v_2^i, \ldots, v_m^i)$. The position vector of the ith particle is given by $X^i = (x_1^i, x_2^i, \ldots, x_m^i)$, and the best position value encountered by each particle at the best time t is $P_t^i = (p_1^i, p_2^i, \ldots, p_m^i)$. The global best particle in the swarm at time t is given by $P_t^g = (p_1^g, p_2^g, \ldots, p_m^g)$. The new position of the particle is determined as

$$x_{t+1}^i = x_t^i + w_t v_t^i + c_1 r_1 (p_t^i - x_t^i) + c_2 r_2 (p_t^g - x_t^i) \qquad (7.23)$$

where
w is the inertia weight
c_1 is the cognitive parameter
c_2 is the social parameter (c_1 and c_2 are together known as acceleration constants)
r_1 and r_2 are random numbers uniformly distributed in the interval [0,1]

The new position of the particle is based on its previous position, distance between the current position and its best determined position, and the collaborative effect of the particles.

7.6.2.1 Watermark Embedding Using PSO

Consider the original image I as a gray-level image of size $I_M \times I_N$, representing the width and height of I, respectively. Let the watermark W be a binary image of size $W_P \times W_Q$, representing the width and height, respectively. The original image is then decomposed into the wavelet representation of l levels. Let $B_i^x(i,j)$ represent the frequency coefficient in the coordinate (i,j), where $l = 0,1,2,3$ denotes the level and $x \in \{0,1,2,3\} \in \{LL,LH,HL,HH\}$ represents the orientation (Tao et al. 2010). The binary image is embedded into the level 3 component of the original image so that the visual quality and robustness are maintained. The steps required to embed the binary image into the original image are as follows:

Step 1: The original image is decomposed into l level subbands, thus obtaining a series of multiresolution fine subshapes $\{HL_i, LH_i, HH_i\}$ with $i = 1,2,3$ and coarse overall shape LL_3. Further, the HL_3 component is decomposed into nonoverlapping blocks NB_k of size 2×2 with $k = 1,2,\dots,(W_P \times W_Q)$.

Step 2: The watermark information is pretreated in order to eliminate the correlation of the watermark pixels, thus enhancing the robustness and security. Affine scrambling is adopted to pretreat according to the equation

$$\begin{pmatrix} u' \\ v' \end{pmatrix} = \begin{pmatrix} a & b \\ c & d \end{pmatrix}\begin{pmatrix} u \\ v \end{pmatrix} + \begin{pmatrix} e \\ f \end{pmatrix}, \quad \text{where} \begin{vmatrix} a & b \\ c & d \end{vmatrix} \neq 0 \tag{7.24}$$

The imperceptibility is enhanced by substituting $\{-1\ 1\}$ for $\{0\ 1\}$ in the process of scrambling. The new watermark $w_i' = w_i p_i$ is then generated on the basis of the binary pseudo-random p_i modulating the watermark, where $p_i = \{-1\,1\}$ and $i \in [0, W_P \times W_Q]$.

Step 3: The watermark bits are embedded into the nonoverlapping blocks NB_k based on experiments.

In each block, $\max[B_3^{HL}(i,j), B_3^{HL}(i+1,j), B_3^{HL}(i,j+1), B_3^{HL}(i+1,j+1)]$ and $\min[B_3^{HL}(i,j), B_3^{HL}(i+1,j), B_3^{HL}(i,j+1), B_3^{HL}(i+1,j+1)]$ are computed. The subband coefficients are then modified according to Equation 7.25:

$$B_3'^{HL}(i,j) = \begin{cases} \max[B_3^{HL}(i,j), B_3^{HL}(i+1,j), B_3^{HL}(i,j+1), B_3^{HL}(i+1,j+1)] + \alpha w_k & \text{if } w_k = 1 \\ \min[B_3^{HL}(i,j), B_3^{HL}(i+1,j), B_3^{HL}(i,j+1), B_3^{HL}(i+1,j+1)] - \alpha w_k & \text{if } w_k = 0 \end{cases} \tag{7.25}$$

where α represents the watermark amplification factor.

Step 4: Apply PSO to obtain the best α value. The amplification factor obtained for optimal watermarking depends on the transparency and robustness measures. Each particle in the swarm represents a possible solution to the problem, and thus it is comprised of a set of amplification factors. The initial solution is obtained in PSO in a random manner based on α generated in [0,1]. The parameter α acts as a weight of each watermarked bit and embeds each such bit into the HL_3 subband components using the DWT transform. Using these amplification factors, the subband coefficients are modified, thereby decomposing the host image and the watermarked images of the current generation. The fitness of each particle is evaluated to obtain the local best particle and the global best particle. The particles are updated according to Equation 7.23. Step 4 is repeated until the termination criterion is reached.

Step 5: After embedding, the inverse wavelet transform is applied to obtain the watermarked image.

7.6.2.2 Watermark Extraction

The reverse of the watermark embedding procedure results in the watermark extraction process as explained as follows:

> **Step 1:** The watermarked image is decomposed into three levels using DWT, thus acquiring a series of high-frequency subbands and a high-energy subband.
>
> **Step 2:** The HL_3 component is decomposed into NB_K nonoverlapping blocks of size 2×2 with $k = 1, 2, \ldots, (W_P \times W_Q)$.
>
> In every block, $x = \max[B_3'^{HL}(i,j), B_3'^{HL}(i+1,j), B_3'^{HL}(i,j+1), B_3'^{HL}(i+1,j+1)]$ and $y = \min[B_3'^{HL}(i,j), B_3'^{HL}(i+1,j), B_3'^{HL}(i,j+1), B_3'^{HL}(i+1,j+1)]$ are computed. Then the average is determined as $z = (x+y)/2$ to obtain the extracted watermark,

$$w_k = \begin{cases} 1, & z \le B_3'^{HL}(i,j) \\ 0, & z > B_3'^{HL}(i,j) \end{cases} \tag{7.26}$$

> **Step 3:** The watermark is extracted and NCC is computed to quantify the correlation between the original watermark and the extracted one. NCC is defined as

$$NCC = \frac{\sum_{k=1}^{W_P \times W_Q} w_k w_k'}{\sqrt{\sum_{k=1}^{W_P \times W_Q} w_k^2 \sum_{k=1}^{W_P \times W_Q} w_k'^2}} \tag{7.27}$$

where w_k and w_k' are the original and the extracted watermarks, respectively.

In order to obtain the optimal results in DIWM, PSO is employed to search for the optimal parameters. The feasibility of the extracted watermark is validated on the basis of the performance metric NCC. NCC values evaluate the objective function as a watermark detection performance index because of its role as a robustness measure. Using PSO, the personal best of each particle is calculated by evaluating the fitness function. The best particle in the entire swarm is the global best, and this is applied in the watermark embedding procedure. PSNR is obtained according to Equation 7.19 in order to validate the perceptual imperceptibility of the watermarked image. PSNR is used as the fitness function to evaluate each particle. A PSNR value greater than 30 dB is perceptually acceptable while evaluating the imperceptibility of the watermarked image.

7.6.3 Hybrid Particle Swarm Optimization

Due to its simple concept, easy implementation, and quick convergence, PSO has gained much attention and has a wide variety of applications in different fields, mainly for various continuous optimization problems (Eberhart and Kennedy 1995). At the same time, the performance of a simple PSO depends on its parameters, and it often suffers from the problem of being trapped in local optima, thereby causing premature convergence. Many studies have been carried out to prevent premature convergence and to balance the exploration and exploitation abilities. In this section, a hybrid of the intelligent techniques of GA and PSO (HPSO) is proposed for DIWM to simultaneously improve security, robustness, and image quality of the watermarked images. The overall structure of the proposed hybrid algorithm is described as follows:

- **Initial population:** In optimization problems, the initial population is usually large enough, to get near-optimal solutions. A set of initial random locations are chosen to insert the watermark bits into the cover image to form the initial population. The chromosomes obtained from the coding part constitute the initial population or particles for the HPSO algorithm. The initial population is formed in a random manner. Filtering, rotation, scaling, and JPEG compression attacks are applied on the watermarked images.

- **Fitness evaluation:** The objective of the HPSO is to evaluate the watermarked image and compute the NCC, and hence for each chromosome the similarity is computed. For every generation, all the chromosomes are evaluated for the fitness function based on Equation 7.28:

$$fitness = sim(w, w') \times 50 + PSNR \tag{7.28}$$

where $sim(w, w')$ is defined as the similarity of w and w', expressed as

$$sim(w, w') = \frac{\sum_{i=1}^{W_L} w.w'}{W_L} \tag{7.29}$$

where

W_L is the length of the original watermark
w and w' are the original and extracted watermarks, respectively

- **Offspring generation:** In the evolutionary process, PSO is used to generate new offspring based on new position of the particle, thus updating the velocity according to

$$x_{t+1}^i = x_t^i + \underbrace{w_t v_t^i + c_1 r_1 (p_t^i - x_t^i) + c_2 r_2 (p_t^g - x_t^i)}_{v_{t+1}^i} \tag{7.30}$$

The best position visited by the kth particle since the first time step is known as the local best position, denoted as *pbest* p_t^i, while *gbest* p_t^g represents the best position that the kth particle and its neighbors have visited from the beginning of the algorithm. During each iteration, the *pbest* and *gbest* are updated based on the objective (fitness) value. If the current value of *pbest* is better, p_t^i and its objective value are updated with the current position. Similarly, if the objective value is better than the objective value of p_t^g, then p_t^g and its objective value are updated with the position and objective value of the current best particle.

As a parallel process, the genetic operators of selection, crossover, and mutation are applied to update the particles according to Equation 7.31:

$$x_{t+1}^i = \hat{x}_t^i \oplus (p_t^i \otimes x_t^i) \oplus (p_t^g \otimes x_t^i) \tag{7.31}$$

Here, the symbol $\otimes$ denotes the crossover operation performed between the current best and the best individual obtained so far. The symbol $\oplus$ represents the optimal or best offspring resulting from $(p_t^i \otimes x_t^i)$, or $(p_t^g \otimes x_t^i)$, or $\hat{x}_t^i$. In Equation 7.31, $(p_t^i \otimes x_t^i)$ denotes the crossover operation between the two parents p_t^i and x_t^i, and $\hat{x}_t^i$ is the mutated offspring of x_t^i. The term x_{t+1}^i represents the new particle obtained from the genetic operators. The particles obtained through GA and PSO are evaluated for their fitness, and the best individual is retained in the population for further generations.

The crossover operation has to maintain the number of bits in each chromosome stable and equal to those of the watermark; hence two-point crossover is applied to generate the offspring. Mutation is performed by dividing a bit string into sections of size N, and random sections are selected to which flip-bit mutation is applied.

The driving parameter involved in the hybridization process is the hybridization factor (HF), which represents the percentage of the individuals in the population evolved

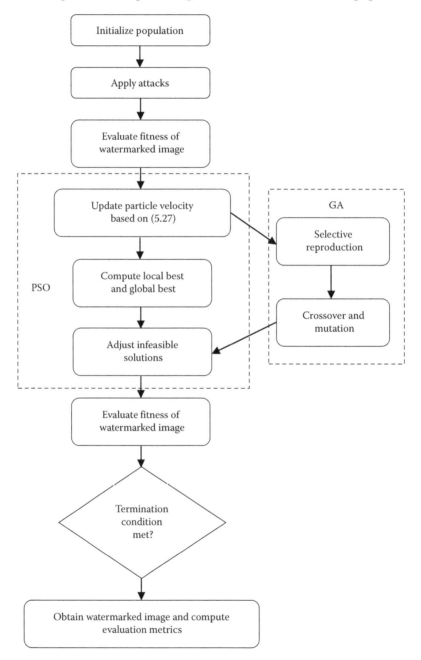

FIGURE 7.4
Flowchart of DIWM using HPSO.

through GA. In simple terms, if HF = 0 then the algorithm is a pure PSO algorithm, and if HF = 1 then the algorithm is a pure GA. HF lies in the range [0, 1] in order to maintain the balance among the individuals evolved through GA and PSO.

- **Stopping condition:** The algorithm terminates with the number of iterations and computation of the best solution.

  The overall operation of HPSO for DIWM is shown in the form of a flowchart in Figure 7.4.

7.7 MATLAB® m-File Snippets for DIWM Using CI Paradigms

Extensive experiments were conducted to prove the validity of the bio-inspired techniques in DIWM. The experiments were carried out to assess the performance of the system from the point of view of both watermark imperceptibility and robustness. In particular, the intelligent system has demonstrated to be resistant to several attacks such as JPEG compression, median filtering, average filtering, Gaussian-noise addition, rotation, scaling, rotation plus scaling, rotation plus scaling plus cropping, and rotation plus scaling plus JPEG compression. The watermark amplification factor α was optimized in the interval [0,1] by the bio-inspired algorithms. The performance of watermarking based on GA, PSO, and HPSO was obtained by analyzing six images (Figure 7.5)—Peppers, Mandrill, Lena, Barbara, Boat, and Cameraman—and the PSNR, MSE, robustness measure, and computational time are evaluated. These images were taken as the cover images, and best.bmp (Figure 7.6) of size 60 × 24 was taken as the

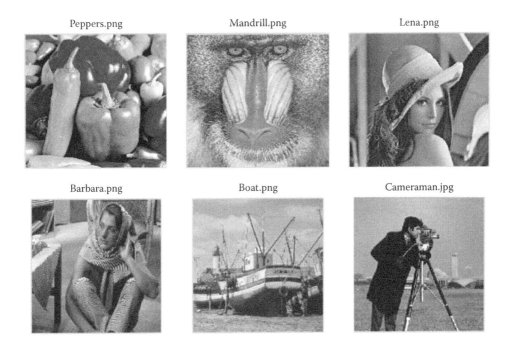

FIGURE 7.5
Standard images used for watermark embedding.

Best

FIGURE 7.6
Digital watermark.

TABLE 7.1

Parameter Setting for Expectation Maximization Algorithm

Parameters	Notation	Value
Input data in the form of image	$X(n,d)$	Image dependent
Size of the image	$[n,d]$	Image dependent
Maximum number of Gaussian components	K	[2,6]
Percentage of the log likelihood difference between two iterations	Tol	0.1
Maximum number of iterations	m_iter	500

watermark image. All experiments were implemented with an Intel i3 CPU, 2.53 GHz, 4 GB RAM PC using MATLAB R2008 software.

The purpose of image segmentation is to partition an image into meaningful regions with respect to a particular application. The segmentation is based on measurements taken from the image and may be the gray-level, color, texture, depth, or motion. The parameters and their settings for segmenting the image using the EM algorithm are shown in Table 7.1. The input image X is read with n observations and d dimensions. The value of k is varied between 2 and 6 based on which the number of Gaussians occur with uniform probability and the log likelihood difference between two successive iterations is set to 0.1. The algorithm is run for a maximum of 500 iterations irrespective of the nature of the input image.

The MATLAB m-file snippet for EM algorithm with the expectation and maximization steps is given here:

```
% Expectation
        prb = distribution(mu,v,p,x);
        scal = sum(prb,2)+eps;
        loglik=sum(h.*log(scal));

% Maximization
        for j=1:k
                pp=h.*prb(:,j)./scal;
                p(j) = sum(pp);
                mu(j) = sum(x.*pp)/p(j);
                vr = (x-mu(j));
                v(j)=sum(vr.*vr.*pp)/p(j)+sml;
        end
```

Using the EM segmentation, the image was segmented into number of homogeneous regions and the parameters such as mean and standard deviation were updated for each segmented regions based on k Gaussians. Dark regions of the segmented image imply that the variance of the images is very high, thus containing more high-frequency components. Figure 7.7 shows the original images and segmented images for different values of k. It is observed that for larger values of k, the segmentation is very coarse; many clusters appear in the images at discrete places.

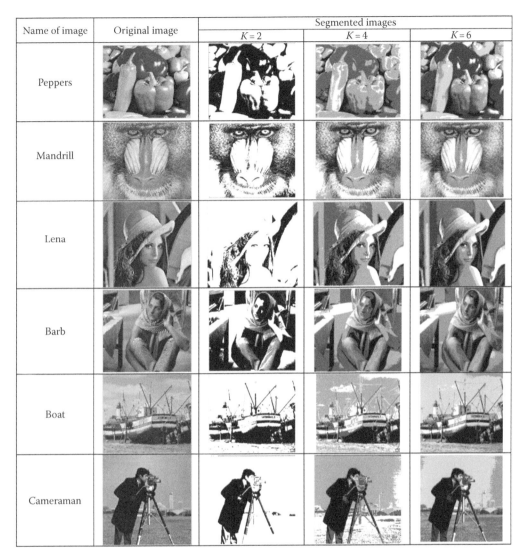

FIGURE 7.7
Segmentation using expectation maximization.

7.7.1 Feature Extraction Using Difference of Gaussian

DoG is a gray-scale image enhancement algorithm to select the feature points from the segmented image. The parameters used in the DoG algorithm and their values are shown in Table 7.3. The Gaussian scale space of an image $I(x, y)$ is the function $G(x, y, \sigma) \overset{\Delta}{=} (g_\sigma * I)(x, y)$, where σ is the scale coordinate discretized in logarithmic steps according to $\sigma = \sigma_0 2^{o+s/S}$, where o is the octave index, s is the scale index, S is the scale resolution, and σ_0 is the scale offset. The scale offset, also known as the smoothing radii, was initially set to 0.5. Two different blurring radii were used for Gaussian blurs with the constraint that they cannot be equal. In case of equal radii values, the resultant image is blank. The scale index s is

TABLE 7.2

Parameter Setting for Difference of Gaussian Algorithm

Parameters	Notation	Value
Minimum octave index	o	−1
Scale resolution	S	3
Scale offset	σ_0	0.5
Initial scale index	s	$[-1,S]$
Scale index during algorithm run	s	$[-1,S+1]$

based on the scale resolution S, which was initially set in the range $[-1, S]$, and during the algorithmic run it was tuned according to the interval $[-1,S+1]$.

The parameters for the DoG algorithm were set based on Table 7.2 and the codes were developed using MATLAB. The segmented image was given as input to the DoG algorithm to extract the features. A code snippet using MATLAB is presented as follows:

```
PSF = fspecial('gaussian',3,0.3);
Blurred = imfilter(I,PSF);
figure;
imshow(Blurred,'DisplayRange',[]);title('Blurred Image');
[m,n]=size(I);
sig=70;
k=5;
for i=1:m
    for j=1:n
        G(i,j)= exp(-(((i^2)+(j^2))/(2*(sig^2))))/(sqrt(2*3.14*(sig^2)));
        G1(i,j)= exp(-(((i^2)+(j^2))/(2*k*(sig^2))))/
        (sqrt(2*3.14*k*(sig^2)));
        g1(i,j)=G(i,j)*Blurred(i,j);
        g2(i,j)=G1(i,j)*Blurred(i,j);
        DOG(i,j)=g2(i,j)-g1(i,j);
    end
end
DOG=double(DOG);
```

This algorithm computes the difference between one blurred version of the original gray-scale image and another less blurred version of the original image, as shown in Figure 7.8. The DoG algorithm was chosen in this experiment to find the extracted points because in each segmented region, one feature point is selected and the circular region centered at the selected feature point with radius will be used for the watermark embedding and detection. After the reference feature points were selected, the rotation- and scaling-invariant properties were assigned to the circular regions centered at the selected feature points.

7.7.2 Orientation and Normalization

The process of orientation and normalization were performed to establish rotation and scaling invariance of the watermarked image. The parameters used in these preprocesses are shown in Table 7.3. The scaling range was selected such that robustness was higher in

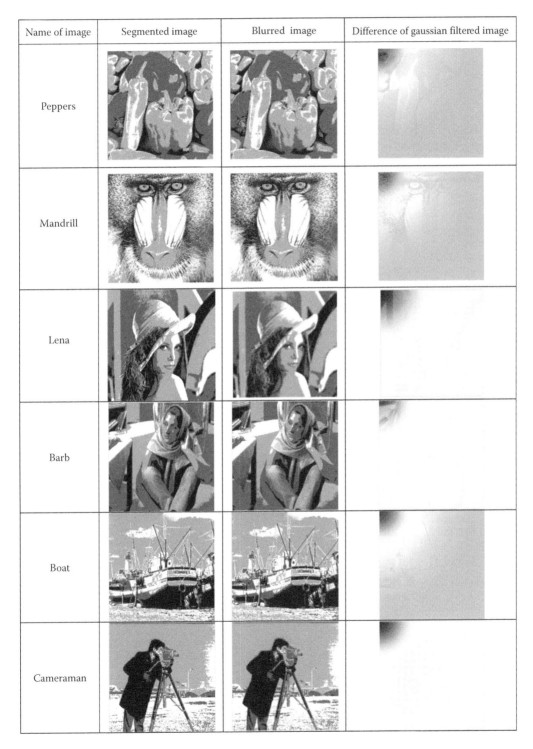

Name of image	Segmented image	Blurred image	Difference of gaussian filtered image
Peppers			
Mandrill			
Lena			
Barb			
Boat			
Cameraman			

FIGURE 7.8
Performance of difference of Gaussian.

TABLE 7.3

Parameters for Orientation and Normalization

Parameters	Value
Scaling range	[0.5, 2]

the low-frequency band and lower in the high-frequency band. The image was rotated in the range [5°, 40°] in counterclockwise direction around its center point.

For different scale values in the range [0.5, 2], the orientations were performed by drawing gradients in small regions. The snippet is shown here:

```
figure('name','Magnitude of Gradient');
[imx,imy]=gaussgradient(fim,0.5);
[imx,imy]
subplot(2,2,1);
imshow(abs(imx)+abs(imy));
title('sigma=0.5');
[imx,imy]=gaussgradient(fim,1.0);
subplot(2,2,2);
imshow(abs(imx)+abs(imy));
title('sigma=1.0');
[imx,imy]=gaussgradient(fim,1.5);
subplot(2,2,3);
imshow(abs(imx)+abs(imy));
title('sigma=1.5');
[imx,imy]=gaussgradient(fim,1.5);
subplot(2,2,4);
imshow(abs(imx)+abs(imy));
title('sigma=2.0');

%draw gradient in a small region with sigma=1.0
[imx,imy]=gaussgradient(fim,1.0);
figure('name','Gradient');
imshow(fim(1:50,1:50),'InitialMagnification','fit');
hold on;
quiver(imx(1:50,1:50),imy(1:50,1:50));
```

The gauss gradient subfunction to generate a 2-D Gaussian kernel is shown as follows:

```
for i=1:size
    for j=1:size
        u=[i-halfsize-1 j-halfsize-1];
        hx(i,j)=gauss(u(1),sigma)*dgauss(u(2),sigma);
    end
end
```

The circular regions were chosen as shown in Figure 7.9 and were made rotation invariant using orientation assignment. In order to achieve orientation invariance, the coordinates of the descriptor and the gradient orientations were rotated relative to the

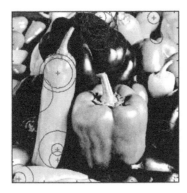

FIGURE 7.9
Circular regions in Peppers, Boat, Lena, and Cameraman images.

keypoint orientation. By finding the image gradients, the key points in the image were extracted and modified to reduce the illumination change.

A technique for normalizing an image against geometric manipulation was implemented, and the purpose was to obtain scaling and rotation invariance for the image during the watermark embedding and extraction phases. Scaling normalization was employed to acquire the scaling invariance for the circular region. It transforms the image into its standard form by translating the origin of the image to its centroid. The normalized form of the images is shown in Figure 7.10. In these results, the normalization effect of each image was evaluated using local mean and standard deviation estimated by Gaussian kernel with $\sigma = 4$.

7.7.3 DWT Watermark Embedding and Extraction

Because of its excellent spatio-frequency localization properties, DWT is suitable for identifying the areas in the host image where a watermark can be embedded effectively. In particular, this property allows the exploitation of the masking effect of the human visual system such that, if a DWT coefficient is modified, only the region corresponding to that coefficient will be modified. In this experiment, the watermark was embedded into the nonoverlapping blocks of the HL_3 subband of the host image with a block size of 8×8. Table 7.4 lists the parameters and their values used in the DWT domain.

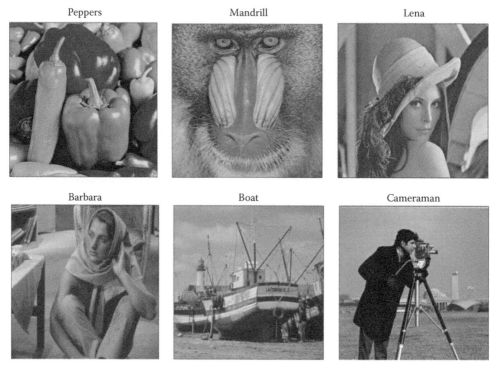

FIGURE 7.10
Normalized images.

TABLE 7.4

Parameters for DWT Watermark Embedding

Parameters	Value
Decomposition	Haar filter
Block size	8
Subband	HL_3
Amplification factor	Optimized by intelligent algorithms

Using the parameters and their settings according to Table 7.4, the MATLAB m-file was developed to embed and extract the watermark. A snippet is presented here:

```
A0 = idwt2(A,cH1,cV1,cD1,'db1');
% A0 = imnoise(A0,'speckle',0.04);
figure(4);
% imshow(uint8(A0))

%%%%%%%%%%%%%%%%%%%%%%%%%%%%%%%%%%
[cA1,cH1,cV1,cD1] = dwt2(A0,'db1');
[x,y]=size(cA1);
cA1=double(cA1);
subplot(1,2,1);
imshow(uint8(A0));
title('received image');
for i=1:x
    for j=1:y
```

Peppers.png   Mandrill.png   Lena.png

Barbara.png   Boat.png   Cameraman.jpg

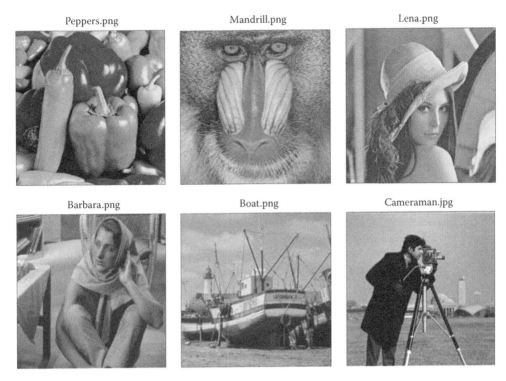

FIGURE 7.11
DWT watermarked images.

```
        s(i,j)=cA1(i,j)-J8(i,j);
          %s(i,j)=s(i,j)/k;
    end
end
s1=imresize(s,[256 256]);
[x,y]=size(s1);
s1=double(s1);
subplot(1,2,2);
imshow((s1));
title('extracted image');
imwrite(s1,'a.bmp')
```

Figure 7.11 shows the set of watermarked images without application of intelligent algorithms and attacks. Decomposition was performed through Haar filter. The watermark was added to the largest coefficients in all bands of details that represented the high and middle frequencies of the image.

7.8 Optimization in Watermarking

The results of the optimization techniques GA, PSO, and HPSO applied to the set of images are explained in this section. The performance of these techniques is analyzed by varying the algorithm-dependent parameters and thus computing the MSE, PSNR, fitness, and computational time.

7.8.1 Genetic Algorithm

A series of experiments were performed by varying several parameters in GA, such as the number of generations, population size, crossover probability, and mutation probability. The choice of parameters was obtained by varying the parameters over a range and computing the MSE, PSNR, fitness, and computational time. Based on the outcome of these values, the GA parameters were fixed. The following sections explain the effect of GA-dependent parameters on watermarking application.

7.8.1.1 Variation in the Number of Generations

With a population size of 120, the number of generations (N_G) was varied from 10 to 40 with an interval of 10 to optimize the watermark amplification factor and thus compute the PSNR, MSE, robustness (NCC), and computational time.

The crossover probability was chosen to be 0.7. The mutation probability was fixed as 0.02 based on previous experiments (Shieh et al. 2004) and maintained constant for variation in the number of generations. From Table 7.5, it is observed that the maximum PSNR and efficient fitness are obtained at 10 generations for Peppers, Mandrill, and Lena images, at 20 generations for Barbara and Boat, and at 30 generation for Cameraman.

TABLE 7.5

Effect of Number of Generations on Images

Images	N_G	MSE (dB)	PSNR (dB)	NCC	Fitness	Time (s)
Peppers	10	**5.9415**	**40.39184**	0.9965	**52.34984**	14.12
	20	6.1214	40.2623	0.9982	52.2407	15.76
	30	6.9732	39.69648	0.9971	51.66168	16.92
	40	7.4712	39.3969	0.9952	51.3393	18.01
Mandrill	10	**5.1172**	**41.04048**	0.9892	**52.91088**	13.9
	20	5.3783	40.82435	0.991	52.71635	14.56
	30	5.6125	40.63924	0.9987	52.62364	15.25
	40	5.9372	40.39499	0.9954	52.33979	15.98
Lena	10	**3.3476**	**42.88347**	0.9978	**54.85707**	13.5
	20	3.9899	42.12118	0.9895	53.99518	13.98
	30	4.1024	42.00042	0.9864	53.83722	14.22
	40	4.2921	41.80411	0.9778	53.53771	15.02
Barbara	10	4.052	42.05411	0.9912	53.94851	13.87
	20	**3.9866**	**42.12478**	0.9945	**54.05878**	14.06
	30	4.0256	42.0825	0.989	53.9505	14.67
	40	4.1244	41.9772	0.9856	53.8044	15.15
Boat	10	4.2378	41.8594	0.9376	53.1106	14.18
	20	**4.0267**	**42.08131**	0.9634	**53.64211**	15.04
	30	4.6432	41.46263	0.9912	53.35703	15.79
	40	4.5433	41.55709	0.9875	53.40709	16.13
Cameraman	10	5.1245	41.03429	0.9877	52.88669	15.45
	20	5.1156	41.04184	0.9823	52.82944	16.66
	30	**5.0123**	**41.13043**	0.9912	**53.02483**	17.13
	40	5.2366	40.94031	0.9891	52.80951	17.99

N_G = Number of generations.

7.8.1.2 Variation in the Population Size

The major issue while applying GA for optimization is choosing the correct size of the population of the encoded chromosomes. The choice of population size (N_P) is a tradeoff between the quality of the solution and the computational cost. A larger population size will maintain a high genetic diversity, thus leading to higher possibility of locating the global optimum but with a high computational cost. In this experiment, the population size was varied in multiples of 4, and the number of generations for the images corresponded to the optimum results obtained from Table 7.5.

The crossover rate was maintained constant at 0.7 and mutation rate at 0.02, and the PSNR, MSE, robustness, and the computational time were evaluated, as shown in Table 7.6. The maximum number of generations for Peppers, Mandrill, and Lena is found to be 10, for Barbara and Boat it is 20, and for Cameraman it is 30. The optimal values are obtained for different images at different population sizes, and these values are carried over for the next set of experiments.

7.8.1.3 Variation in Crossover Rate

At a higher the crossover rate (P_C), a new offspring is added to the population more quickly. If the crossover rate is too high, high-performance strings are eliminated faster so that the

TABLE 7.6

Population Sizes and Its Impact on Images

Images	N_P	MSE (dB)	PSNR (dB)	NCC	Fitness	Time (s)
Peppers	64	6.1425	40.24735	0.9961	52.20055	13.98
($N_G = 10$)	**128**	**5.3214**	**40.87054**	**0.9987**	**52.85494**	14.47
	256	6.1712	40.22711	0.9968	52.18871	15.82
	512	7.1112	39.61137	0.9949	51.55017	17.11
Mandrill	64	5.1342	41.02608	0.9791	52.77528	14.12
($N_G = 10$)	128	5.4534	40.76413	0.9892	52.63453	14.75
	256	**5.0322**	**41.11322**	**0.9987**	**53.09762**	15.17
	512	5.9657	40.37419	0.9945	52.30819	15.88
Lena	**64**	**3.1486**	**43.14963**	**0.9965**	**55.10763**	13.05
($N_G = 10$)	128	3.9724	42.14027	0.9812	53.91467	13.68
	256	4.1128	41.98943	0.9826	53.78063	13.98
	512	4.2821	41.81424	0.9833	53.61384	14.19
Barbara	64	4.154	41.94614	0.9944	53.87894	13.56
($N_G = 20$)	**128**	**3.924**	**42.19351**	**0.9965**	**54.15151**	14.12
	256	4.011	42.09828	0.9823	53.88588	14.34
	512	4.128	41.97341	0.9876	53.82461	15.11
Boat	64	4.2251	41.87243	0.9265	52.99043	14.22
($N_G = 20$)	128	4.3412	41.75471	0.9576	53.24591	14.78
	256	4.0211	42.08735	0.9867	53.92775	15.08
	512	**4.0045**	**42.10532**	**0.9943**	**54.03692**	16.02
Cameraman	64	5.3423	40.85352	0.9827	52.64592	15.12
($N_G = 30$)	128	5.1216	41.03675	0.9809	52.80755	16.72
	256	5.1477	41.01467	0.9897	52.89107	16.99
	512	**5.0366**	**41.10943**	**0.9991**	**53.09863**	17.49

Notes: Optimal values in bold. N_P = Population size; N_G = Number of generations.

TABLE 7.7

Performance Evaluation of Images Based on Crossover Rates

Images	P_C	MSE (dB)	PSNR (dB)	NCC	Fitness	Time (s)
Peppers	0.5	6.0102	40.34191	0.9788	52.08751	14.55
($N_G = 10\ N_P = 128$)	0.6	5.4532	40.76429	0.9823	52.55189	14.72
	0.7	**5.1255**	**41.03344**	**0.9982**	**53.01184**	14.98
	0.8	6.1121	40.2689	0.9856	52.0961	15.12
	0.9	6.4239	40.05282	0.9821	51.83802	15.65
Mandrill	0.5	5.3985	40.80807	0.9734	52.48887	14.67
($N_G = 10\ N_P = 256$)	0.6	5.6623	40.60087	0.9822	52.38727	14.98
	0.7	5.2785	40.9057	0.9991	52.8949	15.98
	0.8	**5.3932**	**40.81234**	**0.9901**	**52.69354**	16.01
	0.9	5.8723	40.44272	0.9828	52.23632	16.55
Lena	0.5	4.0254	42.08271	0.9821	53.86791	14.15
($N_G = 10\ N_P = 64$)	0.6	3.9876	42.12369	0.9856	53.95089	14.21
	0.7	**3.7834**	**42.35198**	**0.9967**	**54.31238**	14.22
	0.8	4.0909	42.01261	0.9837	53.81701	14.08
	0.9	4.3726	41.72341	0.9784	53.46421	14.27
Barbara	0.5	4.1276	41.97383	0.9912	53.86823	14.76
($N_G = 20\ N_P = 128$)	0.6	**3.8769**	**42.24596**	**0.9959**	**54.19676**	14.92
	0.7	4.0986	42.00445	0.9926	53.91565	15.23
	0.8	4.5329	41.56704	0.9892	53.43744	14.12
	0.9	4.8934	41.2347	0.9826	53.0259	14.33
Boat	0.5	4.3415	41.75441	0.9758	53.46401	15.21
($N_G = 20\ N_P = 512$)	0.6	**4.1249**	**41.97667**	**0.9983**	**53.95627**	15.02
	0.7	4.3289	41.76703	0.9856	53.59423	15.62
	0.8	4.7834	41.33344	0.9784	53.07424	15.72
	0.9	4.9167	41.21407	0.9711	52.86727	15.19
Cameraman	0.5	5.2278	40.94761	0.9781	52.68481	17.49
($N_G = 30\ N_P = 512$)	0.6	5.1916	40.97779	0.9856	52.80499	17.12
	0.7	**5.1256**	**41.03336**	**0.9981**	**53.01056**	16.98
	0.8	5.2784	40.90578	0.9897	52.78218	17.16
	0.9	5.3329	40.86117	0.9798	52.61877	17.05

N_P = Population size; N_G = Number of generations; P_C = Crossover rate.

selection cannot produce improvements. A low crossover rate may cause stagnation due to the lower exploration rate. Here, the crossover rate was varied in the range [0.45, 0.95] according to Grefenstette (1986). The number of generations and the population size were chosen from the optimal values obtained from Tables 7.5 and 7.6. The mutation rate was maintained constant at 0.02, and the evaluations were performed. For Peppers, Lena, and Cameraman, the optimal values were obtained for a crossover rate of 0.7, for Mandrill with 0.8, and for Barbara and Boat with 0.6. The evaluation results for all images with different crossover rates are shown in Table 7.7.

7.8.1.4 Variation in Mutation Rate

Mutation probability (P_M) is a very important parameter in mutation processes and decides the rate at which the genes in the chromosome get swapped. A low mutation rate helps to prevent any bit positions from getting stuck to single values, whereas a high mutation rate results in random search.

With the optimal values of population size, number of generations, and crossover rate obtained in the previous experiments (Tables 7.5 through 7.7), the mutation rate was varied and the parameters were evaluated, as shown in Table 7.8. The mutation rate was varied in the range [0.01, 0.2] and GA was run to compute the optimized values of PSNR, NCC, and fitness. From Table 7.8, it can be seen that a mutation rate of 0.02 produced improved results for Peppers, Mandrill, Lena, Barbara, and cameraman and 0.01 for Boat image.

From Tables 7.5–7.8, the optimal values of GA parameters for testing the watermarked image against attacks were determined. The effect of GA parameters on watermarking was tested by varying the number of generations, population size, crossover rate, and mutation rate. Simultaneously, MSE, PSNR, NCC, fitness, and computational time were recorded. During the GA run, the watermark amplification factor was dynamically optimized, and

TABLE 7.8

Effect of Mutation Rate on Images

Images	P_M	MSE (dB)	PSNR (dB)	NCC	Fitness	Time (s)
Peppers	0.01	5.3465	40.85011	0.9879	52.70491	14.95
($N_G = 10$ $N_P = 128$ $P_C = 0.7$)	**0.02**	**5.1289**	**41.03056**	**0.9984**	**53.01136**	14.87
	0.1	5.1782	40.98902	0.9892	52.85942	14.99
	0.15	5.3549	40.84329	0.9823	52.63089	14.76
	0.2	5.8971	40.42442	0.9809	52.19522	15.12
Mandrill	0.01	5.4976	40.72907	0.9854	52.55387	14.87
($N_G = 10$ $N_P = 256$ $P_C = 0.8$)	**0.02**	**5.1287**	**41.03073**	**0.9991**	**53.01993**	14.34
	0.1	5.2267	40.94853	0.9987	52.93293	14.81
	0.15	5.9734	40.36859	0.9789	52.11539	14.92
	0.2	5.8623	40.45012	0.9693	52.08172	14.23
Lena	0.01	3.1274	43.17897	0.9972	55.14537	14.11
($N_G = 10$ $N_P = 64$ $P_C = 0.7$)	**0.02**	**3.0106**	**43.34427**	**0.9991**	**55.33347**	14.94
	0.1	3.2216	43.05009	0.9964	55.00689	14.76
	0.15	3.1415	43.15943	0.9944	55.09223	14.93
	0.2	3.3969	42.81998	0.9895	54.69398	14.11
Barbara	0.01	3.9781	42.13405	0.9873	53.98165	14.19
($N_G = 20$ $N_P = 128$ $P_C = 0.6$)	**0.02**	**3.6742**	**42.47918**	**0.9982**	**54.45758**	14.45
	0.1	3.8862	42.23555	0.9913	54.13115	14.23
	0.15	4.0151	42.09384	0.9876	53.94504	14.82
	0.2	4.1214	41.98036	0.9894	53.85316	14.15
Boat	**0.01**	**4.0124**	**42.09676**	0.9992	**54.08716**	15.43
($N_G = 20$ $N_P = 512$ $P_C = 0.6$)	0.02	4.1214	41.98036	0.9976	53.95156	15.87
	0.1	4.2146	41.88324	0.9961	53.83644	15.12
	0.15	4.3421	41.75381	0.9943	53.68541	15.87
	0.2	4.3989	41.69736	0.9952	53.63976	15.32
Cameraman	0.01	5.2165	40.95701	0.9981	52.93421	16.76
($N_G = 30$ $N_P = 512$ $P_C = 0.7$)	**0.02**	**5.1413**	**41.02007**	**0.9993**	**53.01167**	17.02
	0.1	5.2247	40.95019	0.9976	52.92139	16.22
	0.15	5.2989	40.88895	0.9955	52.83495	16.43
	0.2	5.4012	40.8059	0.9947	52.7423	16.94

N_P = Population size; N_G = Number of generations; P_C = Crossover rate; P_M = Mutation rate.

the optimal value obtained at the stopping condition was determined as 0.12. The parameter values that produced improved MSE and NCC are summarized here:

- Population size (N_P): 128
- Number of generations (N_G): 30
- Crossover rate (P_C): 0.7
- Mutation rate (P_M): 0.02
- Amplification factor (α): 0.12

7.8.2 Particle Swarm Optimization

In PSO, the initial particles are formed by embedding the chosen watermark image bits into the cover image, which comprises the initial population. The particle size in the simulation was chosen as 36. If the particle size is larger, more points can be searched in the search space to determine the global optimal solution, but this increases the number of iterations and hence the computational time. The parameters such as the number of iterations, acceleration constants c_1 and c_2, and the inertia weight w were varied to obtain improved results in terms of MSE, PSNR, NCC, fitness, and computational time. The effect of these parameters on DIWM is explained in this section.

7.8.2.1 Effect of Number of Iterations

With a swarm size of 36, the number of iterative runs (N_I) was varied from 20 to 50 with the interval of 10 to optimize the watermark amplification factor and thus compute the PSNR, MSE, robustness, NCC, and computational time. The acceleration constants c_1 and c_2 were fixed at 1.8, and they were maintained constant during the variation in the number of generations. From Table 7.9, it can be seen that the maximum PSNR and NCC were obtained at the end of 20 iterations for Peppers, Lena, Barbara, Boat, and Cameraman images, while for Mandrill they were obtained at the end of 30 iterations.

7.8.2.2 Effect of Acceleration Constants c_1 and c_2

The cognitive and the social components are important parameters in PSO, and are collectively known as the acceleration constants. The cognitive component represents the personal thinking of each particle, thus encouraging the particles to move toward their own best positions found so far. The social component represents the collaborative effect of the particles, thus pulling the particles toward the global best solution. Hence it is vital to tune these values to obtain the optimal solution. A small value of the acceleration constants result in a slower velocity updating process, thus reducing the convergence speed. On the other hand, for a large value of the acceleration constant, the velocity change is too fast, causing high-speed movement of the particles and thus becoming hard to converge. Hence optimal values of c_1 and c_2 were determined by testing upon different images in the range [1.2, 2] in intervals of 0.2. Table 7.10 shows the effect of the acceleration constants on the chosen images by recording the MSE, PSNR, NCC, fitness value, and the computational time. For an acceleration constant of 1.8, Peppers, Lena, Barbara, Boat, and Cameraman images responded with improved values, while the performance of Mandrill was most efficient with $c_1 = c_2 = 1.6$.

TABLE 7.9

Effect of Number of Iterations on Images

Images	N_I	MSE (dB)	PSNR (dB)	NCC	Fitness	Time (s)
Peppers	**20**	**5.0456**	**41.10168**	**0.9991**	**53.09088**	**12.52**
	30	5.2487	40.93029	0.9985	52.91229	12.78
	40	5.0759	41.07567	0.9985	53.05767	12.77
	50	5.2173	40.95635	0.9899	52.83515	12.65
Mandrill	20	5.3384	40.85669	0.9899	52.73549	12.85
	30	**5.0687**	**41.08184**	**0.9996**	**53.07704**	**12.52**
	40	5.1172	41.04048	0.9993	53.03208	12.44
	50	5.7366	40.54426	0.9855	52.37026	12.81
Lena	**20**	**2.9989**	**43.36118**	**0.9993**	**55.35278**	**12.38**
	30	3.0987	43.21901	0.9998	55.21661	12.69
	40	3.1581	43.13654	0.9985	55.11854	12.95
	50	3.0874	43.23487	0.9981	55.21207	12.74
Barbara	**20**	**3.4578**	**42.7428**	**0.999**	**54.7308**	**12.44**
	30	3.7795	42.35646	0.9986	54.33966	13.05
	40	3.6512	42.50645	0.9954	54.45125	12.85
	50	3.9952	42.11542	0.9886	53.97862	13.09
Boat	**20**	**3.8659**	**42.2583**	**0.9995**	**54.2523**	**13.15**
	30	4.0211	42.08735	0.9982	54.06575	12.77
	40	4.1589	41.94102	0.9977	53.91342	12.49
	50	4.2256	41.87192	0.9958	53.82152	12.53
Cameraman	**20**	**5.0369**	**41.10917**	**0.9992**	**53.09957**	**12.74**
	30	5.1024	41.05306	0.9995	53.04706	12.55
	40	5.1406	41.02067	0.9981	52.99787	12.82
	50	5.1052	41.05068	0.9961	53.00388	12.79

Note: N_I = Number of iterations.

7.8.2.3 Effect of Inertia Weight w

The inertia weight controls the momentum of the particle in the swarm. For large values of the inertia weight, the particles move faster and multiply with the old velocity, thus decreasing the convergence speed. In this experiment, the inertia weight was varied in the range [0.2, 0.8] in intervals of 0.2 to determine the optimal choice of the parameter. Table 7.11 shows the MSE, PSNR, NCC, fitness, and computational time for different values of the inertia weight. It is observed that the optimal results are obtained for an inertia weight of 0.6 for most of the images, including Mandrill, Lena, Boat, and Cameraman. Peppers and Barbara images gave optimal results for an inertia weight of 0.8.

Tables 7.9–7.11 show the effect of the PSO parameters on the test images. The parameter values that resulted in improved outcomes of MSE and NCC were considered the optimal values and used for further tests with attacks. The watermark amplification factor, initially set to a random value in [0,1], was tuned dynamically to an optimal value by PSO. The optimal values of the parameters are given here:

- Number of iterations (N_I): 20
- Acceleration constants ($c_1 = c_2$): 1.8
- Inertia weight (w): 0.6
- Amplification factor (α): 0.096

TABLE 7.10

Impact of Acceleration Constants on Watermarking

Images	$c_1 = c_2$	MSE (dB)	PSNR (dB)	NCC	Fitness	Time (s)
Peppers	1.2	5.1194	41.03861	0.9895	52.91261	11.95
($N_I = 20$)	1.4	5.1056	41.05034	0.9914	52.94714	11.97
	1.6	5.0856	41.06738	0.9945	53.00138	12.06
	1.8	**5.0145**	**41.12853**	**0.9993**	**53.12013**	**11.86**
	2.0	5.0456	41.10168	0.9991	53.09088	12.08
Mandrill	1.2	5.2113	40.96134	0.9934	52.88214	12.77
($N_I = 30$)	1.4	5.1987	40.97186	0.9952	52.91426	12.45
	1.6	**5.0687**	**41.08184**	**0.9996**	**53.07704**	**12.65**
	1.8	5.0985	41.05638	0.9996	53.05158	12.75
	2.0	5.1145	41.04277	0.9984	53.02357	11.82
Lena	1.2	3.0974	43.22083	0.9953	55.16443	12.41
($N_I = 20$)	1.4	2.9921	43.37104	0.9972	55.33744	12.44
	1.6	2.9523	43.4292	0.9983	55.4088	12.42
	1.8	**2.8954**	**43.51372**	**0.9994**	**55.50652**	**12.38**
	2.0	2.9989	43.36118	0.9993	55.35278	12.54
Barbara	1.2	3.5124	42.67476	0.9953	54.61836	12.44
($N_I = 20$)	1.4	3.4985	42.69198	0.9969	54.65478	12.53
	1.6	3.4428	42.76169	0.9975	54.73169	11.96
	1.8	**3.3742**	**42.8491**	**0.9991**	**54.8383**	**12.19**
	2.0	3.4578	42.7428	0.999	54.7308	12.16
Boat	1.2	3.8869	42.23477	0.9973	54.20237	11.87
($N_I = 20$)	1.4	3.8242	42.3054	0.9982	54.2838	11.92
	1.6	3.7972	42.33617	0.9986	54.31937	12.05
	1.8	**3.7584**	**42.38077**	**0.999**	**54.36877**	**11.9**
	2.0	3.8659	42.2583	0.9995	54.2523	12.04
Cameraman	1.2	5.1068	41.04932	0.9951	52.99052	12.16
($N_I = 20$)	1.4	5.0842	41.06858	0.9968	53.03018	12.47
	1.6	5.0993	41.0557	0.9986	53.0389	12.59
	1.8	**5.0247**	**41.1197**	**0.9994**	**53.1125**	**12.74**
	2.0	5.0369	41.10917	0.9992	53.09957	12.54

Note: N_I = Number of iterations; $c_1 = c_2$ = Acceleration constants.

7.8.3 Hybrid Particle Swarm Optimization

In HPSO, the parameters are initialized to the optimal values obtained by the individual runs of GA and PSO algorithms. The size of the initial population was set to 128, and the HPSO algorithm terminated at the end of 30 iterations. The acceleration constants c_1 and c_2 were set to 1.8. The inertia weight w determines the search behavior of the algorithm. Large values for w facilitate searching new locations, whereas small values provide a finer search in the current area. A balance can be established between global and local exploration by choosing the optimal inertia weight. In this experiment, the value of inertia weight was set to 0.6. The watermark amplification factor was set to a random number in [0,1]. The effectiveness of the proposed technique was validated based on the HF. Setting HF = 1 implies that the algorithm is a wholesome GA approach, while HF = 0

TABLE 7.11

Impact of Inertia Weight

Images	w	MSE (dB)	PSNR (dB)	NCC	Fitness	Time (s)
Peppers	0.2	5.1247	41.03412	0.9789	52.78092	12.05
($N_I = 20 > c_1 = c_2 = 1.8$)	0.4	5.1173	41.04039	0.9812	52.81479	12.11
	0.6	5.0983	41.05655	0.9993	53.04815	12.19
	0.8	**5.0145**	**41.12853**	**0.9985**	**53.07813**	**12.22**
Mandrill	0.2	5.2147	40.95851	0.9952	52.90091	11.98
($N_I = 30 \; c_1 = c_2 = 2.0$)	0.4	5.1242	41.03454	0.9967	52.99494	11.85
	0.6	**5.0985**	**41.05638**	**0.9996**	**53.05158**	**12.09**
	0.8	5.1142	41.04303	0.9982	53.02143	11.84
Lena	0.2	2.9975	43.36321	0.9964	55.32001	12.19
($N_I = 20 \; c_1 = c_2 = 1.8$)	0.4	2.9785	43.39083	0.9987	55.37523	12.27
	0.6	**2.9523**	**43.4292**	**0.9994**	**55.422**	**12.06**
	0.8	2.9648	43.41085	0.9989	55.39765	11.91
Barbara	0.2	3.6241	42.5388	0.9964	54.4956	12.19
($N_I = 20 \; c_1 = c_2 = 1.8$)	0.4	3.5234	42.66118	0.9984	54.64198	12.13
	0.6	3.5043	42.68479	0.999	54.67279	11.91
	0.8	**3.4428**	**42.76169**	**0.9992**	**54.75209**	**12.01**
Boat	0.2	3.8745	42.24865	0.9971	54.21385	11.99
($N_I = 20 \; c_1 = c_2 = 1.8$)	0.4	3.7735	42.36336	0.9986	54.34656	11.92
	0.6	**3.7584**	**42.38077**	**0.999**	**54.36877**	**12.06**
	0.8	3.7814	42.35428	0.9988	54.33988	11.89
Cameraman	0.2	5.1157	41.04175	0.9959	52.99255	12.19
($N_I = 20 \; c_1 = c_2 = 1.8$)	0.4	5.1049	41.05093	0.9973	53.01853	11.87
	0.6	**5.0369**	**41.10917**	**0.9992**	**53.09957**	**12.11**
	0.8	5.0495	41.09832	0.9983	53.07792	12.16

Note: N_I = Number of iterations; $c_1 = c_2$ = acceleration constants; w = inertia weight.

results in a wholesome PSO approach. In order to maintain a balance between individuals of PSO and GA in HPSO, HF was set to 0.7. The optimal values of parameter setting are shown here:

- Population size: 128
- Cognitive factor c_1: 1.8
- Social coefficient c_2: 1.8
- Inertia weight w: 0.6
- Number of generations: 30
- Crossover rate: 0.7
- Mutation rate: 0.02

The MSE, PSNR, NCC, fitness, and computational time were computed for the chosen set of watermarked images, as shown in Table 7.12. The watermarking system must embed the watermark in the image such that the visual quality of the image is not perceptibly distorted. Thus, to study the embedding effect, the PSNR was computed. It is observed that the PSNR values are 41.15, 41.13, 43.64, 43.19, and 41.21 dB for the images Peppers, Mandrill, Lena, Barbara, Boat, and Cameraman, respectively. It is also seen that the PSNR

TABLE 7.12

Performance Metrics of Gray-Scale Images Using HPSO

Image	MSE (dB)	PSNR (dB)	NCC	Fitness	Time (s)
Peppers	4.985	41.15415	0.9989	53.14095	11.12
Mandrill	5.013	41.12983	0.9997	53.12623	10.95
Lena	2.812	43.64065	0.9996	55.63585	10.96
Barbara	3.116	43.19483	0.9995	55.18883	11.28
Boat	3.499	42.69136	0.9991	54.68056	11.31
Cameraman	4.924	41.20762	0.9998	53.20522	11.09

of the HPSO watermarking technique for all the images is reasonably high and the artifacts introduced by watermark embedding are almost invisible. However, if an attacker ignores all the refinement bits at this threshold, the obtained PSNR will be less than 30 dB, which is not satisfactorily enough in most cases. The watermark amplification factor using HPSO was found to be 0.082.

7.8.4 Robustness against Watermarking Attacks

The aim of attacking a watermarking system is usually to prevent the watermark embedder (usually the owner or copyright holder) from using the watermark to support his claims. This can be accomplished in two ways: either by rendering the watermark unreadable or by successfully disputing the claim based on the watermark detection result. In general, image processing attacks have to fulfill two rather conflicting requirements: the image quality must not suffer, and the attack must make it impossible for the watermark embedder to successfully detect the mark. To evaluate the performance of the optimization techniques, experiments are conducted by applying several attacks. The common attacks employed to the watermarked image in this work were filtering, addition of Gaussian noise, rotation, scaling, and JPEG compression. The NCC values were calculated between the original watermark and the extracted watermarks (robustness measure) according to Equation 7.27. The GA, PSO, and HPSO procedures were repeated by applying these attacks, and the robustness measure was evaluated.

7.8.4.1 Filtering Attacks

The watermarked image under consideration is subject to several types of filtering attacks such as average filtering, Gaussian filtering, median filtering, and Wiener filtering. The mask for the filter is usually a window that can take various sizes. Average filtering removes the high-frequency components present in the image acting like a low-pass filter. The average filter with a 5×5 mask was applied to the watermark image during the optimization process of GA, PSO, and HPSO to evaluate the robustness measure. While using Gaussian filter attack, the mean was set to 0 and the variance to 1, with a window size 3×3. Median filtering was applied on the watermarked image with a mask size of 2×2 and 3×3, and this preserved the edges while recovering the watermark. Table 7.13 shows the NCC values for different types of filtering attacks. The correlation factor was evaluated based on the similarity between the original watermark and the attacked watermark for all the filtering techniques. The NCC values of Barbara image against Gaussian filtering attack obtained using GA, PSO and HPSO are 0.9816, 0.9817, and 0.99 respectively. An NCC value of 1 indicates that the image is highly robust. In the experiments conducted, it was

TABLE 7.13

Computed Robustness for Filtering Attack

Optimization Technique	Average Filtering	Gaussian Filtering	Median Filtering 2 × 2	Median Filtering 3 × 3
Peppers				
GA	0.974	0.9826	0.9896	0.9894
PSO	0.9834	0.9873	0.9902	0.9897
HPSO	**0.9876**	**0.9944**	**0.9988**	**0.9951**
Mandrill				
GA	0.9723	0.9891	0.9934	0.9926
PSO	0.9787	0.9934	0.9967	0.9945
HPSO	**0.9817**	**0.9947**	**0.998**	**0.999**
Lena				
GA	0.9806	0.9875	0.9927	0.9918
PSO	0.9852	0.9949	0.9974	0.9969
HPSO	**0.9948**	**1**	**0.9977**	**0.9985**
Barbara				
GA	0.9745	0.9816	0.9889	0.9884
PSO	0.9796	0.9817	0.9935	0.9892
HPSO	**0.9808**	**0.99**	**0.9979**	**0.9908**
Boat				
GA	0.9822	0.9796	0.9963	0.9956
PSO	0.9852	0.9879	0.9981	0.9979
HPSO	**0.9898**	**0.9962**	**1**	**0.9992**
Cameraman				
GA	0.9718	0.9879	0.9911	0.9906
PSO	0.9804	0.9915	0.9921	0.9987
HPSO	**0.9875**	**0.9971**	**1**	**0.999**

observed that HPSO resulted in NCC = 1 for image Lena against Gaussian filtering attack and for images Boat and Cameraman against median filtering attack with window size of 2 × 2. Thus, for all the test images, the results show that the proposed HPSO optimization technique has better similarity of extracted watermarks among the compared approaches such as GA and PSO.

7.8.4.2 Additive Noise

A Gaussian noise is added to the watermarked image with zero mean and different variance σ, indicating the percentage of gray levels added into the image. The robustness measure is computed by varying σ in [0.001, 0.5] for GA, PSO, and HPSO algorithms, as indicated in Table 7.14. For instance, in case of Mandrill image against Gaussian noise attack with variance of 0.5, it is observed that the NCC value is 0.8719 using GA, 0.8821 using PSO, and 0.8906 using HPSO. For a Gaussian noise attack with variance $\sigma = 0.01$, the correlation of Lena image is 1.195% higher than GA and 0.66% higher than PSO. Likewise, for all the images against Gaussian noise attack, the correlation values clearly indicate improvement in robustness of HPSO when compared with GA and PSO algorithms.

TABLE 7.14

Evaluated Results for Gaussian Noise Attack

Optimization Technique	Gaussian Noise $\sigma = 0.001$	Gaussian Noise $\sigma = 0.01$	Gaussian Noise $\sigma = 0.1$	Gaussian Noise $\sigma = 0.5$
Peppers				
GA	0.8753	0.8742	0.8617	0.8582
PSO	0.8798	0.8776	0.8625	0.8608
HPSO	**0.8837**	**0.8864**	**0.8655**	**0.8627**
Mandrill				
GA	0.8896	0.8886	0.8746	0.8719
PSO	0.8956	0.8934	0.8856	0.8821
HPSO	**0.8961**	**0.8991**	**0.8936**	**0.8906**
Lena				
GA	0.8934	0.8923	0.8856	0.8824
PSO	0.9031	0.8971	0.8903	0.8882
HPSO	**0.9071**	**0.9031**	**0.9002**	**0.8941**
Barbara				
GA	0.8799	0.8789	0.8701	0.8529
PSO	0.8869	0.8842	0.8786	0.8719
HPSO	**0.8878**	**0.8848**	**0.881**	**0.8798**
Boat				
GA	0.8902	0.8891	0.8806	0.8773
PSO	0.8933	0.8897	0.8889	0.8854
HPSO	**0.903**	**0.8903**	**0.8902**	**0.8884**
Cameraman				
GA	0.8967	0.8959	0.8895	0.8847
PSO	0.9066	0.9027	0.8918	0.8886
HPSO	**0.907**	**0.9029**	**0.8958**	**0.8911**

7.8.4.3 JPEG Compression

JPEG is one of the most widely used lossy compression algorithms, and any watermarking technique should be resilient to some degree of JPEG compression attacks. In general, such lossy compression algorithms discard the redundant and perceptual insignificant information during the coding process, but watermark embedding schemes add invisible information to the image. JPEG compression calculates the visual components based on the relationship with the neighboring pixels in the image. The specific positions to embed the watermark are derived from these visual components which are proportional to the quality levels of the JPEG compression. In practice, it is difficult to choose the minimal JPEG quality factor (QF) for compression. Low values of QF indicate high compression ratio, and vice versa. QF was varied between 20% and 95% for simulation, and the results are shown in Table 7.15. The watermark was detected well even after the image was compressed using a QF of 20%. This is evident from the resultant values of NCC > 0.75 in case of all images using GA, PSO, and HPSO. The correlation factor seems to be high for QF = 95%, indicating that the similarity values prove the closest match between the original watermark and the extracted watermark. These results show that the proposed HPSO technique is better than GA and PSO in terms of the robustness measure against JPEG compression attacks.

TABLE 7.15

JPEG Compression Attack and Robustness Computation

Optimization Technique	JPEG QF = 20%	JPEG QF = 40%	JPEG QF = 70%	JPEG QF = 95%
Peppers				
GA	0.8467	0.8573	0.8856	0.927
PSO	0.8498	0.8584	0.8892	0.9299
HPSO	**0.8565**	**0.8611**	**0.8915**	**0.9372**
Mandrill				
GA	0.8534	0.8587	0.8916	0.9378
PSO	0.8675	0.8634	0.8999	0.9458
HPSO	**0.8759**	**0.8641**	**0.9001**	**0.9521**
Lena				
GA	0.8662	0.8679	0.8835	0.9543
PSO	0.8694	0.8748	0.8855	0.9594
HPSO	**0.8729**	**0.8844**	**0.8953**	**0.9684**
Barbara				
GA	0.8589	0.8645	0.8897	0.9129
PSO	0.8603	0.8681	0.8897	0.9216
HPSO	**0.8632**	**0.8735**	**0.8933**	**0.9248**
Boat				
GA	0.8622	0.8744	0.8959	0.9334
PSO	0.871	0.8817	0.9037	0.9406
HPSO	**0.8724**	**0.8842**	**0.912**	**0.9483**
Cameraman				
GA	0.8563	0.8656	0.8933	0.9452
PSO	0.866	0.8701	0.8977	0.9458
HPSO	**0.8745**	**0.88**	**0.9055**	**0.9513**

7.8.4.4 Rotation

Geometric attacks usually make the watermark detector loose the synchronization information, and one of the major attacks among this group is rotation. The image was rotated by an angle in the counterclockwise direction before extracting the watermark. The rotation angles were 5°, 15°, 30°, and 40° to the right. Then they were rotated back to their original position using bilinear interpolation. While rotating, the black pixels left after rotation in the corners were included to maintain the image size and shape. For larger angles, more black pixels are padded and hence the correlation factor or the robustness measure tends to decrease. For each degree of rotation, the correlation factor was measured between the original watermark and the attacked watermark to determine the degree of similarity. Table 7.16 shows the effect of rotation attack on the robustness measure while applying GA, PSO, and HPSO algorithms. Rotation attack with 5°, 15°, and 30° resulted in NCC > 0.75 for all the images using GA, PSO, and HPSO, whereas for the rotation attack of 40°, it was observed that the robustness of intelligent watermarking algorithms decreased in case of Peppers, Barbara, and Cameraman images since NCC was <0.75.

TABLE 7.16

Robustness Results for Rotation Attack

Optimization Technique	Rotation 5°	Rotation 15°	Rotation 30°	Rotation 40°
Peppers				
GA	0.8934	0.8659	0.7943	0.7157
PSO	0.9021	0.8692	0.8212	0.7287
HPSO	**0.9042**	**0.8762**	**0.824**	**0.7306**
Mandrill				
GA	0.912	0.8854	0.8268	0.7839
PSO	0.9674	0.8923	0.8475	0.7967
HPSO	**0.9696**	**0.8926**	**0.8565**	**0.8055**
Lena				
GA	0.9025	0.8786	0.8215	0.7546
PSO	0.9078	0.8795	0.8288	0.761
HPSO	**0.9083**	**0.886**	**0.8356**	**0.7633**
Barbara				
GA	0.8993	0.8911	0.8137	0.7298
PSO	0.9002	0.8984	0.8138	0.7328
HPSO	**0.9065**	**0.9**	**0.8166**	**0.741**
Boat				
GA	0.8978	0.8897	0.8187	0.7745
PSO	0.9059	0.8929	0.8231	0.7842
HPSO	**0.9153**	**0.8941**	**0.8258**	**0.7866**
Cameraman				
GA	0.8933	0.8887	0.8115	0.7489
PSO	0.8943	0.8906	0.8146	0.7499
HPSO	**0.8967**	**0.9005**	**0.8177**	**0.7507**

7.8.4.5 Scaling

Scaling is generally considered more challenging than other attacks because changing the image size or its orientation even by slight amount can dramatically reduce the receiver's ability to retrieve the watermark. The scaling factors are selected such that the robustness, invisibility, and quality of the extracted watermark are usually maintained high in the low-frequency band and low in the high-frequency band. In this experimental analysis, the watermarked image was scaled by using different scale factors within the range [0.5, 2], and the NCC values were computed using GA, PSO, and HPSO, as shown in Table 7.17. For a scale factor of 1, maximum values of NCC was obtained for all the images using the HPSO algorithm. From Table 7.17, it is seen that the NCC values for GA, PSO, and HPSO are very close to 1. Thus, it can be concluded that the watermark image is extracted with better robustness using HPSO compared to GA and PSO algorithms.

7.8.4.6 Combination of Attacks

The performance of the proposed algorithms was also tested by combining attacks such as rotation, scaling, JPEG compression, and cropping. Applying a combination of attacks minimizes the visual difference, thus making it impossible for the watermark embedder

TABLE 7.17

Experimental Results for Scaling Attack

Optimization Technique	Scale Factor 0.5	Scale Factor 0.75	Scale Factor 1	Scale Factor 1.5	Scale Factor 2
Peppers					
GA	0.9921	0.9953	0.9984	0.9957	0.9911
PSO	0.9935	0.9969	0.9985	0.9962	0.9940
HPSO	**0.9943**	**0.9971**	**0.9989**	**0.9969**	**0.9954**
Mandrill					
GA	0.9925	0.9971	0.9991	0.9973	0.9917
PSO	0.9976	0.9983	0.9996	0.9979	0.9972
HPSO	**0.9969**	**0.9987**	**0.9997**	**0.9980**	**0.9983**
Lena					
GA	0.9936	0.9979	0.9991	0.9967	0.9949
PSO	0.9951	0.9981	0.9994	0.9973	0.9970
HPSO	**0.9973**	**0.9984**	**0.9996**	**0.9981**	**0.9982**
Barbara					
GA	0.9879	0.9965	0.9982	0.9961	0.9934
PSO	0.9888	0.9973	0.9992	0.9975	0.9935
HPSO	**0.9968**	**0.9984**	**0.9995**	**0.9983**	**0.9953**
Boat					
GA	0.9923	0.9957	0.9992	0.9954	0.9949
PSO	0.9940	0.9971	0.999	0.9966	0.9961
HPSO	**0.9970**	**0.9983**	**0.9991**	**0.9979**	**0.9971**
Cameraman					
GA	0.9913	0.9971	0.9993	0.9979	0.9921
PSO	0.9958	0.9987	0.9992	0.9984	0.9947
HPSO	**0.9965**	**0.9988**	**0.9998**	**0.9989**	**0.9968**

to successfully detect the mark. In this application, the attacks were combined as rotation plus scaling, rotation plus scaling plus cropping, and rotation plus scaling plus JPEG compression, and the NCC values were obtained as shown in Table 7.18. Cropping is a lossy operation, which is used with block sizes of 10 and 100 to attack the watermarked image along with geometric attacks such as rotation and scaling. For large block sizes of cropping, for example 100, NCC is found to be very low. However, the NCC values in all the other combinations show that the watermark is capable of surviving during extraction using the proposed HPSO approach.

7.8.5 PSNR Computation

When considering a watermarking scheme, different requirements need to be taken into account. One of the important requirements is the perceptual transparency of the superimposed watermark on the host data. Perceptual transparency implies that the alterations caused by the watermark embedded into the data should not degrade the perceptual quality of the latter. PSNR is one of the most used metrics employed with respect to the host image to test the perceptual transparency of the watermarking algorithm.

TABLE 7.18

Robustness Measure for Images against Combination of Attacks

Technique	Peppers	Mandrill	Lena	Barbara	Boat	Cameraman
Attack: Rotation 5° + Scaling 0.5						
GA	0.7955	0.821	0.8105	0.7986	0.8182	0.8056
PSO	0.7987	0.8493	0.8169	0.8081	0.8243	0.8072
HPSO	**0.803**	**0.8527**	**0.8197**	**0.8094**	**0.8291**	**0.8089**
Attack: Rotation 5° + Scaling 1.1						
GA	0.8836	0.9098	0.8976	0.8895	0.8874	0.8901
PSO	0.8889	0.9197	0.9002	0.8983	0.8933	0.8921
HPSO	**0.8965**	**0.9218**	**0.9072**	**0.9007**	**0.9008**	**0.9012**
Attack: Rotation 5° + Scaling 1.1 + Cropping (Block size = 10)						
GA	0.8765	0.8967	0.8806	0.8745	0.8769	0.884
PSO	0.8773	0.9016	0.8879	0.8772	0.8801	0.8902
HPSO	**0.8826**	**0.9088**	**0.8941**	**0.8863**	**0.8845**	**0.8977**
Attack: Rotation 5° + Scaling 1.1 + Cropping (Block size = 100)						
GA	0.6543	0.7145	0.6982	0.658	0.6659	0.7108
PSO	0.6567	0.7467	0.6989	0.6634	0.6681	0.712
HPSO	**0.6612**	**0.6659**	**0.7013**	**0.6697**	**0.6767**	**0.7151**
Attack: Rotation 5° + Scaling 1.1 + JPEG 95%						
GA	0.8821	0.9085	0.8945	0.8833	0.8859	0.8896
PSO	0.8839	0.9298	0.9032	0.8863	0.8865	0.8986
HPSO	**0.8915**	**0.8868**	**0.9115**	**0.895**	**0.8888**	**0.9026**
Attack: Rotation 5° + Scaling 1.1 + JPEG 60%						
GA	0.8701	0.8971	0.8821	0.874	0.8724	0.8698
PSO	0.8749	0.9121	0.8872	0.8799	0.8772	0.8746
HPSO	**0.881**	**0.875**	**0.8939**	**0.8838**	**0.8796**	**0.8811**

A watermark embedding process is imperceptible if humans cannot differentiate between the original and watermarked images. However, modifications introduced by attacks are noticeable only when the original data is compared with the watermarked data. But this is not possible since the users of the watermarked data are unable to access the original data. Thus blind tests are conducted without the embedded information to assess the perceptual transparency of data embedding procedures. Such tests are performed by computing the PSNR values. For transparency evaluation, PSNR is the most common metric used frequently in watermarking applications (Nguyen et al. 2010).

Filtering attack is one of the most common manipulations in DIWM. The images were tested by applying an average filtering with window size 5 × 5, a Gaussian filtering with window size 3 × 3, and median filtering with window sizes 2 × 2 and 3 × 3. The PSNR values using GA, PSO, and HPSO were calculated for all the images, as shown in Table 7.19. For an average filtering attack of window size 5 × 5, the PSNR value is 36.21 dB using GA, 36.88 dB using PSO, and 37.24 dB using HPSO for the Boat image. It is observed that the PSNR is higher in case of HPSO than with GA and PSO. Likewise, for almost all the images, the PSNR values of HPSO are improved while applying filtering attacks. PSNR was also computed with the application of Gaussian noise with different σ values. High values of PSNR (30–50 dB) resulted in high-quality images, thus minimizing the MSE. The PSNR values obtained against Gaussian noise attack was improved over those against

TABLE 7.19

PSNR Values for Images against Noise and Filtering Attacks

Technique	Peppers	Mandrill	Lena	Barbara	Boat	Cameraman
Attack: Average Filtering 5 × 5						
GA	35.07	34.98	36.92	34.76	36.21	36.13
PSO	35.33	35.81	37.91	35.16	36.88	36.94
HPSO	**35.79**	**36.24**	**38.43**	**36.08**	**37.24**	**37.14**
Attack: Gaussian Filtering 3 × 3						
GA	35.15	35.26	37.24	35.17	36.47	36.77
PSO	35.53	35.27	37.55	36.12	37.43	37.26
HPSO	**36.35**	**35.57**	**38.12**	**36.67**	**37.56**	**37.9**
Attack: Median Filtering 2 × 2						
GA	36.24	35.74	37.81	35.51	36.81	36.99
PSO	36.6	36.61	38.73	35.78	37.66	37.53
HPSO	**37.35**	**36.73**	**39.34**	**36.36**	**38.03**	**37.86**
Attack: Median Filtering 3 × 3						
GA	36.17	35.67	37.76	35.46	36.72	36.92
PSO	36.2	36.27	38.15	35.48	37.67	37.04
HPSO	**36.5**	**37.23**	**38.93**	**35.77**	**37.83**	**37.05**
Attack: Gaussian Noise σ = 0.001						
GA	42.56	43.17	44.47	43.77	43.89	41.78
PSO	42.79	44.08	45.19	44.62	44.66	41.9
HPSO	**43.18**	**44.61**	**45.65**	**45.24**	**44.85**	**42.54**
Attack: Gaussian Noise σ = 0.01						
GA	42.25	42.96	44.36	44.64	43.71	41.56
PSO	42.33	43.1	44.61	44.87	44.07	42.06
HPSO	**42.94**	**43.61**	**44.64**	**45.49**	**45.04**	**42.07**
Attack: Gaussian Noise σ = 0.1						
GA	42.07	42.83	44.24	44.15	43.48	41.32
PSO	43	43.16	44.43	44.84	44.41	41.61
HPSO	**43.3**	**43.29**	**44.59**	**44.9**	**45.38**	**41.89**

filtering attacks. Thus the experiments proved that the proposed intelligent techniques result in perceptually transparent watermarked images.

Images with excessive compression rate are considered unauthentic due to poor quality. QF of JPEG compression is inversely proportional to the compression rate. Table 7.20 summarizes the PSNR values of the extracted watermark against JPEG compression attacks with QF = 20%, 40%, 70%, and 95%. For QF = 70%, the PSNR computed using HPSO for the Boat image is increased by 2.3% over GA and 2.09% over PSO. Likewise, in case of the Peppers image, the PSNR of HPSO (40.55 dB) is improved over those of GA (39.97 dB) and PSO (40.12 dB) for a JPEG attack with QF = 20%. Thus the PSNR values for all images show that the intelligent algorithms perform well in improving the perceptual transparency of the watermark against JPEG compression attacks.

Rotation attacks were applied on the watermarked images to compute the PSNR values between the watermarked images and the attacked watermark images. Here, four levels of rotation of 5°, 15°, 30°, and 40° were applied to the watermarked image. The computed PSNR values shown in Table 7.21 indicate that they are higher for small degrees of rotation,

TABLE 7.20

PSNR Values for Images against JPEG Compression

Technique	Peppers	Mandrill	Lena	Barbara	Boat	Cameraman
Attack: JPEG QF = 20%						
GA	39.97	39.04	41.75	41.02	41.23	39.44
PSO	40.12	39.64	41.8	41.52	41.94	39.6
HPSO	**40.55**	**40.35**	**42.71**	**42.11**	**42.72**	**40.28**
Attack: JPEG QF = 40%						
GA	40.01	39.17	41.89	41.16	41.36	39.56
PSO	40.87	39.28	42.58	42.02	41.64	40.46
HPSO	**40.92**	**40.2**	**42.67**	**42.52**	**42.31**	**41.32**
Attack: JPEG QF = 70%						
GA	40.15	39.34	42.09	41.33	41.65	39.78
PSO	40.36	40.23	42.52	42.22	41.76	40.69
HPSO	**40.81**	**41.06**	**43.02**	**42.48**	**42.65**	**41.09**
Attack: JPEG QF = 95%						
GA	40.3	39.67	42.35	41.62	41.88	39.97
PSO	40.54	40.27	43.14	41.93	41.94	40.71
HPSO	**41.33**	**40.44**	**44.06**	**42.06**	**42.14**	**41.53**

TABLE 7.21

PSNR Values for Images against Rotation Attacks

Technique	Peppers	Mandrill	Lena	Barbara	Boat	Cameraman
Attack: Rotation 5°						
GA	31.54	32.25	33.35	31.65	32.91	31.33
PSO	32.02	33.2	34.3	32.43	33.16	32.04
HPSO	**32.5**	**33.24**	**34.63**	**32.72**	**33.65**	**32.24**
Attack: Rotation 15°						
GA	31.02	31.82	32.94	31.12	32.75	31.16
PSO	31.78	32.13	33.04	31.51	33.06	31.41
HPSO	**31.98**	**32.34**	**34.03**	**32.25**	**33.71**	**31.87**
Attack: Rotation 30°						
GA	30.89	31.34	32.26	30.88	32.6	30.89
PSO	31.78	31.45	32.53	31.03	33.21	31.25
HPSO	**32.64**	**31.78**	**33.1**	**31.94**	**33.38**	**31.65**
Attack: Rotation 40°						
GA	30.62	31.02	32.05	30.75	32.23	30.77
PSO	31.05	31.62	32.26	30.84	32.93	31.25
HPSO	**31.5**	**32.05**	**32.34**	**31.73**	**33.24**	**31.95**

and vice versa. For Lena image, the PSNR value computed by HPSO (34.63 dB) against 5° rotation was higher than that of HPSO (32.34 dB) against 40° rotation. Likewise, for all the other images, it was observed that the PSNR decreased as the rotation angle increased. While comparing the performance of GA, PSO, and HPSO, it was found that the PSNR values of HPSO had improved in all cases of images.

TABLE 7.22

PSNR Values against Combination of Attacks

Technique	Peppers	Mandrill	Lena	Barbara	Boat	Cameraman
Attack: Rotation 5° + Scaling 0.5						
GA	31.06	32.01	33.09	31.21	32.72	31.1
PSO	31.4	32.89	33.24	32.03	33.46	31.47
HPSO	**31.93**	**33.31**	**33.51**	**32.23**	**33.52**	**32.02**
Attack: Rotation 5° + Scaling 1.1						
GA	31.23	32.18	33.17	31.43	32.82	31.21
PSO	31.8	32.87	34.14	31.75	33.33	31.34
HPSO	**32.74**	**33.44**	**34.91**	**32.05**	**33.71**	**32.06**
Attack: Rotation 5° + Scaling 1.1 + Cropping (Block size = 10)						
GA	30.85	31.56	32.13	30.32	31.29	30.23
PSO	31.01	32.1	32.18	31.31	31.4	30.97
HPSO	**31.21**	**32.45**	**32.63**	**32.26**	**32.08**	**31.29**
Attack: Rotation 5° + Scaling 1.1 + Cropping (Block size = 100)						
GA	26.14	27.98	28.55	27.76	26.12	27.52
PSO	27.1	28.04	29.36	27.93	26.58	28.32
HPSO	**27.92**	**28.81**	**29.84**	**28.56**	**27.3**	**28.72**
Attack: Rotation 5° + Scaling 1.1 + JPEG 95%						
GA	31.02	32.01	33.11	31.32	32.75	31.16
PSO	31.51	32.38	33.69	31.46	33.4	32.14
HPSO	**32.09**	**33.1**	**34.17**	**31.59**	**34.17**	**32.84**
Attack: Rotation 5° + Scaling 1.1 + JPEG 60%						
GA	30.89	31.94	33.03	31.29	32.71	31.02
PSO	31.37	32.11	33.84	31.59	32.73	31.35
HPSO	**32.09**	**32.29**	**34.34**	**32.15**	**33.6**	**31.83**

Geometric attacks were combined among themselves and with JPEG compression to obtain the perceptual transparency in terms of PSNR. Combination of such attacks allows the user to obtain better quality results in terms of PSNR. In Table 7.22, the PSNR value of the test images is higher for 5° rotation plus 1.1 scaling attack when compared with 5° rotation plus 0.5 scaling attacks. Similarly, while including cropping, PSNR (<30 dB) is very low for large block sizes, thus failing to achieve perceptual transparency. The rotation plus scaling plus JPEG compression attack proves that the watermarked image is of fine quality as seen by the acceptable range of PSNR values by using the intelligent techniques.

7.8.6 DIWM with Real-Time Color Image

In practical scenarios, it is important to analyze the effect of bio-inspired techniques on real-time camera-captured images. In this section, the Flowers.jpg (Figure 7.12) image taken using a Sony DSC-HX300 camera with an original size of 5184 × 3888 (resized to 512 × 512) was chosen as the cover image and a fingerprint.jpg (Figure 7.13) image of size 256 × 256 as the watermark image. The experiments such as image segmentation, feature extraction, watermark embedding, and extraction using GA, PSO, and HPSO were implemented with an Intel i3 CPU, 2.53 GHz, 4 GB RAM PC using MATLAB R2008 software.

FIGURE 7.12
Real-time color image.

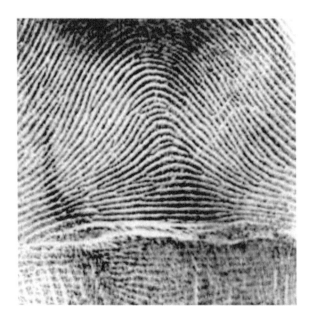

FIGURE 7.13
Digital watermark.

Name of image	Original image	Segmented images		
		$K = 5$	$K = 10$	$K = 15$
Flowers.jpg				

FIGURE 7.14
Segmentation of real-time image using EM algorithm.

7.8.6.1 Image Segmentation Using Expectation Maximization

In order to segment the real-time image automatically, the joint distribution of color, texture, and position features were modeled with a mixture of Gaussians. The EM algorithm estimated the parameters of the Gaussian model, and based on the resulting pixel cluster memberships, the image was segmented. The parameters were set according to Table 7.1. The number of Gaussian components k was varied between 5 and 15, based on which the number of Gaussians occurred with uniform probability. The log likelihood difference between two successive iterations was set to 0.1. The algorithm terminated at the end of 500 iterations. The original image and the segmented images are presented in Figure 7.14. For large values of k, several clusters occur at discrete locations, and hence the segmentation gets coarse.

7.8.6.2 Feature Extraction Using Difference of Gaussian

The DoG algorithm was applied to select the feature points from the segmented image. The number of Gaussian components was chosen as 10 based on the results from the segmentation. The difference between one blurred version of the segmented image and another less blurred version of the segmented image was computed, and the results are presented in Figure 7.15. The blurring radii for the blurred version of segmented images were set to random values in the range [0,1] such that both are not equal.

7.8.6.3 DWT Watermark Embedding and Extraction

The real-time host image was converted into YCbCr channels; the Y channel was then decomposed into three-level wavelet coefficients. The coarsest subband HL_3 was chosen as the target subband for embedding the watermark image in the DWT domain.

Name of image	Blurred version of the segmented image	Less blurred version of the segmented image	Difference of Gaussian filtered image
Flowers			

FIGURE 7.15
Performance of DoG on real-time image.

FIGURE 7.16
DWT watermarked image.

The watermark embedded image for an amplification factor of 0.4 with a PSNR of 21.0198 dB is shown in Figure 7.16. Optimization techniques such as GA, PSO, and HPSO were applied in order to optimize the amplification factor.

7.8.6.4 Genetic Algorithm

Several experiments were conducted by varying the GA parameters, such as number of generations (N_G), crossover probability (P_C), and mutation probability (P_M). The values of PSNR, MSE, NCC, fitness, and computational time were evaluated for the chosen watermarked image by varying the number of generations, crossover rate, and mutation rate, as shown in Table 7.23. For a population size of 256, the number of generations was varied between 10 and 40. The MSE was found to be a minimum (0.6244) at $N_G = 30$, with a PSNR of 50.17617 dB. The effect of different crossover rates of 0.6, 0.7, 0.8, and 0.9 was analyzed with $N_p = 256$ and $N_G = 30$. In this case, MSE and PSNR were improved for $P_C = 0.7$. Likewise, with $N_p = 256$, $N_G = 30$, and $P_C = 0.7$, the MSE, PSNR, NCC, fitness, and time were recorded for different mutation probabilities, and based on the MSE and PSNR values, P_M was set to 0.02.

During the GA run, the watermark amplification factor was dynamically optimized, and the optimal value obtained at stopping condition was found to be 0.012. The optimal choice of GA parameter values were as follows:

- Population size (N_P): 256
- Number of generations (N_G): 30
- Crossover rate (P_C): 0.7
- Mutation rate (P_M): 0.02
- Amplification factor (α): 0.012.

TABLE 7.23

Effect of GA Parameters on Real-Time Image

Parameters	N_G	MSE (dB)	PSNR (dB)	NCC	Fitness	Time (s)
No. of generations (N_G)						
$N_P = 256$	10	0.6379	50.08328	0.9985	62.06528	19.88
	20	0.6921	49.72912	0.9984	61.70992	20.86
	30	**0.6244**	**50.17617**	0.9991	62.16537	20.21
	40	0.6906	49.73854	0.9989	61.72534	21.11
Crossover rate (P_C)						
	P_C	MSE (dB)	PSNR (dB)	NCC	Fitness	Time (s)
$N_P = 256$	0.6	0.7895	49.15728	0.9897	61.03368	20.18
$N_G = 30$	0.7	**0.6854**	**49.77136**	0.9914	61.66816	20.68
	0.8	0.8112	49.03952	0.9862	60.87392	20.44
	0.9	0.8348	48.91498	0.9874	60.76378	20.94
Mutation rate (P_M)						
	P_M	MSE (dB)	PSNR (dB)	NCC	Fitness	Time (s)
$N_P = 256$	0.01	0.842	48.87768	0.9877	60.73008	19.98
$N_G = 30$	0.02	**0.6532**	**49.98034**	0.9912	61.87474	20.17
$P_C = 0.7$	0.05	0.796	49.12167	0.9852	60.94407	20.24
	0.1	0.8357	48.9103	0.9834	60.7111	19.82

7.8.6.5 Particle Swarm Optimization

In PSO, the effects of the number of iterations, acceleration constants, and inertia weight were investigated on the watermarked image. PSNR, MSE, robustness, NCC, and computational time were calculated, and the results are presented in Table 7.24. The swarm size was initially set to 256, and the number of iterations (N_I) was varied from 20 to 50 with the interval of 10, and the minimum MSE and maximum PSNR were obtained at $N_I = 30$. Similarly, minimum MSE and maximum PSNR were achieved with $c_1 = c_2 = 1.6$ and $w = 0.6$.

TABLE 7.24

Effect of PSO Parameters on Real-Time Image

Parameters	N_I	MSE (dB)	PSNR (dB)	NCC	Fitness	Time (s)
Effect of number of iterations (N_I)						
Swarm size = 256	20	0.6214	50.19709	0.9991	62.18629	17.65
	30	**0.6025**	**50.33123**	0.9994	62.32403	17.92
	40	0.6587	49.94393	0.9989	61.93073	18.40
	50	0.6613	49.92682	0.9989	61.91362	18.73
Effect of acceleration constants c_1 and c_2						
	$c_1 = c_2$	MSE (dB)	PSNR (dB)	NCC	Fitness	Time (s)
Swarm size = 256	1.4	0.6043	50.31828	0.9986	62.30148	17.89
$N_I = 30$	1.6	**0.5884**	**50.43408**	0.9995	62.42808	18.09
	1.8	0.6342	50.10854	0.9982	62.08694	18.11
	2	0.6574	49.95251	0.9973	61.92011	17.96
Effect of inertia weight (W)						
	W	MSE (dB)	PSNR (dB)	NCC	Fitness	Time (s)
Swarm size = 256	0.2	0.6377	50.08464	0.9983	62.06424	18.68
$N_I = 30$	0.4	0.6257	50.16714	0.9899	62.04594	18.45
$c_1 = c_2 = 1.6$	0.6	**0.6027**	**50.33141**	0.9991	62.32043	18.78
	0.8	0.6127	50.25832	0.9897	62.13472	18.49

TABLE 7.25

Performance Metrics of Real-Time Color Image Using HPSO

MSE (dB)	PSNR (dB)	NCC	Fitness	Time (s)
0.502	51.12636	0.9997	63.12276	15.54

The watermark amplification factor, which was initially set to a random value in [0,1], was tuned dynamically to an optimal value of 0.006. Based on the results in Table 7.24, the PSO parameters were selected as follows:

- Number of iterations (N_I): 30
- Acceleration constants ($c_1 = c_2$): 1.6
- Inertia weight (w): 0.6
- Amplification factor (α): 0.006.

7.8.6.6 Hybrid Particle Swarm Optimization

The parameters for the HPSO were set based on the results in Tables 7.23 and 7.24. The size of the initial population was set to 256, and the HPSO algorithm terminated at the end of 30 iterations. The acceleration constants c_1 and c_2 were both set to 1.6, and the value of inertia weight was set to 0.6. The parameters for HPSO were as follows:

- Population size: 256
- Cognitive factor c_1: 1.6
- Social coefficient c_2: 1.6
- Inertia weight w: 0.6
- Number of generations: 30
- Crossover rate: 0.7
- Mutation rate: 0.02
- Watermark amplification factor: 0.0002.

In order to maintain a balance between the individuals of PSO and GA in HPSO, the value of HF was set to 0.5. Table 7.25 shows the evaluated metrics obtained for the real-time image using HPSO. The resulting watermark amplification factor was found to be 0.0002. The PSNR value of the real-time image was 51.13 dB, which was reasonably high, and hence the artifacts introduced during embedding were almost invisible.

7.9 Discussion

The comparative analysis of experimental results achieved by GA, PSO, and HPSO in terms of PSNR, NCC, and computational time for gray-scale images and real-time color image is presented in this section.

7.9.1 Perceptual Transparency of Gray-Scale Images

The results of the bio-inspired algorithms used for watermarking are compared against intelligent methods in the literature based on the PSNR values. Table 7.26 compares the PSNR results obtained by GA, PSO, and HPSO techniques for images Lena, Mandrill, and Boat with approaches in the literature such as GA (Sikander et al. 2010), GP (Khan et al. 2004), wavelet perceptual model (WPM) (Khan et al. 2005), and genetic perceptual model (GPM) (Khan et al. 2005). The PSNR values of the proposed intelligent algorithms are higher than those of the GA, GP, WPM, and GPM techniques without applying any attacks. This is due to the fact that the proposed GA, PSO, and HPSO techniques choose the appropriate watermark amplification factor for each DWT coefficient. Moreover, PSO is capable of distributing the watermark in less perceivable regions of the image. Because of this, the PSNR values are higher for PSO and HPSO when compared to that of GA.

The PSNR values of the Lena image subjected to median filtering attacks with different window sizes of 2×2 and 3×3 are compared with approaches in the literature such as PSO (Ishtiaq et al. 2010), GP (Khan et al. 2004), and GA (Sikander et al. 2010), as shown in Table 7.27. It can be seen from the table that for a median filter attack of window size 2×2, the PSNR values of the proposed GA, PSO, and HPSO are 37.81, 38.73, and 39.34 dB, respectively, while the values for PSO (2010), GP (2004), and GA (2010) are 28.7747, 21.61, and 28.76 dB, respectively. It can be seen from these values that the PSNR of HPSO is higher, thus illustrating that the images are perceptually transparent under the effect of filtering attacks.

7.9.2 Robustness Measure of Gray-Scale Images

The performance of the proposed intelligent watermarking techniques was compared with those of algorithms available in the literature such as genetic algorithm-singular value decomposition (GA-SVD) (Jagadeesh et al. 2009) and Intelligent PSO (IPSO) (Wang et al. 2010b) in terms of the robustness measure (NCC), and the results are shown in Table 7.28.

For a scaling attack of 0.5, the proposed PSO (0.9453) and HPSO (0.9762) have higher correlation than IPSO (0.91). During the application of cropping, it is seen that all three proposed intelligent heuristics—GA (0.8913), PSO (0.9079), and HPSO (0.9248)—performed much better than GA-SVD (0.8516) and IPSO (0.76). Similarly, the results obtained from the proposed methods are compared with wavelet-based method (WBM) (Wang et al. 2002a) in terms of NCC due to Gaussian noise, median filtering, and JPEG compression, as shown in Table 7.29. The results of the proposed HPSO method are improved over those of WBM, specifically against median filtering and JPEG compression attacks.

TABLE 7.26

Comparative Analysis of PSNR Values without Attacks

Technique	Comparison of PSNR Values for Gray-Scale Images		
	Lena	Mandrill	Boat
GA (2010)	45.292	39.6354	43.9381
GP (2004)	44.45	36.28	40.94
WPM (2005)	41.5125	36.066	39.1691
GPM (2005)	38.4192	34.408	36.5404
GA	44.47	43.67	43.89
PSO	45.19	44.08	44.66
HPSO	**45.65**	**44.61**	**44.85**

TABLE 7.27

Comparison of PSNR Values with Attacks for Lena Image

Attacks	PSNR Values of Intelligent Algorithms					
	PSO (2010)	GP (2004)	GA (2010)	GA	PSO	HPSO
Median filter 2 × 2	28.7747	21.61	28.76	37.81	38.73	**39.34**
Median filter 3 × 3	33.675	26.52	33.64	37.76	38.15	**38.93**

TABLE 7.28

Comparison of Correlation Values of GA, PSO, and HPSO with GA-SVD and IPSO for Lena Image

Attacks	*NCC* Values Computed by Intelligent Algorithms				
	GA-SVD	IPSO	GA	PSO	HPSO
Median filtering 3 × 3	0.8549	NA	0.9918	0.9969	**0.9985**
Gaussian noise σ = 0.001	NA	**0.94**	0.8934	0.9031	0.9071
Scaling 0.5	NA	0.91	0.9113	0.9453	**0.9762**
JPEG compression 40%	**0.9581**	NA	0.8679	0.8748	0.8844
Cropping 1/4	0.8516	0.76	0.8913	0.9079	**0.9248**
Gaussian filtering 3 × 3	0.6918	NA	0.9875	0.9949	**1.00**
Rotation 30°	NA	0.68	0.8215	0.8288	**0.8356**

TABLE 7.29

Performance Evaluation of GA, PSO, and HPSO and Comparison with WBM

Images	Methods	Correlation of Watermarked Image	Correlation due to Gaussian Noise	Correlation due to Median Filtering	Correlation due to JPEG Compression
Peppers	GA	0.9984	0.8753	0.9894	0.927
	PSO	0.9989	0.8798	0.9902	0.9299
	HPSO	**1.000**	0.8837	**0.9988**	**0.9372**
	WBM	0.986	**0.975**	0.413	0.478
Lena	GA	0.9991	0.8934	0.9918	0.9543
	PSO	0.999	0.9031	0.9974	0.9594
	HPSO	**1.000**	0.9071	**0.9977**	**0.9684**
	WBM	0.982	**0.971**	0.384	0.404
Barbara	GA	0.9982	0.8799	0.9884	0.9129
	PSO	0.9997	0.8869	0.9935	0.9216
	HPSO	**1.000**	0.8878	**0.9979**	**0.9248**
	WBM	0.987	**0.971**	0.501	0.671
Mandrill	GA	0.999	0.8896	0.9926	0.9378
	PSO	**1.000**	0.8956	0.9967	0.9458
	HPSO	**1.000**	0.8961	**0.998**	**0.9521**
	WBM	0.988	0.974	0.367	0.661

7.9.3 PSNR and NCC of Real-Time Color Image

To evaluate the performance of the optimization techniques on real-time camera-captured color images, several experiments were conducted by applying common attacks such as filtering, addition of Gaussian noise, rotation, scaling, and JPEG compression. Performance metrics such as NCC and PSNR were evaluated, and are shown in Table 7.30. For Gaussian noise attack with variance $\sigma = 0.001$, the PSNR values of the real-time color image is 49.88 dB for GA, 50.06 dB for PSO, and 51.06 dB for HPSO. Likewise, for JPEG compression attack with a QF of 70%, the value of PSNR is found to be 47.5 dB for GA, 47.93 dB for PSO, and 48.43 dB for HPSO. In addition, PSNR was also evaluated for all combination of attacks as listed in table, and it is observed that the PSNR of HPSO is higher than those of GA and PSO.

The NCC values for JPEG compression attack with QF = 95% are 0.9582, 0.9622, and 0.9749 for GA, PSO, and HPSO, respectively. Similarly, for combination of attacks such as rotation 5° + scaling 1.1 + cropping (block size = 10), the NCC values obtained by GA, PSO, and HPSO are 0.8824, 0.8938, and 0.901 respectively. From these results, it can be concluded that for all attacks presented, the NCC values of HPSO are improved relative to those of GA and PSO.

TABLE 7.30

PSNR and NCC for Real-Time Color Image against Attacks

Type of Attack	GA		PSO		HPSO	
	NCC	PSNR	NCC	PSNR	NCC	PSNR
Average filtering	0.9877	42.33	0.9941	43.32	0.9985	43.84
Gaussian filtering	0.9942	42.65	0.9957	42.96	0.9971	43.53
Median filtering 2 × 2	0.9981	43.22	0.9984	44.14	0.9987	44.75
Median filtering 3 × 3	0.9982	43.17	0.9989	43.56	0.9988	44.34
Gaussian noise $\sigma = 0.001$	0.9007	49.88	0.9058	50.06	0.9095	51.06
Gaussian noise $\sigma = 0.01$	0.8959	49.77	0.9018	49.84	0.9039	50.05
Gaussian noise $\sigma = 0.1$	0.887	49.65	0.8945	49.84	0.9023	50
JPEG $QF = 20\%$	0.8703	47.16	0.8791	47.21	0.8737	48.12
JPEG $QF = 40\%$	0.8761	47.3	0.8779	47.99	0.8902	48.08
JPEG $QF = 70\%$	0.8838	47.5	0.8868	47.93	0.9008	48.43
JPEG $QF = 95\%$	0.9582	47.76	0.9622	48.55	0.9749	49.47
Rotation 5°	0.9087	38.76	0.9172	39.71	0.9131	40.04
Rotation 15°	0.8864	38.35	0.8849	38.45	0.8918	39.44
Rotation 30°	0.8268	37.67	0.8294	37.94	0.8378	38.51
Rotation 40°	0.757	37.46	0.7641	37.67	0.7671	37.75
Rotation 5° + Scaling 0.5	0.8182	38.5	0.8242	38.65	0.8269	38.92
Rotation 5° + Scaling 1.1	0.907	38.58	0.9052	39.55	0.9082	40.32
Rotation 5° + Scaling 1.1 + Cropping (Block size = 10)	0.8824	37.54	0.8938	37.59	0.901	38.04
Rotation 5° + Scaling 1.1 + Cropping (Block size = 100)	0.7054	33.96	0.7076	34.77	0.7065	35.25
Rotation 5° + Scaling 1.1 + JPEG 95%	0.9042	38.52	0.9112	39.1	0.9182	39.58
Rotation 5° + Scaling 1.1 + JPEG 60%	0.8836	38.44	0.891	39.25	0.8947	39.75

TABLE 7.31

Comparative Analysis of Computational Time

Technique	Computational Time (s)						Real Time Color Image
	Peppers	Mandrill	Lena	Barbara	Boat	Cameraman	
GA	11.86	12.65	12.38	12.19	11.9	12.74	20.17
PSO	12.22	12.09	12.06	12.01	12.06	12.11	18.78
HPSO	11.12	10.95	10.96	11.28	11.31	11.09	15.54

7.9.4 Computational Efficiency

Simulations were carried out on an Intel dual-core processor with a speed of 2.53 GHz. The time taken for embedding and extracting the watermark for the test images is presented in Table 7.31. The computational overhead is related to the population size and the number of iterations. In all the three optimization algorithms, there is a steep increase in the execution time when the population size increases beyond 256. Thus the optimal tradeoff for population size is based on the computational time. The computational time is recorded as shown in Table 7.31 for a population size of 128 for gray-scale images and 256 for the real-time color image. No significant improvements in PSNR and MSE are observed for an increase in the number of generations beyond 30 in both GA and PSO. Hence the results in terms of computational time are recorded, as shown in Table 7.31, at the end of 30 generations. For the Peppers image, the computational time taken by HPSO (11.12 s) is improved over those GA (11.86 s) and PSO (12.22 s). Likewise, for the real-time camera-captured color image, the computational time of HPSO (15.54 s) was reduced compares to those of GA (20.17 s) and PSO (18.78 s). Hence, it is concluded that the time required to optimize the watermark amplification factor using HPSO is less than those of GA and PSO. It can also be concluded that the HPSO algorithm minimizes the chances of getting trapped in local minima, thus converging faster than GA and PSO.

7.10 Advantages of CI Paradigms

The algorithms with test cases for DIWM problem in terms of optimal solution, perceptual transparency, robustness, and computational time are presented in Table 7.32.

In the DIWM problem, considering all the six gray-scale images and one real-time color image, the computational time of HPSO is found to be improved over GA by 14.15% and over PSO by 11.04%. The merits and demerits of the optimization techniques in terms of the computational time are as follows:

- GA: Since linearization of the problem is not required while using GA, it is applied to solve DIWM. In a noisy environment, convergence is difficult in GA, as a result of which the computational time increases.

- PSO: The algorithm ensures stable convergence with respect to the DIWM problem, but the computational time is high since the algorithm gets trapped in local minima.

TABLE 7.32

Evaluation of Algorithms for DIWM

Advantages	Performance of Algorithms
Optimal solution	HPSO produced an optimal watermark amplification factor for all the six gray-scale images and one real-time camera-captured color image
Perceptual transparency	HPSO resulted in improved PSNR for all images over GA and PSO with and without geometric and nongeometric attacks
Robustness	Improvement in terms of NCC values for all the images computed by HPSO shows that the watermarks are detected effectively with attacks such as average filtering, Gaussian filtering, median filtering, Gaussian noise, JPEG compression, rotation, scaling, and cropping
Computational time	For all gray-scale images such as Peppers, Mandrill, Lena, Barbara, Boat, and Cameraman and real-time color image, it is observed that the time required to optimize the watermark amplification factor using HPSO is less than that for GA and PSO

- HPSO: PSO has a strong ability to determine the optimal solution, but at times it has a disadvantage of getting stuck in local optimum. Thus a combination of PSO with GA eliminates this disadvantage and hence increases the computational efficiency.

7.11 Summary

In this chapter, intelligent DIWMs based on GA, PSO, and HPSO were proposed. The host image was made to undergo a set of preprocessing stages such as segmentation, feature extraction, orientation assignment, and normalization. This increased the probability of choosing efficient blocks for watermark embedding and extraction in the DWT domain. Such type of image modeling provides a better guidance to adaptively adjust the watermark embedding strength, also known as the watermark amplification factor. The amplification factor was optimized by the proposed intelligent algorithms GA, PSO, and HPSO so that an optimal DWT subband coefficient was chosen for embedding the watermark. In addition, since the coefficients were randomly selected from different subbands, this led to an increase in the security against unauthorized access to remove the watermark or to detect the existence of the watermark.

The efficiency of the proposed algorithms on DIWM was validated on a set of six gray-scale images and one real-time, camera-captured color image. Perceptual transparency and robustness are important evaluation criteria in judging any digital watermarking algorithm. These parameters were determined by computing the PSNR and the NCC values for the images using GA, PSO, and HPSO, thus preserving the quality of the image. High values of PSNR and NCC imply that the watermarked image and the cover image are more similar to each other. In general, an increase in PSNR decreases the NCC value. But a proper balance is maintained between the PSNR and NCC values due to the application of intelligent algorithms such as GA, PSO, and HPSO.

An extensive analysis was carried out by applying several geometric and nongeometric attacks on the watermarked images, thus computing the PSNR and NCC values. For almost all attacks on gray-scale images and the real-time, camera-captured color image, it was observed that the NCC values were greater than 0.75 and PSNR greater than 30 dB, thus proving the effectiveness of the intelligent algorithms GA, PSO, and HPSO in DIWM. Based on the NCC values, it was observed that the HPSO algorithm was nearly robust against geometric and nongeometric attacks and also for a combination of these attacks when compared with GA and PSO. The experimental results of HPSO based on PSNR showed that the watermarked images were perceptually equal to the original images and that the watermarks were still detectable against attacks such as average filtering, Gaussian filtering, median filtering, Gaussian noise, JPEG compression, rotation, scaling, and cropping.

The computational time required for intelligent watermarking of gray-scale images and the real-time color image using HPSO was found to be less than those of GA and PSO, thereby increasing the speed of operation and making HPSO more suitable for real-time applications. From the observations listed in this chapter, it can be concluded that the proposed HPSO watermarking strategy offers an improved degree of authentication, in turn protecting the watermark against unauthorized access. In future, approaches such as differential evolution and multiobjective optimization can be investigated for DIWM and their compared with the obtained results here. In addition, the proposed techniques can be applied to video watermarking applications, thus providing a high degree of authentication in the transfer of video information over the Internet.

References

Abu-Errub, A. and Al-Haj, A., Performance optimization of discrete wavelets transform based image watermarking using genetic algorithms, *Journal of Computer Science*, 4(10), 834–841, 2008.

Alghoniemy, M. and Tewfik, A.H., Geometric distortion correction through image normalization, in *Proceedings of the IEEE International Conference on Multimedia and Expo: ICME 2000*, New York, July 30–August 2, 2000, Vol. 3, pp. 1291–1294.

Al-Haj, A., Combined DWT-DCT digital watermarking, *Journal of Computer Science*, 3(9), 740–746, 2007.

Ali, K.H., Kasirun, Z.M., Hamid, A.J., Gazi, M.A., and Zaidan, A.A., On the accuracy of hiding information metrics: Counterfeit protection for education and important certificates, *International Journal of the Physical Sciences*, 5(7), 1054–1062, 2010.

Aslantas, V., Ozer, S., and Ozturk, S., Improving the performance of DCT-based fragile watermarking using intelligent optimization algorithms, *Optics Communication*, 282(14), 2806–2817, 2009.

Barni, M., Bartolini, F., and Piva, A., Improved wavelet-based watermarking through pixel-wise masking, *IEEE Transactions on Image Processing*, 10(5), 1–21, 2001.

Belongie, S., Carson, C., Greenspan, H., and Malik, J., Color and texture-based image segmentation using EM and its application to content-based image retrieval, in *Proceedings of the Sixth IEEE International Conference on Computer Vision 1998*, Bombay, India, January 4–7, 1998, pp. 675–682.

Boato, G., Conotter, V., and De Natale, F.G.B., GA-based robustness evaluation method for digital image watermarking, in *Digital Watermarking*, Y.Q. Shi, H.-J. Kim, and K. Stefan, Eds. Springer-Verlag, New York, 2008, pp. 294–307.

Boato, G., Conotter, V., and De Natale, F.G.B., Watermarking robustness evaluation based on perceptual quality via genetic algorithms, *IEEE Transactions on Information Forensics and Security*, 4(2), 207–216, 2009.

Burt, P.J. and Adelson, E., The Laplacian Pyramid as a compact image code, *IEEE Transactions on Communications*, 31(4), 532–540, 1983.

Chen, Q., Kotani, K., Lee, F., and Ohmi, T., Scale-invariant feature extraction by VQ-based local image descriptor, in *Proceedings of the International Conference on Computational Intelligence for Modelling, Control and Automation*, Vienna, Austria, December 10–12, 2008, pp. 1217–1222.

Christian, R. and Jean-Luc, D., A survey of watermarking algorithms for image authentication, *EURASIP Journal on Applied Signal Processing*, 2002(6), 613–621, 2002.

Comaniciu, D. and Meer, P., Mean shift: A robust approach toward feature space analysis, *IEEE Transactions on Pattern Analysis and Machine Intelligence*, 24(5), 603–619, 2002.

Cong, J., Affine invariant watermarking algorithm using feature matching,*Digital Signal Processing*, 16(3), 247–254, 2006.

Cox, I., Miller, M., and Bloom, J., *Digital Watermarking*. San Mateo, CA: Morgan Kaufmann, 2002.

Dempster, A., Laird, N., and Rubin, D., Maximum likelihood from incomplete data via the EM algorithm, *Journal of the Royal Statistical Society: Series B*, 39(1), 1–38, 1977.

Do, C.B. and Batzoglou, S., What is the expectation maximization algorithm?, *Nature Biotechnology*, 26(8), 897–899, 2008.

Eberhart, R. and Kennedy, J., A new optimizer using particles Swarm theory, *Proceedings of the Sixth IEEE International Symposium on Micro Machine and Human Science: MHS '95*, Nagoya, Japan, pp. 39–43, 1995.

Emek, S. and Pazarci, M., A cascade DWT-DCT based digital watermarking scheme, in *Proceedings of the 13th European Signal Processing Conference: EUSIPCO'05*, Antalya, Turkey, September 4–8, 2005, pp. 492–495.

Felzenszwalb, P. and Huttenlocher, D., Efficient graph-based image segmentation, *International Journal of Computer Vision*, 59(2), 167–181, 2004.

Fu, Y.G. and Shen, R.M., Color image watermarking scheme based on linear discriminant analysis, *Computer Standards and Interfaces*, 30(3), 115–120, 2008.

Goldberg, D.E., *Genetic Algorithms in Search Optimization and Machine Learning*. Boston, MA: Addison Wesley Longman Publishing Co., 1989.

Grefenstette, J.J., Optimization of control parameters for genetic algorithms, *IEEE Transactions on Systems, Man and Cybernetics*, 16(1), 122–128, 1986.

Gunjal, B.L. and Mali, S.N., Comparative performance analysis of digital image watermarking scheme in DWT and DWT-FWHT-SVD domains, in *India Conference (INDICON), 2014 Annual IEEE*, pp. 1–6. IEEE, 2014.

Hartley, H., Maximum likelihood estimation from incomplete data, *Biometrics*, 14(2), 174–194, June 1958.

Hernandez, J.R., Amado, M., and Perez-Gonzalez, F., DCT-domain watermarking techniques for still images: Detector performance analysis and a new structure, *IEEE Transactions on Image Processing*, 9(1), 55–68, 2000.

Ishtiaq, M., Sikandar, B., Jaffar, M.A., and Khan, A., Adaptive watermark strength selection using particle swarm optimization, *ICIC Express Letters*, 4(5), 1–6, 2010.

Jagadeesh, B., Kumar, S.S., and Rajeswari, K.R., A genetic algorithm based oblivious image watermarking scheme using singular value decomposition, in *Proceedings of the IEEE International Conference on Networks and Communications: NETCOM '09*, Chennai, India, December 27–29, 2009, pp. 224–229.

Ketcham, M. and Vongpradhip, S., Intelligent audio watermarking using genetic algorithm in DWT domain, *International Journal of Intelligent Technology*, 2(2), 135–140, 2007.

Khan, A., Mirza A.M., and Majid, A., Optimizing perceptual shaping of a digital watermark using genetic programming, *Iranian Journal of Electrical and Computer Engineering*, 3(2), 1251–1260, 2004.

Khan, A., Mirza, A.M., and Majid, A., Intelligent perceptual shaping of a digital watermark: Exploiting characteristics of human visual system, *International Journal of Knowledge-Based Intelligent Engineering Systems*, 9, 1–11, 2005.

Kim, H., Lee, H.-Y., and Lee, H.-K., Robust image watermarking using local invariant features, *Optical Engineering*, 45(3), 1–11, 2006.

Lee, H.-Y. and Lee, H.-K., Copyright protection through feature-based watermarking using scale-invariant keypoints, in *Proceedings of the International Conference on Consumer Electronics: ICCE '06*, January 7–11, 2006, pp. 225–226.

Lee, Z.J., Lin, S.W., Su, S.F., and Lin, C.Y., A hybrid watermarking technique applied to digital images, *Applied Soft Computing*, 8(1), 798–808, 2008.

Lin, P.L., Hsieh, C.K., and Huang, P.W., A hierarchical digital watermarking method for image tamper detection and recovery, *Pattern Recognition*, 38(12), 2519–2529, 2005.

Lowe, D.G., Distinctive image features from scale-invariant keypoints, *International Journal on Computer Vision*, 60(2), 91–110, 2004.

Mairgiotis, A., Chantas, G., Galatsanos, N., Blekas, K., and Yang, Y., New detectors for watermarks with unknown power based on student-t image priors, in *Proceedings of the IEEE 9th Workshop on Multimedia Signal Processing: MMSP 2007*, Crete, Greece, October 1–3, 2007, pp. 353–356.

Mohamed, F.K. and Abbes, R., RST robust watermarking schema based on image normalization and DCT decomposition, *Malaysian Journal of Computer Science*, 20(1), 77–90, 2007.

Nguyen, P.B., Luong, M., and Beghdadi, A., Statistical analysis of image quality metrics for watermark transparency assessment, in *Proceedings of the Conference on Advances in Multimedia Information Processing: PCM 2010*, Shanghai, China, September 21–24, 2010, pp. 685–696.

Pan, J.S., Huang, H.C., and Jain, L.C., *Intelligent Watermarking Techniques (Innovative Intelligence)*, World Scientific Press, 2004.

Piva, A., Barni, M., Bartolini, F., and Cappellini, V., Mask building for perceptually hiding frequency embedded watermarks, in *Proceedings of the 5th IEEE International Conference on Image Processing ICIP'98*, Chicago, IL, October 4–7, 1998, pp. 450–454.

Potdar, V., Han, S., and Chang, E., A survey of digital image watermarking techniques, in *Proceedings of the 3rd IEEE-International Conference on Industrial Informatics*, Perth, Australia, August 10–12, 2005, pp. 709–716.

Revathy, K., Applying EM algorithm for segmentation of textured images, in *Proceedings of the World Congress on Engineering (WCE 2007)*, London, U.K., July 2–4, 2007, Vol. 1, pp. 702–707.

Shia, X.H., Lianga, Y.C., Leeb, H.P., Lub, C., and Wanga, L.M., An improved GA and a novel PSO-GA based hybrid algorithm, *Information Processing*, 93(5), 255–261, 2005.

Shieh, C.S., Huang, H.C., Pan, J.S., and Wang, F.H., Genetic watermarking based on transform domain technique, *Pattern Recognition*, 37(3), 555–565, 2004.

Sikander, B., Ishtiaq, M., Jaffar, M.A., and Mirza, A.M., Adaptive digital image watermarking of images using genetic algorithm, in *Proceedings of the IEEE International Conference on Information Science and Applications: ICISA 2010*, Seoul, Korea, April 21–23, 2010, pp. 1–8.

Sun, Z. and Ma, J., DWT-domain watermark detection using Gaussian mixture model with automated model selection, in *International Symposium on Computer Network and Multimedia Technology: CNMT 2009*, Wuhan, January 18–20, 2009, pp. 1–4.

Sviatoslav, V., Frederic, D., and Thierry, P., Content adaptive watermarking based on a stochastic multiresolution image modeling, in *Proceedings of the Tenth European Signal Processing Conference: EUSIPCO 2000*, Tampere, Finland, September 5–8, 2000, pp. 5–8.

Tang, C.W. and Hang, H.M., A feature-based robust digital image watermarking scheme, *IEEE Transactions on Signal Processing*, 51(4), 950–959, 2003.

Tao, H., Zain, J.M., Abd Alla, A.N., and Hongwu, Q., An implementation of digital image watermarking based on particle swarm optimization, *Communications in Computer and Information Science*, 87(2), 314–320, 2010.

Vallabha, H., *Multiresolution Watermark Based on Wavelet Transform for Digital images*, Technical Report, Cranes Software International Ltd, Bangalore, India, 2003.

Voloshynovskiy, S., Pereira, S., Iquise, V., and Pun, T., Attack modeling: Towards a second generation watermarking benchmark, *Journal of Signal Processing*, 80(6), 1177–1214, 2001.

Wang, M.S. and Chen, W.C., A hybrid DWT-SVD copyright protection scheme based on k-means clustering and visual cryptography, *Computer Standards and Interfaces*, 31(4), 757–762, 2009.

Wang, W., Men, A., and Yang, B., A feature-based semi-fragile watermarking scheme in DWT domain, in *Proceedings of the 2nd IEEE International Conference on Network Infrastructure and Digital Content*, Beijing, China, September 24–26, 2010a, pp. 768–772.

Wang, Y., Doherty, J.F., and Van Dyck, R.E., A wavelet-based watermarking algorithm for ownership verification of digital images, *IEEE Transactions on Image Processing*, 11(2), 77–88, 2002a.

Wang, Y.-R., Lin, W.-H., and Yang, L., An intelligent PSO watermarking, in *Proceedings of the International Conference on Machine Learning and Cybernetics: ICMLC*, Qingdao, China, July 11–14, 2010b, pp. 2555–2558.

Wang, Z., Bovik, A.C., and Lu, L., Why is image quality assessment so difficult?, in *Proceedings of the IEEE International Conference on Acoustics, Speech and Signal Processing: ICASSP*, Orlando, FL, May 13–17, 2002b, Vol. 4, pp. 3313–3316.

Watson, A.B., Visual optimization of DCT quantization matrices for individual images, in *Proceedings of AIAA Computing in Aerospace 9*, San Diego, CA, October 19–21, 1993, pp. 286–291.

Wei, Z., Li, H., and Dai, J., Image watermarking based on genetic algorithm, in *Proceedings of the IEEE International Conference on Multimedia and Expo: ICME '06*, Toronto, Canada, July 9–12, 2006, pp. 1117–1120.

Wu, Y., Agrawal, D., and Abbadi, A.E., A comparison of DFT and DWT based similarity search in time-series databases, in *Proceedings of the Ninth International Conference on Information and Knowledge Management: CIKM'00*, McClean, VA, November 6–11, 2000, pp. 488–495.

Zheng, D. and Zhao, J., A rotation invariant feature and image normalization based image watermarking algorithm, in *Proceedings of the IEEE International Conference on Multimedia and Expo 2007*, Beijing, China, July 2–5, 2007, pp. 2098–2101.

Zheng, D., Wang, S., and Zhao, J., RST invariant image watermarking algorithm with mathematical modeling and analysis of the watermarking processes, *IEEE Transactions on Image Processing*, 18(5), 1055–1068, 2009.

Appendix A: Unit Commitment and Economic Load Dispatch Test Systems

In this appendix, the specifications of the test systems used in the UC-ELD problem are discussed. The load demand during 24 h, generator characteristics, and the transmission loss coefficients of the IEEE 30-bus system (6-unit system), 10-unit test system, Indian utility 75-bus system (15-unit system), and the 20-unit test system are explained. The IEEE 30-bus system data were obtained online (http://www.ece.mtu.edu/faculty/ljbohman/peec/Dig_Rsor.htm) from the Power Engineering Digital Educational Resource of the IEEE Power and Energy Society. The Indian utility 75-bus system data were obtained online (http://www.wseas.us/e-library/transactions/power/2009/32-168.pdf) from the Uttar Pradesh State Electricity Board (UPSEB).

A.1 Case Study I: 6-Unit Test System

The 6-unit test system chosen in this thesis is the IEEE 30-bus system (Labbi and Attous 2010) in which the cost coefficients of the generating units, generating capacity of each unit and transmission, loss matrix and 24 h power demand requirements are specified. The test system comprises 6 generators, 41 transmission lines, and 30 buses. The IEEE 30-bus system has a minimum generation capacity of 117 MW and a maximum generation capacity of 435 MW. The load demand for 24 h, characteristics of the 6-unit test system, and the transmission loss matrix are detailed in Tables A.1 through A.3, respectively.

A.2 Case Study II: 10-Unit Test System

The second case study consists of a 10-unit test system (Park et al. 2010). The input data include the generator limits, fuel cost coefficients, transmission loss matrix, and load profile for 24 h. The minimum generating capacity of the system is 690 MW and the maximum generating capacity is 2358 MW. The load profile, generator characteristics, and transmission loss coefficients are given in Tables A.4 through A.6, respectively.

A.3 Case Study III: 15-Unit Test System

The Indian utility 75-bus UPSEB system with 15 generating units is chosen as the test system for the analysis of commitment and optimal dispatch. The 75-bus Indian system has 15 generators (at buses 1–15) and 98 transmission lines (including 24 transformers) (Prabhakar et al. 2009). The load demand, characteristics of generators, and loss coefficients are given in Tables A.7 through A.9, respectively.

TABLE A.1

24 h Load Demand for 6-Unit Test System

Hour	1	2	3	4	5	6	7	8	9	10	11	12
Load (MW)	166	196	229	267	283.4	272	246	213	192	161	147	160
Hour	13	14	15	16	17	18	19	20	21	22	23	24
Load (MW)	170	185	208	232	246	241	236	225	204	182	161	131

TABLE A.2

Generator Characteristics of 6-Unit Test System

Units	Parameters				
	a ($/W – h²)	b ($/W – h)	c ($)	Min Power (MW)	Max Power (MW)
1	0.00375	2	0	50	200
2	0.01750	1.75	0	20	80
3	0.06250	1	0	15	50
4	0.00834	3.25	0	10	35
5	0.02500	3	0	10	30
6	0.02500	3	0	12	40

TABLE A.3

Transmission Loss Coefficients for 6-Unit Test System

B_{mn}	Loss Coefficient Matrix					
	0.000218	0.000103	0.000009	−0.00001	0.000002	0.000027
	0.000103	0.000181	0.000004	−0.000015	0.000002	0.00003
	0.000009	0.000004	0.000417	−0.000131	−0.000153	−0.000107
	−0.00014	0.000015	−0.000131	0.000221	0.000094	0.00005
	0.000002	0.000002	−0.000153	0.000094	0.000243	0
	0.000027	0.00003	−0.000107	0.00005	0	0.000358

TABLE A.4

Load Demand for 10-Unit Test System

Hour	1	2	3	4	5	6	7	8	9	10	11	12
Load (MW)	1036	1110	1258	1406	1480	1628	1702	1776	1924	2072	2146	2220
Hour	13	14	15	16	17	18	19	20	21	22	23	24
Load (MW)	2072	1924	1776	1554	1480	1628	1776	2072	1924	1628	1332	1184

TABLE A.5

Generator Characteristics of 10-Unit Test System

Units	Parameters				
	a ($/W – h²)	b ($/W – h)	c ($)	Min Power (MW)	Max Power (MW)
1	0.00043	21.60	958.2	150	470
2	0.00063	21.05	1313.6	135	460
3	0.00039	20.81	604.97	73	340
4	0.0007	23.9	471.6	60	300
5	0.00079	21.62	480.29	73	243
6	0.00056	17.87	601.75	57	160
7	0.00211	16.51	502.7	20	130
8	0.0048	23.23	639.4	47	120
9	0.10908	19.58	455.6	20	80
10	0.00951	22.54	692.4	55	55

TABLE A.6

Transmission Loss Coefficients for 10-Unit Test System

$B_{mn} = 10^{-3}*$	Loss Coefficient Matrix									
	8.7	0.43	–4.61	0.36	0.32	–0.66	0.96	–1.6	0.8	–0.1
	0.43	0.83	–0.97	0.22	0.75	–0.28	5.04	1.7	0.54	7.2
	–4.61	–0.97	9	–2	0.63	3	1.7	–4.3	3.1	–2
	0.36	0.22	–2	5.3	0.47	2.62	–1.96	2.1	0.67	1.8
	0.32	0.75	0.63	0.47	8.6	–0.8	0.37	0.72	–0.9	0.69
	–0.66	–0.28	3	2.62	–0.8	11.8	–4.9	0.3	3	–3
	0.96	5.04	1.7	–1.96	0.37	–4.9	8.24	–0.9	5.9	–0.6
	–1.6	1.7	–4.3	2.1	0.72	0.3	–0.9	1.2	–0.96	0.56
	0.8	0.54	3.1	0.67	–0.9	3	5.9	–0.96	0.93	–0.3
	–0.1	7.2	–2	1.8	0.69	–3	–0.6	0.56	–0.3	0.99

Note: * implies that the values given in the table need to be multiplied by 10^{-3} etc.

TABLE A.7

Load Demand for 15-Unit Test System

Hour	1	2	3	4	5	6	7	8	9	10	11	12
Load (MW)	3352	3384	3437	3489	3659	3849	3898	3849	3764	3637	3437	3384
Hour	13	14	15	16	17	18	19	20	21	22	23	24
Load (MW)	3357	3394	3616	3828	3828	3786	3659	3458	3394	3334	3329	3348

TABLE A.8

Generator Characteristics of 15-Unit Test System

Units	Parameters				
	a ($/W − h²)	b ($/W − h)	c ($)	Min Power (MW)	Max Power (MW)
1	0.0008	0.814	0	100	1500
2	0.0014	1.3804	0	100	300
3	0.0016	1.5662	0	40	200
4	0.0016	1.6069	0	40	170
5	0.0016	1.5662	0	2	240
6	0.0018	1.7422	0	1	120
7	0.0018	1.7755	0	1	100
8	0.0018	1.7422	0	20	100
9	0.0012	1.1792	0	60	570
10	0.0017	1.6947	0	30	250
11	0.0016	1.6208	0	40	200
12	0.0004	0.4091	0	80	1300
13	0.0007	0.677	0	50	900
14	0.0015	1.491	0	10	150
15	0.001	1.0025	0	20	454

TABLE A.9

Transmission Loss Coefficients for 15-Unit Test System

Loss Coefficient Matrix														
$B_{mn} = 10^{-5*}$ 1.4	1.2	0.7	−0.1	−0.3	−0.1	−0.1	−0.1	−0.3	−0.5	−0.3	−0.2	0.4	0.3	−0.1
1.2	1.5	1.3	0	−0.5	−0.2	0	0.1	−0.2	−0.4	−0.4	0	0.4	1	−0.2
0.7	1.3	7.6	−0.1	−1.3	−0.9	−0.1	0	−0.8	−1.2	−1.7	0	−2.6	11.1	−2.8
−0.1	0	−0.1	3.4	−0.7	−0.4	1.1	5	2.9	3.2	−1.1	0	0.1	0.1	−2.6
−0.3	−0.5	−1.3	−0.7	9	1.4	−0.3	−1.2	−1	−1.3	0.7	−0.2	−0.2	−2.4	−0.3
−0.1	−0.2	−0.9	−0.4	1.4	1.6	0	−0.6	−0.5	−0.8	1.1	−0.1	−0.2	−1.7	0.3
−0.1	0	−0.1	1.1	−0.3	0	1.5	1.7	1.5	0.9	−0.5	0.7	0	−0.2	−0.8
−0.1	0.1	0	5	−1.2	−0.6	1.7	16.8	8.2	7.9	−2.3	−3.6	0.1	0.5	−7.8
−0.3	−0.2	−0.8	2.9	−1	−0.5	1.5	8.2	12.9	11.6	−2.1	−2.5	0.7	−1.2	−7.2
−0.5	−0.4	−1.2	3.2	−1.3	−0.8	0.9	7.9	11.6	20	−2.7	−3.4	0.9	−1.1	−8.8
−0.3	−0.4	−1.7	−1.1	0.7	1.1	−0.5	−2.3	−2.1	−2.7	14	0.1	0.4	−3.8	16.8
−0.2	0	0	0	−0.2	−0.1	0.7	−3.6	−2.5	−3.4	0.1	5.4	−0.1	−0.4	2.8
0.4	0.4	−2.6	0.1	−0.2	−0.2	0	0.1	0.7	0.9	0.4	−0.1	10.3	−10.1	2.8
0.3	1	11.1	0.1	−2.4	−1.7	−0.2	0.5	−1.2	−1.1	−3.8	−0.4	−10.1	57.8	−9.4

A.4 Case Study IV: 20-Unit Test System

The fourth case study considered in this thesis consists of 20 generating units. The data are generated based on the reference obtained from Abookazemi et al. (2009). The power limits and cost coefficients of the 10-unit system data are duplicated to obtain the 20-unit system data. The hourly load demand of the 10-unit system is doubled to obtain the hourly profile of the 20-unit system. The load profile, transmission loss coefficients, fuel cost coefficients, and maximum and minimum power are presented in Tables A.10 through A.12, respectively.

TABLE A.10

Load Demand for 20-Unit Test System

Hour	1	2	3	4	5	6	7	8	9	10	11	12
Load (MW)	2072	2220	2516	2812	2960	3256	3404	3552	3848	4144	4292	4440
Hour	13	14	15	16	17	18	19	20	21	22	23	24
Load (MW)	4144	3848	3552	3108	2960	3256	3552	4144	3848	3256	2664	2368

TABLE A.11

Generator Characteristics of 20-Unit Test System

	Parameters				
Units	a ($/W – h²)	b ($/W – h)	c ($)	Min Power (MW)	Max Power (MW)
1	0.00043	21.60	958.2	150	470
2	0.00063	21.05	1313.6	135	460
3	0.00039	20.81	604.97	73	340
4	0.0007	23.9	471.6	60	300
5	0.00079	21.62	480.29	73	243
6	0.00056	17.87	601.75	57	160
7	0.00211	16.51	502.7	20	130
8	0.0048	23.23	639.4	47	120
9	0.10908	19.58	455.6	20	80
10	0.00951	22.54	692.4	55	55
11	0.00043	21.60	958.2	150	470
12	0.00063	21.05	1313.6	135	460
13	0.00039	20.81	604.97	73	340
14	0.0007	23.9	471.6	60	300
15	0.00079	21.62	480.29	73	243
16	0.00056	17.87	601.75	57	160
17	0.00211	16.51	502.7	20	130
18	0.0048	23.23	639.4	47	120
19	0.10908	19.58	455.6	20	80
20	0.00951	22.54	692.4	55	55

TABLE A.12

Transmission Loss Coefficients for 20-Unit Test System

Loss Coefficient Matrix

$B_{mn} = 10^{-3}*$

8.7	0.43	-4.61	0.36	0.32	-0.66	0.96	-1.6	0.8	-0.1	8.7	0.43	-4.61	0.36	0.32	-0.66	0.96	-1.6	0.8	-0.1
0.43	0.83	-0.97	0.22	0.75	-0.28	5.04	1.7	0.54	7.2	0.43	0.83	-0.97	0.22	0.75	-0.28	5.04	1.7	0.54	7.2
-4.61	-0.97	9	-2	0.63	3	1.7	-4.3	3.1	-2	-4.61	-0.97	9	-2	0.63	3	1.7	-4.3	3.1	-2
0.36	0.22	-2	5.3	0.47	2.62	-1.96	2.1	0.67	1.8	0.36	0.22	-2	5.3	0.47	2.62	-1.96	2.1	0.67	1.8
0.32	0.75	0.63	0.47	8.6	-0.8	0.37	0.72	-0.9	0.69	0.32	0.75	0.63	0.47	8.6	-0.8	0.37	0.72	-0.9	0.69
-0.66	-0.28	3	2.62	-0.8	11.8	-4.9	0.3	3	-3	-0.66	-0.28	3	2.62	-0.8	11.8	-4.9	0.3	3	-3
0.96	5.04	1.7	-1.96	0.37	-4.9	8.24	-0.9	5.9	-0.6	0.96	5.04	1.7	-1.96	0.37	-4.9	8.24	-0.9	5.9	-0.6
-1.6	1.7	-4.3	2.1	0.72	0.3	-0.9	1.2	-0.96	0.56	-1.6	1.7	-4.3	2.1	0.72	0.3	-0.9	1.2	-0.96	0.56
0.8	0.54	3.1	0.67	-0.9	3	5.9	-0.96	0.93	-0.3	0.8	0.54	3.1	0.67	-0.9	3	5.9	-0.96	0.93	-0.3
-0.1	7.2	-2	1.8	0.69	-3	-0.6	0.56	-0.3	0.99	-0.1	7.2	-2	1.8	0.69	-3	-0.6	0.56	-0.3	0.99
8.7	0.43	-4.61	0.36	0.32	-0.66	0.96	-1.6	0.8	-0.1	8.7	0.43	-4.61	0.36	0.32	-0.66	0.96	-1.6	0.8	-0.1
0.43	0.83	-0.97	0.22	0.75	-0.28	5.04	1.7	0.54	7.2	0.43	0.83	-0.97	0.22	0.75	-0.28	5.04	1.7	0.54	7.2
-4.61	-0.97	9	-2	0.63	3	1.7	-4.3	3.1	-2	-4.61	-0.97	9	-2	0.63	3	1.7	-4.3	3.1	-2
0.36	0.22	-2	5.3	0.47	2.62	-1.96	2.1	0.67	1.8	0.36	0.22	-2	5.3	0.47	2.62	-1.96	2.1	0.67	1.8
0.32	0.75	0.63	0.47	8.6	-0.8	0.37	0.72	-0.9	0.69	0.32	0.75	0.63	0.47	8.6	-0.8	0.37	0.72	-0.9	0.69
-0.66	-0.28	3	2.62	-0.8	11.8	-4.9	0.3	3	-3	-0.66	-0.28	3	2.62	-0.8	11.8	-4.9	0.3	3	-3
0.96	5.04	1.7	-1.96	0.37	-4.9	8.24	-0.9	5.9	-0.6	0.96	5.04	1.7	-1.96	0.37	-4.9	8.24	-0.9	5.9	-0.6
-1.6	1.7	-4.3	2.1	0.72	0.3	-0.9	1.2	-0.96	0.56	-1.6	1.7	-4.3	2.1	0.72	0.3	-0.9	1.2	-0.96	0.56
0.8	0.54	3.1	0.67	-0.9	3	5.9	-0.96	0.93	-0.3	0.8	0.54	3.1	0.67	-0.9	3	5.9	-0.96	0.93	-0.3
-0.1	7.2	-2	1.8	0.69	-3	-0.6	0.56	-0.3	0.99	-0.1	7.2	-2	1.8	0.69	-3	-0.6	0.56	-0.3	0.99

References

Abookazemi, K., Mustafa, M.W., and Ahmad, H., Structured genetic algorithm technique for unit commitment problem, *International Journal of Recent Trends in Engineering*, 1(3), 135–139, 2009.

Labbi, Y. and Attous, B.D., A hybrid GA–PS method to solve the economic load dispatch problem, *Journal of Theoretical and Applied Information Technology*, 15(1), 61–68, 2010.

Park, J.-B., Jeong, Y.-W., Shin, J.-R., and Lee, K.Y., An improved particle swarm optimization for nonconvex economic dispatch problems, *IEEE Transactions on Power Systems*, 25(1), 156–166, February 2010.

Prabhakar, K.S., Palanisamy, K., Jacob, R.I., and Kothari, D.P., Security constrained UCP with operational and power flow constraints, *International Journal of Recent Trends in Engineering*, 1(3), 106–114, 2009.

Appendix B: Harmonic Reduction— MATLAB®/Simulink® Models

A set of MATLAB®/Simulink® models used for implementing harmonic reduction in Chapter 3 is presented in this appendix (Figures B.1 through B.10).

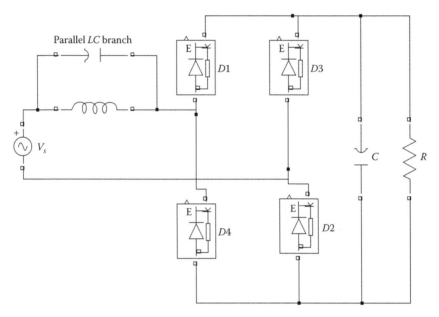

FIGURE B.1
Two-pulse drive rectifier section 1.

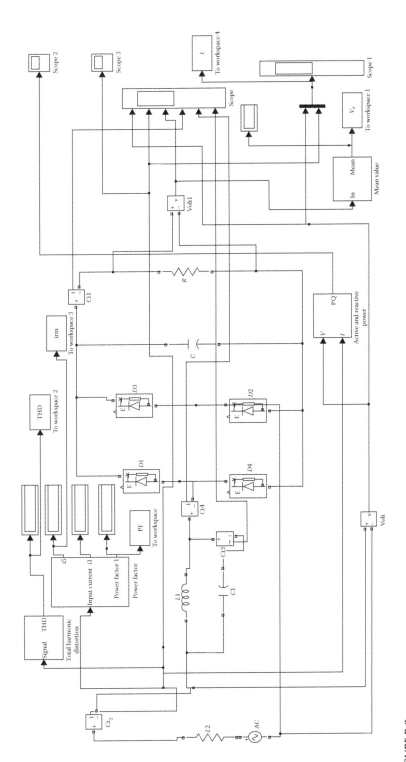

FIGURE B.2

Power and control structure for six-pulse drive.

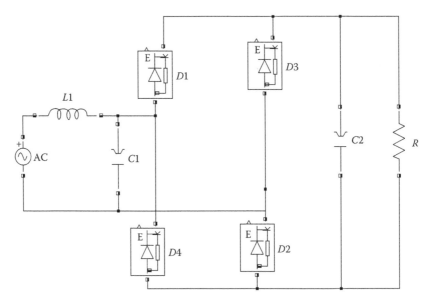

FIGURE B.3
Two-pulse drive rectifier section 2.

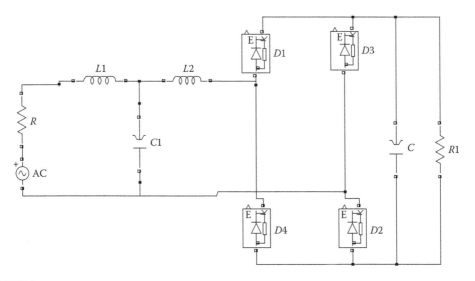

FIGURE B.4
Diode bridge rectifier with series LC filter.

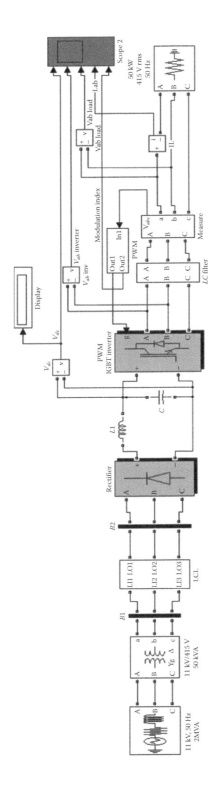

FIGURE B.5

Simulation model of the six-pulse drive and filter model.

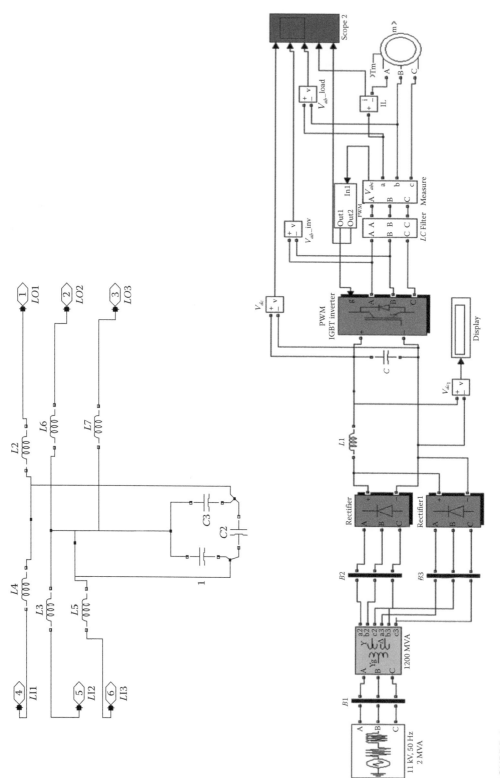

FIGURE B.6
Simulation model of the 12-pulse drive.

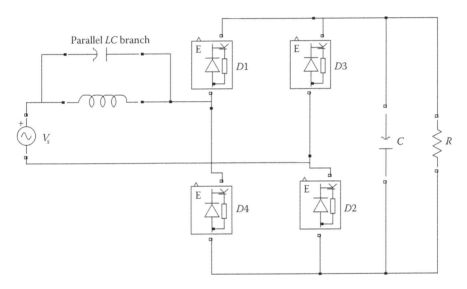

FIGURE B.7
Parallel LC connected in series with the source.

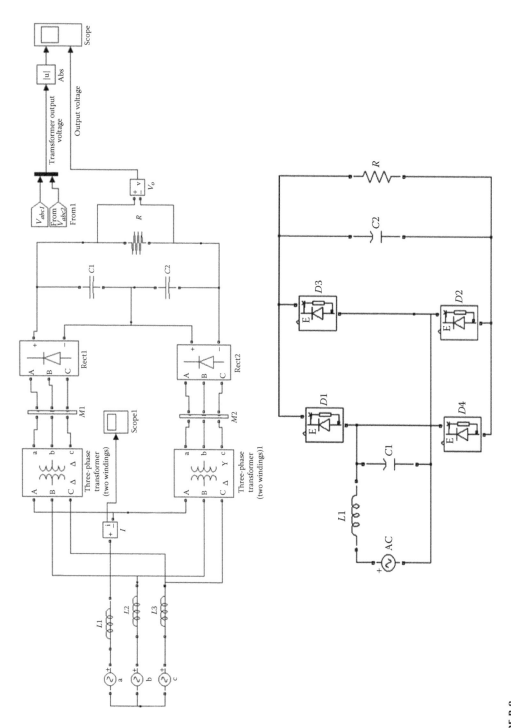

FIGURE B.8

Full bridge rectifier circuit with shunt LC filter.

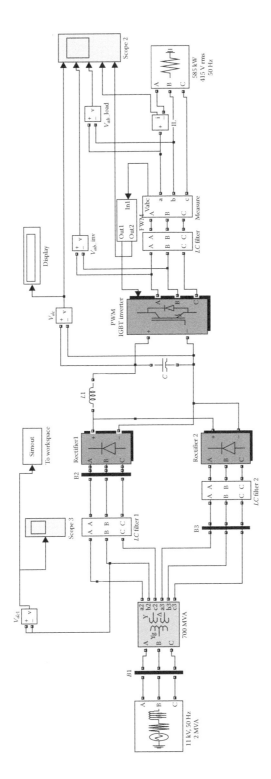

FIGURE B.9
Drive power structure.

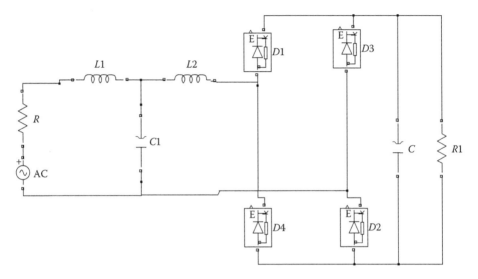

FIGURE B.10
Full bridge rectifier circuit with LCL filter.

Appendix C: MATLAB®/Simulink® Functions—An Overview

MATLAB® is a high-performance language for technical computing. It integrates computation, visualization, and programming in an easy-to-use environment where problem and solution are expressed in familiar mathematical notation. MATLAB is an interactive system whose basic data element is an array that does not require dimensioning. The name MATLAB stands for matrix laboratory. The combination of analysis capabilities, flexibility, and power graphics makes MATLAB the premier software package for electrical engineers. It is the mainstay of the mathematical department software lineup and also available for PCs and Macintoshes. MATLAB was originally written to provide easy access to matrix software developed by the LINPACK and EISPACK projects.

Today MATLAB engines incorporate the LINPACK and Basic Linear Algebra Subprograms (BLAS) libraries, embedding the state-of-the-art software for matrix computation. It is extremely powerful, simple to use, and can be found in most research and engineering environments. It gets more powerful and specialized with the addition of "toolboxes," additional functions added to the MATLAB environment by various developers for specific tasks or fields. In this section, the MATLAB functions used in fuzzy algorithm, genetic algorithm (GA), particle swarm optimization (PSO), ant colony optimization, and hybrid genetic algorithm are listed and their definitions are provided.

- *fuzzy*—Basic fuzzy inference system (FIS) editor.
- *mfedit*—Membership function editor.
- *ruleedit*—Rule editor and parser.
- *ruleview*—Rule viewer and fuzzy inference diagram.
- *Dsigmf*—Difference of two-sigmoid membership functions.
- *trapmf*—Trapezoidal membership function.
- *trimf*—Triangular membership function.
- *zmfZ*—Shaped curve membership function.
- *addmf*—Add a membership function to an FIS.
- *addrule*—Add rule to FIS.
- *addvar*—Add variable to FIS.
- *defuzz*—Defuzzify membership function.
- *newfis*—Create new FIS.
- *rmmf*—Remove membership function from FIS.
- *rmvar*—Remove variable from FIS.
- *setfis*—Set fuzzy system properties
- *showrule*—Display FIS rules.
- *writefis*—Save FIS to disk.

- *fuzblock*—Simulink® fuzzy logic library.
- *sffis*—Fuzzy inference S-function for Simulink.
- *ga*—Implement the GA at the command line to minimize an objective function.
- *gaoptimget*—Get values of a GA options structure.
- *gaoptimset*—Create a GA options structure.
- *gatool*—Open the GA tool.
- *population type* (Population Type)—Specifies the data type of the input to the fitness function. Population type can be set to any one of the following:
  - *Double Vector* (double Vector)—Use this option if the individuals in the population have type double.
  - *Bit string* (bitstring)—Use this option if the individuals in the population are bit strings.
  - *Custom* (custom)—Use this option to create a population whose data type is neither of the preceding.
  - *Creation function* (CreationFcn)—Specifies the function that creates the initial population for GA.
  - *Mutation function* (MutationFcn)—This option is used to specify Mutation function.
  - *Crossover function* (CrossoverFcn)—Specifies the function that performs the crossover.
  - Set options = *gaoptimset*(CreationFcn, @myfun);—Get values of a GA options structure function.
  - Population = *myfun*(GenomeLength, FitnessFcn, options)—The input arguments to the function are
    - *Genomelength*—Number of independent variables for the fitness function
    - *FitnessFcn*—Fitness function
    - *options*—Options structure
- The function returns Population, the initial population for the GA.
- *Genomelength*—Number of independent variables for the fitness function.
- *FitnessFcn*—Fitness function options.
- *Options structure Off* (off)—Only the final answer is displayed.
- *Iterative* (iter)—Information is displayed at each iteration.
- *Diagnose* (diagnose)—Information is displayed at each iteration. In addition, options that are changed from the defaults are listed.
- *Final* (final)—The outcome of the GA (successful or unsuccessful), the reason for stopping, and the final point.
- *Hybrid Function Option*—A hybrid function is another minimization function that runs after the GA terminates. Hybrid function is specified in Hybrid function (*HybridFcn*) options.
- *fminsearch* (@fminsearch)—Uses the MATLAB function fminsearch
- *patternsearch* (@patternsearch)—To find the minimum of a function using a pattern search.

- *fminunc* (@fminunc—Uses the Optimization Toolbox function fminunc.
- *Generations* (Generations)—Specifies the maximum number of iterations the GA will perform. The default is 100.
- *Time limit* (TimeLimit)—Specifies the maximum time in seconds the GA runs before stopping.
- *Fitness limit* (FitnessLimit)—The algorithm stops if the best fitness value is less than or equal to the value of Fitness limit.
- *Stall generations* (StallGenLimit)—The algorithm stops if there is no improvement in the best fitness value for the number of generations specified by Stall generations.
- *Stall time* (StallTimeLimit)—The algorithm stops if there is no improvement in the best fitness value for an interval of time in seconds specified by Stall time
- *PSOparams*—PSO parameters
- *P(1)*—Epochs between updating display, default = 100; if 0, no display.
- *P(2)*—Maximum number of iterations (epochs) to train, default = 2000.
- *P(3)*—Population size, default = 24.
- *P(4)*—Acceleration const 1 (local best influence), default = 2.
- *P(5)*—Acceleration const 2 (global best influence), default = 2.
- *P(6)*—Initial inertia weight, default = 0.9.
- *P(7)*—Final inertia weight, default = 0.4.
- *P(8)*—Epoch when inertial weight at final value, default = 1500.
- *P(9)*—Minimum global error gradient, if abs(Gbest(i+1)-Gbest(i)) < gradient over certain length of epochs, terminate run, default = 1e-25.
- *P(10)*—Epochs before error gradient criterion terminates run, default = 150, if the SSE does not change over 250 epochs then exit.
- *P(11)*—Error goal, if NaN then unconstrained min or max, default = NaN.
- *P(12)*—Type flag (which kind of PSO to use) 0 = Common PSO w/intertia (default) 1, 2 = Trelea types 1,2 3 = Clerc's Constricted PSO, Type 1'
- *P(13)*—PSOseed, default = 0

  = 0 for initial random positions

  = 1 for initial particles as user input

C.1 MATLAB® Commands

`matlab\general`—General purpose commands.

`matlab\ops`—Operators and special characters.

`matlab\lang`—Language constructs and debugging.

`matlab\elmat`—Elementary matrices and matrix manipulation.

`matlab\specmat`—Specialized matrices.

`matlab\elfun`—Elementary math functions.

`matlab\specfun`—Specialized math functions.

`matlab\matfun`—Matrix function-numerical linear algebra.

`matlab\datafun`—Data analysis and Fourier transform functions.

`matlab\polyfun`—Polynomial and interpolation functions.

`matlab\graphics`—General purpose graphics functions.

`matlab\color`—Color control and lighting model functions.

`matlab\sounds`—Sound processing functions.

`matlab\strfun`—Character string functions.

`matlab\iofun`—Low-level file I/O functions.

`matlab\demos`—The MATLAB Expo and other demonstrations.

C.2 Simulink®

Simulink is an interactive environment for modeling, analyzing, and simulating a wide variety of dynamic systems. It supports linear and nonlinear systems, modeled in continuous time, sampled time, or a hybrid of the two. Simulink provides a graphical user interface for constructing block diagram models using "drag and drop" operations. A system is configured in terms of block diagram representation from a library of standard components. A system block diagram representation is built easily, and the simulation results are displayed quickly. Simulation algorithms and parameters can be changed in the middle of a simulation with intuitive results, thus providing the user with a real world. Simulink is particularly useful for studying the effects of nonlinearities on the behavior of the system, and as such it is also an ideal research tool. The key features of Simulink are as follows:

- Interactive simulations with live display.
- A comprehensive block library for creating linear, nonlinear, discrete, or hybrid multi-input/output systems.
- Seven integration methods for fixed steps and variable-step systems.
- Unlimited hierarchical model structure.
- Scalar and vector connections.
- Mask facility for creating custom blocks and block libraries.

Simulink provides an open architecture that allows extension of the simulation environment:

- The model parameters can be changed either interactively or in batch mode while simulation is running.
- Custom blocks and block libraries can be created with own icons and user interface from MATLAB, FORTRAN, or C code.
- C code from Simulink models can be generated for embedded applications and rapid prototyping of control systems.

- Hierarchical models can be created by grouping blocks into systems.
- Simulink provides immediate access to the mathematical, graphical, and programming capabilities of MATLAB.
- Analysis of data, automation procedures, and optimization of parameters can be done directly in Simulink.
- The advanced design and analysis capabilities of the tool boxes can be executed from within a simulation using the mask facility in Simulink.
- The Simulink block library can be executed with special purpose block sets.

C.3 Simulation Parameters and Solver

Parameters from the simulation menu are used for setting the simulation parameters and selecting the solver. Simulink displays the simulation parameters dialog box, which uses three "pages": solver, work space I/O, and diagnostics to manage simulation parameters.

C.3.1 Solver Page

When the solver tab is selected, the solver page is displayed and it allows the following operations:

- Set the start and stop times: the start time and the stop time for the simulation can be changed by entering new values in the start time and stop time fields. The default start time is 0.0 s, and the default stop time is 10.0 s.
- Choose the solver and specify solver parameters: the default solver provides accurate and efficient results for most of the problems. There are two types of solvers: variable-step solver and fixed-step solver. Variable solvers can modify their step sizes during simulation. These are ode45, ode23, ode113, ode15s, ode23s, discrete, and the default is ode45. For variable-step solvers, the maximum and initial step size parameters can be varied. For fixed-step solvers, ode5, ode4, ode3, ode2, ode1, and discrete can be chosen.
- The output options area of the dialog box enables the control of the output generated by simulation.

C.3.2 Workspace I/O Page

The workspace I/O page manages the input from the MATLAB workspace and the output to the MATLAB workspace. The functions of workspace include the following:

1. Loading of input from the workspace can be specified either as MATLAB command or as matrix for the import blocks.
2. Saving the output to the workspace-return variables can be specified by selecting the time, state, and/or output check boxes in the save to workspace area.

C.3.3 Diagnostic Page

The diagnostic page allows the selection of level of warming messages displayed during a simulation.

C.4 Simulation Parameters Dialog Box

The following section summarizes the actions performed by the dialog box buttons, which appear on the bottom of each dialog box page.

C.4.1 Button Action

- *Apply*: Applies the current parameter values and keeps the dialog box open; during a simulation, the parameter values are applied immediately.
- *Revert*: Changes the parameter values back to the values they had when the dialog box was most recently opened and applies the parameter.
- *Help*: Displays help text for the dialog box page.
- *Close*: Applies the parameter values and closes the dialog box.

During a simulation, the parameter values are applied immediately. To stop the simulation, choose *stop* from the simulation menu. The keyboard shortcut for stopping a simulation is CTRL – T. We can suspend the running simulation by choosing *pause* from the simulation menu; when we select *pause*, the menu item changes to *continue*. We can proceed with a suspended simulation by choosing *continue*.

C.5 Block Diagram Construction

The easy-to-use pull-down menus allow us to create a Simulink block diagram, or open an existing file, perform the simulation, and make any modifications. Basically, one has to specify the model of the system (state space, discrete, transfer functions, nonlinear codes, etc), the input (source) to the system, and where the output (sink) of the simulation of the system will go. When putting blocks together into a model, add the blocks to the model window before adding the lines that connect them. The block diagram is constructed to represent the digital simulator of the single- and two-area interconnected power system. It is clear from the block diagram that each area has its own control on its operation. There are three main blocks: speed governor, turbine, and power system. An intelligent controller is incorporated in each block for obtaining zero steady-state error for the deviations of both the system frequency and tie-line power. The block diagrams can be simulated for different load and regulation values during the course of simulation.

Appendix D: Instances of Job-Shop Scheduling Problems

The list of benchmark instances with the best known makespan and problem size is presented in this appendix. The benchmark instances for solving the job-shop scheduling problems (JSSPs) are the test vectors listing the machine number and processing time for each step of the job. Each instance consists of the number of jobs and the number of machines, along with the upper and lower bound of the best known makespan. In this thesis work, 162 instances were tested, and the specifications are listed in Tables D.1 through D.4. These instances were taken from the online resources of the University of Brunel, London, United Kingdom (http://people.brunel.ac.uk/).

D.1 Specifications of JSSP Instances

The details of contribution of the instances used in this thesis are as follows:

- Instances ABZ5–ABZ9 are contributed by Adams et al. (1988).
- Instances FT06, FT10, and FT20 are obtained from Fisher and Thompson (1963).
- Instances ORB01–ORB10 are contributed by Applegate and Cook (1991).
- Twenty instances, SWV01–SWV20, are reported by Storer et al. (1992).
- Instances YN1–YN4 are developed by Storer et al. (1992).
- Forty benchmark instances LA01–LA40 are contributed by Lawrence (1984).
- Taillard (1993) presented a set of 80 instances denoted by TA01–TA80, with job sizes varying between 15 and 100 and machine size varying between 15 and 20.

TABLE D.1

Specifications of Group A Instances

Instance	No. of Jobs × No. of Machines	Makespan Lower Bound	Makespan Upper Bound	Instance	No. of Jobs × No. of Machines	Makespan Lower Bound	Makespan Upper Bound
ABZ5	10 × 10		1234	SWV04	20 × 10	1450	1470
ABZ6	10 × 10		943	SWV05	20 × 10	1424	1424
ABZ7	20 × 15		656	SWV06	20 × 15	1591	1675
ABZ8	20 × 15	645	665	SWV07	20 × 15	1446	1594
ABZ9	20 × 15	661	679	SWV08	20 × 15	1640	1755
FT06	6 × 6		55	SWV09	20 × 15	1604	1661
FT10	10 × 10		930	SWV10	20 × 15	1631	1743
FT20	20 × 5		1165	SWV11	50 × 10		2983
ORB01	10 × 10		1059	SWV12	50 × 10	2972	2979
ORB02	10 × 10		888	SWV13	50 × 10		3104
ORB03	10 × 10		1005	SWV14	50 × 10		2968
ORB04	10 × 10		1005	SWV15	50 × 10	2885	2886
ORB05	10 × 10		887	SWV16	50 × 10		2924
ORB06	10 × 10		1010	SWV17	50 × 10		2794
ORB07	10 × 10		397	SWV18	50 × 10		2852
ORB08	10 × 10		899	SWV19	50 × 10		2843
ORB09	10 × 10		934	SWV20	50 × 10		2823
ORB10	10 × 10		944	YN1	20 × 20	836	884
SWV01	20 × 10		1407	YN2	20 × 20	861	904
SWV02	20 × 10		1475	YN3	20 × 20	827	892
SWV03	20 × 10	1369	1398	YN4	20 × 20	918	968

TABLE D.2

Specifications of Group B Instances

Instance	No. of Jobs × No. of Machines	Makespan Upper Bound	Instance	No. of Jobs × No. of Machines	Makespan Upper Bound
LA01	10 × 5	666	LA21	15 × 10	1046
LA02	10 × 5	655	LA22	15 × 10	927
LA03	10 × 5	597	LA23	15 × 10	1032
LA04	10 × 5	590	LA24	15 × 10	935
LA05	10 × 5	593	LA25	15 × 10	977
LA06	15 × 5	926	LA26	20 × 10	1218
LA07	15 × 5	890	LA27	20 × 10	1235
LA08	15 × 5	863	LA28	20 × 10	1216
LA09	15 × 5	951	LA29	20 × 10	1157
LA10	15 × 5	958	LA30	20 × 10	1355
LA11	20 × 5	1222	LA31	30 × 10	1784
LA12	20 × 5	1039	LA32	30 × 10	1850
LA13	20 × 5	1150	LA33	30 × 10	1719
LA14	20 × 5	1292	LA34	30 × 10	1721
LA15	20 × 5	1207	LA35	30 × 10	1888
LA16	10 × 10	945	LA36	15 × 15	1268
LA17	10 × 10	784	LA37	15 × 15	1397
LA18	10 × 10	848	LA38	15 × 15	1196
LA19	10 × 10	842	LA39	15 × 15	1233
LA20	10 × 10	902	LA40	15 × 15	1222

TABLE D.3

Specifications of Group C Instances

Instance	No. of Jobs × No. of Machines	Makespan Lower Bound	Makespan Upper Bound	Instance	No. of Jobs × No. of Machines	Makespan Lower bound	Makespan Upper bound
TA01	15 × 15		1231	TA21	20 × 20	1539	1644
TA02	15 × 15		1244	TA22	20 × 20	1511	1600
TA03	15 × 15		1218	TA23	20 × 20	1472	1557
TA04	15 × 15		1175	TA24	20 × 20	1602	1647
TA05	15 × 15		1224	TA25	20 × 20	1504	1595
TA06	15 × 15		1238	TA26	20 × 20	1539	1645
TA07	15 × 15		1227	TA27	20 × 20	1616	1680
TA08	15 × 15		1217	TA28	20 × 20	1591	1614
TA09	15 × 15		1274	TA29	20 × 20	1514	1625
TA10	15 × 15		1241	TA30	20 × 20	1473	1584
TA11	20 × 15	1323	1361	TA31	30 × 15		1764
TA12	20 × 15	1351	1367	TA32	30 × 15	1774	1796
TA13	20 × 15	1282	1342	TA33	30 × 15	1778	1793
TA14	20 × 15		1345	TA34	30 × 15	1828	1829
TA15	20 × 15	1304	1340	TA35	30 × 15		2007
TA16	20 × 15	1302	1360	TA36	30 × 15		1819
TA17	20 × 15		1462	TA37	30 × 15	1771	1778
TA18	20 × 15	1369	1396	TA38	30 × 15		1673
TA19	20 × 15	1297	1335	TA39	30 × 15		1795
TA20	20 × 15	1318	1351	TA40	30 × 15	1631	1674

TABLE D.4

Specifications of Group D Instances

Instance	No. of Jobs × No. of Machines	Makespan Lower Bound	Makespan Upper Bound	Instance	No. of Jobs × No. of Machines	Makespan Lower Bound	Makespan Upper Bound
TA41	30 × 20	1859	2006	TA61	50 × 20		2868
TA42	30 × 20	1867	1945	TA62	50 × 20		2869
TA43	30 × 20	1809	1848	TA63	50 × 20		2755
TA44	30 × 20	1927	1983	TA64	50 × 20		2702
TA45	30 × 20	1997	2000	TA65	50 × 20		2725
TA46	30 × 20	1940	2008	TA66	50 × 20		2845
TA47	30 × 20	1789	1897	TA67	50 × 20		2825
TA48	30 × 20	1912	1945	TA68	50 × 20		2784
TA49	30 × 20	1915	1966	TA69	50 × 20		3071
TA50	30 × 20	1807	1924	TA70	50 × 20		2995
TA51	50 × 15		2760	TA71	100 × 20		5464
TA52	50 × 15		2756	TA72	100 × 20		5181
TA53	50 × 15		2717	TA73	100 × 20		5568
TA54	50 × 15		2839	TA74	100 × 20		5339
TA55	50 × 15		2679	TA75	100 × 20		5392
TA56	50 × 15		2781	TA76	100 × 20		5342
TA57	50 × 15		2943	TA77	100 × 20		5436
TA58	50 × 15		2885	TA78	100 × 20		5394
TA59	50 × 15		2655	TA79	100 × 20		5358
TA60	50 × 15		2723	TA80	100 × 20		5183

References

Adams, J., Balas, E., and Zawack, D., The shifting bottleneck procedure for job shop scheduling, *Management Science*, 34, 391–401, 1988.

Applegate, D. and Cook, W., A computational study of the job-shop scheduling problem, *ORSA Journal on Computing*, 3, 149–156, 1991.

Fisher, H. and Thompson, G.L., *Probabilistic Learning Combinations of Local Job-Shop Scheduling Rules*, Englewood Cliffs, NJ: Prentice-Hill, 1963.

Lawrence, S., Resource constrained project scheduling: An experimental investigation of heuristic scheduling techniques, Supplement PA1984, Graduate School of Industrial Administration, Carnegie Mellon University, Pittsburgh, 1984.

Storer, R.H., Wu, S.D., and Vaccari, R., New search spaces for sequencing instances with application to job shop scheduling, *Management Science* 38(10), 1495–1509, 1992.

Taillard, E.D., Benchmarks for basic scheduling problems, *European Journal of Operational Research*, 64(2), 108–117, 1993.

Appendix E: MDVRP Instances

A company may have several depots from which it can serve its customers. If the customers are clustered around depots, then the distribution problem should be modeled as a set of independent vehicle routing problems. However, if the customers and the depots are intermingled, then a multi-depot vehicle routing problem (MDVRP) should be solved. MDVRP requires the assignment of customers to depots with a fleet of vehicles based at each depot. Each vehicle originates from one depot, services the customers assigned to that depot, and returns to the same depot. The objective of the problem is to service all customers while minimizing the number of vehicles and travel distance. The formal descriptions of the MDVRP along with the specifications of the benchmark instances chosen in this research are presented in this appendix.

The formal description for the MDVRP can be stated as follows:

- *Objective*: The objective is to minimize the vehicle fleet and the sum of travel time, and the total demand of commodities must be served from several depots.

- *Feasibility*: A solution is feasible if each route satisfies the standard vehicle routing problem (VRP) constraints and begins and ends at the same depot.

- *Formulation*: The VRP is extended to multiple depots, with the vertex set $V = \{v_1, v_2, \ldots, v_n\} \cup V_0$ where $V_0 = \{v_{01}, v_{02}, \ldots, v_{0d}\}$ is the vertex representing the depots. A route i is defined by $R_i = \{d, v_1, \ldots, v_m, d\}$ with $d \in V_0$, where m is the number of vehicles, and d is a vector of customer demands. The cost of a route is calculated as $C(R_i) = \sum_{i=0}^{m} c_{i,i+1} + \sum_{i=1}^{m} \delta_i$, where δ_i is the service time required to unload all goods.

A set of five Cordeau's instances—such as p01, p02, p03, p04, and p06—is taken as the benchmark problem for solving an MDVRP in this thesis work. These instances were taken from the online resources of the University of Malaga, Spain (http://neo.lcc.uma.es/radi-aeb/WebVRP/). In this section, the specifications of the instances are described with the information provided as follows:

The first line of data denotes:

$$\text{type } m \; n \; t$$

where
    type = 0 (VRP)
            1 (PVRP)
            2 (MDVRP)
            3 (SDVRP)
            4 (VRPTW)
            5 (PVRPTW)
            6 (MDVRPTW)
            7 (SDVRPTW)
    m = number of vehicles
    n = number of customers
    t = number of days (PVRP), depots (MDVRP) or vehicle types (SDVRP)

The next t lines contain, for each day (or depot or vehicle type), the following information:

$$D \, Q$$

where
    D = maximum duration of a route
    Q = maximum load of a vehicle

The next lines contain, for each customer, the following information:

$$i \; x \; y \; d \; q \; f \; a \; \text{list} \; e \; l$$

where
    i = customer number
    x = x-coordinate
    y = y-coordinate
    d = service duration
    q = demand
    f = frequency of visit
    a = number of possible visit combinations
    list = list of all possible visit combinations
    e = beginning of time window (earliest time for start of service), if any
    l = end of time window (latest time for start of service), if any

A.E.1 Cordeau's p01 Instance

The p01 instance has 50 customers and 4 depots, and the number of vehicles in each depot is 8. Each vehicle capacity is 80, and the demand of each customer is as follows:

2	8	50	4						
0	80								
0	80								
0	80								
0	80								
1	37	52	0	7	1	4	1	2	4	8
2	49	49	0	30	1	4	1	2	4	8
3	52	64	0	16	1	4	1	2	4	8

(Continued)

4	20	26	0	9	1	4	1	2	4	8
5	40	30	0	21	1	4	1	2	4	8
6	21	47	0	15	1	4	1	2	4	8
7	17	63	0	19	1	4	1	2	4	8
8	31	62	0	23	1	4	1	2	4	8
9	52	33	0	11	1	4	1	2	4	8
10	51	21	0	5	1	4	1	2	4	8
11	42	41	0	19	1	4	1	2	4	8
12	31	32	0	29	1	4	1	2	4	8
13	5	25	0	23	1	4	1	2	4	8
14	12	42	0	21	1	4	1	2	4	8
15	36	16	0	10	1	4	1	2	4	8
16	52	41	0	15	1	4	1	2	4	8
17	27	23	0	3	1	4	1	2	4	8
18	17	33	0	41	1	4	1	2	4	8
19	13	13	0	9	1	4	1	2	4	8
20	57	58	0	28	1	4	1	2	4	8
21	62	42	0	8	1	4	1	2	4	8
22	42	57	0	8	1	4	1	2	4	8
23	16	57	0	16	1	4	1	2	4	8
24	8	52	0	10	1	4	1	2	4	8
25	7	38	0	28	1	4	1	2	4	8
26	27	68	0	7	1	4	1	2	4	8
27	30	48	0	15	1	4	1	2	4	8
28	43	67	0	14	1	4	1	2	4	8
29	58	48	0	6	1	4	1	2	4	8
30	58	27	0	19	1	4	1	2	4	8
31	37	69	0	11	1	4	1	2	4	8
32	38	46	0	12	1	4	1	2	4	8
33	46	10	0	23	1	4	1	2	4	8
34	61	33	0	26	1	4	1	2	4	8
35	62	63	0	17	1	4	1	2	4	8
36	63	69	0	6	1	4	1	2	4	8
37	32	22	0	9	1	4	1	2	4	8
38	45	35	0	15	1	4	1	2	4	8
39	59	15	0	14	1	4	1	2	4	8
40	5	6	0	7	1	4	1	2	4	8
41	10	17	0	27	1	4	1	2	4	8
42	21	10	0	13	1	4	1	2	4	8
43	5	64	0	11	1	4	1	2	4	8
44	30	15	0	16	1	4	1	2	4	8
45	39	10	0	10	1	4	1	2	4	8
46	32	39	0	5	1	4	1	2	4	8
47	25	32	0	25	1	4	1	2	4	8
48	25	55	0	17	1	4	1	2	4	8
49	48	28	0	18	1	4	1	2	4	8
50	56	37	0	10	1	4	1	2	4	8

A.E.2 Cordeau's p02 Instance

Fifty customers and 4 depots are available in the p02 instance and each depot has 4 vehicles. The capacity of each vehicle is 100, and the demand of each customer is as follows:

2	5	50	4							
0	100									
0	100									
0	100									
0	100									
1	37	52	0	7	1	4	1	2	4	8
2	49	49	0	30	1	4	1	2	4	8
3	52	64	0	16	1	4	1	2	4	8
4	20	26	0	9	1	4	1	2	4	8
5	40	30	0	21	1	4	1	2	4	8
6	21	47	0	15	1	4	1	2	4	8
7	17	63	0	19	1	4	1	2	4	8
8	31	62	0	23	1	4	1	2	4	8
9	52	33	0	11	1	4	1	2	4	8
10	51	21	0	5	1	4	1	2	4	8
11	42	41	0	19	1	4	1	2	4	8
12	31	32	0	29	1	4	1	2	4	8
13	5	25	0	23	1	4	1	2	4	8
14	12	42	0	21	1	4	1	2	4	8
15	36	16	0	10	1	4	1	2	4	8
16	52	41	0	15	1	4	1	2	4	8
17	27	23	0	3	1	4	1	2	4	8
18	17	33	0	41	1	4	1	2	4	8
19	13	13	0	9	1	4	1	2	4	8
20	57	58	0	28	1	4	1	2	4	8
21	62	42	0	8	1	4	1	2	4	8
22	42	57	0	8	1	4	1	2	4	8
23	16	57	0	16	1	4	1	2	4	8
24	8	52	0	10	1	4	1	2	4	8
25	7	38	0	28	1	4	1	2	4	8
26	27	68	0	7	1	4	1	2	4	8
27	30	48	0	15	1	4	1	2	4	8
28	43	67	0	14	1	4	1	2	4	8
29	58	48	0	6	1	4	1	2	4	8
30	58	27	0	19	1	4	1	2	4	8
31	37	69	0	11	1	4	1	2	4	8
32	38	46	0	12	1	4	1	2	4	8
33	46	10	0	23	1	4	1	2	4	8
34	61	33	0	26	1	4	1	2	4	8
35	62	63	0	17	1	4	1	2	4	8
36	63	69	0	6	1	4	1	2	4	8
37	32	22	0	9	1	4	1	2	4	8

(Continued)

38	45	35	0	15	1	4	1	2	4	8
39	59	15	0	14	1	4	1	2	4	8
40	5	6	0	7	1	4	1	2	4	8
41	10	17	0	27	1	4	1	2	4	8
42	21	10	0	13	1	4	1	2	4	8
43	5	64	0	11	1	4	1	2	4	8
44	30	15	0	16	1	4	1	2	4	8
45	39	10	0	10	1	4	1	2	4	8
46	32	39	0	5	1	4	1	2	4	8
47	25	32	0	25	1	4	1	2	4	8
48	25	55	0	17	1	4	1	2	4	8
49	48	28	0	18	1	4	1	2	4	8
50	56	37	0	10	1	4	1	2	4	8

A.E.3 Cordeau's p03 Instance

The p03 problem instance has 75 customers and 5 depots, and the number of vehicles in each depot is 7. Each vehicle capacity is 140, and the demand of each customer is as follows:

2	7	75	5							
0	140									
0	140									
0	140									
0	140									
0	140									
1	22	22	0	18	1	5	1	2	4	8	16
2	36	26	0	26	1	5	1	2	4	8	16
3	21	45	0	11	1	5	1	2	4	8	16
4	45	35	0	30	1	5	1	2	4	8	16
5	55	20	0	21	1	5	1	2	4	8	16
6	33	34	0	19	1	5	1	2	4	8	16
7	50	50	0	15	1	5	1	2	4	8	16
8	55	45	0	16	1	5	1	2	4	8	16
9	26	59	0	29	1	5	1	2	4	8	16
10	40	66	0	26	1	5	1	2	4	8	16
11	55	65	0	37	1	5	1	2	4	8	16
12	35	51	0	16	1	5	1	2	4	8	16
13	62	35	0	12	1	5	1	2	4	8	16
14	62	57	0	31	1	5	1	2	4	8	16
15	62	24	0	8	1	5	1	2	4	8	16
16	21	36	0	19	1	5	1	2	4	8	16
17	33	44	0	20	1	5	1	2	4	8	16
18	9	56	0	13	1	5	1	2	4	8	16

(Continued)

19	62	48	0	15	1	5	1	2	4	8	16
20	66	14	0	22	1	5	1	2	4	8	16
21	44	13	0	28	1	5	1	2	4	8	16
22	26	13	0	12	1	5	1	2	4	8	16
23	11	28	0	6	1	5	1	2	4	8	16
24	7	43	0	27	1	5	1	2	4	8	16
25	17	64	0	14	1	5	1	2	4	8	16
26	41	46	0	18	1	5	1	2	4	8	16
27	55	34	0	17	1	5	1	2	4	8	16
28	35	16	0	29	1	5	1	2	4	8	16
29	52	26	0	13	1	5	1	2	4	8	16
30	43	26	0	22	1	5	1	2	4	8	16
31	31	76	0	25	1	5	1	2	4	8	16
32	22	53	0	28	1	5	1	2	4	8	16
33	26	29	0	27	1	5	1	2	4	8	16
34	50	40	0	19	1	5	1	2	4	8	16
35	55	50	0	10	1	5	1	2	4	8	16
36	54	10	0	12	1	5	1	2	4	8	16
37	60	15	0	14	1	5	1	2	4	8	16
38	47	66	0	24	1	5	1	2	4	8	16
39	30	60	0	16	1	5	1	2	4	8	16
40	30	50	0	33	1	5	1	2	4	8	16
41	12	17	0	15	1	5	1	2	4	8	16
42	15	14	0	11	1	5	1	2	4	8	16
43	16	19	0	18	1	5	1	2	4	8	16
44	21	48	0	17	1	5	1	2	4	8	16
45	50	30	0	21	1	5	1	2	4	8	16
46	51	42	0	27	1	5	1	2	4	8	16
47	50	15	0	19	1	5	1	2	4	8	16
48	48	21	0	20	1	5	1	2	4	8	16
49	12	38	0	5	1	5	1	2	4	8	16
50	15	56	0	22	1	5	1	2	4	8	16
51	29	39	0	12	1	5	1	2	4	8	16
52	54	38	0	19	1	5	1	2	4	8	16
53	55	57	0	22	1	5	1	2	4	8	16
54	67	41	0	16	1	5	1	2	4	8	16
55	10	70	0	7	1	5	1	2	4	8	16
56	6	25	0	26	1	5	1	2	4	8	16
57	65	27	0	14	1	5	1	2	4	8	16
58	40	60	0	21	1	5	1	2	4	8	16
59	70	64	0	24	1	5	1	2	4	8	16
60	64	4	0	13	1	5	1	2	4	8	16
61	36	6	0	15	1	5	1	2	4	8	16
62	30	20	0	18	1	5	1	2	4	8	16
63	20	30	0	11	1	5	1	2	4	8	16
64	15	5	0	28	1	5	1	2	4	8	16
65	50	70	0	9	1	5	1	2	4	8	16
66	57	72	0	37	1	5	1	2	4	8	16

(Continued)

67	45	42	0	30	1	5	1	2	4	8	16
68	38	33	0	10	1	5	1	2	4	8	16
69	50	4	0	8	1	5	1	2	4	8	16
70	66	8	0	11	1	5	1	2	4	8	16
71	59	5	0	3	1	5	1	2	4	8	16
72	35	60	0	1	1	5	1	2	4	8	16
73	27	24	0	6	1	5	1	2	4	8	16
74	40	20	0	10	1	5	1	2	4	8	16
75	40	37	0	20	1	5	1	2	4	8	16

A.E.4 Cordeau's p04 Instance

In the p04 instance, 100 customers and 2 depots are available, and each depot has 12 vehicles. The capacity of each vehicle in every depot is 100.

2	12	100	2					
0	100							
0	100							
1	41	49	0	10	1	2	1	2
2	35	17	0	7	1	2	1	2
3	55	45	0	13	1	2	1	2
4	55	20	0	19	1	2	1	2
5	15	30	0	26	1	2	1	2
6	25	30	0	3	1	2	1	2
7	20	50	0	5	1	2	1	2
8	10	43	0	9	1	2	1	2
9	55	60	0	16	1	2	1	2
10	30	60	0	16	1	2	1	2
11	20	65	0	12	1	2	1	2
12	50	35	0	19	1	2	1	2
13	30	25	0	23	1	2	1	2
14	15	10	0	20	1	2	1	2
15	30	5	0	8	1	2	1	2
16	10	20	0	19	1	2	1	2
17	5	30	0	2	1	2	1	2
18	20	40	0	12	1	2	1	2
19	15	60	0	17	1	2	1	2
20	45	65	0	9	1	2	1	2
21	45	20	0	11	1	2	1	2
22	45	10	0	18	1	2	1	2
23	55	5	0	29	1	2	1	2
24	65	35	0	3	1	2	1	2
25	65	20	0	6	1	2	1	2

(Continued)

26	45	30	0	17	1	2	1	2
27	35	40	0	16	1	2	1	2
28	41	37	0	16	1	2	1	2
29	64	42	0	9	1	2	1	2
30	40	60	0	21	1	2	1	2
31	31	52	0	27	1	2	1	2
32	35	69	0	23	1	2	1	2
33	53	52	0	11	1	2	1	2
34	65	55	0	14	1	2	1	2
35	63	65	0	8	1	2	1	2
36	2	60	0	5	1	2	1	2
37	20	20	0	8	1	2	1	2
38	5	5	0	16	1	2	1	2
39	60	12	0	31	1	2	1	2
40	40	25	0	9	1	2	1	2
41	42	7	0	5	1	2	1	2
42	24	12	0	5	1	2	1	2
43	23	3	0	7	1	2	1	2
44	11	14	0	18	1	2	1	2
45	6	38	0	16	1	2	1	2
46	2	48	0	1	1	2	1	2
47	8	56	0	27	1	2	1	2
48	13	52	0	36	1	2	1	2
49	6	68	0	30	1	2	1	2
50	47	47	0	13	1	2	1	2
51	49	58	0	10	1	2	1	2
52	27	43	0	9	1	2	1	2
53	37	31	0	14	1	2	1	2
54	57	29	0	18	1	2	1	2
55	63	23	0	2	1	2	1	2
56	53	12	0	6	1	2	1	2
57	32	12	0	7	1	2	1	2
58	36	26	0	18	1	2	1	2
59	21	24	0	28	1	2	1	2
60	17	34	0	3	1	2	1	2
61	12	24	0	13	1	2	1	2
62	24	58	0	19	1	2	1	2
63	27	69	0	10	1	2	1	2
64	15	77	0	9	1	2	1	2
65	62	77	0	20	1	2	1	2
66	49	73	0	25	1	2	1	2
67	67	5	0	25	1	2	1	2
68	56	39	0	36	1	2	1	2
69	37	47	0	6	1	2	1	2
70	37	56	0	5	1	2	1	2
71	57	68	0	15	1	2	1	2
72	47	16	0	25	1	2	1	2
73	44	17	0	9	1	2	1	2

(Continued)

74	46	13	0	8	1	2	1	2
75	49	11	0	18	1	2	1	2
76	49	42	0	13	1	2	1	2
77	53	43	0	14	1	2	1	2
78	61	52	0	3	1	2	1	2
79	57	48	0	23	1	2	1	2
80	56	37	0	6	1	2	1	2
81	55	54	0	26	1	2	1	2
82	15	47	0	16	1	2	1	2
83	14	37	0	11	1	2	1	2
84	11	31	0	7	1	2	1	2
85	16	22	0	41	1	2	1	2
86	4	18	0	35	1	2	1	2
87	28	18	0	26	1	2	1	2
88	26	52	0	9	1	2	1	2
89	26	35	0	15	1	2	1	2
90	31	67	0	3	1	2	1	2
91	15	19	0	1	1	2	1	2
92	22	22	0	2	1	2	1	2
93	18	24	0	22	1	2	1	2
94	26	27	0	27	1	2	1	2
95	25	24	0	20	1	2	1	2
96	22	27	0	11	1	2	1	2
97	25	21	0	12	1	2	1	2
98	19	21	0	10	1	2	1	2
99	20	26	0	9	1	2	1	2
100	18	18	0	17	1	2	1	2

A.E.5 Cordeau's p06 Instance

The numbers of customers and depots in the p06 instance are 100 and 3. Each depot has 10 vehicles, and the capacity of each vehicle is 100.

2	10	100	3					
0	100							
0	100							
0	100							
1	41	49	0	10	1	3	1	2	4
2	35	17	0	7	1	3	1	2	4
3	55	45	0	13	1	3	1	2	4
4	55	20	0	19	1	3	1	2	4
5	15	30	0	26	1	3	1	2	4
6	25	30	0	3	1	3	1	2	4

(Continued)

7	20	50	0	5	1	3	1	2	4
8	10	43	0	9	1	3	1	2	4
9	55	60	0	16	1	3	1	2	4
10	30	60	0	16	1	3	1	2	4
11	20	65	0	12	1	3	1	2	4
12	50	35	0	19	1	3	1	2	4
13	30	25	0	23	1	3	1	2	4
14	15	10	0	20	1	3	1	2	4
15	30	5	0	8	1	3	1	2	4
16	10	20	0	19	1	3	1	2	4
17	5	30	0	2	1	3	1	2	4
18	20	40	0	12	1	3	1	2	4
19	15	60	0	17	1	3	1	2	4
20	45	65	0	9	1	3	1	2	4
21	45	20	0	11	1	3	1	2	4
22	45	10	0	18	1	3	1	2	4
23	55	5	0	29	1	3	1	2	4
24	65	35	0	3	1	3	1	2	4
25	65	20	0	6	1	3	1	2	4
26	45	30	0	17	1	3	1	2	4
27	35	40	0	16	1	3	1	2	4
28	41	37	0	16	1	3	1	2	4
29	64	42	0	9	1	3	1	2	4
30	40	60	0	21	1	3	1	2	4
31	31	52	0	27	1	3	1	2	4
32	35	69	0	23	1	3	1	2	4
33	53	52	0	11	1	3	1	2	4
34	65	55	0	14	1	3	1	2	4
35	63	65	0	8	1	3	1	2	4
36	2	60	0	5	1	3	1	2	4
37	20	20	0	8	1	3	1	2	4
38	5	5	0	16	1	3	1	2	4
39	60	12	0	31	1	3	1	2	4
40	40	25	0	9	1	3	1	2	4
41	42	7	0	5	1	3	1	2	4
42	24	12	0	5	1	3	1	2	4
43	23	3	0	7	1	3	1	2	4
44	11	14	0	18	1	3	1	2	4
45	6	38	0	16	1	3	1	2	4
46	2	48	0	1	1	3	1	2	4
47	8	56	0	27	1	3	1	2	4
48	13	52	0	36	1	3	1	2	4
49	6	68	0	30	1	3	1	2	4
50	47	47	0	13	1	3	1	2	4
51	49	58	0	10	1	3	1	2	4
52	27	43	0	9	1	3	1	2	4
53	37	31	0	14	1	3	1	2	4
54	57	29	0	18	1	3	1	2	4

(Continued)

55	63	23	0	2	1	3	1	2	4
56	53	12	0	6	1	3	1	2	4
57	32	12	0	7	1	3	1	2	4
58	36	26	0	18	1	3	1	2	4
59	21	24	0	28	1	3	1	2	4
60	17	34	0	3	1	3	1	2	4
61	12	24	0	13	1	3	1	2	4
62	24	58	0	19	1	3	1	2	4
63	27	69	0	10	1	3	1	2	4
64	15	77	0	9	1	3	1	2	4
65	62	77	0	20	1	3	1	2	4
66	49	73	0	25	1	3	1	2	4
67	67	5	0	25	1	3	1	2	4
68	56	39	0	36	1	3	1	2	4
69	37	47	0	6	1	3	1	2	4
70	37	56	0	5	1	3	1	2	4
71	57	68	0	15	1	3	1	2	4
72	47	16	0	25	1	3	1	2	4
73	44	17	0	9	1	3	1	2	4
74	46	13	0	8	1	3	1	2	4
75	49	11	0	18	1	3	1	2	4
76	49	42	0	13	1	3	1	2	4
77	53	43	0	14	1	3	1	2	4
78	61	52	0	3	1	3	1	2	4
79	57	48	0	23	1	3	1	2	4
80	56	37	0	6	1	3	1	2	4
81	55	54	0	26	1	3	1	2	4
82	15	47	0	16	1	3	1	2	4
83	14	37	0	11	1	3	1	2	4
84	11	31	0	7	1	3	1	2	4
85	16	22	0	41	1	3	1	2	4
86	4	18	0	35	1	3	1	2	4
87	28	18	0	26	1	3	1	2	4
88	26	52	0	9	1	3	1	2	4
89	26	35	0	15	1	3	1	2	4
90	31	67	0	3	1	3	1	2	4
91	15	19	0	1	1	3	1	2	4
92	22	22	0	2	1	3	1	2	4
93	18	24	0	22	1	3	1	2	4
94	26	27	0	27	1	3	1	2	4
95	25	24	0	20	1	3	1	2	4
96	22	27	0	11	1	3	1	2	4
97	25	21	0	12	1	3	1	2	4
98	19	21	0	10	1	3	1	2	4
99	20	26	0	9	1	3	1	2	4
100	18	18	0	17	1	3	1	2	4

Appendix F: Image Watermarking Metrics and Attacks

Besides designing digital watermarking methods, an important issue is to verify the proper evaluation of images against attacks. The evaluation process is based on a standard set of metrics, which in turn allow the user to judge the quality of a watermark after subjecting it to a set of attacks. In this appendix, a list of pixel-based metrics and possible watermark attacks are presented.

F.1 Pixel-Based Metrics

The watermark robustness depends directly on the embedding strength, which in turn influences the visual degradation of the image. For fair benchmarking and performance evaluation, the visual degradation due to the embedding is an important and unfortunately often neglected issue. The most popular pixel-based distortion metrics are presented as follows.

Most distortion measures or quality metrics used in visual information processing belong to the group of *difference distortion measures*. Table F.1 lists the difference distortion measures based on the difference between the original, undistorted and the modified, distorted signal. In the table, $I_{m,n}$ represents a pixel whose coordinates are (m, n) in the original, undistorted image, and $\bar{I}_{m,n}$ represents a pixel whose coordinates are (m, n) in the watermarked image, and in the Laplacian Mean Square Error, the term $\nabla^2 I_{m,n}$ is given by $\nabla^2 I_{m,n} = I_{m+1,n} + I_{m-1,n} + I_{m,n+1} + I_{m,n-1} - 4I_{m,n}$.

Table F.2 shows distortion measures based on the correlation between the original and the distorted signals. Among the metrics listed in Table F.1, the most popular distortion measures in the field of image and video coding and compression are the *signal-to-noise ratio* (SNR), and the *peak-signal-to-noise ratio* (PSNR). They are usually measured in *decibels* (dB), SNR (dB) = $10 \log_{10}$ (SNR). Their popularity is very likely due to the simplicity of the metric. However, it is well known that these difference distortion metrics are not correlated with human visual system (HVS). This might be a problem for their application in digital watermarking since sophisticated watermarking methods exploit the HVS in one way or the other. Using the aforementioned metrics to quantify the distortion caused by a watermarking process might, therefore, result in misleading quantitative distortion measurements. Furthermore, correlation distortion metrics are usually applied to the luminance and chrominance channels of images. If the watermarking methods work in the same color space, for example luminance modification, this does not pose a problem. On the contrary, if the methods use different color spaces, these metrics are not suitable.

Apart from the difference distortion and correlation distortion metrics, Table F.3 presents miscellaneous metrics such as the structural content (SC), global sigma signal-to-noise ratio (GSSNR), sigma signal-to-noise ratio (SSNR), sigma-to-error ratio (SER), and the histogram similarity (HS). GSSNR, SSNR, and SER require the division of the original and watermarked images into B blocks of P pixels (e.g., 4×4 pixels).

TABLE F.1

Difference Distortion Metrics

S. No.	Distortion Type	Metric
1	Maximum difference	$MD = \max\limits_{m,n} \left| I_{m,n} - \bar{I}_{m,n} \right|$
2	Average absolute difference	$AD = \dfrac{1}{MN} \sum\limits_{m,n} \left| I_{m,n} - \bar{I}_{m,n} \right|$
3	Normalized average absolute difference	$NAD = \sum\limits_{m,n} \left| I_{m,n} - \bar{I}_{m,n} \right| \Big/ \sum\limits_{m,n} \left| I_{m,n} \right|$
4	Mean square error	$MSE = \dfrac{1}{MN} \sum\limits_{m,n} \left(I_{m,n} - \bar{I}_{m,n} \right)^2$
5	Normalized mean square error	$NMSE = \sum\limits_{m,n} \left(I_{m,n} - \bar{I}_{m,n} \right)^2 \Big/ \sum\limits_{m,n} \left(I_{m,n} \right)^2$
6	L^P-Norm	$L^P = \left(\dfrac{1}{MN} \sum\limits_{m,n} \left| I_{m,n} - \bar{I}_{m,n} \right|^P \right)^{1/P}$
7	Laplacian mean square error	$LMSE = \sum\limits_{m,n} \left(\nabla^2 I_{m,n} - \nabla^2 \bar{I}_{m,n} \right)^2 \Big/ \sum\limits_{m,n} \left(\nabla^2 I_{m,n} \right)$
8	Signal-to-noise ratio	$SNR = \sum\limits_{m,n} \left(I_{m,n} \right)^2 \Big/ \sum\limits_{m,n} \left(I_{m,n} - \bar{I}_{m,n} \right)^2$
9	Peak-signal-to-noise ratio	$PSNR = MN \max\limits_{m,n} I_{m,n}{}^2 \Big/ \sum\limits_{m,n} \left(I_{m,n} - \bar{I}_{m,n} \right)^2$
10	Image fidelity	$IF = 1 - \sum\limits_{m,n} \left(I_{m,n} - \bar{I}_{m,n} \right) \Big/ \sum\limits_{m,n} \left(I_{m,n} \right)$

TABLE F.2

Correlation Distortion Metrics

S. No.	Distortion Type	Metric
1	Normalized cross-correlation	$NC = \sum\limits_{m,n} I_{m,n} \bar{I}_{m,n} \Big/ \sum\limits_{m,n} I_{m,n}^2$
2	Correlation quality	$CQ = \sum\limits_{m,n} I_{m,n} \bar{I}_{m,n} \Big/ \sum\limits_{m,n} I_{m,n}$

TABLE F.3

Miscellaneous Metrics

S. No.	Distortion Type	Metric
1	Structural content	$SC = \sum_{m,n} I_{m,n}^2 \Big/ \sum_{m,n} \bar{I}_{m,n}^2$
2	Global sigma signal-to-noise ratio	$GSSNR = \sum_b \sigma_b^2 \Big/ \sum_b (\sigma_b - \bar{\sigma}_b)^2$, where $$\sigma_b = \sqrt{\frac{1}{P}\sum_{block\,b} I_{m,n}^2 - \left(\frac{1}{P}\sum_{block\,b} I_{m,n}\right)^2}$$
3	Sigma signal-to-noise ratio	$SSNR = \frac{1}{P}\sum_b SSNR_b$, where $SSNR_b = 10\log_{10}\dfrac{\sigma_b^2}{(\sigma_b - \bar{\sigma}_b)^2}$
4	Sigma-to-error ratio	$SER_b = \dfrac{\sigma_b^2}{\frac{1}{P}\sum_{block\,b}(I_{m,n} - \bar{I}_{m,n})^2}$
5	Histogram similarity	$HS = \sum_{c=0}^{255}\left|f_I(c) - f_{\bar{I}}(c)\right|$, where $f_I(c)$ is the relative frequency of level c in a 255 levels image

F.2 Possible Attacks on Watermarks

In practical situations, the watermarked image may either be altered due to necessity or accidentally during transmission; such kind of process is referred to as attacks on the watermarked image. Even in such situations, the watermarking system should be able to detect and extract the watermark. Such distortions introduce a deprivation on the performance of the system. In image processing, several kinds of malicious attacks result in a partial or total destruction of watermarked image. The attacks are classified broadly into the following categories:

- *Active attacks*: The watermark is removed completely, or it is made undetectable. This is one of the major issues in copyright protection and image authentication.
- *Passive attacks*: An attempt is made to identify the existence of the watermark. This is mainly required in communication applications to obtain the knowledge of the presence of the watermark.
- *Collusion attacks*: The collusive attack is similar to active attacks but differs in terms of removing the watermark. Multiple copies of the watermark data are generated to construct a new copy without any watermark. Major role involves in authentication applications involving fingerprints.
- *Forgery attacks*: Rather than removing a watermark, a new watermark is embedded into the image, thus replacing the destructed one and making the corrupted image as genuine.

We introduce some of the best known attacks classified into geometric and nongeometric attacks, which may either be intentional or unintentional, depending on the application:

- *JPEG compression*: JPEG is currently one of the most widely used compression algorithms for images, and any watermarking system should be resilient to certain degree of compression.

- *Geometric attacks*: The geometric attacks are easy to implement, but they tend to convert the existing watermarking algorithms as ineffective ones. These geometric attacks are able to destroy the synchronization in the watermarked bitsteam, which is vital for most of the watermarking techniques. Some of the most common geometric attacks are discussed as follows:

  - *Horizontal flip*: Many images can be flipped without losing any value. Although resilience to flipping is usually straightforward, only very few systems survive it.

  - *Rotation*: Small angle rotation, often in combination with cropping, does not usually change the commercial value of the image but can make the watermark undetectable. Rotations are used to realign horizontal features of an image, and it is certainly the first modification applied to an image after it has been scanned.

  - *Cropping*: In some cases, infringers are just interested by the central part of the copyrighted material. Moreover, several websites use image segmentation, which is the basis of the "mosaic" attack. This is, of course, an extreme case of cropping.

  - *Scaling*: Scaling happens when a printed image is scanned or when a high-resolution digital image is used for electronic applications such as web publishing. Scaling can be divided into two groups: uniform and nonuniform scaling. Uniform scaling is the same in horizontal and vertical direction. Nonuniform scaling uses different scaling factors in horizontal and vertical directions (change of aspect ratio). Very often digital watermarking methods are resilient only to uniform scaling.

  - *Deletion of lines or columns*: This attack is very efficient against any straightforward implementation of spread–spectrum techniques in the spatial domain. Removing k samples at regular intervals in a pseudo-random sequence (−1; 1) (hence shifting the next ones) typically divides the amplitude of the cross-correlation peak with the original sequence by k.

  - *Generalized geometrical transformations*: A generalized geometrical transformation is a combination of nonuniform scaling, rotation, and shearing.

  - *Random geometric distortions (StirMark)*: These distortions do not survive and should be considered unacceptably easy to break.

  - *Geometric distortions with JPEG*: Rotation and scaling alone are not enough; they should be also tested in combination with JPEG compression. Since most artists will first apply the geometric transformation and then save the image in a compressed format, it makes sense to test the robustness of watermarking system to geometric transformation followed by compression. However, an exhaustive test should also include the contrary since it might be tried by willful infringers. It is difficult to chose a minimal quality factor for JPEG as artifacts appear quickly. However, experience from professionals shows that quality factors down to 70% are reasonable.

- *Enhancement techniques*
  - *Low-pass filtering*: This includes linear and nonlinear filters. Frequently used filters include median, Gaussian, and standard average filters.
  - *Sharpening*: Sharpening functions belong to the standard functionalities of photo edition software. These filters can be used as an effective attack on some watermarking schemes because they are very effective at detecting high-frequency noise introduced by some digital watermarking software.
  - *Histogram modification*: This includes histogram stretching or equalization, which are sometimes used to compensate poor lighting conditions.
  - *Gamma correction*: This type of attack is used frequently to enhance images or adapt images for display.
  - *Color quantization*: This is mostly applied when pictures are converted to the graphics interchange format (GIF) extensively used for web publishing. Color quantization is very often accompanied by dithering, which diffuses the error of the quantization.
  - *Restoration*: This technique is usually designed to reduce the effects of specific degradation processes but could also be used without prior knowledge of the noise introduced by the watermarking system.
  - *Noise addition*: Additive noise and uncorrelated multiplicative noise have been largely addressed in the communication theory and signal processing theory literature. Authors often claim that their copyright marking techniques survive this kind of noise, but the maximum level of acceptable noise is not revealed.
  - *Printing–scanning*: This process introduces geometrical as well as noise-like distortions.
  - *Statistical averaging and collusion*: Given two or more copies of the same image but with different marks, it should not be possible to remove the marks by averaging these images or by taking small parts of all images and reassembling them.
  - *Over-marking*: In this case, the attacker needs special access to the marking software. Current commercial implementations will refuse to add a watermark if another is already embedded. However, manufacturers have full access to the marking software and can perform this test without any difficulty.
  - *Oracle attack*: When a public decoder is available, an attacker can remove a mark by applying small changes to the image until the decoder cannot find it anymore. One could also make the decoding process computationally expensive. However, neither approach is really satisfactory in the absence of tamper-resistant hardware.

Index